Extended
Mathematics
for Cambridge **IGCSE**

David Rayner

OXFORD

OXFORD
UNIVERSITY PRESS

Great Clarendon Street, Oxford OX2 6DP

Oxford University Press is a department of the University of Oxford.
It furthers the University's objective of excellence in research,
scholarship, and education by publishing worldwide in

Oxford   New York

Auckland   Cape Town   Dar es Salaam   Hong Kong   Karachi
Kuala Lumpur   Madrid   Melbourne   Mexico City   Nairobi
New Delhi   Shanghai   Taipei   Toronto

With offices in

Argentina   Austria   Brazil   Chile   Czech Republic   France   Greece
Guatemala   Hungary   Italy   Japan   Poland   Portugal   Singapore
South Korea   Switzerland   Thailand   Turkey   Ukraine   Vietnam

Oxford is a registered trade mark of Oxford University Press
in the UK and in certain other countries

British Library Cataloguing in Publication Data

Data available

ISBN: 978-0-19-913874-6   98515

10 9 8 7 6 5 4 3

Printed in Great Britain by Bell and Bain Ltd, Glasgow

Paper used in the production of this book is a natural, recyclable product made
from wood grown in sustainable forests. The manufacturing process conforms to
the environmental regulations to the country of origin.

Acknowledgments
The publishers would like to thank University of Cambridge Local Examinations
Syndicate for kind permission to reproduce past paper questions.

University of Cambridge Local Examinations Syndicate bears no responsibility for
the example answers to questions taken from its past question papers which are
contained in this publication.

We are grateful to the following for permission to reproduce copyright material:

Cover photo by Phaif/Dreamstime

p1: PhotoSpin, Inc/Alamy; p13: Photocreo Michal Bednarek/Shutterstock; p51:
Lebrecht Music & Arts/Corbis; p92: Science Photo Library; p124: Science Photo
Library; p137: Photocreo Michal Bednarek/Shutterstock; p162: Science Source/
Science Photo Library; p192: Science Source/Science Photo Library; p223: Science
Source/Science Photo Library; p256: Mary Evans Picture Library/IDA KAR; p291:
INTERFOTO/Alamy; p325: Pictorial Press Ltd/Alamy; p367: Emilio Segre Visual
Archives/American Institute Of Physics/Science Photo Library

Contents

Contents

Oxford University Press is a worldwide publisher of educational textbooks and we have made full use of our international reputation to bring you the most authoritative text for the Cambridge IGCSE syllabus.

About this book

This book is designed to provide the best preparation for your Cambridge IGCSE examination. It is written by a very popular and successful author, and is endorsed by University of Cambridge International Examinations so you can be sure it covers everything you need to know.

Finding your way around

To get the most out of this book when studying or revising, use the:
- **Contents list** to help you find the appropriate units.
- **Index** to find key words so you can turn to any concept straight away.

Exercises and exam questions

There are literally thousands of questions in this book, providing ample opportunities to practise the skills and techniques required in the exam.

- **Worked examples and comprehensive exercises** are one of the main features of the book. The examples show you the important skills and techniques required. The exercises are carefully graded, starting from the basics and going up to exam standard, allowing you to practise the skills and techniques.
- **Revision exercises** at the end of each unit allow you to bring together all your knowledge on a particular topic and encourage regular revision.
- **Examination exercises** at the end of each unit consist of questions from past Cambridge IGCSE papers.
- **Specimen exam papers** athere are two papers, corresponding to the papers you will take at the end of your course: Paper 2 and Paper 4. They give you the opportunity to practise for the real thing.
- **Revision section:** Unit 12 contains multiple choice questions to provide an extra opportunity to revise, making sure you are completely ready for your exam.
- **Answers to numerical problems** are at the end of the book so you can check your progress.

Investigations

Unit 11 provides plenty of ideas to help you gain the special skills required for the Investigation paper. Remember that you can only gain by taking this optional paper - you cannot lose marks - so it is worth developing these skills.

Links to curriculum content

At the start of each unit you will find a list of objectives that are covered in the unit. These objectives are drawn from the Core and Supplement sections of the Cambridge IGCSE syllabus.

1 NUMBER

Karl Friedrich Gauss (1777–1855) was the son of a German labourer and is thought by many to have been the greatest all-round mathematician of all time. He considered that his finest discovery was the method for constructing a regular seventeen-sided polygon. This was not of the slightest use outside the world of mathematics, but was a great achievement of the human mind. Gauss would not have understood the modern view held by many that mathematics must somehow be 'useful' to be worthy of study.

1 Identify and use natural numbers, integers, prime numbers, square numbers, common factors and common multiples, rational and irrational numbers; continue a given number sequence; recognise patterns in sequences and relationships between different sequences, generalise to simple algebraic statements

6 Use the standard form $A \times 10^n$

7 Use the four rules for calculations with whole numbers, decimal fractions and vulgar fractions

8 Make estimates, give approximations and round off answers to reasonable accuracy

9 Obtain appropriate upper and lower bounds to solutions of simple problems

10 Demonstrate an understanding of ratio, direct and inverse proportion and common measures of rate; divide a quantity in a given ratio; use scales in practical situations; calculate average speed

11 Calculate percentage increase or decrease; carry out calculations involving reverse percentages

12 Use an electronic calculator efficiently

15 Calculate using money and convert from one currency to another

16 Solve problems on simple interest and compound interest

1.1 Arithmetic

Decimals

Example

Evaluate: (a) $7{\cdot}6 + 19$ (b) $3{\cdot}4 - 0{\cdot}24$ (c) $7{\cdot}2 \times 0{\cdot}21$

(d) $0{\cdot}84 \div 0{\cdot}2$ (e) $3{\cdot}6 \div 0{\cdot}004$

(a) $\begin{array}{r} 7{\cdot}6 \\ + 19{\cdot}0 \\ \hline 26{\cdot}6 \\ \hline \end{array}$

(b) $\begin{array}{r} 3{\cdot}40 \\ - 0{\cdot}24 \\ \hline 3{\cdot}16 \\ \hline \end{array}$

(c) $\begin{array}{r} 7{\cdot}2 \\ \times 0{\cdot}21 \\ \hline 72 \\ 1440 \\ \hline 1{\cdot}512 \\ \hline \end{array}$

No decimal points in the working, '3 figures after the points in the question *and* in the answer'.

(d) $0{\cdot}84 \div 0{\cdot}2 = 8{\cdot}4 \div 2$

$\begin{array}{r} 4{\cdot}2 \\ 2\overline{)8{\cdot}4} \end{array}$ Multiply both numbers by 10 so that we can divide by a whole number.

(e) $3{\cdot}6 \div 0{\cdot}004 = 3600 \div 4$

$= 900$

Exercise 1

Evaluate the following without a calculator:

1. $7{\cdot}6 + 0{\cdot}31$

2. $15 + 7{\cdot}22$

3. $7{\cdot}004 + 0{\cdot}368$

4. $0{\cdot}06 + 0{\cdot}006$

5. $4{\cdot}2 + 42 + 420$

6. $3{\cdot}84 - 2{\cdot}62$

7. $11{\cdot}4 - 9{\cdot}73$

8. $4{\cdot}61 - 3$

9. $17 - 0{\cdot}37$

10. $8{\cdot}7 + 19{\cdot}2 - 3{\cdot}8$

11. $25 - 7{\cdot}8 + 9{\cdot}5$

12. $3{\cdot}6 - 8{\cdot}74 + 9$

13. $20{\cdot}4 - 20{\cdot}399$

14. $2{\cdot}6 \times 0{\cdot}6$

15. $0{\cdot}72 \times 0{\cdot}04$

16. $27{\cdot}2 \times 0{\cdot}08$

17. $0{\cdot}1 \times 0{\cdot}2$

18. $(0{\cdot}01)^2$

19. $2{\cdot}1 \times 3{\cdot}6$

20. $2{\cdot}31 \times 0{\cdot}34$

21. $0{\cdot}36 \times 1000$

22. $0{\cdot}34 \times 100\,000$

23. $3{\cdot}6 \div 0{\cdot}2$

24. $0{\cdot}592 \div 0{\cdot}8$

25. $0{\cdot}1404 \div 0{\cdot}06$

26. $3{\cdot}24 \div 0{\cdot}002$

27. $0{\cdot}968 \div 0{\cdot}11$

28. $600 \div 0{\cdot}5$

29. $0{\cdot}007 \div 4$

30. $2640 \div 200$

31. $1100 \div 5{\cdot}5$

32. $(11 + 2{\cdot}4) \times 0{\cdot}06$

33. $(0{\cdot}4)^2 \div 0{\cdot}2$

34. $77 \div 1000$

35. $(0{\cdot}3)^2 \div 100$

36. $(0{\cdot}1)^4 \div 0{\cdot}01$

37. $\dfrac{92 \times 4{\cdot}6}{2{\cdot}3}$

38. $\dfrac{180 \times 4}{36}$

39. $\dfrac{0{\cdot}55 \times 0{\cdot}81}{4{\cdot}5}$

40. $\dfrac{63 \times 600 \times 0{\cdot}2}{360 \times 7}$

Exercise 2

1. A maths teacher bought 40 calculators at $8·20 each and a number of other calculators costing $2·95 each. In all she spent $387. How many of the cheaper calculators did she buy?

2. At a temperature of 20°C the common amoeba reproduces by splitting in half every 24 hours. If we start with a single amoeba how many will there be after (a) 8 days, (b) 16 days?

3. Copy and complete.

$$3^2 + 4^2 + 12^2 = 13^2$$
$$5^2 + 6^2 + 30^2 = 31^2$$
$$6^2 + 7^2 + \quad = $$
$$x^2 + \quad + \quad = $$

You can find out about square numbers on page 5.

4. Find all the missing digits in these multiplications.

(a) $\begin{array}{r} 5* \\ 9\times \\ \hline **6 \\ \hline \end{array}$
(b) $\begin{array}{r} *7 \\ *\times \\ \hline 4*6 \\ \hline \end{array}$
(c) $\begin{array}{r} 5* \\ *\times \\ \hline 1*4 \\ \hline \end{array}$

5. Pages 6 and 27 are on the same (double) sheet of a newspaper. What are the page numbers on the opposite side of the sheet? How many pages are there in the newspaper altogether?

6. Use the numbers 1, 2, 3, 4, 5, 6, 7, 8, 9 once each and in their natural order to obtain an answer of 100. You may use only the operations $+, -, \times, \div$.

7. The ruler below has eleven marks and can be used to measure lengths from one unit to twelve units.

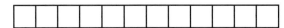

Design a ruler which can be used to measure all the lengths from one unit to twelve units but this time put the minimum possible number of marks on the ruler.

8. Each packet of washing powder carries a token and four tokens can be exchanged for a free packet. How many free packets will I receive if I buy 64 packets?

9. Put three different numbers in the circles so that when you add the numbers at the end of each line you always get a square number.

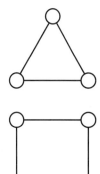

10. Put four different numbers in the circles so that when you add the numbers at the end of each line you always get a square number.

11. A group of friends share a bill for $13·69 equally between them. How many were in the group?

Fractions

Common fractions are added or subtracted from one another directly only when they have a common denominator.

Example

Evaluate: (a) $\frac{3}{4} + \frac{2}{5}$ (b) $2\frac{3}{8} - 1\frac{5}{12}$ (c) $\frac{2}{5} \times \frac{6}{7}$ (d) $2\frac{2}{5} \div 6$

(a) $\frac{3}{4} + \frac{2}{5} = \frac{15}{20} + \frac{8}{20}$

$= \frac{23}{20}$

$= 1\frac{3}{20}$

(b) $2\frac{3}{8} - 1\frac{5}{12} = \frac{19}{8} - \frac{17}{12}$

$= \frac{57}{24} - \frac{34}{24}$

$= \frac{23}{24}$

(c) $\frac{2}{5} \times \frac{6}{7} = \frac{12}{35}$

(d) $2\frac{2}{5} \div 6 = \frac{12}{5} \div \frac{6}{1}$

$= \frac{\overset{2}{\cancel{12}}}{5} \times \frac{1}{\cancel{6}_1} = \frac{2}{5}$

> **Remember**
> The order of operations follows the BODMAS rule:
> Brackets then powers Of then
> Divide then
> Multiply then
> Add then
> Subtract.

Exercise 3

Evaluate and simplify your answer.

1. $\frac{3}{4} + \frac{4}{5}$
2. $\frac{1}{3} + \frac{1}{8}$
3. $\frac{5}{6} + \frac{6}{9}$
4. $\frac{3}{4} - \frac{1}{3}$
5. $\frac{3}{5} - \frac{1}{3}$

6. $\frac{1}{2} - \frac{2}{5}$
7. $\frac{2}{3} \times \frac{4}{5}$
8. $\frac{1}{7} \times \frac{5}{6}$
9. $\frac{5}{8} \times \frac{12}{13}$
10. $\frac{1}{3} \div \frac{4}{5}$

11. $\frac{3}{4} \div \frac{1}{6}$
12. $\frac{5}{6} \div \frac{1}{2}$
13. $\frac{3}{8} + \frac{1}{5}$
14. $\frac{3}{8} \times \frac{1}{5}$
15. $\frac{3}{8} \div \frac{1}{5}$

16. $1\frac{3}{4} - \frac{2}{3}$
17. $1\frac{3}{4} \times \frac{2}{3}$
18. $1\frac{3}{4} \div \frac{2}{3}$
19. $3\frac{1}{2} + 2\frac{3}{5}$
20. $3\frac{1}{2} \times 2\frac{3}{5}$

21. $3\frac{1}{2} \div 2\frac{3}{5}$
22. $\left(\frac{3}{4} - \frac{2}{3}\right) \div \frac{3}{4}$
23. $\left(\frac{3}{5} + \frac{1}{3}\right) \times \frac{5}{7}$
24. $\dfrac{\frac{3}{8} - \frac{1}{5}}{\frac{7}{10} - \frac{2}{3}}$
25. $\dfrac{\frac{2}{3} + \frac{1}{5}}{\frac{3}{4} - \frac{1}{3}}$

26. Arrange the fractions in order of size:

   (a) $\frac{7}{12}, \frac{1}{2}, \frac{2}{3}$ (b) $\frac{3}{4}, \frac{2}{3}, \frac{5}{6}$ (c) $\frac{1}{3}, \frac{17}{24}, \frac{5}{8}, \frac{3}{4}$ (d) $\frac{5}{6}, \frac{8}{9}, \frac{11}{12}$

27. Find the fraction which is mid-way between the two fractions given:

   (a) $\frac{2}{5}, \frac{3}{5}$ (b) $\frac{5}{8}, \frac{7}{8}$ (c) $\frac{2}{3}, \frac{3}{4}$ (d) $\frac{1}{3}, \frac{4}{9}$ (e) $\frac{4}{15}, \frac{1}{3}$ (f) $\frac{3}{8}, \frac{11}{24}$

28. In the equation below all the asterisks stand for the same number. What is the number?

   $$\left[\frac{*}{*} - \frac{*}{6} = \frac{*}{30}\right]$$

29. When it hatches from its egg, the shell of a certain crab is 1 cm across. When fully grown the shell is approximately 10 cm across. Each new shell is one-third bigger than the previous one. How many shells does a fully grown crab have during its life?

30. Glass A contains 100 ml of water and glass B contains 100 ml of juice.

   A 10 ml spoonful of juice is taken from glass B and mixed thoroughly with the water in glass A. A 10 ml spoonful of the mixture from A is returned to B. Is there now more juice in the water or more water in the juice?

A

B

10 ml

Fractions and decimals

A decimal is simply a fraction expressed in tenths, hundredths etc.

Example

Change   (a) $\frac{7}{8}$ to a decimal   (b) 0.35 to a fraction   (c) $\frac{1}{3}$ to a decimal

(a) $\frac{7}{8}$, divide 8 into 7   (b) $0.35 = \frac{35}{100} = \frac{7}{20}$ (c) $\frac{1}{3}$, divide 3 into 1

$$\frac{7}{8} = 0.875 \quad 8\overline{)7.000}^{\,0.875}$$

$$\frac{1}{3} = 0.\dot{3} \ (0.3 \text{ recurring})$$

$$3\overline{)1.0^1 0^1 0^1 000}^{\,0.3\,3\,3\,3}$$

Exercise 4

In questions **1** to **24**, change the fractions to decimals.

1. $\frac{1}{4}$ 2. $\frac{2}{5}$ 3. $\frac{4}{5}$ 4. $\frac{3}{4}$ 5. $\frac{1}{2}$ 6. $\frac{3}{8}$

7. $\frac{9}{10}$ 8. $\frac{5}{8}$ 9. $\frac{5}{12}$ 10. $\frac{1}{6}$ 11. $\frac{2}{3}$ 12. $\frac{5}{6}$

13. $\frac{2}{7}$ 14. $\frac{3}{7}$ 15. $\frac{4}{9}$ 16. $\frac{5}{11}$ 17. $1\frac{1}{5}$ 18. $2\frac{5}{8}$

19. $2\frac{1}{3}$ 20. $1\frac{7}{10}$ 21. $2\frac{3}{16}$ 22. $2\frac{2}{7}$ 23. $2\frac{6}{7}$ 24. $3\frac{19}{100}$

In questions **25** to **40**, change the decimals to fractions and simplify.

25. 0.2 26. 0.7 27. 0.25 28. 0.45

29. 0.36 30. 0.52 31. 0.125 32. 0.625

33. 0.84 34. 2.35 35. 3.95 36. 1.05

37. 3.2 38. 0.27 39. 0.007 40. $0.000\,11$

Evaluate, giving the answer to 2 decimal places:

41. $\frac{1}{4} + \frac{1}{3}$ 42. $\frac{2}{3} + 0.75$ 43. $\frac{8}{9} - 0.24$ 44. $\frac{7}{8} + \frac{5}{9} + \frac{2}{11}$

45. $\frac{1}{3} \times 0.2$ 46. $\frac{5}{8} \times \frac{1}{4}$ 47. $\frac{8}{11} \div 0.2$ 48. $\left(\frac{4}{7} - \frac{1}{3}\right) \div 0.4$

Arrange the numbers in order of size (smallest first).

49. $\frac{1}{3}, 0.33, \frac{4}{15}$ 50. $\frac{2}{7}, 0.3, \frac{4}{9}$ 51. $0.71, \frac{7}{11}, 0.705$ 52. $\frac{4}{13}, 0.3, \frac{5}{18}$

1.2 Number facts and sequences

Number facts

- An *integer* is a whole number. e.g. 2, −3, ...
- A *prime* number is divisible only by itself and by 1.
  e.g. 2, 3, 5, 7, 11, 13, ...
- The *multiples* of 12 are 12, 24, 36, 48, ...
- The *factors* of 12 are 1, 2, 3, 4, 6, 12.
- A *square number* is the result of multiplying a number by itself.
  e.g. $5 \times 5 = 25$ so 25 is a square number.
- A *cube number* is the result of multiplying a number by itself three
  times. e.g. $5 \times 5 \times 5 = 125$, so 125 is a cube number.

Exercise 5

1. Which of the following are prime numbers?

3, 11, 15, 19, 21, 23, 27, 29, 31, 37, 39, 47, 51, 59, 61, 67, 72, 73, 87, 99

2. Write down the first five multiples of the following numbers:

(a) 4   (b) 6   (c) 10   (d) 11   (e) 20

3. Write down the first six multiples of 4 and of 6. What are the first two *common* multiples of 4 and 6? [i.e. multiples of both 4 and 6]

4. Write down the first six multiples of 3 and of 5. What is the lowest common multiple of 3 and 5?

5. Write down all the factors of the following:

(a) 6   (b) 9   (c) 10   (d) 15   (e) 24   (f) 32

6. (a) Is 263 a prime number?
By how many numbers do you need to divide 263 so that you can find out?

(b) Is 527 a prime number?

(c) Suppose you used a computer to find out if 1147 was a prime number. Which numbers would you tell the computer to divide by?

7. Make six prime numbers using the digits 1, 2, 3, 4, 5, 6, 7, 8, 9 once each.

Rational and irrational numbers

- A rational number can always be written exactly in the form $\dfrac{a}{b}$ where a and b are whole numbers.

$\frac{3}{7}$ $\qquad\qquad$ $1\frac{1}{2} = \frac{3}{2}$ $\qquad\qquad$ $5 \cdot 14 = \frac{257}{50}$ $\qquad\qquad$ $0 \cdot \dot{6} = \frac{2}{3}$
All these are rational numbers.

- An irrational number cannot be written in the form $\dfrac{a}{b}$.

$\sqrt{2}, \sqrt{5}, \pi, \sqrt[3]{2}$ are all irrational numbers.
- In general $\sqrt{n}$ is irrational unless n is a square number.

In this triangle the length of the hypotenuse is *exactly* $\sqrt{5}$.
On a calculator, $\sqrt{5} = 2 \cdot 236068$. This value of $\sqrt{5}$ is *not* exact and is correct to only 6 decimal places.

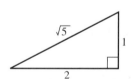

Exercise 6

1. Which of the following numbers are rational?

(a) $\dfrac{\pi}{2}$

(b) $\sqrt{5}$

(c) $(\sqrt{17})^2$

(d) $\sqrt{3}$

(e) $3 \cdot 14$

(f) $\dfrac{\sqrt{12}}{\sqrt{3}}$

(g) π^2

(h) $3^{-1} + 3^{-2}$

(i) $7^{-\frac{1}{2}}$

(j) $\dfrac{22}{7}$

(k) $\sqrt{2} + 1$

(l) $\sqrt{2 \cdot 25}$

2. (a) Write down any rational number between 4 and 6.

(b) Write down any irrational number between 4 and 6.

(c) Find a rational number between $\sqrt{2}$ and $\sqrt{3}$.

(d) Write down any rational number between π and $\sqrt{10}$.

3. (a) For each shape state whether the *perimeter* is rational or irrational.

(b) For each shape state whether the *area* is rational or irrational.

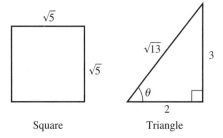

Square   Triangle

4. The diagram shows a circle of radius 3 cm drawn inside a square. Write down the exact value of the following and state whether the answer is rational or not:

(a) the circumference of the circle

(b) the diameter of the circle

(c) the area of the square

(d) the area of the circle

(e) the shaded area.

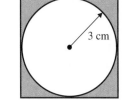

5. Think of two *irrational* numbers x and y such that $\dfrac{x}{y}$ is a *rational* number.

6. Explain the difference between a rational number and an irrational number.

7. (a) Is it possible to multiply a rational number and an irrational number to give an answer which is rational?

(b) Is it possible to multiply two irrational numbers together to give a rational answer?

(c) If either or both are possible, give an example.

Sequences, the *n*th term

Exercise 7

Write down each sequence and find the next two numbers.

1. 2, 6, 10, 14	**2.** 2, 9, 16, 23	**3.** 95, 87, 79, 71
4. 13, 8, 3, −2	**5.** 7, 9, 12, 16	**6.** 20, 17, 13, 8
7. 1, 2, 4, 7, 11	**8.** 1, 2, 4, 8	**9.** 55, 49, 42, 34
10. 10, 8, 5, 1	**11.** −18, −13, −9, −6	**12.** 120, 60, 30, 15
13. 27, 9, 3, 1	**14.** 162, 54, 18, 6	**15.** 2, 5, 11, 20
16. 1, 4, 20, 120	**17.** 2, 3, 1, 4, 0	**18.** 720, 120, 24, 6

Look at the sequence which starts 5, 9, 13, 17, ...

What is the 10th number in the sequence?
What is the nth number in the sequence? [n stands for any whole number.]
Is there a formula so that we can easily find the 50th or 100th number
in the sequence?

The 10th term in the sequence is $(4 \times 10) + 1 = 41$
The 50th term in the sequence is $(4 \times 50) + 1 = 201$
The 1000th term in the sequence is $(4 \times 1000) + 1 = 4001$
The nth term in the sequence is $(4 \times n) + 1 = 4n + 1$

Exercise 8

1. Write down each sequence and select the correct formula for the nth
term from the list given.

$\boxed{11n}$ $\boxed{10n}$ $\boxed{2n}$ $\boxed{n^2}$ $\boxed{10^n}$ $\boxed{3n}$ $\boxed{100n}$ $\boxed{n^3}$

(a) 2, 4, 6, 8, ...   (b) 10, 20, 30, 40, ...
(c) 3, 6, 9, 12, ...   (d) 11, 22, 33, 44, ...
(e) 100, 200, 300, 400, ...   (f) $1^2, 2^2, 3^2, 4^2, ...$
(g) 10, 100, 1000, 10000, ...   (h) $1^3, 2^3, 3^3, 4^3, ...$

2. Look at the sequence: 5, 8, 13, 20, ...
Decide which of the following is the correct expression for the nth
term of the sequence.

$\boxed{4n + 1}$ $\boxed{3n + 2}$ $\boxed{n^2 + 4}$

3. Write down the first five terms of the sequence whose nth term is
$2n + 7$.

4. Write down the first five terms of the sequence whose nth term is

(a) $n + 2$ (b) $5n$ (c) $10n - 1$ (d) $100 - 3n$ (e) $\dfrac{1}{n}$ (f) n^2

Finding the nth term

- In an *arithmetic* sequence the difference between successive terms is
  always the same number.
  Here are some arithmetic sequences: A 5, 7, 9, 11, 13
                                        B 12, 32, 52, 72, 92
                                        C 20, 17, 14, 11, 8

- The expression for the nth term of an arithmetic sequence is always of
  the form $an + b$.
  The *difference* between successive terms is equal to the number a.
  The number b can be found by looking at the terms.
  Look at sequences A, B and C above.
  For sequence A, the nth term $= 2n + b$ [the terms go up by 2]
  For sequence B, the nth term $= 20n + b$ [the terms go up by 20]

For sequence C, the nth term $= -3n + b$ [the terms go up by -3]

Look at each sequence and find the value of b in each case.
For example in sequence A: when $n = 1$, $2 \times 1 + b = 5$
$$\text{so } b = 3$$
The nth term in sequence A is $2n + 3$.

Exercise 9

In questions **1** to **18** find a formula for the nth term of the sequence.

1. 5, 9, 13, 17, ...

2. 7, 10, 13, 16, ...

3. 4, 9, 14, 19, ...

4. 6, 10, 14, 18, ...

5. 5, 8, 11, 14, ...

6. 25, 22, 19, 16, ...

7. 5, 10, 15, 20, ...

8. 2, 4, 8, 16, 32, ...

9. $(1 \times 3), (2 \times 4), (3 \times 5), \ldots$

10. $\frac{1}{2}, \frac{2}{3}, \frac{3}{4}, \frac{4}{5}, \ldots$

11. 7, 14, 21, 28, ...

12. 1, 4, 9, 16, 25, ...

13. $\frac{5}{1^2}, \frac{5}{2^2}, \frac{5}{3^2}, \frac{5}{4^2}, \ldots$

14. $\frac{3}{1}, \frac{4}{2}, \frac{5}{3}, \frac{6}{4}, \ldots$

15. 3, 7, 11, 15, ...

16. 5, 7, 9, 11, ...

17. 7, 5, 3, 1, ...

18. $-5, -1, 3, 7, \ldots$

19. Write down each sequence and then find the nth term.
(a) 8, 10, 12, 14, 16, ...
(b) 3, 7, 11, 15, ...
(c) 8, 13, 18, 23, ...

20. Write down each sequence and write the nth term.
(a) 11, 19, 27, 35, ...
(b) $2\frac{1}{2}, 4\frac{1}{2}, 6\frac{1}{2}, 8\frac{1}{2}, \ldots$
(c) $-7, -4, -1, 2, 5, \ldots$

21. Here is a sequence of shapes made from sticks

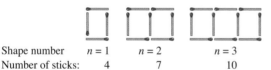

Shape number	$n = 1$	$n = 2$	$n = 3$
Number of sticks:	4	7	10

The number of sticks makes the sequence 4, 7, 10, 13, ...
(a) Find an expression for the nth term in the sequence.
(b) How many sticks are there in shape number 1000?

1.3 Approximations and estimation

Example

(a) $7 \cdot 8126 = 8$ to the nearest whole number
   ↑ This figure is '5 or more'.

(b) $7 \cdot 8126 = 7 \cdot 81$ to three significant figures
    ↑ This figure is not '5 or more'.

(c) $7 \cdot 8126 = 7 \cdot 813$ to three decimal places
     ↑ This figure is '5 or more'.

(d) $0 \cdot 078\,126 = 0 \cdot 0781$ to three significant figures.
     ↑ 7 is the first significant figure.

(e) $3596 = 3600$ to two significant figures.
   ↑ This figure is '5 or more'.

Exercise 10

Write the following numbers correct to:
(a) the nearest whole number   (b) three significant figures   (c) two decimal places.

1. $8 \cdot 174$	**2.** $19 \cdot 617$	**3.** $20 \cdot 041$	**4.** $0 \cdot 814\,52$	**5.** $311 \cdot 14$
6. $0 \cdot 275$	**7.** $0 \cdot 007\,47$	**8.** $15 \cdot 62$	**9.** $900 \cdot 12$	**10.** $3 \cdot 555$
11. $5 \cdot 454$	**12.** $20 \cdot 961$	**13.** $0 \cdot 0851$	**14.** $0 \cdot 5151$	**15.** $3 \cdot 071$

Write the following numbers correct to one decimal place.

16. $5 \cdot 71$	**17.** $0 \cdot 7614$	**18.** $11 \cdot 241$	**19.** $0 \cdot 0614$	**20.** $0 \cdot 0081$	**21.** $11 \cdot 12$

Measurements and bounds

Measurement is approximate

Example 1

A length is measured as 145 cm to the nearest cm.

The actual length could be anything from $144 \cdot 5$ cm to $145 \cdot 49999 \ldots$ cm using the normal convention which is to round up a figure of 5 or more. Clearly $145 \cdot 49999 \ldots$ is effectively $145 \cdot 5$ and we say the *upper bound* is $145 \cdot 5$.

The *lower bound* is $144 \cdot 5$.

As an inequality we can write $144 \cdot 5 \leqslant \text{length} < 145 \cdot 5$

The upper limit often causes confusion. We use $145 \cdot 5$ as the upper bound simply because it is *inconvenient* to work with $145 \cdot 49999 \ldots$

Example 2

When measuring the length of a page in a book, you might say the length is 437 mm to the nearest mm.

In this case the actual length could be anywhere from 436·5 mm to 437·5 mm. We write 'length is between 436·5 mm and 437·5 mm'.

In both Examples 1 and 2, the measurement expressed to a given unit is in *possible error* of *half* a *unit*.

Example 3

(a) If you say your mass is 57 kg to the nearest kg, your mass could actually be anything from 56·5 kg to 57·5 kg.

> The 'unit' is 1 so 'half a unit' is 0·5.

(b) If your brother's mass was measured on more sensitive scales and the result was 57·2 kg, his actual mass could be from 57·15 kg to 57·25 kg.

> The 'unit' is 0·1 so 'half a unit' is 0·05.

(c) The mass of a butterfly might be given as 0·032 g. The actual mass could be from 0·0315 g to 0·0325 g.

> The 'unit' is 0·001 so 'half a unit' is 0·005.

Here are some further examples:

Measurement	Lower bound	Upper bound
The diameter of a CD is 12 cm to the nearest cm.	11·5 cm	12·5 cm
The mass of a coin is 6·2 g to the nearest 0·1 g.	6·15 g	6·25 g
The length of a fence is 330 m to the nearest 10 m.	325 m	335 m

Exercise 11

1. In a DIY store the height of a door is given as 195 cm to the nearest cm. Write down the upper bound for the height of the door.

2. A vet measures the mass of a goat at 37 kg to the nearest kg. What is the least possible mass of the goat?

3. A cook's scales measure mass to the nearest 0·1 kg. What is the upper bound for the mass of a chicken which the scales say has a mass of 3·2 kg?

4. A surveyor using a laser beam device can measure distances to the nearest 0·1 m. What is the least possible length of a warehouse which he measures at 95·6 m?

5. In the county sports Stefan was timed at 28·6 s for the 200 m. What is the upper bound for the time she could have taken?

6. Copy and complete the table.

	Measurement	Lower bound	Upper bound
(a)	temperature in a fridge $= 2\,°C$ to the nearest degree		
(b)	mass of an acorn $= 2\!\cdot\!3\,g$ to 1 d.p.		
(c)	length of telephone cable $= 64\,m$ to nearest m		
(d)	time taken to run $100\,m = 13\!\cdot\!6\,s$ to nearest $0\!\cdot\!1\,s$		

7. The length of a telephone is measured as 193 mm, to the nearest mm. The length lies between:

A	B	C
192 and 194 mm	192·5 and 193·5 mm	188 and 198 mm

8. The mass of a suitcase is 35 kg, to the nearest kg. The mass lies between:

A	B	C
30 and 40 kg	34 and 36 kg	34·5 and 35·5 kg

9. Adra and Leila each measure a different worm and they both say that their worm is 11 cm long to the nearest cm.
(a) Does this mean that both worms are the same length?
(b) If not, what is the maximum possible difference in the length of the two worms?

10. To the nearest cm, the length l of a stapler is 12 cm. As an inequality we can write $11\!\cdot\!5 \leqslant l < 12\!\cdot\!5$.

For parts (a) to (j) you are given a measurement. Write the possible values using an inequality as above.

(a) mass $= 17\,kg$ (2 s.f.)
(b) $d = 256\,km$ (3 s.f.)
(c) length $= 2\!\cdot\!4\,m$ (1 d.p.)
(d) $m = 0\!\cdot\!34\,grams$ (2 s.f.)
(e) $v = 2\!\cdot\!04\,m/s$ (2 d.p.)
(f) $x = 12\!\cdot\!0\,cm$ (1 d.p.)
(g) $T = 81\!\cdot\!4\,°C$ (1 d.p.)
(h) $M = 0\!\cdot\!3\,kg$ (1 s.f.)
(i) mass $= 0\!\cdot\!7\,tonnes$ (1 s.f.)
(j) $n = 52\,000$ (nearest thousand)

11. A card measuring 11·5 cm long (to the nearest 0·1 cm) is to be posted in an envelope which is 12 cm long (to the nearest cm). Can you guarantee that the card will fit inside the envelope? Explain your answer.

11.5 cm

12 cm

Exercise 12

1. The sides of the triangle are measured correct to the nearest cm.
   (a) Write down the upper bounds for the lengths of the three sides.
   (b) Work out the maximum possible perimeter of the triangle.

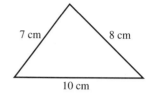

2. The dimensions of a photo are measured correct to the nearest cm.
   Work out the minimum possible area of the photo.

3. In this question the value of a is either exactly 4 or 5, and the value
   of b is either exactly 1 or 2. Work out:

   (a) the maximum value of $a + b$ (b) the minimum value of $a + b$
   (c) the maximum value of ab (d) the maximum value of $a - b$

   (e) the minimum value of $a - b$ (f) the maximum value of $\dfrac{a}{b}$

   (g) the minimum value of $\dfrac{a}{b}$ (h) the maximum value of $a^2 - b^2$.

4. If $p = 7\,\text{cm}$ and $q = 5\,\text{cm}$, both to the nearest cm, find:
   (a) the largest possible value of $p + q$
   (b) the smallest possible value of $p + q$
   (c) the largest possible value of $p - q$
   (d) the largest possible value of $\dfrac{p^2}{q}$.

5. If $a = 3{\cdot}1$ and $b = 7{\cdot}3$, correct to 1 decimal place, find the largest
   possible value of:
   (i) $a + b$ (ii) $b - a$

6. If $x = 5$ and $y = 7$ to one significant figure, find the largest and
   smallest possible values of:

   (i) $x + y$ (ii) $y - x$ (iii) $\dfrac{x}{y}$

7. In the diagram, ABCD and EFGH are rectangles with
   $AB = 10\,\text{cm}$, $BC = 7\,\text{cm}$, $EF = 7\,\text{cm}$ and $FG = 4\,\text{cm}$,
   all figures accurate to the nearest cm.
   Find the largest possible value of the shaded area.

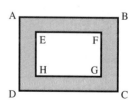

8. When a voltage V is applied to a resistance R the power
   consumed P is given by $P = \dfrac{V^2}{R}$.
   If you measure V as $12{\cdot}2$ and R as $2{\cdot}6$, correct to 1 d.p., calculate
   the smallest possible value of P.

Estimation

You should check that the answer to a calculation is 'about the right size'.

Example

Estimate the value of $\dfrac{57 \cdot 2 \times 110}{2 \cdot 146 \times 46 \cdot 9}$, correct to one significant figure.

We have approximately, $\dfrac{\cancel{50} \times 100}{2 \times \cancel{50}} \approx 50$ ——————————— On a calculator the value is 62·52 (to 4 significant figures).

Exercise 13

In this exercise there are 25 questions, each followed by three possible answers. Decide (by estimating) which answer is correct.

1. $7 \cdot 2 \times 9 \cdot 8$ [52·16, 98·36, 70·56]
2. $2 \cdot 03 \times 58 \cdot 6$ [118·958, 87·848, 141·116]
3. $23 \cdot 4 \times 19 \cdot 3$ [213·32, 301·52, 451·62]
4. $313 \times 107 \cdot 6$ [3642·8, 4281·8, 33 678·8]
5. $6 \cdot 3 \times 0 \cdot 098$ [0·6174, 0·0622, 5·98]
6. $1200 \times 0 \cdot 89$ [722, 1068, 131]
7. $0 \cdot 21 \times 93$ [41·23, 9·03, 19·53]
8. $88 \cdot 8 \times 213$ [18 914·4, 1693·4, 1965·4]
9. $0 \cdot 04 \times 968$ [38·72, 18·52, 95·12]
10. $0 \cdot 11 \times 0 \cdot 089$ [0·1069, 0·095 9, 0·009 79]
11. $13 \cdot 92 \div 5 \cdot 8$ [0·52, 4·2, 2·4]
12. $105 \cdot 6 \div 9 \cdot 6$ [8·9, 11, 15]
13. $8405 \div 205$ [4·6, 402, 41]
14. $881 \cdot 1 \div 99$ [4·5, 8·9, 88]
15. $4 \cdot 183 \div 0 \cdot 89$ [4·7, 48, 51]
16. $6 \cdot 72 \div 0 \cdot 12$ [6·32, 21·2, 56]
17. $20 \cdot 301 \div 1010$ [0·0201, 0·211, 0·0021]
18. $0 \cdot 288\,96 \div 0 \cdot 0096$ [312, 102·1, 30·1]
19. $0 \cdot 143 \div 0 \cdot 11$ [2·3, 1·3, 11·4]
20. $159 \cdot 65 \div 515$ [0·11, 3·61, 0·31]
21. $(5 \cdot 6 - 0 \cdot 21) \times 39$ [389·21, 210·21, 20·51]
22. $\dfrac{17 \cdot 5 \times 42}{2 \cdot 5}$ [294, 504, 86]
23. $(906 + 4 \cdot 1) \times 0 \cdot 31$ [473·21, 282·131, 29·561]
24. $\dfrac{543 + 472}{18 \cdot 1 + 10 \cdot 9}$ [65, 35, 85]
25. $\dfrac{112 \cdot 2 \times 75 \cdot 9}{6 \cdot 9 \times 5 \cdot 1}$ [242, 20·4, 25·2]

1.4 Standard form

When dealing with either very large or very small numbers, it is not convenient to write them out in full in the normal way. It is better to use standard form. Most calculators represent large and small numbers in this way.

The number $a \times 10^n$ is in standard form when $1 \leqslant a < 10$ and n is a positive or negative integer.

Example

Write the following numbers in standard form:

(a) $2000 = 2 \times 1000 = 2 \times 10^3$

(b) $150 = 1 \cdot 5 \times 100 = 1 \cdot 5 \times 10^2$

(c) $0 \cdot 0004 = 4 \times \dfrac{1}{10\,000} = 4 \times 10^{-4}$

Exercise 14

Write the following numbers in standard form:

1. 4000	2. 500	3. 70 000	4. 60	5. 2400	6. 380
7. 46 000	8. 46	9. 900 000	10. 2560	11. 0·007	12. 0·0004
13. 0·0035	14. 0·421	15. 0·000 055	16. 0·01	17. 564 000	18. 19 million

19. The population of China is estimated at 1 100 000 000. Write this in standard form.

20. The mass of a hydrogen atom is 0.000 000 000 000 000 000 000 001 67 grams. Write this mass in standard form.

21. The area of the surface of the Earth is about 510 000 000 km^2. Express this in standard form.

22. An atom is 0.000 000 000 25 cm in diameter. Write this in standard form.

23. Avogadro's number is 602 300 000 000 000 000 000 000. Express this in standard form.

24. The speed of light is 300 000 km/s. Express this speed in cm/s in standard form.

> **Hint**
> 1 km = 1000 m
> 1 m = 100 cm

25. A very rich man leaves his fortune of $\$3 \cdot 6 \times 10^8$ to be divided between his 100 grandchildren. How much does each child receive? Give the answer in standard form.

Example

Work out $1500 \times 8\,000\,000$.

$$1500 \times 8\,000\,000 = (1 \cdot 5 \times 10^3) \times (8 \times 10^6)$$
$$= 12 \times 10^9$$
$$= 1 \cdot 2 \times 10^{10}$$

Notice that we multiply the numbers and the powers of 10 separately.

Exercise 15

In questions **1** to **12** give the answer in standard form.

1. 5000×3000
2. $60\,000 \times 5000$
3. $0 \cdot 000\,07 \times 400$
4. $0 \cdot 0007 \times 0 \cdot 000\,01$
5. $8000 \div 0 \cdot 004$
6. $(0 \cdot 002)^2$
7. $150 \times 0 \cdot 0006$
8. $0 \cdot 000\,033 \div 500$
9. $0 \cdot 007 \div 20\,000$
10. $(0 \cdot 0001)^4$
11. $(2000)^3$
12. $0 \cdot 005\,92 \div 8000$

13. If $a = 512 \times 10^2 \qquad b = 0 \cdot 478 \times 10^6 \qquad c = 0 \cdot 0049 \times 10^7$
    arrange a, b and c in order of size (smallest first).

14. If the number $2 \cdot 74 \times 10^{15}$ is written out in full, how many zeros follow the 4?

15. If the number $7 \cdot 31 \times 10^{-17}$ is written out in full, how many zeros would there be between the decimal point and the first significant figure?

16. If $x = 2 \times 10^5$ and $y = 3 \times 10^{-3}$ correct to one significant figure, find the greatest and least possible values of:

    (i) xy (ii) $\dfrac{x}{y}$

> **Remember**
> The limits of accuracy of 2 to one significant figure are $1 \cdot 5$ to $2 \cdot 5$.

17. Oil flows through a pipe at a rate of $40\,\mathrm{m}^3/\mathrm{s}$. How long will it take to fill a tank of volume $1 \cdot 2 \times 10^5\,\mathrm{m}^3$?

18. Given that $L = 2\sqrt{\dfrac{a}{k}}$, find the value of L in standard form when $a = 4 \cdot 5 \times 10^{12}$ and $k = 5 \times 10^7$.

19. (a) The number 10 to the power 100 (10 000 sexdecillion) is called a 'Googol'! If it takes $\frac{1}{5}$ second to write a zero and $\frac{1}{10}$ second to write a 'one', how long would it take to write the number 100 'Googols' in full?
    (b) The number 10 to the power of a 'Googol' is called a 'Googolplex'. Using the same speed of writing, how long in years would it take to write 1 'Googolplex' in full? You may assume that your pen has enough ink.

1.5 Ratio and proportion

The word 'ratio' is used to describe a fraction. If the *ratio* of a boy's height to his father's height is $4 : 5$, then he is $\frac{4}{5}$ as tall as his father.

Example 1

Change the ratio $2:5$ into the form

(a) $1:n$ (b) $m:1$

(a) $2:5 = 1:\frac{5}{2}$ (b) $2:5 = \frac{2}{5}:1$

         $= 1:2\cdot5$ $= 0\cdot4:1$

Example 2

Divide \$60 between two people A and B in the ratio $5:7$.

Consider \$60 as 12 equal parts (i.e. $5 + 7$). Then A receives 5 parts and B receives 7 parts.

$\therefore$ A receives $\frac{5}{12}$ of \$60 $=$ \$25

      B receives $\frac{7}{12}$ of \$60 $=$ \$35

Example 3

Divide 200 kg in the ratio $1:3:4$.

The parts are $\frac{1}{8}$, $\frac{3}{8}$ and $\frac{4}{8}$ (of 200 kg). i.e. 25 kg, 75 kg and 100 kg.

Exercise 16

In questions **1** to **8** express the ratios in the form $1:n$.

1. $2:6$ **2.** $5:30$ **3.** $2:100$ **4.** $5:8$
5. $4:3$ **6.** $8:3$ **7.** $22:550$ **8.** $45:360$

In questions **9** to **12** express the ratios in the form $n:1$.

9. $12:5$ **10.** $5:2$ **11.** $4:5$ **12.** $2:100$

In questions **13** to **18** divide the quantity in the ratio given.

13. \$40; $(3:5)$ **14.** \$120; $(3:7)$ **15.** 250 m; $(14:11)$
16. \$117; $(2:3:8)$ **17.** 180 kg; $(1:5:6)$ **18.** 184 minutes; $(2:3:3)$

19. When \$143 is divided in the ratio $2:4:5$, what is the difference between the largest share and the smallest share?

20. Divide 180 kg in the ratio $1:2:3:4$.

21. Divide \$4000 in the ratio $2:5:5:8$.

22. If $\frac{5}{8}$ of the children in a school are boys, what is the ratio of boys to girls?

23. A man and a woman share a prize of \$1000 between them in the ratio $1:4$. The woman shares her part between herself, her mother and her daughter in the ratio $2:1:1$. How much does her daughter receive?

24. A man and a woman share a sum of money in the ratio $3:2$. If the sum of money is doubled, in what ratio should they divide it so that the man still receives the same amount?

25. In a herd of x cattle, the ratio of the number of bulls to cows is $1:6$. Find the number of bulls in the herd in terms of x.

26. If $x : 3 = 12 : x$, calculate the positive value of x.

27. If $y : 18 = 8 : y$, calculate the positive value of y.

28. $400 is divided between Kas, Jaspar and Jae so that Kas has twice as much as Jaspar and Jaspar has three times as much as Jae. How much does Jaspar receive?

29. A cake of mass 550 g has three ingredients: flour, sugar and raisins. There is twice as much flour as sugar and one and a half times as much sugar as raisins. How much flour is there?

30. A brother and sister share out their collection of 5000 stamps in the ratio $5:3$. The brother then shares his stamps with two friends in the ratio $3:1:1$, keeping most for himself. How many stamps do each of his friends receive?

Proportion

The majority of problems where proportion is involved are usually solved by finding the value of a unit quantity.

Example 1

If a wire of length 2 metres costs $10, find the cost of a wire of length 35 cm.

200 cm costs 1000 cents

$\therefore$ 1 cm costs $\frac{1000}{200}$ cents $= 5$ cents

$\therefore$ 35 cm costs 5×35 cents $= 175$ cents

$\qquad\qquad\qquad\qquad\qquad = \$1{\cdot}75$

Example 2

Eight men can dig a hole in 4 hours. How long will it take five men to dig the same size hole?

   8 men take 4 hours

   1 man would take 32 hours

   5 men would take $\frac{32}{5}$ hours $= 6$ hours 24 minutes.

Exercise 17

1. Five cans of cola cost $1·20. Find the cost of seven cans.

2. A man earns $140 in a 5-day week. What is his pay for 3 days?

3. Three people build a wall in 10 days. How long would it take five people?

4. Nine fruit juice bottles contain $4\frac{1}{2}$ litres of fruit juice between them. How much juice do five bottles hold?

5. A car uses 10 litres of petrol in 75 km. How far will it go on 8 litres?

6. A wire 11 cm long has a mass of 187 g. What is the mass of 7 cm of this wire?

7. A shopkeeper can buy 36 toys for $20·52. What will he pay for 120 toys?

8. A ship has sufficient food to supply 600 passengers for 3 weeks. How long would the food last for 800 people?

9. The cost of a phone call lasting 3 minutes 30 seconds was 52·5 cents. At this rate, what was the cost of a call lasting 5 minutes 20 seconds?

10. 80 machines can produce 4800 identical pens in 5 hours. At this rate
(a) how many pens would one machine produce in one hour?
(b) how many pens would 25 machines produce in 7 hours?

11. Three men can build a wall in 10 hours. How many men would be needed to build the wall in $7\frac{1}{2}$ hours?

12. If it takes 6 men 4 days to dig a hole 3 metres deep, how long will it take 10 men to dig a hole 7 metres deep?

13. Find the cost of 1 km of pipe at 7 cents for every 40 cm.

14. A wheel turns through 90 revolutions per minute. How many degrees does it turn through in 1 second?

15. Find the cost of 200 grams of flour at $6 per kilogram.

> **Remember**
> 1 kg = 1000 g

16. The height of One Kansas City Place is 623 feet. Express this height to the nearest metre using 1 m = 3·281 feet.

17. A floor is covered by 800 tiles measuring 10 cm². How many square tiles of side 8 cm would be needed to cover the same floor?

18. A battery has enough energy to operate eight toy bears for 21 hours. For how long could the battery operate 15 toy bears?

19. An engine has enough fuel to operate at full power for 20 minutes. For how long could the engine operate at 35% of full power?

20. A large drum, when full, contains 260 kg of oil of density 0·9 g/cm³. What mass of petrol, of density 0·84 g/cm³, can be contained in the drum?

21. A wall can be built by 6 men working 8 hours per day in 5 days. How many days will it take 4 men to build the wall if they work only 5 hours per day?

Foreign exchange

Money is changed from one currency into another using the method of proportion.

Exchange rate for US dollars ($):

Country	Rate of exchange
Argentina (pesos)	$1 = 3·79 ARS
Australia (dollar)	$1 = 1·13 AUD
Euro (euros)	$1 = €0·70 EUR
India (rupees)	$1 = 46·50 INR
Japan (yen)	$1 = 91·20 JPY
Kuwait (dinar)	$1 = 0·29 KWD
UK (pounds)	$1 = £0·63 GBP

Example

Convert: (a) $22.50 to dinars   (b) € 300 to dollars.

(a) $1 = 0·29 dinars (KWD)

so $22·50 = 0·29 × 22·50 KWD

= 6·53 KWD

(b)   €0·70 = $1

so $€1 = \dfrac{1}{0·70}$

so €300 = $\$\dfrac{1}{0·70} × 300$

= $428.57

Exercise 18

Give your answers correct to two decimal places. Use the exchange rates given in the table.

1. Change the number of dollars into the foreign currency stated.
   (a) $20 [euros]   (b) $70 [pounds]   (c) $200 [pesos]
   (d) $1·50 [rupees]   (e) $2·30 [yen]   (f) 90c [dinars]

2. Change the amount of foreign currency into dollars.
   (a) € 500   (b) £2500   (c) 7·50 rupees
   (d) 900 dinars   (e) 125·24 pesos   (f) 750 AUD

3. A CD costs £9·50 in London and $9·70 in Chicago. How much cheaper, in British money, is the CD when bought in the US?

4. An MP3 player costs € 20·46 in Spain and £12·60 in the UK. Which is the cheaper in dollars, and by how much?

5. The monthly rent of a flat in New Delhi is 32 860 rupees. How much is this in euros?

6. A Persian kitten is sold in several countries at the prices given below.

Kuwait	150 dinars
France	550 euros
Japan	92 000 yen

   Write out in order a list of the prices converted into GBP.

7. An Australian man on holiday in Germany finds that his wallet contains 700 AUD. If he changes the money at a bank how many euros will he receive?

Map scales

You can use proportion to work out map scales. First you need to know these metric equivalents:

1 km = 1000 m   km means kilometre
1 m = 100 cm   m means metre
1 cm = 10 mm   cm means centimetre
                       mm means millimetre

Example

A map is drawn to a scale of 1 to 50 000. Calculate:

(a) the length of a road which appears as 3 cm long on the map.
(b) the length on the map of a lake which is 10 km long.

(a) 1 cm on the map is equivalent to 50 000 cm on the Earth.

$\therefore$ 1 cm $\equiv$ 50 000 cm
$\therefore$ 1 cm $\equiv$ 500 m
$\therefore$ 1 cm $\equiv$ 0·5 km

so 3 cm $\equiv$ 3 $\times$ 0·5 km = 1·5 km.
The road is 1·5 km long.

(b)   0·5 km $\equiv$ 1 cm
$\therefore$ 1 km $\equiv$ 2 cm
$\therefore$ 10 km $\equiv$ 2 $\times$ 10 cm
               = 20 cm

The lake appears 20 cm long on the map.

Exercise 19

1. Find the actual length represented on a drawing by
   (a) 14 cm   (b) 3·2 cm
   (c) 0·71 cm   (d) 21·7 cm

when the scale is 1 cm to 5 m.

2. Find the length on a drawing that represents
   (a) 50 m   (b) 35 m
   (c) 7·2 m   (d) 28·6 m

when the scale is 1 cm to 10 m.

3. If the scale is $1:10\,000$, what length will $45\,cm$ on the map represent:

(a) in cm   (b) in m   (c) in km?

4. On a map of scale $1:100\,000$, the distance between De'aro and Debeka is $12\cdot3\,cm$. What is the actual distance in km?

5. On a map of scale $1:15\,000$, the distance between Noordwijk aan Zee and Katwijk aan Zee is $31\cdot4\,cm$. What is the actual distance in km?

6. If the scale of a map is $1:10\,000$, what will be the length on this map of a road which is $5\,km$ long?

7. The distance from Hong Kong to Shenzhen is $32\,km$. How far apart will they be on a map of scale $1:50\,000$?

8. The 17th hole at the famous St Andrews golf course is $420\,m$ in length. How long will it appear on a plan of the course of scale $1:8000$?

An area involves two dimensions multiplied together and hence the scale is multiplied *twice*.

For example, if the linear scale is $\frac{1}{100}$, then the area scale is $\frac{1}{100} \times \frac{1}{100} = \frac{1}{10\,000}$.

You can use a diagram to help:
If a scale is $1:50\,000$
then $2\,cm \equiv 1\,km$

3 cm

An area of $6\,cm^2$ can be thought of as: | $6\,cm^2$ | 2 cm

1·5 km

so the equivalent area using the scale is: | $1.5\,km^2$ | 1 km

Exercise 20

1. The scale of a map is $1:1000$. What are the actual dimensions of a rectangle which appears as $4\,cm$ by $3\,cm$ on the map? What is the area on the map in cm^2? What is the actual area in m^2?

2. The scale of a map is $1:100$. What area does $1\,cm^2$ on the map represent? What area does $6\,cm^2$ represent?

3. The scale of a map is $1:20\,000$. What area does $8\,cm^2$ represent?

4. The scale of a map is $1:1000$. What is the area, in cm^2, on the map of a lake of area $5000\,m^2$?

5. The scale of a map is $1\,cm$ to $5\,km$. A farm is represented by a rectangle measuring $1{\cdot}5\,cm$ by $4\,cm$. What is the actual area of the farm?

6. On a map of scale $1\,cm$ to $250\,m$ the area of a car park is $3\,cm^2$. What is the actual area of the car park in hectares? (1 hectare $= 10\,000\,m^2$)

7. The area of the playing surface at the Olympic Stadium in Beijing is $\frac{3}{5}$ of a hectare. What area will it occupy on a plan drawn to a scale of $1:500$?

8. On a map of scale $1:20\,000$ the area of a forest is $50\,cm^2$. On another map the area of the forest is $8\,cm^2$. Find the scale of the second map.

1.6 Percentages

Percentages are simply a convenient way of expressing fractions or decimals. '50% of \$60' means $\frac{50}{100}$ of \$60, or more simply $\frac{1}{2}$ of \$60. Percentages are used very frequently in everyday life and are misunderstood by a large number of people. What are the implications if 'inflation falls from 10% to 8%'? Does this mean prices will fall?

Example
(a) Change 80% to a fraction.
(b) Change $\frac{3}{8}$ to a percentage.
(c) Change 8% to a decimal.

(a) $80\% = \dfrac{80}{100} = \dfrac{4}{5}$

(b) $\dfrac{3}{8} = \left(\dfrac{3}{8} \times \dfrac{100}{1}\right)\% = 37\frac{1}{2}\%$

(c) $8\% = \dfrac{8}{100} = 0{\cdot}08$

Exercise 21
1. Change to fractions:
    (a) 60%   (b) 24%   (c) 35%   (d) 2%

2. Change to percentages:
    (a) $\frac{1}{4}$ (b) $\frac{1}{10}$ (c) $\frac{7}{8}$
    (d) $\frac{1}{3}$ (e) $0{\cdot}72$ (f) $0{\cdot}31$

3. Change to decimals:
   (a) 36%   (b) 28%   (c) 7%
   (d) 13·4%   (e) $\frac{3}{5}$ (f) $\frac{7}{8}$

4. Arrange in order of size (smallest first):
   (a) $\frac{1}{2}$; 45%; 0·6   (b) 0·38; $\frac{6}{16}$; 4%
   (c) 0·111; 11%; $\frac{1}{9}$ (d) 32%; 0·3; $\frac{1}{3}$

5. The following are marks obtained in various tests. Convert them to percentages.
   (a) 17 out of 20   (b) 31 out of 40   (c) 19 out of 80
   (d) 112 out of 200   (e) $2\frac{1}{2}$ out of 25   (f) $7\frac{1}{2}$ out of 20

Example 1

A car costing $400 is reduced in price by 10%. Find the new price.

$$10\% \text{ of } \$2400 = \frac{10}{100} \times \frac{2400}{1}$$

$$= \$240$$

New price of car $= \$(2400 - 240)$
$$= \$2160$$

Example 2

After a price increase of 10% a television set costs $286.
What was the price before the increase?

The price before the increase is 100%.

∴   110% of old price = \$286

∴   1% of old price $= \$\dfrac{286}{110}$

∴   100% of old price $= \$\dfrac{286}{110} \times \dfrac{100}{1}$

   Old price of TV = \$260

Exercise 22

1. Calculate:
   (a) 30% of $50   (b) 45% of 2000 kg
   (c) 4% of $70   (d) 2·5% of 5000 people

2. In a sale, a jacket costing $40 is reduced by 20%. What is the sale price?

3. The charge for a telephone call costing 12 cents is increased by 10%. What is the new charge?

4. In peeling potatoes 4% of the mass of the potatoes is lost as 'peel'. How much is *left* for use from a bag containing 55 kg?

5. Work out, to the nearest cent:
(a) 6·4% of $15·95   (b) 11·2% of $192·66
(c) 8·6% of $25·84   (d) 2·9% of $18·18

6. Find the total bill:
5 golf clubs at $18·65 each
60 golf balls at $16·50 per dozen
1 bag at $35·80
Sales tax at 15% is added to the total cost.

7. In 2010 a club has 250 members who each pay $95 annual subscription. In 2011 the membership increases by 4% and the annual subscription is increased by 6%. What is the total income from subscriptions in 2011?

8. In Thailand the population of a town is 48 700 men and 1600 women. What percentage of the total population are men?

9. In South Korea there are 21 280 000 licensed vehicles on the road. Of these, 16 486 000 are private cars. What percentage of the licensed vehicles are private cars?

10. A quarterly telephone bill consists of $19·15 rental plus 4·7 cents for each dialled unit. Tax is added at 15%. What is the total bill for Bryndis who used 915 dialled units?

11. 70% of Hassan's collection of goldfish died. If he has 60 survivors, how many did he have originally?

12. The average attendance at Parma football club fell by 7% in 2010. If 2030 fewer people went to matches in 2010, how many went in 2009?

13. When heated an iron bar expands by 0·2%. If the increase in length is 1 cm, what is the original length of the bar?

14. In the last two weeks of a sale, prices are reduced first by 30% and then by a *further* 40% of the new price. What is the final sale price of a shirt which originally cost $15?

15. During a Grand Prix car race, the tyres on a car are reduced in mass by 3%. If their mass is 388 kg at the end of the race, what was their mass at the start?

16. Over a period of 6 months, a colony of rabbits increases in number by 25% and then by a further 30%. If there were originally 200 rabbits in the colony, how many were there at the end?

17. A television costs $270·25 including 15% tax. How much of the cost is tax?

18. The cash price for a car was $7640. Gurtaj bought the car on the following terms: 'A deposit of 20% of the cash price and 36 monthly payments of $191·60'. Calculate the total amount Gurtaj paid.

Percentage increase or decrease

In the next exercise use the formulae:

$$\text{Percentage profit} = \frac{\text{Actual profit}}{\text{Original price}} \times \frac{100}{1}$$

$$\text{Percentage loss} = \frac{\text{Actual loss}}{\text{Original price}} \times \frac{100}{1}$$

Example 1

A radio is bought for $16 and sold for $20. What is the percentage profit?

$$\text{Actual profit} = \$4$$

$$\therefore \quad \text{Percentage profit} = \frac{4}{16} \times \frac{100}{1} = 25\%$$

The radio is sold at a 25% profit.

Example 2

A car is sold for $2280, at a loss of 5% on the cost price. Find the cost price.

Do *not* calculate 5% of $2280!

The loss is 5% of the cost price.

$$\therefore \quad 95\% \text{ of cost price} = \$2280$$

$$1\% \text{ of cost price} = \$\frac{2280}{95}$$

$$\therefore \quad 100\% \text{ of cost price} = \$\frac{2280}{95} \times \frac{100}{1}$$

$$\text{Cost price} = \$2400$$

Exercise 23

1. The first figure is the cost price and the second figure is the selling price. Calculate the percentage profit or loss in each case.
   (a) $20, $25   (b) $400, $500   (c) $60, $54
   (d) $9000, $10 800   (e) $460, $598   (f) $512, $550·40
   (g) $45, $39·60   (h) 50c, 23c

Hint
c is the symbol for cents.
100c = $1

2. A car dealer buys a car for $500, and then sells it for $640. What is the percentage profit?

3. A damaged carpet which cost $180 when new, is sold for $100. What is the percentage loss?

4. During the first four weeks of her life, a baby girl increases her mass from 3·2 kg to 4·7 kg. What percentage increase does this represent? (Give your answer to 3 sig. fig.)

5. When tax is added to the cost of a lipstick, its price increases from $16·50 to $18·48. What is the rate at which tax is charged?

6. The price of a sports car is reduced from $30 000 to $28 400. What percentage reduction is this?

7. Find the *cost* price of the following:
   (a) selling price $55, profit 10%   (b) selling price $558, profit 24%
   (c) selling price $680, loss 15%   (d) selling price $11·78, loss 5%

8. A sari is sold for $60 thereby making a profit of 20% on the cost price. What was the cost price?

9. A pair of jeans is sold for $15, thereby making a profit of 25% on the cost price. What was the cost price?

10. A book is sold for $5·40, at a profit of 8% on the cost price. What was the cost price?

11. A can of worms is sold for 48c, incurring a loss of 20%. What was the cost price?

12. A car was sold for $1430, thereby making a loss of 35% on the cost price. What was the cost price?

13. If an employer reduces the working week from 40 hours to 35 hours, with no loss of weekly pay, calculate the percentage increase in the hourly rate of pay.

14. The rental for a television set changed from $80 per year to $8 per month. What is the percentage increase in the yearly rental?

15. A greengrocer sells a melon at a profit of $37\frac{1}{2}\%$ on the price he pays for it. What is the ratio of the cost price to the selling price?

16. Given that $G = ab$, find the percentage increase in G when both a and b increase by 10%.

17. Given that $T = \dfrac{kx}{y}$, find the percentage increase in T when k, x and y all increase by 20%.

Simple interest

When a sum of money P is invested for T years at $R\%$ interest per annum (each year), then the interest gained I is given by:

$$I = \frac{P \times R \times T}{100}$$

This is known as simple interest.

Example

Joel invests $400 for 6 months at 5%.
Work out the simple interest gained.

$P = \$400 \qquad R = 5 \qquad T = 0.5 \qquad$ (6 months is half a year)

so $\quad I = \dfrac{400 \times 5 \times 0.5}{100}$

$\qquad I = \$10$

Exercise 24

1. Calculate:

    (a) the simple interest on $1200 for 3 years at 6% per annum
    (b) the simple interest on $700 at 8·25% per annum for 2 years
    (c) the length of time for $5000 to earn $1000 if invested at 10% per annum
    (d) the length of time for $400 to earn $160 if invested at 8% per annum.

2. Khalid invests $6750 at 8·5% per annum. How much interest has he earned and what is the total amount in his account after 4 years?

3. Petra invests $10 800. After 4 years she has earned $3240 in interest. At what annual rate of interest did she invest her money?

Compound interest

Suppose a bank pays a fixed interest of 10% on money in deposit accounts. A man puts $500 in the bank.

After one year he has
$\quad 500 + 10\%$ of $500 = \$550$

After two years he has
$\quad 550 + 10\%$ of $550 = \$605$

$\qquad$ [Check that this is $1.10^2 \times 500$]

After three years he has
$\quad 605 + 10\%$ of $605 = \$665.50$

$\qquad$ [Check that this is $1.10^3 \times 500$]

In general after n years the money in the bank will be $\$(1.10^n \times 500)$

Exercise 25

1. A bank pays interest of 9% on money in deposit accounts. Carme puts $2000 in the bank. How much has she after (a) one year, (b) two years, (c) three years?

2. A bank pays interest of 11%. Mamuru puts $5000 in the bank. How much has he after (a) one year, (b) three years, (c) five years?

3. A student gets a grant of $10 000 a year. Assuming her grant is increased by 7% each year, what will her grant be in four years time?

4. Isoke's salary in 2010 is $30 000 per year. Every year her salary is increased by 5%.
   In 2011 her salary will be $30\,000 \times 1{\cdot}05$ $= \$31\,500$
   In 2012 her salary will be $30\,000 \times 1{\cdot}05 \times 1{\cdot}05$ $= \$33\,075$
   In 2013 her salary will be $30\,000 \times 1{\cdot}05 \times 1{\cdot}05 \times 1{\cdot}05 = \$34\,728{\cdot}75$
   And so on.
   (a) What will her salary be in 2014?
   (b) What will her salary be in 2016?

5. The rental price of a dacha was $9000. At the end of each month the price is increased by 6%.
   (a) Find the price of the house after 1 month.
   (b) Find the price of the house after 3 months.
   (c) Find the price of the house after 10 months.

6. Assuming an average inflation rate of 8%, work out the probable cost of the following items in 10 years:
   (a) motor bike $6500
   (b) iPod $340
   (c) car $50 000

7. A new scooter is valued at $15 000. At the end of each year its value is reduced by 15% of its value at the start of the year. What will it be worth after 3 years?

8. The population of an island increases by 10% each year. After how many years will the original population be doubled?

9. A bank pays interest of 11% on $6000 in a deposit account. After how many years will the money have trebled?

10. A tree grows in height by 21% per year. It is 2 m tall after one year. After how many more years will the tree be over 20 m tall?

11. Which is the better investment over ten years:
        $20 000 at 12% compound interest
    or $30 000 at 8% compound interest?

1.7 Speed, distance and time

Calculations involving these three quantities are simpler when the speed is *constant*. The formulae connecting the quantities are as follows:

(a) distance = speed × time

(b) speed = $\dfrac{\text{distance}}{\text{time}}$

(c) time = $\dfrac{\text{distance}}{\text{speed}}$

A helpful way of remembering these formulae is to write the letters D, S and T in a triangle,

thus: to find D, cover D and we have ST

                    to find S, cover S and we have $\dfrac{D}{T}$

to find T, cover T and we have $\dfrac{D}{S}$

Great care must be taken with the units in these questions.

Example 1

A man is running at a speed of 8 km/h for a distance of 5200 metres. Find the time taken in minutes.

$$5200 \text{ metres} = 5\cdot2 \text{ km}$$

$$\text{time taken in hours} = \left(\frac{D}{S}\right) = \frac{5\cdot2}{8}$$

$$= 0\cdot65 \text{ hours}$$

$$\text{time taken in minutes} = 0\cdot65 \times 60$$

$$= 39 \text{ minutes}$$

Example 2

Change the units of a speed of 54 km/h into metres per second.

$$54 \text{ km/hour} = 54\,000 \text{ metres/hour}$$

$$= \frac{54\,000}{60} \text{ metres/minute}$$

$$= \frac{54\,000}{60 \times 60} \text{ metres/second}$$

$$= 15 \text{ m/s}$$

Exercise 26

1. Find the time taken for the following journeys:
   (a) 100 km at a speed of 40 km/h
   (b) 250 miles at a speed of 80 miles per hour
   (c) 15 metres at a speed of 20 cm/s (answer in seconds)
   (d) 10^4 metres at a speed of 2·5 km/h

2. Change the units of the following speeds as indicated:
   (a) 72 km/h into m/s
   (b) 108 km/h into m/s
   (c) 300 km/h into m/s
   (d) 30 m/s into km/h
   (e) 22 m/s into km/h
   (f) 0·012 m/s into cm/s
   (g) 9000 cm/s into m/s
   (h) 600 miles/day into miles per hour
   (i) 2592 miles/day into miles per second

3. Find the speeds of the bodies which move as follows:
   (a) a distance of 600 km in 8 hours
   (b) a distance of 31·64 km in 7 hours
   (c) a distance of 136·8 m in 18 seconds
   (d) a distance of 4×10^4 m in 10^{-2} seconds
   (e) a distance of 5×10^5 cm in 2×10^{-3} seconds
   (f) a distance of 10^8 mm in 30 minutes (in km/h)
   (g) a distance of 500 m in 10 minutes (in km/h)

4. Find the distance travelled (in metres) in the following:
   (a) at a speed of 55 km/h for 2 hours
   (b) at a speed of 40 km/h for $\frac{1}{4}$ hour
   (c) at a speed of 338·4 km/h for 10 minutes
   (d) at a speed of 15 m/s for 5 minutes
   (e) at a speed of 14 m/s for 1 hour
   (f) at a speed of 4×10^3 m/s for 2×10^{-2} seconds
   (g) at a speed of 8×10^5 cm/s for 2 minutes

5. A car travels 60 km at 30 km/h and then a further 180 km at 160 km/h. Find:
   (a) the total time taken
   (b) the average speed for the whole journey.

6. A cyclist travels 25 kilometres at 20 km/h and then a further 80 kilometres at 25 km/h. Find:
   (a) the total time taken
   (b) the average speed for the whole journey.

7. A swallow flies at a speed of 50 km/h for 3 hours and then at a speed of 40 km/h for a further 2 hours. Find the average speed for the whole journey.

8. A runner ran two laps around a 400 m track. She completed the first lap in 50 seconds and then decreased her speed by 5% for the second lap. Find:
   (a) her speed on the first lap
   (b) her speed on the second lap
   (c) her total time for the two laps
   (d) her average speed for the two laps.

9. An airliner flies 2000 km at a speed of 1600 km/h and then returns due to bad weather at a speed of 1000 km/h. Find the average speed for the whole trip.

10. A train travels from A to B, a distance of 100 km, at a speed of 20 km/h. If it had gone two and a half times as fast, how much earlier would it have arrived at B?

11. Two men running towards each other at 4 m/s and 6 m/s respectively are one kilometre apart. How long will it take before they meet?

12. A car travelling at 90 km/h is 500 m behind another car travelling at 70 km/h in the same direction. How long will it take the first car to catch the second?

13. How long is a train which passes a signal in twenty seconds at a speed of 108 km/h?

14. A train of length 180 m approaches a tunnel of length 620 m. How long will it take the train to pass completely through the tunnel at a speed of 54 km/h?

15. An earthworm of length 15 cm is crawling along at 2 cm/s. An ant overtakes the worm in 5 seconds. How fast is the ant walking?

16. A train of length 100 m is moving at a speed of 50 km/h. A horse is running alongside the train at a speed of 56 km/h. How long will it take the horse to overtake the train?

17. A car completes a journey at an average speed of 40 km/h. At what speed must it travel on the return journey if the average speed for the complete journey (out and back) is 60 km/h?

Mixed problems

Exercise 27

1. Fill in the blank spaces in the table so that each row contains equivalent values.

fraction	decimal	percentage
	0·28	
		64%
$\frac{5}{8}$		

2. An engine pulls four identical carriages. The engine is $\frac{2}{3}$ the length of a carriage and the total length of the train is 86·8 m. Find the length of the engine.

3. A cake is made from the ingredients listed below.

> 500 g flour, 450 g butter, 470 g sugar,
> 1·8 kg mixed fruit, 4 eggs (70 g each)

The cake loses 12% of its mass during cooking. What is its final mass?

4. Abdul left his home at 7.35 a.m. and drove at an average speed of 45 km/h arriving at the airport at 8.50 a.m. How far is his home from the airport?

5. Tuwile's parents have agreed to lend him 60% of the cost of buying a car. If Tuwile still has to find $328 himself, how much does the car cost?

6. Which bag of potatoes is the better value:
Bag A, 6 kg for $4·14 or
Bag B, 2·5 kg for $1·80?

7. An aeroplane was due to take off from Madrid airport at 18:42 but it was 35 min late. During the flight, thanks to a tail wind, the plane made up the time and in fact landed 16 min before its scheduled arrival time of 00:05. (Assume that the plane did not cross any time zones on its journey.)
(a) What time did the aeroplane take off?
(b) What time did it land?

8. A 20 cent coin is 1·2 mm thick. What is the value of a pile of 20 cent coins which is 21·6 cm high?

9. Work out $\frac{3}{5} + 0·12 + 6\%$ of 10.

Exercise 28

1. Find the distance travelled by light in one hour, given that the speed of light is 300 000 kilometres per second.
Give the answer in kilometres in standard form.

2. When the lid is left off an ink bottle, the ink evaporates at a rate of $2·5 \times 10^{-6} \, cm^3/s$. A full bottle contains $36 \, cm^3$ of ink. How long, to the nearest day, will it take for all the ink to evaporate?

3. Convert 3·35 hours into hours and minutes.

> **Remember**
> There are 60 minutes in 1 hour.

4. When I think of a number, multiply it by 6 and subtract 120, my answer is −18. What was my original number?

5. The cost of advertising in a local paper for one week is:

> 28 cents per word plus 75 cents

(a) What is the cost of an advertisement of 15 words for one week?
(b) What is the greatest number of words in an advertisement costing up to $8 for one week?
(c) If an advertisement is run for two weeks, the cost for the second week is reduced by 30%. Calculate the total cost for an advertisement of 22 words for two weeks.

6. Bronze is made up of zinc, tin and copper in the ratio $1:4:95$. A bronze statue contains 120 g of tin. Find the quantities of the other two metals required and the total mass of the statue.

Exercise 29

1. In the diagram $\frac{5}{6}$ of the circle is shaded and $\frac{2}{3}$ of the triangle is shaded. What is the ratio of the area of the circle to the area of the triangle?

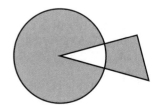

2. Find the exact answer to the following by first working out a rough answer and then using the information given.
Do *not* use a calculator.
(a) If $142\cdot3 \times 98\cdot5 = 14016\cdot55$ find $140\cdot1655 \div 14\cdot23$
(b) If $76\cdot2 \times 8\cdot6 = 655\cdot32$ find $6553\cdot2 \div 86$
(c) If $22\cdot3512 \div 0\cdot268 = 83\cdot4$ find $8340 \times 26\cdot8$
(d) If $1\cdot6781 \div 17\cdot3 = 0\cdot097$ find $9700 \times 0\cdot173$

3. A sales manager reports an increase of 28% in sales this year compared to last year.
The increase was $70 560.
What were the sales last year?

4. Small cubes of side 1 cm are stuck together to form a large cube of side 4 cm. Opposite faces of the large cube are painted the same colour, but adjacent faces are different colours. The three colours used are red, black and green.
(a) How many small cubes have just one red and one green face?
(b) How many small cubes are painted on one face only?
(c) How many small cubes have one red, one green and one black face?
(d) How many small cubes have no faces painted?

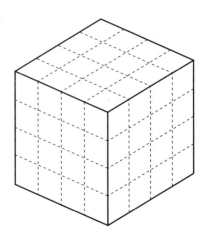

5. A bullet travels at a speed of 3×10^4 cm/s. Work out the length of time in seconds taken for the bullet to hit a target 54 m away.

6. A sewing machine cost $162·40 after a price increase of 16%. Find the price before the increase.

7. To get the next number in a sequence you double the previous number and subtract two.
The fifth number in the sequence is 50.
Find the first number.

8. A code uses 1 for A, 2 for B, 3 for C and so on up to 26 for Z. Coded words are written without spaces to confuse the enemy, so 18 could be AH or R. Decode the following message.

    208919 919 1 2251825 199121225 31545

9. A coach can take 47 passengers. How many coaches are needed to transport 1330 passengers?

1.8 Calculator

In this book, the keys are described thus:

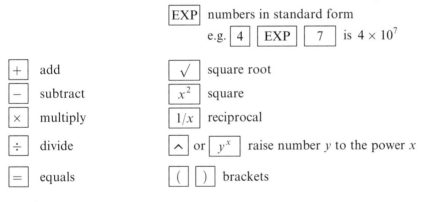

| EXP | numbers in standard form |

e.g. $\boxed{4}$ $\boxed{\text{EXP}}$ $\boxed{7}$ is 4×10^7

$\boxed{+}$ add $\boxed{\sqrt{}}$ square root

$\boxed{-}$ subtract $\boxed{x^2}$ square

$\boxed{\times}$ multiply $\boxed{1/x}$ reciprocal

$\boxed{\div}$ divide $\boxed{\wedge}$ or $\boxed{y^x}$ raise number y to the power x

$\boxed{=}$ equals $\boxed{(}\boxed{)}$ brackets

Using the $\boxed{\text{ANS}}$ button

The $\boxed{\text{ANS}}$ button can be used as a 'short term memory'.
It holds the answer from the previous calculation.

Example
Evaluate the following to 4 significant figures:

(a) $\dfrac{5}{1·2 - 0·761}$ (b) $\left(\dfrac{1}{0·084}\right)^4$ (c) $\sqrt[3]{[3·2 \times (1·7 - 1·64)]}$

(a) Find the bottom line first.

$$\boxed{5}\ \boxed{\div}\ \boxed{(}\ \boxed{1\cdot2}\ \boxed{-}\ \boxed{0\cdot761}\ \boxed{)}\ \boxed{=}$$

or

$$\boxed{1\cdot2}\ \boxed{-}\ \boxed{0\cdot761}\ \boxed{\text{EXE}}\ \boxed{5}\ \boxed{\div}\ \boxed{\text{ANS}}\ \boxed{\text{EXE}}$$

The calculator reads 11·38952164

∴ Answer = 11·39 (to four sig. fig.)

Note: The $\boxed{\text{EXE}}$ button works the same as the $\boxed{=}$ button.

(b) $\left(\dfrac{1}{0\cdot084}\right)^4$

$$\boxed{1}\ \boxed{\div}\ \boxed{0\cdot084}\ \boxed{=}\ \boxed{\wedge}\ \boxed{4}\ \boxed{=}\ \text{or}\ \boxed{0\cdot084}\ \boxed{1/x}\ \boxed{y^x}\ \boxed{4}\ \boxed{=}$$

Answer 20 090 (to four sig. fig.)

(c) $\sqrt[3]{[3\cdot2(1\cdot7-1\cdot64)]}$

$$\boxed{3\cdot2}\ \boxed{\times}\ \boxed{(}\ \boxed{1\cdot7}\ \boxed{-}\ \boxed{1\cdot64}\ \boxed{)}\ \boxed{=}\ \boxed{\sqrt[3]{\ }}\ \boxed{=}$$

or

$$\boxed{1\cdot7}\ \boxed{-}\ \boxed{1\cdot64}\ \boxed{=}\ \boxed{\times}\ \boxed{3\cdot2}\ \boxed{=}$$

$$\boxed{y^x}\ \boxed{0\cdot333\,333}\ \boxed{=}$$

Answer 0·5769 (to four sig. fig.)

Note: To find a cube root, raise to the power $\frac{1}{3}$, or as a decimal 0·333 ...

Exercise 30

Use a calculator to evaluate the following, giving the answers to
4 significant figures:

1. $\dfrac{7\cdot351 \times 0\cdot764}{1\cdot847}$

2. $\dfrac{0\cdot0741 \times 14\,700}{0\cdot746}$

3. $\dfrac{0\cdot0741 \times 9\cdot61}{23\cdot1}$

4. $\dfrac{417\cdot8 \times 0\cdot008\,41}{0\cdot073\,24}$

5. $\dfrac{8\cdot41}{7\cdot601 \times 0\cdot008\,47}$

6. $\dfrac{4\cdot22}{1\cdot701 \times 5\cdot2}$

7. $\dfrac{9\cdot61}{17\cdot4 \times 1\cdot51}$

8. $\dfrac{8\cdot71 \times 3\cdot62}{0\cdot84}$

9. $\dfrac{0\cdot76}{0\cdot412 - 0\cdot317}$

10. $\dfrac{81\cdot4}{72\cdot6 + 51\cdot92}$

11. $\dfrac{111}{27\cdot4 + 2960}$

12. $\dfrac{27\cdot4 + 11\cdot61}{5\cdot9 - 4\cdot763}$

13. $\dfrac{6\cdot51 - 0\cdot1114}{7\cdot24 + 1\cdot653}$

14. $\dfrac{5\cdot71 + 6\cdot093}{9\cdot05 - 5\cdot77}$

15. $\dfrac{0\cdot943 - 0\cdot788}{1\cdot4 - 0\cdot766}$

16. $\dfrac{2\cdot6}{1\cdot7} + \dfrac{1\cdot9}{3\cdot7}$

17. $\dfrac{8\cdot06}{5\cdot91} - \dfrac{1\cdot594}{1\cdot62}$

18. $\dfrac{4\cdot7}{11\cdot4 - 3\cdot61} + \dfrac{1\cdot6}{9\cdot7}$

19. $\dfrac{3\cdot74}{1\cdot6 \times 2\cdot89} - \dfrac{1}{0\cdot741}$

20. $\dfrac{1}{7\cdot2} - \dfrac{1}{14\cdot6}$

21. $\dfrac{1}{0.961} \times \dfrac{1}{0.412}$

22. $\dfrac{1}{7} + \dfrac{1}{13} - \dfrac{1}{8}$

23. $4.2\left(\dfrac{1}{5.5} - \dfrac{1}{7.6}\right)$

24. $\sqrt{(9.61 + 0.1412)}$

25. $\sqrt{\left(\dfrac{8.007}{1.61}\right)}$

26. $(1.74 + 9.611)^2$

27. $\left(\dfrac{1.63}{1.7 - 0.911}\right)^2$

28. $\left(\dfrac{9.6}{2.4} - \dfrac{1.5}{0.74}\right)^2$

29. $\sqrt{\left(\dfrac{4.2 \times 1.611}{9.83 \times 1.74}\right)}$

30. $(0.741)^3$

31. $(1.562)^5$

32. $(0.32)^3 + (0.511)^4$

33. $(1.71 - 0.863)^6$

34. $\left(\dfrac{1}{0.971}\right)^4$

35. $\sqrt[3]{(4.714)}$

36. $\sqrt[3]{(0.9316)}$

37. $\sqrt[3]{\left(\dfrac{4.114}{7.93}\right)}$

38. $\sqrt[4]{(0.8145 - 0.799)}$

39. $\sqrt[5]{(8.6 \times 9.71)}$

40. $\sqrt[3]{\left(\dfrac{1.91}{4.2 - 3.766}\right)}$

41. $\left(\dfrac{1}{7.6} - \dfrac{1}{18.5}\right)^3$

42. $\dfrac{\sqrt{(4.79)} + 1.6}{9.63}$

43. $\dfrac{(0.761)^2 - \sqrt{(4.22)}}{1.96}$

44. $\sqrt[3]{\left(\dfrac{1.74 \times 0.761}{0.0896}\right)}$

Example

Work out the following to 4 significant figures.

(a) $2 \times 10^5 - 1.734 \times 10^4$

(b) $(3.6 \times 10^{-4})^2$

(a) $\boxed{2}$ $\boxed{\text{EXP}}$ $\boxed{5}$ $\boxed{-}$ $\boxed{1.734}$ $\boxed{\text{EXP}}$ $\boxed{4}$ $\boxed{=}$

Answer 182 700 (to 4 s.f.)

(b) $\boxed{3.6}$ $\boxed{\text{EXP}}$ $\boxed{-}$ $\boxed{4}$ $\boxed{x^2}$

Answer 1.296×10^{-7}

Exercise 31

1. $\left(\dfrac{8.6 \times 1.71}{0.43}\right)^3$

2. $\dfrac{9.61 - \sqrt{(9.61)}}{9.61^2}$

3. $\dfrac{9.6 \times 10^4 \times 3.75 \times 10^7}{8.88 \times 10^6}$

4. $\dfrac{8.06 \times 10^{-4}}{1.71 \times 10^{-6}}$

5. $\dfrac{3.92 \times 10^{-7}}{1.884 \times 10^{-11}}$

6. $\left(\dfrac{1.31 \times 2.71 \times 10^5}{1.91 \times 10^4}\right)^5$

7. $\left(\dfrac{1}{9.6} - \dfrac{1}{9.99}\right)^{10}$

8. $\dfrac{\sqrt[3]{(86.6)}}{\sqrt[4]{(4.71)}}$

9. $\dfrac{23.7 \times 0.0042}{12.48 - 9.7}$

10. $\dfrac{0.482 + 1.6}{0.024 \times 1.83}$

11. $\dfrac{8.52 - 1.004}{0.004 - 0.0083}$

12. $\dfrac{1.6 - 0.476}{2.398 \times 41.2}$

13. $\left(\dfrac{2.3}{0.791}\right)^7$

14. $\left(\dfrac{8.4}{28.7 - 0.47}\right)^3$

15. $\left(\dfrac{5.114}{7.332}\right)^5$

16. $\left(\dfrac{4.2}{2.3} + \dfrac{8.2}{0.52}\right)^3$

17. $\dfrac{1}{8.2^2} - \dfrac{3}{19^2}$

18. $\dfrac{100}{11^3} + \dfrac{100}{12^3}$

19. $\dfrac{7.3 - 4.291}{2.6^2}$

20. $\dfrac{9.001 - 8.97}{0.95^3}$

21. $\dfrac{10.1^2 + 9.4^2}{9.8}$

22. $(3.6 \times 10^{-8})^2$

23. $(8.24 \times 10^4)^3$

24. $(2.17 \times 10^{-3})^3$

25. $(7.095 \times 10^{-6})^{\frac{1}{3}}$

26. $\sqrt[3]{\left(\dfrac{4.7}{2.3^2}\right)}$

Checking answers

Here are five calculations, followed by sensible checks.

(a) $22 \cdot 2 \boxed{\div} 6 = 3 \cdot 7$ check $3 \cdot 7 \boxed{\times} 6 = 22 \cdot 2$

(b) $31 \cdot 7 \boxed{-} 4 \cdot 83 = 26 \cdot 87$ check $26 \cdot 87 \boxed{+} 4 \cdot 83 = 31 \cdot 7$

(c) $42 \cdot 8 \boxed{\times} 30 = 1284$ check $1284 \boxed{\div} 30 = 42 \cdot 8$

(d) $\sqrt{17} = 4 \cdot 1231$ check $4 \cdot 1231^2$

(e) $3 \cdot 7 + 17 \cdot 6 + 13 \cdot 9$ check $13 \cdot 9 + 17 \cdot 6 + 3 \cdot 7$
                                      (add in reverse order)

Calculations can also be checked by rounding numbers to a given number of significant figures.

(f) $\dfrac{6 \cdot 1 \times 32 \cdot 6}{19 \cdot 3} = 10 \cdot 3$ (to 3 s.f.)

Check this answer by rounding each number to 1 significant figure and estimating.

$$\frac{6 \cdot 1 \times 32 \cdot 6}{19 \cdot 3} \approx \frac{6 \times 30}{20} = \frac{180}{20} = 9$$

Remember
'$\approx$' means 'approximately equal to'

This is close to $10 \cdot 3$
so the actual answer probably is $10 \cdot 3$

Exercise 32

1. Use a calculator to work out the following then check the answers as indicated.

(a) $92 \cdot 5 \times 20 = \boxed{}$ Check $\boxed{} \div 20 = \boxed{}$

(b) $14 \times 328 = \boxed{}$ Check $\boxed{} \div 328 = \boxed{}$

(c) $63 - 12 \cdot 6 = \boxed{}$ Check $\boxed{} + 12 \cdot 6 = \boxed{}$

(d) $221 \cdot 2 \div 7 = \boxed{}$ Check $\boxed{} \times 7 = \boxed{}$

(e) $384 \cdot 93 \div 9 \cdot 1 = \boxed{}$ Check $\boxed{} \times 9 \cdot 1 = \boxed{}$

(f) $13 \cdot 71 + 25 \cdot 8 = \boxed{}$ Check $\boxed{} - 25 \cdot 8 = \boxed{}$

(g) $95 \cdot 4 \div 4 \cdot 5 = \boxed{}$ Check $\boxed{} \times 4 \cdot 5 = \boxed{}$

(h) $8 \cdot 2 + 3 \cdot 1 + 19 \cdot 6 + 11 \cdot 5 = \boxed{}$ Check $11 \cdot 5 + 19 \cdot 6 + 3 \cdot 1 + 8 \cdot 2 = \boxed{}$

(i) $\sqrt{39} = \boxed{}$ Check $\boxed{}^2 = 39$

(j) $3\cdot17 + 2\cdot06 + 8\cdot4 + 16 = \boxed{}$ Check $16 + 8\cdot4 + 2\cdot06 + 3\cdot17 = \boxed{}$

2. The numbers below are rounded to 1 significant figure to *estimate* the answer to each calculation. Match each question below to the correct estimated answer.

A $\boxed{21\cdot9 \times 1\cdot01}$ P $\boxed{10}$

B $\boxed{\dfrac{19\cdot82^2}{(18\cdot61 + 22\cdot3)}}$ Q $\boxed{5}$

C $\boxed{7\cdot8 \times 1\cdot01}$ R $\boxed{0\cdot5}$

D $\boxed{\dfrac{\sqrt{98\cdot7}}{8\cdot78 + 11\cdot43}}$ S $\boxed{8}$

E $\boxed{\dfrac{21\cdot42 + 28\cdot6}{18\cdot84 - 8\cdot99}}$ T $\boxed{20}$

3. Do *not* use a calculator.
$281 \times 36 = 10\,116$
Work out
(a) $10\,116 \div 36$ (b) $10\,116 \div 281$ (c) $28\cdot1 \times 3\cdot6$

4. Mavis is paid a salary of \$49 620 per year. Work out a rough estimate for her weekly pay. (Give your answer correct to one significant figure.)

5. In 2011, the population of France was 61 278 514 and the population of Greece was 9 815 972. Roughly how many times bigger is the population of France compared to the population of Greece? (Hint: round the numbers to 1 significant figure.)

6. *Estimate*, correct to 1 significant figure:

(a) $41\cdot56 \div 7\cdot88$

(b) $\dfrac{5\cdot13 \times 18\cdot777}{0\cdot952}$

(c) $\dfrac{1}{5}$ of £14 892

(d) $\dfrac{0\cdot0974 \times \sqrt{104}}{1\cdot03}$

(e) 52% of 0·394 kg

(f) $\dfrac{6\cdot84^2 + 0\cdot983}{5\cdot07^2}$

(g) $\dfrac{2848\cdot7 + 1024\cdot8}{51\cdot2 - 9\cdot98}$

(h) $\dfrac{2}{3}$ of £3 124

(i) $18\cdot13 \times (3\cdot96^2 + 2\cdot07^2)$

1.9 Using a spreadsheet on a computer

This section is written for use with Microsoft Excel. Other spreadsheet programs work in a similar way.

Select Microsoft Excel from the desk top.

A spreadsheet appears on your screen as a grid with rows numbered 1, 2, 3, 4, ... and the columns lettered A, B, C, D, ...
The result should be a window like the one below.

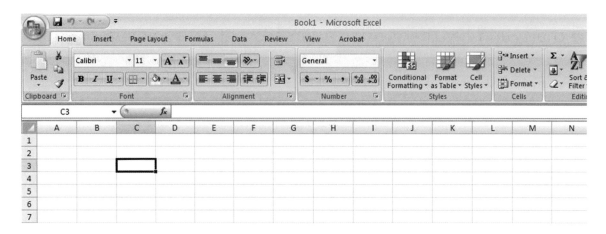

Cell The spaces on the spreadsheet are called cells. Individual cells are referred to as A1, B3, F9, like grid references. Cells may contain *labels*, *values* or *formulae*. The current cell has a black border.

Label Any words, headings or messages used to help the layout and organisation of the spreadsheet.

Value A number placed in a cell. It may be used as input to a calculation.

Tasks 1, 2 and 3 are written for you to become familiar with how the main functions of a spreadsheet program work. Afterwards there are sections on different topics where spreadsheets can be used.

Task 1. To generate the whole numbers from 1 to 10 in column A.

    (a) In cell A1 type '1' and press *Return*. This will automatically take you to the cell below. [NOTE that you must use the *Return* button and not the arrow keys to move down the column.]

    (b) In cell A2 type the formula '= A1 + 1' and press *Return*. [NOTE that the = sign is needed before any formula.]

    (c) We now want to copy the formula in A2 down column A as far as A10. Click on A2 again and put the arrow in the bottom right corner of cell A2 (a + sign will appear) and drag down to A10.

Task 2. To generate the odd numbers in column B.

    (a) In B1 type '1' (press *Return*).

    (b) In B2 type the formula ' $= B1 + 2$' (press *Return*).

    (c) Click in B2 and copy the formula down column B as far as B10.

Task 3. To generate the first 15 square numbers.

    (a) As before generate the numbers from 1 to 15 in cells A1 to A15.

    (b) In B1 put the formula ' $= A1 * A1$' and press *Return*.

    (c) Click in B1 and copy the formula down as far as B15.

Pie charts and bar charts using a spreadsheet on a computer

Example

Display the data about the activities in one day.

Enter the headings: *Sleep* in A1, *School* in B1 etc. [Use the *tab* key to move across the page.]

Enter the data: 8 in A2, 7 in B2 etc.

	A	B	C	D	E	F	G	H	I
1	Sleep	School	Tv	Eating	Homework	Other			
2	8	7	1.5	1	1.5	5			
3									
4									

Now highlight all the cells from A1 to F2. [Click on A1 and drag across to F2.]

Click on the (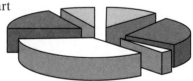) Chart wizard on the toolbar.

Select 'pie' and then choose one of the examples displayed. Follow the on-screen prompts.

Alternatively, for a bar chart, select 'charts' after clicking on the chart wizard. Proceed as above.

You will be able to display your charts with various '3D' effects, possibly in colour. This approach is recommended when you are presenting data that you have collected as part of an investigation.

Scatter graphs on a computer

Example

Plot a scatter graph showing the marks of 10 students in Maths and Science.

Enter the headings: *Maths* in A1, *Science* in B1
Enter the data as shown.

Now highlight all the cells from A2 to B11.
[Click on A1 and drag across and down to B11.]

Click on the ( ) Chart wizard on the toolbar.

Select XY (Scatter) and select the picture which looks like a
scatter graph.

Follow the on-screen prompts.

On 'Titles' enter: Chart title: Maths/Science results
                    Value (X) axis: Maths
                    Value (Y) axis: Science

	A	B
1	**Maths**	**Science**
2	**23**	**30**
3	**45**	**41**
4	**73**	**67**
5	**35**	**74**
6	**67**	**77**
7	**44**	**50**
8	**32**	**41**
9	**66**	**55**
10	**84**	**70**
11	**36**	**32**

Experiment with 'Axes', 'Gridlines', 'Legend' and 'Data Labels'.

Task   Enter the data on a spreadsheet and print a scatter graph.
           What does each scatter graph show?

(a)

Height	Armspan
162	160
155	151
158	157
142	144
146	148
165	163
171	167
148	150
150	147

(b)

Temperature	Sales
23	7
18	14
7	23
20	9
4	30
12	19
15	15
18	15
10	20

Revision exercise 1A

1. Evaluate, without a calculator:
   (a) $148 \div 0 \cdot 8$
   (b) $0 \cdot 024 \div 0 \cdot 000\,16$
   (c) $(0 \cdot 2)^2 \div (0 \cdot 1)^3$
   (d) $2 - \frac{1}{2} - \frac{1}{3} - \frac{1}{4}$

   (e) $1\frac{3}{4} \times 1\frac{3}{5}$
   (f) $\dfrac{1\frac{1}{6}}{1\frac{2}{3} + 1\frac{1}{4}}$

2. On each bounce, a ball rises to $\frac{4}{5}$ of its previous height. To what
   height will it rise after the third bounce, if dropped from a height of
   250 cm?

3. A man spends $\frac{1}{3}$ of his salary on accommodation and $\frac{2}{5}$ of the remainder on food. What fraction is left for other purposes?

4. $a = \frac{1}{2}$, $b = \frac{1}{4}$. Which one of the following has the greatest value?

(i) ab (ii) $a + b$ (iii) $\dfrac{a}{b}$ (iv) $\dfrac{b}{a}$ (v) $(ab)^2$

5. Express 0·054 73:
   (a) correct to three significant figures
   (b) correct to three decimal places
   (c) in standard form.

6. Evaluate $\frac{2}{3} + \frac{4}{7}$, correct to three decimal places.

7. Evaluate the following and give the answer in standard form:

(a) $3600 \div 0{\cdot}000\,12$ (b) $\dfrac{3{\cdot}33 \times 10^4}{9 \times 10^{-1}}$ (c) $(30\,000)^3$

8. (a) \$143 is divided in the ratio $2 : 3 : 6$; calculate the smallest share.
   (b) A prize is divided between three people X, Y and Z. If the ratio of X's share to Y's share is $3 : 1$ and Y's share to Z's share is $2 : 5$, calculate the ratio of X's share to Z's share.
   (c) If $a : 3 = 12 : a$, calculate the positive value of a.

9. Labour costs, totalling \$47·25, account for 63% of a car repair bill. Calculate the total bill.

10. (a) Convert to percentages:
     (i) 0·572   (ii) $\frac{7}{8}$
   (b) Express 2·6 kg as a percentage of 6·5 kg.
   (c) In selling a red herring for 92c, a fishmonger makes a profit of 15%. Find the cost price of the fish.

11. The length of a rectangle is decreased by 25% and the width is increased by 40%. Calculate the percentage change in the area of the rectangle.

12. (a) What sum of money, invested at 9% interest per year, is needed to provide an income of \$45 per year?
   (b) A particle increases its speed from 8×10^5 m/s to $1{\cdot}1 \times 10^6$ m/s. What is the percentage increase?

13. A family on holiday in France exchanged \$450 for euros when the exchange rate was 1·41 euros to the dollar. They spent 500 euros and then changed the rest back into dollars, by which time the exchange rate had become 1·46 euros to the dollar. How much did the holiday cost?   (Answer in dollars.)

14. Given that

$$t = 2\pi \sqrt{\left(\dfrac{l}{g}\right)},$$

find the value of t, to three sig. fig., when $l = 2{\cdot}31$ and $g = 9{\cdot}81$

15. A map is drawn to a scale of 1 : 10 000. Find:
  (a) the distance between two railway stations which appear on the map 24 cm apart.
  (b) the area, in square kilometres, of a lake which has an area of 100 cm² on the map.

16. A map is drawn to a scale of 1 : 2000. Find:
  (a) the actual distance between two points, which appear 15 cm apart on the map.
  (b) the length on the map of a road, which is 1·2 km in length.
  (c) the area on the map of a field, with an actual area of 60 000 m².

17. (a) On a map, the distance between two points is 16 cm. Calculate the scale of the map if the actual distance between the points is 8 km.
  (b) On another map, two points appear 1·5 cm apart and are in fact 60 km apart. Calculate the scale of the map.

18. (a) A house is bought for $20 000 and sold for $24 400. What is the percentage profit?
  (b) A piece of fish, initially of mass 2·4 kg, is cooked and subsequently has mass 1·9 kg. What is the percentage loss in mass?
  (c) An article is sold at a 6% loss for $225·60. What was the cost price?

19. (a) Convert into metres per second:
    (i) 700 cm/s   (ii) 720 km/h   (iii) 18 km/h
  (b) Convert into kilometres per hour:
    (i) 40 m/s   (ii) 0·6 m/s

20. (a) Calculate the speed (in metres per second) of a slug which moves a distance of 30 cm in 1 minute.
  (b) Calculate the time taken for a bullet to travel 8 km at a speed of 5000 m/s.
  (c) Calculate the distance flown, in a time of four hours, by a pigeon which flies at a speed of 12 m/s.

21. A motorist travelled 200 km in five hours. Her average speed for the first 100 km was 50 km/h. What was her average speed for the second 100 kilometres?

22. 1   3   8   9   10
  From these numbers, write down:
  (a) the prime number, (Note: 1 is not a prime number)
  (b) a multiple of 5,
  (c) two square numbers,
  (d) two factors of 32.
  (e) Find two numbers m and n from the list such that $m = \sqrt{n}$ and $n = \sqrt{81}$.
  (f) If each of the numbers in the list can be used once, find p, q, r, s, t such that $(p + q)r = 2(s + t) = 36$.

23. The value of t is given by

$$t = 2\pi\sqrt{\left(\frac{2 \cdot 31^2 + 0 \cdot 9^2}{2 \cdot 31 \times 9 \cdot 81}\right)}.$$

Without using a calculator, and using suitable approximate values for the numbers in the formula, find an estimate for the value of t. (To earn the marks in this question you must show the various stages of your working.)

24. Throughout his life Baichu's heart has beat at an average rate of 72 beats per minute. Baichu is sixty years old. How many times has his heart beat during his life? Give the answer in standard form correct to two significant figures.

25. Estimate the answer correct to one significant figure. Do not use a calculator.
   (a) $(612 \times 52) \div 49 \cdot 2$ (b) $(11 \cdot 7 + 997 \cdot 1) \times 9 \cdot 2$

   (c) $\sqrt{\left(\frac{91 \cdot 3}{10 \cdot 1}\right)}$ (d) $\pi\sqrt{(5 \cdot 2^2 + 18 \cdot 2)}$

26. Evaluate the following using a calculator:
   (give answers to 4 s.f.)

   (a) $\dfrac{0 \cdot 74}{0 \cdot 81 \times 1 \cdot 631}$ (b) $\sqrt{\left(\dfrac{9 \cdot 61}{8 \cdot 34 - 7 \cdot 41}\right)}$

   (c) $\left(\dfrac{0 \cdot 741}{0 \cdot 8364}\right)^4$ (d) $\dfrac{8 \cdot 4 - 7 \cdot 642}{3 \cdot 333 - 1 \cdot 735}$

27. Evaluate the following and give the answers to three significant figures:

   (a) $\sqrt[3]{(9 \cdot 61 \times 0 \cdot 0041)}$ (b) $\left(\dfrac{1}{9 \cdot 5} - \dfrac{1}{11 \cdot 2}\right)^3$

   (c) $\dfrac{15 \cdot 6 \times 0 \cdot 714}{0 \cdot 0143 \times 12}$ (d) $\sqrt[4]{\left(\dfrac{1}{5 \times 10^3}\right)}$

28. The edges of a cube are all increased by 10%. What is the percentage increase in the volume?

Examination exercise 1B

1. Calculate $\dfrac{5^2}{2^5}$

   (a) giving your answer as a fraction,   [1]
   (b) giving your answer as a decimal.   [1]

Cambridge IGCSE Mathematics 0580
Paper 2 Q1 June 2005

2. Work out the value of $1 + \dfrac{2}{3 + \frac{4}{5+6}}$. [2]

Cambridge IGCSE Mathematics 0580
Paper 22 Q3 November 2008

3. Write down
   (a) an irrational number, [1]
   (b) a prime number between 60 and 70. [1]

Cambridge IGCSE Mathematics 0580
Paper 2 Q9 June 2007

4. At 05 06 Mr Ho bought 850 fish at a fish market for $2·62 each.
   95 minutes later he sold them all to a supermarket for $2·86 each.
   (a) What was the time when he sold the fish? [1]
   (b) Calculate his total profit. [1]

Cambridge IGCSE Mathematics 0580
Paper 21 Q3 June 2009

5. Write down the next term in each of the following sequences.
   (a) 8.2, 6.2, 4.2, 2.2, 0.2, ... [1]
   (b) 1, 3, 6, 10, 15, ... [1]

Cambridge IGCSE Mathematics 0580
Paper 2 Q4 November 2005

6. (a) The formula for the nth term of the sequence
   1, 5, 14, 30, 55, 91, ... is $\dfrac{n(n+1)(2n+1)}{6}$.
   Find the 15th term. [1]
   (b) The nth term of the sequence 17, 26, 37, 50, 65, ... is $(n+3)^2 + 1$.
   Write down the formula for the nth term of the sequence 26, 37,
   50, 65, 82, ... [1]

Cambridge IGCSE Mathematics 0580
Paper 22 Q4 June 2008

7. To raise money for charity, Jalaj walks 22 km, correct to the nearest
   kilometre, every day for 5 days.
   (a) Complete the statement in the answer space for the distance,
   d km, he walks in one day. [2]
   (b) He raises $1.60 for every kilometre that he walks.
   Calculate the least amount of money that he raises at the end
   of the 5 days. [1]

Cambridge IGCSE Mathematics 0580
Paper 2 Q7 June 2005

8. The distance between Singapore and Sydney is 6300 km correct to
   the nearest 100 km.
   A businessman travelled from Singapore to Sydney and then back
   to Singapore.

He did this six times in a year.
Between what limits is the total distance he travelled?
Write your answer as ....km $\leqslant$ total distance travelled $<$ ....km.   [2]

Cambridge IGCSE Mathematics 0580
Paper 2 Q9 June 2006

9. A rectangle has sides of length 6·1 cm and 8·1 cm correct to
   1 decimal place.
   Calculate the upper bound for the area of the rectangle as
   accurately as possible.   [2]

Cambridge IGCSE Mathematics 0580
Paper 21 Q7 November 2008

10. In 2005 there were 9 million bicycles in Beijing, correct to the
    nearest million.
    The average distance travelled by each bicycle in one day was
    6.5 km correct to one decimal place.
    Work out the upper bound for the **total** distance travelled by all the
    bicycles in one day.   [2]

Cambridge IGCSE Mathematics 0580
Paper 21 Q6 June 2009

11. The mass of the Earth is $\dfrac{1}{95}$ of the mass of the planet Saturn.

    The mass of the Earth is $5 \cdot 97 \times 10^{24}$ kilograms.
    Calculate the mass of the planet Saturn, giving your answer in
    standard form, correct to 2 significant figures.   [3]

Cambridge IGCSE Mathematics 0580
Paper 2 Q10 November 2005

12. Maria, Carolina and Pedro receive $800 from their grandmother in
    the ratio
    $$\text{Maria : Carolina : Pedro} = 7 : 5 : 4.$$

    (a) Calculate how much money each receives.   [3]

    (b) Maria spends $\dfrac{2}{7}$ of her money and then invests the rest for two

        years at 5% per year simple interest.
        How much money does Maria have at the end of the two years?   [3]

    (c) Carolina spends all of her money on a hi-fi set and two years
        later sells it at a loss of 20%.
        How much money does Carolina have at the end of the two
        years?   [2]

(d) Pedro spends some of his money and at the end of the two years
he has $100.
Write down and simplify the ratio of the amounts of money
Maria, Carolina and Pedro have at the end of the two years.   [2]
(e) Pedro invests his $100 for two years at a rate of 5% per year
compound interest.
Calculate how much money he has at the end of these two
years.   [2]

Cambridge IGCSE Mathematics 0580
Paper 4 Q1 November 2006

13. In 2004 Colin had a salary of $7200.
    (a) This was an increase of 20% on his salary in 2002.
        Calculate his salary in 2002.   [2]
    (b) In 2006 his salary increased to $8100.
        Calculate the percentage increase from 2004 to 2006.   [2]

Cambridge IGCSE Mathematics 0580
Paper 2 Q16 June 2006

14. A student played a computer game 500 times and won 370 of these
    games.
    He then won the next x games and lost none.
    He has now won 75% of the games he has played.
    Find the value of x.   [4]

Cambridge IGCSE Mathematics 0580
Paper 21 Q17 June 2008

15.

NORTH EASTERN BANK	SOUTH WESTERN BANK
SAVINGS ACCOUNT	SAVINGS ACCOUNT
5%	4.9%
Per Year	Per Year
Simple Interest	Compound Interest

Kalid and his brother have $2000 each to invest for 3 years.
(a) North Eastern Bank advertises savings with **simple** interest at
    5% per year.
    Kalid invests his money in this bank.
    How much money will he have at the end of 3 years?   [2]
(b) South Western Bank advertises savings with **compound** interest
    at 4.9% per year.
    Kalid's brother invests his money in this bank.
    At the end of 3 years, how much **more** money will he have than Kalid?   [3]

Cambridge IGCSE Mathematics 0580
Paper 2 Q22 June 2007

16. Use your calculator to work out
(a) $\sqrt{(7 + 6 \times 243^{0.2})}$, [1]
(b) $2 - \tan 30° \times \tan 60°$. [1]

Cambridge IGCSE Mathematics 0580
Paper 2 Q3 November 2006

17. Hassan sells fruit and vegetables at the market.
(a) The mass of fruit and vegetables he sells is in the ratio
fruit : vegetables = 5 : 7.
Hassan sells 1.33 **tonnes** of vegetables.
How many **kilograms** of fruit does he sell? [3]
(b) The amount of money Hassan receives from selling fruit and
vegetables is in the ratio
fruit : vegetables = 9 : 8.
Hassan receives a **total** of $765 from selling fruit and vegetables.
Calculate how much Hassan receives from selling fruit. [2]
(c) Calculate the average price of Hassan's fruit, in dollars per
kilogram. [2]
(d) (i) Hassan sells oranges for $0.35 per kilogram.
He reduces this price by 40%.
Calculate the new price per kilogram. [2]
(ii) The price of $0.35 per kilogram of oranges is an increase of
25% on the previous day's price.
Calculate the previous day's price. [2]

Cambridge IGCSE Mathematics 0580
Paper 4 Q1 June 2005

18. (a) The scale of a map is 1 : 20 000 000.
On the map, the distance between Cairo and Addis Ababa is
12 cm.
(i) Calculate the distance, in kilometres, between Cairo and
Addis Ababa. [2]
(ii) On the map the area of a desert region is 13 square
centimetres.
Calculate the actual area of this desert region, in square
kilometres. [2]

(b) (i) The actual distance between Cairo and Khartoum is 1580 km.
On a different map this distance is represented by 31.6 cm.
Calculate, in the form 1 : n, the scale of this map. [2]
(ii) A plane flies the 1580 km from Cairo to Khartoum.
It departs from Cairo at 1155 and arrives in Khartoum at
1403.
Calculate the average speed of the plane, in kilometres per hour. [4]

Cambridge IGCSE Mathematics 0580
Paper 4 Q1 June 2007

19. $1 + 2 + 3 + 4 + 5 + \ldots + n = \frac{n(n+1)}{2}$

    (a) (i) Show that this formula is true for the sum of the first 8
          natural numbers.   [2]
        (ii) Find the sum of the first 400 natural numbers.   [1]
    (b) (i) Show that $2 + 4 + 6 + 8 + \ldots + 2n = n(n + 1)$.   [1]
        (ii) Find the sum of the first 200 even numbers.   [1]
       (iii) Find the sum of the first 200 odd numbers.   [1]
    (c) (i) Use the formula at the beginning of the question to find the
          sum of the first $2n$ natural numbers.   [1]
        (ii) Find a formula, in its simplest form, for
          $1 + 3 + 5 + 7 + 9 + \ldots + (2n - 1)$.
          Show your working.   [2]

Cambridge IGCSE Mathematics 0580
Paper 4 Q10 November 2008

20. Each year a school organises a concert.
    (a) (i) In 2004 the cost of organising the concert was $385.
          In 2005 the cost was 10% less than in 2004.
          Calculate the cost in 2005.   [2]
        (ii) The cost of $385 in 2004 was 10% more than the cost in
          2003.
          Calculate the cost in 2003.   [2]
    (b) (i) In 2006 the number of tickets sold was 210.
          The ratio
          Number of adult tickets : Number of student tickets was 23 : 19.
          How many adult tickets were sold?   [2]
        (ii) Adult tickets were $2·50 each and student tickets were $1·50
          each.
          Calculate the **total** amount **received** from selling the tickets. [2]
       (iii) In 2006 the cost of organising the concert was $410.
          Calculate the percentage profit in 2006.   [2]
    (c) In 2007, the number of tickets sold was again 210.
      Adult tickets were $2·60 each and student tickets were $1·40
      each.
      The total amount received from selling the 210 tickets was $480.
      How many student tickets were sold?   [4]

Cambridge IGCSE Mathematics 0580
Paper 4 Q1 November 2007

2 ALGEBRA 1

Isaac Newton (1642–1727) is thought by many to have been one of the greatest intellects of all time. He went to Trinity College Cambridge in 1661 and by the age of 23 he had made three major discoveries: the nature of colours, the calculus and the law of gravitation. He used his version of the calculus to give the first satisfactory explanation of the motion of the Sun, the Moon and the stars. Because he was extremely sensitive to criticism, Newton was always very secretive, but he was eventually persuaded to publish his discoveries in 1687.

3   Use directed numbers in practical situations

20   Substitute numbers for words and letters in formulae; construct and transform more complicated formulae and equations

21   Manipulate directed numbers; expand products of algebraic expressions; factorise expressions

24   Solve simple linear equations in one unknown; solve simultaneous linear equations in two unknowns; solve quadratic equations by factorisation, completing the square or use of the formula

2.1 Negative numbers

- If the weather is very cold and the temperature is 3 degrees below zero, it is written $-3°$.

- If a golfer is 5 under par for his round, the scoreboard will show -5.

- On a bank statement if someone is \$55 overdrawn [or 'in the red'] it would appear as $-\$55$.

The above are examples of the use of negative numbers.

An easy way to begin calculations with negative numbers is to think about changes in temperature:

(a) Suppose the temperature is $-2°$ and it rises by $7°$.
   The new temperature is $5°$.
   We can write $-2 + 7 = 5$.

(b) Suppose the temperature is $-3°$ and it falls by $6°$.
The new temperature is $-9°$.
We can write $-3 - 6 = -9$.

Exercise 1

In Questions **1** to **12** move up or down the thermometer to find the new temperature.

1. The temperature is $+8°$ and it falls by $3°$.

2. The temperature is $-8°$ and it rises by $4°$.

3. The temperature is $+4°$ and it falls by $5°$.

4. The temperature is $-3°$ and it rises by $7°$.

5. The temperature is $+2°$ and it falls by $6°$.

6. The temperature is $+4°$ and it rises by $8°$.

7. The temperature is $-1°$ and it falls by $6°$.

8. The temperature is $+9°$ and it falls by $14°$.

9. The temperature is $-5°$ and it rises by $1°$.

10. The temperature is $-13°$ and it rises by $13°$.

11. Some land in Bangladesh is below sea level.
Here are the heights, above sea level, of five villages.
A $1\,m$ **B** $-4\,m$ **C** $21\,m$ **D** $-2\,m$ **E** $-1\cdot5\,m$
(a) Which village is safest from flooding?
(b) Which village is most at risk from serious flooding?

12. A diver is below the surface of the water at $-15\,m$. He dives down by $6\,m$, then rises $4\,m$.
Where is he now?

2.2 Directed numbers

To add two directed numbers with the same sign, find the sum of the numbers and give the answer the same sign.

Example 1

$+3 + (+5) = +3 + 5 = +8$
$-7 + (-3) = -7 - 3 = -10$
$-9\cdot1 + (-3\cdot1) = 9\cdot1 - 3\cdot1 = -12\cdot2$
$-2 + (-1) + (-5) = (-2 - 1) - 5$
$\qquad\qquad\qquad = -3 - 5$
$\qquad\qquad\qquad = -8$

To add two directed numbers with different signs, find the difference between the numbers and give the answer the sign of the larger number.

Example 2

$+7 + (-3) = +7 - 3 \ = +4$
$+9 + (-12) = +9 - 12 = 3$
$-8 + (+4) = -8 + 4 \ = -4$

To subtract a directed number, change its sign and add.

Example 3

$$+7 - (+5) \ = +7 - 5 \ = +2$$
$$+7 - (-5) \ = +7 + 5 \ = +12$$
$$-8 - (+4) \ = -8 - 4 \ = -12$$
$$-9 - (-11) = -9 + 11 = +2$$

Exercise 2

1. $+7 + (+6)$
2. $+11 + (+200)$
3. $-3 + (-9)$
4. $-7 + (-24)$
5. $-5 + (-61)$
6. $+0{\cdot}2 + (+5{\cdot}9)$
7. $+5 + (+4{\cdot}1)$
8. $-8 + (-27)$
9. $+17 + (+1{\cdot}7)$
10. $-2 + (-3) + (-4)$
11. $-7 + (+4)$
12. $+7 + (-4)$
13. $-9 + (+7)$
14. $+16 + (-30)$
15. $+14 + (-21)$
16. $-7 + (+10)$
17. $-19 + (+200)$
18. $+7{\cdot}6 + (-9{\cdot}8)$
19. $-1{\cdot}8 + (+10)$
20. $-7 + (+24)$
21. $+7 - (+5)$
22. $+9 - (+15)$
23. $-6 - (+9)$
24. $-9 - (+5)$
25. $+8 - (+10)$
26. $-19 - (-7)$
27. $-10 - (+70)$
28. $-5{\cdot}1 - (+8)$
29. $-0{\cdot}2 - (+4)$
30. $+5{\cdot}2 - (-7{\cdot}2)$
31. $-4 + (-3)$
32. $+6 - (-2)$
33. $+8 + (-4)$
34. $-4 - (+6)$
35. $+7 - (-4)$
36. $+6 + (-2)$
37. $+10 - (+30)$
38. $+19 - (+11)$
39. $+4 + (-7) + (-2)$
40. $-3 - (+2) + (-5)$
41. $-17 - (-1) + (-10)$
42. $-5 + (-7) - (+9)$
43. $+9 + (-7) - (-6)$
44. $-7 - (-8)$
45. $-10{\cdot}1 + (-10{\cdot}1)$
46. $-75 - (-25)$
47. $-204 - (+304)$
48. $-7 + (-11) - (+11)$
49. $+17 - (+17)$
50. $-6 + (-7) - (+8)$
51. $+7 + (-7{\cdot}1)$
52. $-11 - (-4) + (+3)$
53. $-2 - (-8{\cdot}7)$
54. $+7 + (-11) + (+5)$
55. $-610 + (-240)$
56. $-7 - (-3) - (-8)$
57. $+9 - (-6) + (-9)$
58. $-1 - (-5) + (-8)$
59. $-2{\cdot}1 + (-9{\cdot}9)$
60. $-47 - (-16)$

When two directed numbers with the same sign are multiplied together,
the answer is positive.
- $+7 \times (+3) = +21$
- $-6 \times (-4) = +24$

When two directed numbers with different signs are multiplied together,
the answer is negative.
- $-8 \times (+4) = -32$
- $+7 \times (-5) = -35$
- $-3 \times (+2) \times (+5) = -6 \times (+5) = -30$

When dividing directed numbers, the rules are the same as in multiplication.
- $-70 \div (-2) = +35$
- $+12 \div (-3) = -4$
- $-20 \div (+4) = -5$

Exercise 3

1. $+2 \times (-4)$
2. $+7 \times (+4)$
3. $-4 \times (-3)$
4. $-6 \times (-4)$
5. $-6 \times (-3)$
6. $+5 \times (-7)$
7. $-7 \times (-7)$
8. $-4 \times (+3)$
9. $+0{\cdot}5 \times (-4)$
10. $-1\frac{1}{2} \times (-6)$
11. $-8 \div (+2)$
12. $+12 \div (+3)$

13. $+36 \div (-9)$	**14.** $-40 \div (-5)$	**15.** $-70 \div (-1)$	**16.** $-56 \div (+8)$
17. $-\frac{1}{2} \div (-2)$	**18.** $-3 \div (+5)$	**19.** $+0\cdot1 \div (-10)$	**20.** $-0\cdot02 \div (-100)$
21. $-11 \times (-11)$	**22.** $-6 \times (-1)$	**23.** $+12 \times (-50)$	**24.** $-\frac{1}{2} \div (+\frac{1}{2})$
25. $-600 \div (+30)$	**26.** $-5\cdot2 \div (+2)$	**27.** $+7 \times (-100)$	**28.** $-6 \div (-\frac{1}{3})$
29. $100 \div (-0\cdot1)$	**30.** -8×-80	**31.** $-3 \times (-2) \times (-1)$	**32.** $+3 \times (-7) \times (+2)$
33. $+0\cdot4 \div (-1)$	**34.** $-16 \div (+40)$	**35.** $+0\cdot2 \times (-1000)$	**36.** $-7 \times (-5) \times (-1)$
37. $-14 \div (+7)$	**38.** $-7 \div (-14)$	**39.** $+1\frac{1}{4} \div (-5)$	**40.** $-6 \times (-\frac{1}{2}) \times (-30)$

Exercise 4

1. $-7 + (-3)$	**2.** $-6 - (-7)$	**3.** $-4 \times (-3)$	**4.** $-4 \times (+7)$
5. $4 - (+6)$	**6.** $-4 \times (-4)$	**7.** $+6 \div (-2)$	**8.** $+8 - (-6)$
9. $-7 \times (+4)$	**10.** $-8 \div (-2)$	**11.** $+10 \div (-60)$	**12.** (-3^2)
13. $40 - (+70)$	**14.** $-6 \times (-4)$	**15.** $(-1)^5$	**16.** $-8 \div (+4)$
17. $+10 \times (-3)$	**18.** $-7 \times (-1)$	**19.** $+10 + (-7)$	**20.** $+12 - (-4)$
21. $+100 + (-7)$	**22.** $-60 \times (-40)$	**23.** $-20 \div (-2)$	**24.** $(-1)^{20}$
25. $6 - (+10)$	**26.** $-6 \times (+4) \times (-2)$	**27.** $+8 \div (-8)$	**28.** $0 \times (-6)$
29. $(-2)^3$	**30.** $+100 - (-70)$	**31.** $+18 \div (-6)$	**32.** $(-1)^{12}$
33. $-6 - (-7)$	**34.** $(-2)^2 + (-4)$	**35.** $+8 - (-7)$	**36.** $+7 + (-2)$
37. $-6 \times (+0\cdot4)$	**38.** $-3 \times (-6) \times (-10)$	**39.** $(-2)^2 + (+1)$	**40.** $+6 - (+1000)$
41. $(-3)^2 - 7$	**42.** $-12 \div \frac{1}{4}$	**43.** $-30 \div -\frac{1}{2}$	**44.** $5 - (+7) + (-0\cdot5)$
45. $(-2)^5$	**46.** $0 \div (-\frac{1}{5})$	**47.** $(-0\cdot1)^2 \times (-10)$	**48.** $3 - (+19)$
49. $2\cdot1 + (-6\cdot4)$	**50.** $(-\frac{1}{2})^2 \div (-4)$		

2.3 Formulae

When a calculation is repeated many times it is often helpful to use a formula. Publishers use a formula to work out the selling price of a book based on the production costs and the expected sales of the book.

Exercise 5

1. The final speed v of a car is given by the formula $v = u + at$.
   [u = initial speed, a = acceleration, t = time taken]
   Find v when $u = 15$ m/s, $a = 0\cdot2$ m/s^2, $t = 30$ s.

2. The time period T of a simple pendulum is given by the formula

   $T = 2\pi\sqrt{\left(\frac{l}{g}\right)}$, where l is the length of the pendulum and g is the

   gravitational acceleration. Find T when $l = 0.65$ m, $g = 9.81$ m/s^2
   and $\pi = 3\cdot142$.

3. The total surface area A of a cone is related to the radius r and the slant height l by the formula $A = \pi r(r + l)$. Find A when $r = 7$ cm and $l = 11$ cm.

4. The sum S of the squares of the integers from 1 to n is given by
$S = \frac{1}{6}n(n+1)(2n+1)$. Find S when $n = 12$.

5. The acceleration a of a train is found using the formula $a = \dfrac{v^2 - u^2}{2s}$.
Find a when $v = 20\,\text{m/s}$, $u = 9\,\text{m/s}$ and $s = 2\cdot5\,\text{m}$.

6. Einstein's famous equation relating energy, mass and the speed of
light is $E = mc^2$. Find E when $m = 0\cdot0001\,\text{kg}$ and $c = 3 \times 10^8\,\text{m/s}$.

7. The distance s travelled by an accelerating rocket is given by
$s = ut + \frac{1}{2}at^2$. Find s when $u = 3\,\text{m/s}$, $t = 100\,\text{s}$ and $a = 0\cdot1\,\text{m/s}^2$.

8. Find a formula for the area of
the shape opposite, in terms
of a, b and c.

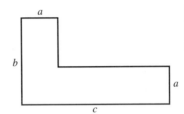

You can find out more about
area in Unit 3 on page 92.

9. Find a formula for the length of the shaded part below, in terms of
p, q and r.

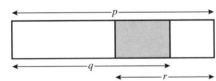

10. A fish lays brown eggs or white eggs and it likes to lay them in a
certain pattern. Each brown egg is surrounded by six white eggs.
Here there are 3 brown eggs and 14 white eggs.
(a) How many eggs does it lay altogether if it lays 200 brown eggs?
(b) How many eggs does it lay altogether if it lays n brown eggs?

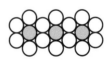

11. In the diagrams below the rows of black tiles are surrounded by
white tiles.

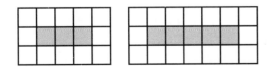

Find a formula for the number of white tiles which would be
needed to surround a row of n black tiles.

Example

When $a = 3$, $b = -2$, $c = 5$, find the value of:

(a)	$3a + b$	(b)	$ac + b^2$	(c)	$\dfrac{a+c}{b}$	(d)	$a(c - b)$

(a) $3a + b$	(b) $ac + b^2$	(c) $\dfrac{a+c}{b}$	(d) $a(c - b)$
$= (3 \times 3) + (-2)$	$= (3 \times 5) + (-2)^2$	$= \dfrac{3+5}{-2}$	$= 3[5 - (-2)]$
$= 9 - 2$	$= 15 + 4$	$= \dfrac{8}{-2}$	$= 3[7]$
$= 7$	$= 19$	$= -4$	$= 21$

Notice that working *down* the page is often easier to follow.

Exercise 6

Evaluate the following:
For questions **1** to **12**, $a = 3$, $c = 2$, $e = 5$.

1. $3a - 2$	**2.** $4c + e$	**3.** $2c + 3a$	**4.** $5e - a$
5. $e - 2c$	**6.** $e - 2a$	**7.** $4c + 2e$	**8.** $7a - 5e$
9. $c - e$	**10.** $10a + c + e$	**11.** $a + c - e$	**12.** $a - c - e$

For questions **13** to **24**, $h = 3$, $m = -2$, $t = -3$.

13. $2m - 3$	**14.** $4t + 10$	**15.** $3h - 12$	**16.** $6m + 4$
17. $9t - 3$	**18.** $4h + 4$	**19.** $2m - 6$	**20.** $m + 2$
21. $3h + m$	**22.** $t - h$	**23.** $4m + 2h$	**24.** $3t - m$

For questions **25** to **36**, $x = -2$, $y = -1$, $k = 0$.

25. $3x + 1$	**26.** $2y + 5$	**27.** $6k + 4$	**28.** $3x + 2y$
29. $2k + x$	**30.** xy	**31.** xk	**32.** $2xy$
33 $2(x + k)$	**34.** $3(k + y)$	**35.** $5x - y$	**36.** $3k - 2x$

$2x^2$ means $2(x^2)$.
$(2x)^2$ means 'work out $2x$ and *then* square it'.
$-7x$ means $-7(x)$.
$-x^2$ means $-1(x^2)$.

Example

When $x = -2$, find the value of:
(a) $2x^2 - 5x$ 　　　　　　　　(b) $(3x)^2 - x^2$

(a) $2x^2 - 5x = 2(-2)^2 - 5(-2)$	(b) $(3x)^2 - x^2 = (3 \times -2)^2 - 1(-2)^2$
$= 2(4) + 10$	$= (-6)^2 - 1(4)$
$= 18$	$= 36 - 4$
	$= 32$

Exercise 7

If $x = -3$ and $y = 2$, evaluate the following:

1. x^2
2. $3x^2$
3. y^2
4. $4y^2$
5. $(2x)^2$
6. $2x^2$
7. $10 - x^2$
8. $10 - y^2$
9. $20 - 2x^2$
10. $20 - 3y^2$
11. $5 + 4x$
12. $x^2 - 2x$
13. $y^2 - 3x^2$
14. $x^2 - 3y$
15. $(2x)^2 - y^2$
16. $4x^2$
17. $(4x)^2$
18. $1 - x^2$
19. $y - x^2$
20. $x^2 + y^2$
21. $x^2 - y^2$
22. $2 - 2x^2$
23. $(3x)^2 + 3$
24. $11 - xy$
25. $12 + xy$
26. $(2x)^2 - (3y)^2$
27. $2 - 3x^2$
28. $y^2 - x^2$
29. $x^2 + y^3$
30. $\dfrac{x}{y}$
31. $10 - 3x$
32. $2y^2$
33. $25 - 3y$
34. $(2y)^2$
35. $-7 + 3x$
36. $-8 + 10y$
37. $(xy)^2$
38. xy^2
39. $-7 + x^2$
40. $17 + xy$
41. $-5 - 2x^2$
42. $10 - (2x)^2$
43. $x^2 + 3x + 5$
44. $2x^2 - 4x + 1$
45. $\dfrac{x^2}{y}$

Example

When $a = -2$, $b = 3$, $c = -3$, evaluate:

(a) $\dfrac{2a(b^2 - a)}{c}$ (b) $\sqrt{(a^2 + b^2)}$

(a) $(b^2 - a) = 9 - (-2)$
$$= 11$$
$$\therefore \quad \frac{2a(b^2 - a)}{c} = \frac{2 \times (-2) \times (11)}{-3}$$
$$= 14\frac{2}{3}$$

(b) $a^2 + b^2 = (-2)^2 + (3)^2$
$$= 4 + 9$$
$$= 13$$
$$\sqrt{(a^2 + b^2)} = \sqrt{13}$$

Exercise 8

Evaluate the following:
In questions **1** to **16**, $a = 4$, $b = -2$, $c = -3$.

1. $a(b + c)$
2. $a^2(b - c)$
3. $2c(a - c)$
4. $b^2(2a + 3c)$
5. $c^2(b - 2a)$
6. $2a^2(b + c)$
7. $2(a + b + c)$
8. $3c(a - b - c)$
9. $b^2 + 2b + a$
10. $c^2 - 3c + a$
11. $2b^2 - 3b$
12. $\sqrt{(a^2 + c^2)}$
13. $\sqrt{(ab + c^2)}$
14. $\sqrt{(c^2 - b^2)}$
15. $\dfrac{b^2}{a} + \dfrac{2c}{b}$
16. $\dfrac{c^2}{b} + \dfrac{4b}{a}$

In questions **17** to **32**, $k = -3$, $m = 1$, $n = -4$.

17. $k^2(2m - n)$
18. $5m\sqrt{(k^2 + n^2)}$
19. $\sqrt{(kn + 4m)}$
20. $kmn(k^2 + m^2 + n^2)$
21. $k^2m^2(m - n)$
22. $k^2 - 3k + 4$
23. $m^3 + m^2 + n^2 + n$
24. $k^3 + 3k$
25. $m(k^2 - n^2)$
26. $m\sqrt{(k - n)}$
27. $100k^2 + m$
28. $m^2(2k^2 - 3n^2)$
29. $\dfrac{2k + m}{k - n}$
30. $\dfrac{kn - k}{2m}$
31. $\dfrac{3k + 2m}{2n - 3k}$
32. $\dfrac{k + m + n}{k^2 + m^2 + n^2}$

In questions **33** to **48**, $w = -2$, $x = 3$, $y = 0$, $z = -\frac{1}{2}$.

33. $\dfrac{w}{z} + x$

34. $\dfrac{w + x}{z}$

35. $y\left(\dfrac{x + z}{w}\right)$

36. $x^2(z + wy)$

37. $x\sqrt{(x + wz)}$

38. $w^2\sqrt{(z^2 + y^2)}$

39. $2(w^2 + x^2 + y^2)$

40. $2x(w - z)$

41. $\dfrac{z}{w} + x$

42. $\dfrac{z + w}{x}$

43. $\dfrac{x + w}{z^2}$

44. $\dfrac{y^2 - w^2}{xz}$

45. $z^2 + 4z + 5$

46. $\dfrac{1}{w} + \dfrac{1}{z} + \dfrac{1}{x}$

47. $\dfrac{4}{z} + \dfrac{10}{w}$

48. $\dfrac{yz - xw}{xz - w}$

49. Find $K = \sqrt{\left(\dfrac{a^2 + b^2 + c^2 - 2c}{a^2 + b^2 + 4c}\right)}$ if $a = 3$, $b = -2$, $c = -1$.

50. Find $W = \dfrac{kmn(k + m + n)}{(k + m)(k + n)}$ if $k = \frac{1}{2}$, $m = -\frac{1}{3}$, $n = \frac{1}{4}$.

2.4 Brackets and simplifying

A term outside a bracket multiplies each of the terms inside the bracket.
This is the *distributive law*.

Example 1
$3(x - 2y) = 3x - 6y$

Example 2
$2x(x - 2y + z) = 2x^2 - 4xy + 2xz$

Example 3
$7y - 4(2x - 3) = 7y - 8x + 12$

In general,
   numbers can be added to numbers
      x's can be added to x's
      y's can be added to y's
      x^2's can be added to x^2's

But they must not be mixed.

Example 4
$2x + 3y + 3x^2 + 2y - x = x + 5y + 3x^2$

Example 5
$7x + 3x(2x - 3) = 7x + 6x^2 - 9x$
$\qquad\qquad\qquad = 6x^2 - 2x$

Exercise 9

Simplify as far as possible:

1. $3x + 4y + 7y$
2. $4a + 7b - 2a + b$
3. $3x - 2y + 4y$
4. $2x + 3x + 5$
5. $7 - 3x + 2 + 4x$
6. $5 - 3y - 6y - 2$
7. $5x + 2y - 4y - x^2$
8. $2x^2 + 3x + 5$
9. $2x - 7y - 2x - 3y$
10. $4a + 3a^2 - 2a$
11. $7a - 7a^2 + 7$
12. $x^2 + 3x^2 - 4x^2 + 5x$
13. $\dfrac{3}{a} + b + \dfrac{7}{a} - 2b$
14. $\dfrac{4}{x} - \dfrac{7}{y} + \dfrac{1}{x} + \dfrac{2}{y}$
15. $\dfrac{m}{x} + \dfrac{2m}{x}$
16. $\dfrac{5}{x} - \dfrac{7}{x} + \dfrac{1}{2}$
17. $\dfrac{3}{a} + b + \dfrac{2}{a} + 2b$
18. $\dfrac{n}{4} - \dfrac{m}{3} - \dfrac{n}{2} + \dfrac{m}{3}$
19. $x^3 + 7x^2 - 2x^3$
20. $(2x)^2 - 2x^2$
21. $(3y)^2 + x^2 - (2y)^2$
22. $(2x)^2 - (2y)^2 - (4x)^2$
23. $5x - 7x^2 - (2x)^2$
24. $\dfrac{3}{x^2} + \dfrac{5}{x^2}$

Remove the brackets and collect like terms:

25. $3x + 2(x + 1)$
26. $5x + 7(x - 1)$
27. $7 + 3(x - 1)$
28. $9 - 2(3x - 1)$
29. $3x - 4(2x + 5)$
30. $5x - 2x(x - 1)$
31. $7x + 3x(x - 4)$
32. $4(x - 1) - 3x$
33. $5x(x + 2) + 4x$
34. $3x(x - 1) - 7x^2$
35. $3a + 2(a + 4)$
36. $4a - 3(a - 3)$
37. $3ab - 2a(b - 2)$
38. $3y - y(2 - y)$
39. $3x - (x + 2)$
40. $7x - (x - 3)$
41. $5x - 2(2x + 2)$
42. $3(x - y) + 4(x + 2y)$
43. $x(x - 2) + 3x(x - 3)$
44. $3x(x + 4) - x(x - 2)$
45. $y(3y - 1) - (3y - 1)$
46. $7(2x + 2) - (2x + 2)$
47. $7b(a + 2) - a(3b + 3)$
48. $3(x - 2) - (x - 2)$

Two brackets

Example 1

$(x + 5)(x + 3) = x(x + 3) + 5(x + 3)$
$\qquad\qquad\quad = x^2 + 3x + 5x + 15$
$\qquad\qquad\quad = x^2 + 8x + 15$

Example 2

$(2x - 3)(4y + 3) = 2x(4y + 3) - 3(4y + 3)$
$\qquad\qquad\qquad\quad = 8xy + 6x - 12y - 9$

Example 3

$3(x + 1)(x - 2) = 3[x(x - 2) + 1(x - 2)]$
$\qquad\qquad\qquad = 3[x^2 - 2x + x - 2]$
$\qquad\qquad\qquad = 3x^2 - 3x - 6$

Exercise 10

Remove the brackets and simplify:

1. $(x + 1)(x + 3)$
2. $(x + 3)(x + 2)$
3. $(y + 4)(y + 5)$
4. $(x - 3)(x + 4)$
5. $(x + 5)(x - 2)$
6. $(x - 3)(x - 2)$

7. $(a-7)(a+5)$	**8.** $(z+9)(z-2)$	**9.** $(x-3)(x+3)$
10. $(k-11)(k+11)$	**11.** $(2x+1)(x-3)$	**12.** $(3x+4)(x-2)$
13. $(2y-3)(y+1)$	**14.** $(7y-1)(7y+1)$	**15.** $(3x-2)(3x+2)$
16. $(3a+b)(2a+b)$	**17.** $(3x+y)(x+2y)$	**18.** $(2b+c)(3b-c)$
19. $(5x-y)(3y-x)$	**20.** $(3b-a)(2a+5b)$	**21.** $2(x-1)(x+2)$
22. $3(x-1)(2x+3)$	**23.** $4(2y-1)(3y+2)$	**24.** $2(3x+1)(x-2)$
25. $4(a+2b)(a-2b)$	**26.** $x(x-1)(x-2)$	**27.** $2x(2x-1)(2x+1)$
28. $3y(y-2)(y+3)$	**29.** $x(x+y)(x+z)$	**30.** $3z(a+2m)(a-m)$

Be careful with an expression like $(x-3)^2$. It is not x^2-9 or even x^2+9.

$$(x-3)^2 = (x-3)(x-3)$$
$$= x(x-3) - 3(x-3)$$
$$= x^2 - 6x + 9$$

Another common mistake occurs with an expression like $4-(x-1)^2$.

$$4-(x-1)^2 = 4 - 1(x-1)(x-1)$$
$$= 4 - 1(x^2 - 2x + 1)$$
$$= 4 - x^2 + 2x - 1$$
$$= 3 + 2x - x^2$$

Exercise 11

Remove the brackets and simplify:

1. $(x+4)^2$	**2.** $(x+2)^2$	**3.** $(x-2)^2$
4. $(2x+1)^2$	**5.** $(y-5)^2$	**6.** $(3y+1)^2$
7. $(x+y)^2$	**8.** $(2x+y)^2$	**9.** $(a-b)^2$
10. $(2a-3b)^2$	**11.** $3(x+2)^2$	**12.** $(3-x)^2$
13. $(3x+2)^2$	**14.** $(a-2b)^2$	**15.** $(x+1)^2 + (x+2)^2$
16. $(x-2)^2 + (x+3)^2$	**17.** $(x+2)^2 + (2x+1)^2$	**18.** $(y-3)^2 + (y-4)^2$
19. $(x+2)^2 - (x-3)^2$	**20.** $(x-3)^2 - (x+1)^2$	**21.** $(y-3)^2 - (y+2)^2$
22. $(2x+1)^2 - (x+3)^2$	**23.** $3(x+2)^2 - (x+4)^2$	**24.** $2(x-3)^2 - 3(x+1)^2$

2.5 Linear equations

● If the x term is negative, take it to the other side, where it becomes positive.

Example 1

$$4 - 3x = 2$$
$$4 = 2 + 3x$$
$$2 = 3x$$
$$\frac{2}{3} = x$$

- If there are x terms on both sides, collect them on one side.

Example 2

$$2x - 7 = 5 - 3x$$
$$2x + 3x = 5 + 7$$
$$5x = 12$$
$$x = \frac{12}{5} = 2\tfrac{2}{5}$$

- If there is a fraction in the x term, multiply out to simplify the equation.

Example 3

$$\frac{2x}{3} = 10$$
$$2x = 30$$
$$x = \frac{30}{2} = 15$$

Exercise 12

Solve the following equations:

1. $2x - 5 = 11$ **2.** $3x - 7 = 20$ **3.** $2x + 6 = 20$ **4.** $5x + 10 = 60$

5. $8 = 7 + 3x$ **6.** $12 = 2x - 8$ **7.** $-7 = 2x - 10$ **8.** $3x - 7 = -10$

9. $12 = 15 + 2x$ **10.** $5 + 6x = 7$ **11.** $\dfrac{x}{5} = 7$ **12.** $\dfrac{x}{10} = 13$

13. $7 = \dfrac{x}{2}$ **14.** $\dfrac{x}{2} = \dfrac{1}{3}$ **15.** $\dfrac{3x}{2} = 5$ **16.** $\dfrac{4x}{5} = -2$

17. $7 = \dfrac{7x}{3}$ **18.** $\dfrac{3}{4} = \dfrac{2x}{3}$ **19.** $\dfrac{5x}{6} = \dfrac{1}{4}$ **20.** $-\dfrac{3}{4} = \dfrac{3x}{5}$

21. $\dfrac{x}{2} + 7 = 12$ **22.** $\dfrac{x}{3} - 7 = 2$ **23.** $\dfrac{x}{5} - 6 = -2$ **24.** $4 = \dfrac{x}{2} - 5$

25. $10 = 3 + \dfrac{x}{4}$ **26.** $\dfrac{a}{5} - 1 = -4$ **27.** $100x - 1 = 98$ **28.** $7 = 7 + 7x$

29. $\dfrac{x}{100} + 10 = 20$ **30.** $1000x - 5 = -6$ **31.** $-4 = -7 + 3x$ **32.** $2x + 4 = x - 3$

33. $x - 3 = 3x + 7$ **34.** $5x - 4 = 3 - x$ **35.** $4 - 3x = 1$ **36.** $5 - 4x = -3$

37. $7 = 2 - x$ **38.** $3 - 2x = x + 12$ **39.** $6 + 2a = 3$ **40.** $a - 3 = 3a - 7$

41. $2y - 1 = 4 - 3y$ **42.** $7 - 2x = 2x - 7$ **43.** $7 - 3x = 5 - 2x$ **44.** $8 - 2y = 5 - 5y$

45. $x - 16 = 16 - 2x$ **46.** $x + 2 = 3{\cdot}1$ **47.** $-x - 4 = -3$ **48.** $-3 - x = -5$

49. $-\dfrac{x}{2} + 1 = -\dfrac{1}{4}$ **50.** $-\dfrac{3}{5} + \dfrac{x}{10} = -\dfrac{1}{5} - \dfrac{x}{5}$

Example

$$x - 2(x - 1) = 1 - 4(x + 1)$$
$$x - 2x + 2 = 1 - 4x - 4$$
$$x - 2x + 4x = 1 - 4 - 2$$
$$3x = -5$$
$$x = -\frac{5}{3}$$

Exercise 13

Solve the following equations:

1. $x + 3(x + 1) = 2x$
2. $1 + 3(x - 1) = 4$
3. $2x - 2(x + 1) = 5x$
4. $2(3x - 1) = 3(x - 1)$
5. $4(x - 1) = 2(3 - x)$
6. $4(x - 1) - 2 = 3x$
7. $4(1 - 2x) = 3(2 - x)$
8. $3 - 2(2x + 1) = x + 17$
9. $4x = x - (x - 2)$
10. $7x = 3x - (x + 20)$
11. $5x - 3(x - 1) = 39$
12. $3x + 2(x - 5) = 15$
13. $7 - (x + 1) = 9 - (2x - 1)$
14. $10x - (2x + 3) = 21$
15. $3(2x + 1) + 2(x - 1) = 23$
16. $5(1 - 2x) - 3(4 + 4x) = 0$
17. $7x - (2 - x) = 0$
18. $3(x + 1) = 4 - (x - 3)$
19. $3y + 7 + 3(y - 1) = 2(2y + 6)$
20. $4(y - 1) + 3(y + 2) = 5(y - 4)$
21. $4x - 2(x + 1) = 5(x + 3) + 5$
22. $7 - 2(x - 1) = 3(2x - 1) + 2$
23. $10(2x + 3) - 8(3x - 5) + 5(2x - 8) = 0$
24. $2(x + 4) + 3(x - 10) = 8$
25. $7(2x - 4) + 3(5 - 3x) = 2$
26. $10(x + 4) - 9(x - 3) - 1 = 8(x + 3)$
27. $5(2x - 1) - 2(x - 2) = 7 + 4x$
28. $6(3x - 4) - 10(x - 3) = 10(2x - 3)$
29. $3(x - 3) - 7(2x - 8) - (x - 1) = 0$
30. $5 + 2(x + 5) = 10 - (4 - 5x)$
31. $6x + 30(x - 12) = 2(x - 1\frac{1}{2})$
32. $3(2x - \frac{2}{3}) - 7(x - 1) = 0$
33. $5(x - 1) + 17(x - 2) = 2x + 1$
34. $6(2x - 1) + 9(x + 1) = 8(x - 1\frac{1}{4})$
35. $7(x + 4) - 5(x + 3) + (4 - x) = 0$
36. $0 = 9(3x + 7) - 5(x + 2) - (2x - 5)$
37. $10(2 \cdot 3 - x) - 0 \cdot 1(5x - 30) = 0$
38. $8(2\frac{1}{2}x - \frac{3}{4}) - \frac{1}{4}(1 - x) = \frac{1}{2}$
39. $(6 - x) - (x - 5) - (4 - x) = -\dfrac{x}{2}$
40. $10\left(1 - \dfrac{x}{10}\right) - (10 - x) - \dfrac{1}{100}(10 - x) = 0 \cdot 05$

Example

$$(x + 3)^2 = (x + 2)^2 + 3^2$$
$$(x + 3)(x + 3) = (x + 2)(x + 2) + 9$$
$$x^2 + 6x + 9 = x^2 + 4x + 4 + 9$$
$$6x + 9 = 4x + 13$$
$$2x = 4$$
$$x = 2$$

Exercise 14

Solve the following equations:

1. $x^2 + 4 = (x + 1)(x + 3)$
2. $x^2 + 3x = (x + 3)(x + 1)$
3. $(x + 3)(x - 1) = x^2 + 5$
4. $(x + 1)(x + 4) = (x - 7)(x + 6)$
5. $(x - 2)(x + 3) = (x - 7)(x + 7)$
6. $(x - 5)(x + 4) = (x + 7)(x - 6)$

7. $2x^2 + 3x = (2x - 1)(x + 1)$

8. $(2x - 1)(x - 3) = (2x - 3)(x - 1)$

9. $x^2 + (x + 1)^2 = (2x - 1)(x + 4)$

10. $x(2x + 6) = 2(x^2 - 5)$

11. $(x + 1)(x - 3) + (x + 1)^2 = 2x(x - 4)$

12. $(2x + 1)(x - 4) + (x - 2)^2 = 3x(x + 2)$

13. $(x + 2)^2 - (x - 3)^2 = 3x - 11$

14. $x(x - 1) = 2(x - 1)(x + 5) - (x - 4)^2$

15. $(2x + 1)^2 - 4(x - 3)^2 = 5x + 10$

16. $2(x + 1)^2 - (x - 2)^2 = x(x - 3)$

17. The area of the rectangle shown exceeds the area of the square by $2\,\text{cm}^2$. Find x.

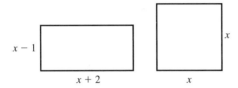

18. The area of the square exceeds the area of the rectangle by $13\,\text{m}^2$. Find y.

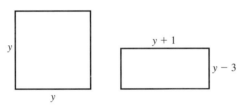

> **Remember**
> For a rectangle:
> Area = length × width

19. The area of the square is half the area of the rectangle. Find x.

When solving equations involving fractions, multiply both sides of the equation by a suitable number to eliminate the fractions.

Example 1

$\dfrac{5}{x} = 2$

$5 = 2x$ (multiply both sides by x)

$\dfrac{5}{2} = x$

Example 2

$$\frac{x+4}{4} = \frac{2x-1}{3} \qquad \qquad \ldots(A)$$

$$12\frac{(x+3)}{4} = 12\frac{(2x-1)}{3}$$

(multiply both sides by 12)

$$\therefore \quad 3(x+3) = 4(2x-1) \qquad \qquad \ldots(B)$$
$$3x+9 = 8x-4$$
$$13 = 5x$$
$$\frac{13}{5} = x$$
$$x = 2\tfrac{3}{5}$$

Note: It is possible to go straight from line (A) to line (B) by 'cross-multiplying'.

Example 3

$$\frac{5}{(x-1)} + 2 = 12$$
$$\frac{5}{(x-1)} = 10$$
$$5 = 10(x-1)$$
$$5 = 10x - 10$$
$$15 = 10x$$
$$\frac{15}{10} = x$$
$$x = 1\tfrac{1}{2}$$

Exercise 15

Solve the following equations:

1. $\dfrac{7}{x} = 21$

2. $30 = \dfrac{6}{x}$

3. $\dfrac{5}{x} = 3$

4. $\dfrac{9}{x} = -3$

5. $11 = \dfrac{5}{x}$

6. $-2 = \dfrac{4}{x}$

7. $\dfrac{x}{4} = \dfrac{3}{2}$

8. $\dfrac{x}{3} = 1\tfrac{1}{4}$

9. $\dfrac{x+1}{3} = \dfrac{x-1}{4}$

10. $\dfrac{x+3}{2} = \dfrac{x-4}{5}$

11. $\dfrac{2x-1}{3} = \dfrac{x}{2}$

12. $\dfrac{3x+1}{5} = \dfrac{2x}{3}$

13. $\dfrac{8-x}{2} = \dfrac{2x+2}{5}$

14. $\dfrac{x+2}{7} = \dfrac{3x+6}{5}$

15. $\dfrac{1-x}{2} = \dfrac{3-x}{3}$

16. $\dfrac{2}{x-1} = 1$

17. $\dfrac{x}{3} + \dfrac{x}{4} = 1$

18. $\dfrac{x}{3} + \dfrac{x}{2} = 4$

19. $\dfrac{x}{2} - \dfrac{x}{5} = 3$

20. $\dfrac{x}{3} = 2 + \dfrac{x}{4}$

21. $\dfrac{5}{x-1} = \dfrac{10}{x}$

22. $\dfrac{12}{2x-3}=4$

23. $2=\dfrac{18}{x+4}$

24. $\dfrac{5}{x+5}=\dfrac{15}{x+7}$

25. $\dfrac{9}{x}=\dfrac{5}{x-3}$

26. $\dfrac{4}{x-1}=\dfrac{10}{3x-1}$

27. $\dfrac{-7}{x-1}=\dfrac{14}{5x+2}$

28. $\dfrac{4}{x+1}=\dfrac{7}{3x-2}$

29. $\dfrac{x+1}{2}+\dfrac{x-1}{3}=\dfrac{1}{6}$

30. $\dfrac{1}{3}(x+2)=\dfrac{1}{5}(3x+2)$

31. $\dfrac{1}{2}(x-1)-\dfrac{1}{6}(x+1)=0$

32. $\dfrac{1}{4}(x+5)-\dfrac{2x}{3}=0$

33. $\dfrac{4}{x}+2=3$

34. $\dfrac{6}{x}-3=7$

35. $\dfrac{9}{x}-7=1$

36. $-2=1+\dfrac{3}{x}$

37. $4-\dfrac{4}{x}=0$

38. $5-\dfrac{6}{x}=-1$

39. $7-\dfrac{3}{2x}=1$

40. $4+\dfrac{5}{3x}=-1$

41. $\dfrac{9}{2x}-5=0$

42. $\dfrac{x-1}{5}-\dfrac{x-1}{3}=0$

43. $\dfrac{x-1}{4}-\dfrac{2x-3}{5}=\dfrac{1}{20}$

44. $\dfrac{4}{1-x}=\dfrac{3}{1+x}$

45. $\dfrac{x+1}{4}-\dfrac{x}{3}=\dfrac{1}{12}$

46. $\dfrac{2x+1}{8}-\dfrac{x-1}{3}=\dfrac{5}{24}$

2.6 Problems solved by linear equations

- Let the unknown quantity be x (or any other letter) and state the units (where appropriate).
- Express the given statement in the form of an equation.
- Solve the equation for x and give the answer in *words*. (Do not finish by writing '$x=3$'.)
- Check your solution using the problem (not your equation).

Example 1

The sum of three consecutive whole numbers is 78. Find the numbers.

(a) Let the smallest number be x; then the other numbers are $(x+1)$ and $(x+2)$.

(b) Form an equation:
   $x+(x+1)+(x+2)=78$

(c) Solve: $3x=75$
              $x=25$
   In words:
   The three numbers are 25, 26 and 27.

(d) Check: $25+26+27=78$

Example 2

The length of a rectangle is three times the width. If the perimeter is 36 cm, find the width.

(a) Let the width of the rectangle be x cm.
   Then the length of the rectangle is $3x$ cm.

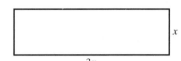

(b) Form an equation.
   $x + 3x + x + 3x = 36$

(c) Solve:　$8x = 36$
   $$x = \frac{36}{8}$$
   $$x = 4 \cdot 5$$
   In words:
   The width of the rectangle is 4·5 cm.

(d) Check:　If width $= 4 \cdot 5$ cm
   　　　　　length $= 13 \cdot 5$ cm
   　　　perimeter $= 36$ cm

Exercise 16

Solve each problem by forming an equation. The first questions are easy but should still be solved using an equation, in order to practise the method:

1. The sum of three consecutive numbers is 276. Find the numbers.

2. The sum of four consecutive numbers is 90. Find the numbers.

3. The sum of three consecutive odd numbers is 177. Find the numbers.

4. Find three consecutive even numbers which add up to 1524.

5. When a number is doubled and then added to 13, the result is 38. Find the number.

6. When a number is doubled and then added to 24, the result is 49. Find the number.

7. When 7 is subtracted from three times a certain number, the result is 28. What is the number?

8. The sum of two numbers is 50. The second number is five times the first. Find the numbers.

9. Two numbers are in the ratio $1:11$ and their sum is 15. Find the numbers.

10. The length of a rectangle is twice the width.
   If the perimeter is 20 cm, find the width.

11. The width of a rectangle is one third of the length. If the perimeter is 96 cm, find the width.

12. If AB is a straight line, find x.
(The angles on a straight line add to 180°.)

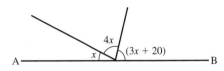

13. If the perimeter of the triangle is 22 cm, find the length of the shortest side.

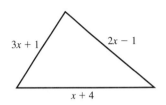

14. If the perimeter of the rectangle is 34 cm, find x.

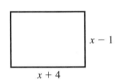

15. The difference between two numbers is 9.
Find the numbers, if their sum is 46.

16. The three angles in a triangle are in the ratio $1:3:5$. Find them.

17. The three angles in a triangle are in the ratio $3:4:5$. Find them.

18. The product of two consecutive odd numbers is 10 more than the square of the smaller number. Find the smaller number.

19. The product of two consecutive even numbers is 12 more than the square of the smaller number. Find the numbers.

20. The sum of three numbers is 66. The second number is twice the first and six less than the third. Find the numbers.

21. The sum of three numbers is 28. The second number is three times the first and the third is 7 less than the second. What are the numbers?

22. David's mass is 5 kg less than John's, who in turn is 8 kg lighter than Paul. If their total mass is 197 kg, how heavy is each person?

23. Nilopal is 2 years older than Devjan who is 7 years older than Sucha. If their combined age is 61 years, find the age of each person.

24. Kimiya has four times as many marbles as Ramneet. If Kimiya gave 18 to Ramneet they would have the same number. How many marbles has each?

25. Mukat has five times as many books as Usha.
If Mukat gave 16 books to Usha, they would each have the same number. How many books did each girl have?

26. The result of trebling a number is the same as adding 12 to it. What is the number?

27. Find the area of the rectangle if the perimeter is 52 cm.
(The perimeter is the distance around the edge of the rectangle.)

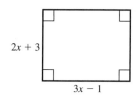

$2x + 3$

$3x - 1$

28. The result of multiplying a number by 3 and subtracting 5 is the same as doubling the number and adding 9. What is the number?

29. Two girls have $76 between them. If the first gave the second $7 they would each have the same amount of money. How much did each girl have?

30. A tennis racket costs $12 more than a hockey stick. If the price of the two is $31, find the cost of the tennis racket.

Example

A man goes out at 16:42 and arrives at a post box, 6 km away, at 17:30. He walked part of the way at 5 km/h and then, realising the time, he ran the rest of the way at 10 km/h. How far did he have to run?

● Let the distance he ran be x km. Then the distance he walked $= (6 - x)$ km.

● Time taken to walk $(6 - x)$ km at 5 km/h $= \dfrac{(6 - x)}{5}$ hours.

Time taken to run x km at 10 km/h $= \dfrac{x}{10}$ hours.

Total time taken $= 48$ minutes

$$= \frac{4}{5} \text{ hour}$$

$$\therefore \quad \frac{(6 - x)}{5} + \frac{x}{10} = \frac{4}{5}$$

● Multiply by 10:
$2(6 - x) + x = 8$
$12 - 2x + x = 8$
$\qquad\quad 4 = x$
He ran a distance of 4 km.

● Check:
Time to run 4 km $= \dfrac{4}{10} = \dfrac{2}{5}$ hour.

Time to walk 2 km $= \dfrac{2}{5}$ hour.

Total time taken $= \left(\dfrac{2}{5} + \dfrac{2}{5}\right) \text{h} = \dfrac{4}{5} \text{ h}$

Exercise 17

1. Every year a man is paid $500 more than the previous year. If he receives $17 800 over four years, what was he paid in the first year?

2. Samir buys x cans of soda at 30 cents each and $(x + 4)$ cans of soda at 35 cents each. The total cost was $3·35. Find x.

3. The length of a straight line ABC is 5 m.
   If $AB : BC = 2 : 5$, find the length of AB.

4. The opposite angles of a cyclic quadrilateral are $(3x + 10)°$ and $(2x + 20)°$. Find the angles.

> Opposite angles of a cyclic quadrilateral add to 180°.

5. The interior angles of a hexagon are in the ratio $1 : 2 : 3 : 4 : 5 : 9$. Find the angles. This is an example of a concave hexagon. Try to sketch the hexagon.

> Interior angles of a hexagon add to 720°.

6. A man is 32 years older than his son. Ten years ago he was three times as old as his son was then. Find the present age of each.

7. Mahmoud runs to a marker and back in 15 minutes. His speed on the way to the marker is 5 m/s and his speed on the way back is 4 m/s. Find the distance to the marker.

8. A car completes a journey in 10 minutes. For the first half of the distance the speed was 60 km/h and for the second half the speed was 40 km/h. How far is the journey?

9. A lemming runs from a point A to a cliff at 4 m/s, jumps over the edge at B and falls to C at an average speed of 25 m/s. If the total distance from A to C is 500 m and the time taken for the journey is 41 seconds, find the height BC of the cliff.

10. A bus is travelling with 48 passengers. When it arrives at a stop, x passengers get off and 3 get on. At the next stop half the passengers get off and 7 get on. There are now 22 passengers. Find x.

11. A bus is travelling with 52 passengers. When it arrives at a stop, y passengers get off and 4 get on. At the next stop one-third of the passengers get off and 3 get on. There are now 25 passengers. Find y.

12. Mr Lee left his fortune to his 3 sons, 4 daughters and his wife. Each son received twice as much as each daughter and his wife received $6000, which was a quarter of the money. How much did each son receive?

13. In a regular polygon with n sides each interior angle is $180 - \dfrac{360}{n}$ degrees. How many sides does a polygon have if each interior angle is 156°?

14. A sparrow flies to see a friend at a speed of 4 km/h. His friend is out, so the sparrow immediately returns home at a speed of 5 km/h. The complete journey took 54 minutes. How far away does his friend live?

15. Consider the equation $an^2 = 182$ where a is any number between 2 and 5 and n is a positive integer. What are the possible values of n?

16. Consider the equation $\dfrac{k}{x} = 12$ where k is any number between

20 and 65 and x is a positive integer. What are the possible values of x?

2.7 Simultaneous equations

To find the value of two unknowns in a problem, *two* different equations must be given that relate the unknowns to each other. These two equations are called *simultaneous* equations.

Substitution method

This method is used when one equation contains a unit quantity of one of the unknowns, as in equation [2] of the example below.

Example

$$3x - 2y = 0 \qquad \qquad \ldots [1]$$
$$2x + y = 7 \qquad \qquad \ldots [2]$$

(a) Label the equations so that the working is made clear.
(b) In *this* case, write y in terms of x from equation [2].
(c) Substitute this expression for y in equation [1] and solve to find x.
(d) Find y from equation [2] using this value of x.

$$2x + y = 7 \qquad \qquad \ldots [2]$$
$$y = 7 - 2x$$

Substituting in [1]
$$3x - 2(7 - 2x) = 0$$
$$3x - 14 + 4x = 0$$
$$7x = 14$$
$$x = 2$$

Substituting in [2]
$$2 \times 2 + y = 7$$
$$y = 3$$

The solutions are $x = 2$, $y = 3$.

These values of x and y are the only pair which simultaneously satisfy *both* equations.

Exercise 18

Use the substitution method to solve the following:

1. $2x + y = 5$
$x + 3y = 5$

2. $x + 2y = 8$
$2x + 3y = 14$

3. $3x + y = 10$
$x - y = 2$

4. $2x + y = -3$
$x - y = -3$

5. $4x + y = 14$
$x + 5y = 13$

6. $x + 2y = 1$
$2x + 3y = 4$

7. $2x + y = 5$
$3x - 2y = 4$

8. $2x + y = 13$
$5x - 4y = 13$

9. $7x + 2y = 19$
$x - y = 4$

10. $b - a = -5$
$a + b = -1$

11. $a + 4b = 6$
$8b - a = -3$

12. $a + b = 4$
$2a + b = 5$

13. $3m = 2n - 6\frac{1}{2}$
$4m + n = 6$

14. $2w + 3x - 13 = 0$
$x + 5w - 13 = 0$

15. $x + 2(y - 6) = 0$
$3x + 4y = 30$

16. $2x = 4 + z$
$6x - 5z = 18$

17. $3m - n = 5$
$2m + 5n = 7$

18. $5c - d - 11 = 0$
$4d + 3c = -5$

It is useful, at this point to revise the operations of addition and subtraction with negative numbers.

Example

Simplify:

(a) $-7 + -4 = -7 - 4 = -11$

(b) $-3x + (-4x) = -3x - 4x = -7x$

(c) $4y - (-3y) = 4y + 3y = 7y$

(d) $3a + (-3a) = 3a - 3a = 0$

Exercise 19

Evaluate:

1. $7 + (-6)$

2. $8 + (-11)$

3. $5 - (+7)$

4. $6 - (-9)$

5. $-8 + (-4)$

6. $-7 - (-4)$

7. $10 + (-12)$

8. $-7 - (+4)$

9. $-10 - (+11)$

10. $-3 - (-4)$

11. $4 - (+4)$

12. $8 - (-7)$

13. $-5 - (+5)$

14. $-7 - (-10)$

15. $16 - (+10)$

16. $-7 - (+4)$

17. $-6 - (-8)$

18. $10 - (+5)$

19. $-12 + (-7)$

20. $7 + (-11)$

Simplify:

21. $3x + (-2x)$

22. $4x + (-7x)$

23. $6x - (+2x)$

24. $10y - (+6y)$

25. $6y - (-3y)$

26. $7x + (-4x)$

27. $-5x + (-3x)$

28. $-3x - (-7x)$

29. $5x - (+3x)$

30. $-7y - (-10y)$

Elimination method

Use this method when the first method is unsuitable (some prefer to use it for every question).

Example 1

$$x + 2y = 8 \qquad \ldots [1]$$
$$2x + 3y = 14 \qquad \ldots [2]$$

(a) Label the equations so that the working is made clear.
(b) Choose an unknown in one of the equations and multiply the equations by a factor or factors so that this unknown has the same coefficient in both equations.
(c) Eliminate this unknown from the two equations by subtracting them, then solve for the remaining unknown.
(d) Substitute in the first equation and solve for the eliminated unknown.

$$x + 2y = 8 \qquad \ldots [1]$$

$$[1] \times 2 \quad 2x + 4y = 16 \qquad \ldots [3]$$
$$2x + 3y = 14 \qquad \ldots [2]$$

Subtract [2] from [3]
$$y = 2$$

Substituting in [1]
$$x + 2 \times 2 = 8$$
$$x = 8 - 4$$
$$x = 4$$

The solutions are $x = 4$, $y = 2$.

Example 2

$$2x + 3y = 5 \qquad \ldots [1]$$
$$5x - 2y = -16 \qquad \ldots [2]$$

$$[1] \times 5 \quad 10x + 15y = 25 \qquad \ldots [3]$$
$$[2] \times 2 \quad 10x - 4y = -32 \qquad \ldots [4]$$

$$[3] - 4 \quad 15y - (-4y) = 25 - (-32)$$
$$19y = 57$$
$$y = 3$$

Substitute in [1]
$$2x + 3 \times 3 = 5$$
$$2x = 5 - 9 = -4$$
$$x = -2$$

The solutions are $x = -2$, $y = 3$.

Exercise 20

Use the elimination method to solve the following:

1. $2x + 5y = 24$
$4x + 3y = 20$

2. $5x + 2y = 13$
$2x + 6y = 26$

3. $3x + y = 11$
$9x + 2y = 28$

4. $x + 2y = 17$
$8x + 3y = 45$

5. $3x + 2y = 19$
$x + 8y = 21$

6. $2a + 3b = 9$
$4a + b = 13$

7. $2x + 3y = 11$
$3x + 4y = 15$

8. $3x + 8y = 27$
$4x + 3y = 13$

9. $2x + 7y = 17$
$5x + 3y = -1$

10. $5x + 3y = 23$
$2x + 4y = 12$

11. $7x + 5y = 32$
$3x + 4y = 23$

12. $3x + 2y = 4$
$4x + 5y = 10$

13. $3x + 2y = 11$
$2x - y = -3$

14. $3x + 2y = 7$
$2x - 3y = -4$

15. $x + 2y = -4$
$3x - y = 9$

16. $5x - 7y = 27$
$3x - 4y = 16$

17. $3x - 2y = 7$
$4x + y = 13$

18. $x - y = -1$
$2x - y = 0$

19. $y - x = -1$
$3x - y = 5$

20. $x - 3y = -5$
$2y + 3x + 4 = 0$

21. $x + 3y - 7 = 0$
$2y - x - 3 = 0$

22. $3a - b = 9$
$2a + 2b = 14$

23. $3x - y = 9$
$4x - y = -14$

24. $x + 2y = 4$
$3x + y = 9\frac{1}{2}$

25. $2x - y = 5$
$\dfrac{x}{4} + \dfrac{y}{3} = 2$

26. $3x - y = 17$
$\dfrac{x}{5} + \dfrac{y}{2} = 0$

27. $3x - 2y = 5$
$\dfrac{2x}{3} + \dfrac{y}{2} = -\dfrac{7}{9}$

28. $2x = 11 - y$
$\dfrac{x}{5} - \dfrac{y}{4} = 1$

29. $4x - 0.5y = 12.5$
$3x + 0.8y = 8.2$

30. $0.4x + 3y = 2.6$
$x - 2y = 4.6$

2.8 Problems solved by simultaneous equations

Example

A motorist buys 24 litres of petrol and 5 litres of oil for $10·70, while another motorist buys 18 litres of petrol and 10 litres of oil for $12·40. Find the cost of 1 litre of petrol and 1 litre of oil at this garage.

Let cost of 1 litre of petrol be x cents.
Let cost of 1 litre of oil be y cents.

We have, $24x + 5y = 1070$ . . . [1]
$18x + 10y = 1240$ . . . [2]

(a) Multiply [1] by 2,
$48x + 10y = 2140$ . . . [3]

(b) Subtract [2] from [3],
$30x = 900$
$x = 30$

(c) Substitute $x = 30$ into equation [2]
$18(30) + 10y = 1240$
$10y = 1240 - 540$
$10y = 700$
$y = 70$

1 litre of petrol costs 30 cents and
1 litre of oil costs 70 cents.

Exercise 21

Solve each problem by forming a pair of simultaneous equations:

1. Find two numbers with a sum of 15 and a difference of 4.

2. Twice one number added to three times another gives 21. Find the numbers, if the difference between them is 3.

3. The average of two numbers is 7, and three times the difference between them is 18. Find the numbers.

4. The line, with equation $y + ax = c$, passes through the points (1, 5) and (3, 1). Find a and c.
   Hint: For the point (1, 5) put $x = 1$ and $y = 5$ into $y + ax = c$, etc.

5. The line $y = mx + c$ passes through (2, 5) and (4, 13). Find m and c.

6. The curve $y = ax^2 + bx$ passes through (2, 0) and (4, 8). Find a and b.

7. A gardener buys fifty carrot seeds and twenty lettuce seeds for $1·10 and her mother buys thirty carrot seeds and forty lettuce seeds for $1·50. Find the cost of one carrot seed and one lettuce seed.

8. A shop owner can buy either two televisions and three video-recorders for $1750 or four televisions and one video-recorder for $1250. Find the cost of one of each.

9. Half the difference between two numbers is 2. The sum of the greater number and twice the smaller number is 13. Find the numbers.

10. A bird can lay either white or brown eggs. Three white eggs and two brown eggs have a mass of 13 grams, while five white eggs and four brown eggs have a mass of 24 grams. Find the mass of a brown egg and of a white egg.

11. A tortoise makes a journey in two parts; it can either walk at 4 cm/s or crawl at 3 cm/s. If the tortoise walks the first part and crawls the second, it takes 110 seconds. If it crawls the first part and walks the second, it takes 100 seconds. Find the lengths of the two parts of the journey.

12. A cyclist completes a journey of 500 m in 22 seconds, part of the way at 10 m/s and the remainder at 50 m/s. How far does she travel at each speed?

13. A bag contains forty coins, all of them either 2 cent or 5 cent coins. If the value of the money in the bag is $1.55, find the number of each kind.

> **Remember**
> 100 cents = $1

14. A slot machine takes only 10 cent and 50 cent coins and contains a total of twenty-one coins altogether. If the value of the coins is $4·90, find the number of coins of each value.

15. Thirty tickets were sold for a concert, some at 60 cents and the rest at $1. If the total raised was $22, how many had the cheaper tickets?

16. The wage bill for five male and six female workers is $6700, while the bill for eight men and three women is $6100. Find the wage for a man and the wage for a woman.

17. A fish can swim at 14 m/s in the direction of the current and at 6 m/s against it. Find the speed of the current and the speed of the fish in still water.

18. If the numerator and denominator of a fraction are both decreased by 1 the fraction becomes $\frac{2}{3}$. If the numerator and denominator are both increased by 1 the fraction becomes $\frac{3}{4}$. Find the original fraction.

19. The denominator of a fraction is 2 more than the numerator. If both denominator and numerator are increased by 1 the fraction becomes $\frac{2}{3}$. Find the original fraction.

20. In three years' time a pet mouse will be as old as his owner was four years ago. Their present ages total 13 years. Find the age of each now.

21. Find two numbers where three times the smaller number exceeds the larger by 5 and the sum of the numbers is 11.

22. A straight line passes through the points (2, 4) and (−1, −5). Find its equation.

23. A spider can walk at a certain speed and run at another speed. If she walks for 10 seconds and runs for 9 seconds she travels 85 m. If she walks for 30 seconds and runs for 2 seconds she travels 130 m. Find her speeds of walking and running.

24. A wallet containing $40 has three times as many $1 notes as $5 notes. Find the number of each kind.

25. At the present time a man is four times as old as his son. Six years ago he was 10 times as old. Find their present ages.

26. A submarine can travel at 25 knots with the current and at 16 knots against it. Find the speed of the wind and the speed of the submarine in still water.

27. The curve $y = ax^2 + bx + c$ passes through the points (1, 8), (0, 5) and (3, 20). Find the values of a, b and c and hence the equation of the curve.

28. The curve $y = ax^2 + bx + c$ passes through the points (1, 4), (−2, 19) and (0, 5). Find the equation of the curve.

29. The curve $y = ax^2 + bx + c$ passes through (1, 8), (−1, 2) and (2, 14). Find the equation of the curve.

30. The curve $y = ax^2 + bx + c$ passes through (2, 5), (3, 12) and (−1, −4). Find the equation of the curve.

2.9 Factorising

Earlier in this section we expanded expressions such as $x(3x - 1)$ to give $3x^2 - x$.

The reverse of this process is called *factorising*.

Example

Factorise: (a) $4x + 4y$ (b) $x^2 + 7x$ (c) $3y^2 - 12y$ (d) $6a^2b - 10ab^2$

(a) 4 is common to $4x$ and $4y$.
$\quad\therefore\quad 4x + 4y = 4(x + y)$

(b) x is common to x^2 and $7x$.
$\quad\therefore\quad x^2 + 7x = x(x + 7)$
The factors are x and $(x + 7)$.

(c) $3y$ is common.
$\quad\therefore\quad 3y^2 - 12y = 3y(y - 4)$

(d) $2ab$ is common.
$\quad\therefore\quad 6a^2b - 10ab^2 = 2ab(3a - 5b)$

Exercise 22

Factorise the following expressions completely:

1. $5a + 5b$
2. $7x + 7y$
3. $7x + x^2$
4. $y^2 + 8y$
5. $2y^2 + 3y$
6. $6y^2 - 4y$
7. $3x^2 - 21x$
8. $16a - 2a^2$
9. $6c^2 - 21c$
10. $15x - 9x^2$
11. $56y - 21y^2$
12. $ax + bx + 2cx$
13. $x^2 + xy + 3xz$
14. $x^2y + y^3 + z^2y$
15. $3a^2b + 2ab^2$
16. $x^2y + xy^2$
17. $6a^2 + 4ab + 2ac$
18. $ma + 2bm + m^2$
19. $2kx + 6ky + 4kz$
20. $ax^2 + ay + 2ab$
21. $x^2k + xk^2$
22. $a^3b + 2ab^2$
23. $abc - 3b^2c$
24. $2a^2e - 5ae^2$
25. $a^3b + ab^3$
26. $x^3y + x^2y^2$
27. $6xy^2 - 4x^2y$
28. $3ab^3 - 3a^3b$
29. $2a^3b + 5a^2b^2$
30. $ax^2y - 2ax^2z$
31. $2abx + 2ab^2 + 2a^2b$
32. $ayx + yx^3 - 2y^2x^2$

Example 1

Factorise $ah + ak + bh + bk$.

(a) Divide into pairs, $ah + ak \quad + bh + bk$.

(b) a is common to the first pair.
b is common to the second pair.
$a(h + k) + b(h + k)$

(c) $(h + k)$ is common to both terms.
Thus we have $(h + k)(a + b)$

Example 2

Factorise $6mx - 3nx + 2my - ny$.

(a) $6mx - 3nx \quad + 2my - ny$

(b) $= 3x(2m - n) + y(2m - n)$

(c) $= (2m - n)(3x + y)$

Exercise 23

Factorise the following expressions:

1. $ax + ay + bx + by$

2. $ay + az + by + bz$

3. $xb + xc + yb + yc$

4. $xh + xk + yh + yk$

5. $xm + xn + my + ny$

6. $ah - ak + bh - bk$

7. $ax - ay + bx - by$

8. $am - bm + an - bn$

9. $hs + ht + ks + kt$

10. $xs - xt + ys - yt$

11. $ax - ay - bx + by$

12. $xs - xt - ys + yt$

13. $as - ay - xs + xy$

14. $hx - hy - bx + by$

15. $am - bm - an + bn$

16. $xk - xm - kz + mz$

17. $2ax + 6ay + bx + 3by$

18. $2ax + 2ay + bx + by$

19. $2mh - 2mk + nh - nk$

20. $2mh + 3mk - 2nh - 3nk$

21. $6ax + 2bx + 3ay + by$

22. $2ax - 2ay - bx + by$

23. $x^2 a + x^2 b + ya + yb$

24. $ms + 2mt^2 - ns - 2nt^2$

Quadratic expressions

Example 1

Factorise $x^2 + 6x + 8$.

(a) Find two numbers which multiply to give 8 and add up to 6.
In this case the numbers are 4 and 2.

(b) Put these numbers into brackets.
So $x^2 + 6x + 8 = (x + 4)(x + 2)$

Example 2

Factorise (a) $x^2 + 2x - 15$
           (b) $x^2 - 6x + 8$

(a) Two numbers which multiply to give -15 and add up to $+2$ are -3 and 5.

$$\therefore \quad x^2 + 2x - 15 = (x - 3)(x + 5)$$

(b) Two numbers which multiply to give $+8$ and add up to -6 are -2 and -4.

$$\therefore \quad x^2 - 6x + 8 = (x - 2)(x - 4)$$

Exercise 24

Factorise the following:

1. $x^2 + 7x + 10$

2. $x^2 + 7x + 12$

3. $x^2 + 8x + 15$

4. $x^2 + 10x + 21$

5. $x^2 + 8x + 12$

6. $y^2 + 12y + 35$

7. $y^2 + 11y + 24$

8. $y^2 + 10y + 25$

9. $y^2 + 15y + 36$

10. $a^2 - 3a - 10$

11. $a^2 - a - 12$

12. $z^2 + z - 6$

13. $x^2 - 2x - 35$

14. $x^2 - 5x - 24$

15. $x^2 - 6x + 8$

16. $y^2 - 5y + 6$

17. $x^2 - 8x + 15$

18. $a^2 - a - 6$

19. $a^2 + 14a + 45$

20. $b^2 - 4b - 21$

21. $x^2 - 8x + 16$

22. $y^2 + 2y + 1$

23. $y^2 - 3y - 28$

24. $x^2 - x - 20$

25. $x^2 - 8x - 240$

26. $x^2 - 26x + 165$

27. $y^2 + 3y - 108$

28. $x^2 - 49$

29. $x^2 - 9$

30. $x^2 - 16$

Example

Factorise $3x^2 + 13x + 4$.

(a) Find two numbers which multiply to give 12 and add up to 13.
In this case the numbers are 1 and 12.

(b) Split the '$13x$' term,
$3x^2 + x + 12x + 4$

(c) Factorise in pairs,
$x(3x + 1) + 4(3x + 1)$

(d) $(3x + 1)$ is common,
$(3x + 1)(x + 4)$

Exercise 25

Factorise the following:

1. $2x^2 + 5x + 3$	**2.** $2x^2 + 7x + 3$	**3.** $3x^2 + 7x + 2$
4. $2x^2 + 11x + 12$	**5.** $3x^2 + 8x + 4$	**6.** $2x^2 + 7x + 5$
7. $3x^2 - 5x - 2$	**8.** $2x^2 - x - 15$	**9.** $2x^2 + x - 21$
10. $3x^2 - 17x - 28$	**11.** $6x^2 + 7x + 2$	**12.** $12x^2 + 23x + 10$
13. $3x^2 - 11x + 6$	**14.** $3y^2 - 11y + 10$	**15.** $4y^2 - 23y + 15$
16. $6y^2 + 7y - 3$	**17.** $6x^2 - 27x + 30$	**18.** $10x^2 + 9x + 2$
19. $6x^2 - 19x + 3$	**20.** $8x^2 - 10x - 3$	**21.** $12x^2 + 4x - 5$
22. $16x^2 + 19x + 3$	**23.** $4a^2 - 4a + 1$	**24.** $12x^2 + 17x - 14$
25. $15x^2 + 44x - 3$	**26.** $48x^2 + 46x + 5$	**27.** $64y^2 + 4y - 3$
28. $120x^2 + 67x - 5$	**29.** $9x^2 - 1$	**30.** $4a^2 - 9$

The difference of two squares

$x^2 - y^2 = (x - y)(x + y)$
Remember this result.

Example

Factorise (a) $4a^2 - b^2$
(b) $3x^2 - 27y^2$

(a) $4a^2 - b^2 = (2a)^2 - b^2$
$= (2a - b)(2a + b)$

(b) $3x^2 - 27y^2 = 3(x^2 - 9y^2)$
$= 3[x^2 - (3y)^2]$
$= 3(x - 3y)(x + 3y)$

Exercise 26

Factorise the following:

1. $y^2 - a^2$	**2.** $m^2 - n^2$	**3.** $x^2 - t^2$	**4.** $y^2 - 1$
5. $x^2 - 9$	**6.** $a^2 - 25$	**7.** $x^2 - \dfrac{1}{4}$	**8.** $x^2 - \dfrac{1}{9}$

9. $4x^2 - y^2$

10. $a^2 - 4b^2$

11. $25x^2 - 4y^2$

12. $9x^2 - 16y^2$

13. $x^2 - \dfrac{y^2}{4}$

14. $9m^2 - \dfrac{4}{9}n^2$

15. $16t^2 - \dfrac{4}{25}s^2$

16. $4x^2 - \dfrac{z^2}{100}$

17. $x^3 - x$

18. $a^3 - ab^2$

19. $4x^3 - x$

20. $8x^3 - 2xy^2$

21. $12x^3 - 3xy^2$

22. $18m^3 - 8mn^2$

23. $5x^2 - 1\tfrac{1}{4}$

24. $50a^3 - 18ab^2$

25. $12x^2y - 3yz^2$

26. $36a^3b - 4ab^3$

27. $50a^5 - 8a^3b^2$

28. $36x^3y - 225xy^3$

Evaluate the following:

29. $81^2 - 80^2$

30. $102^2 - 100^2$

31. $225^2 - 215^2$

32. $1211^2 - 1210^2$

33. $723^2 - 720^2$

34. $3 \cdot 8^2 - 3 \cdot 7^2$

35. $5 \cdot 24^2 - 4 \cdot 76^2$

36. $1234^2 - 1235^2$

37. $3 \cdot 81^2 - 3 \cdot 8^2$

38. $540^2 - 550^2$

39. $7 \cdot 68^2 - 2 \cdot 32^2$

40. $0 \cdot 003^2 - 0 \cdot 002^2$

2.10 Quadratic equations

So far, we have met linear equations which have one solution only. Quadratic equations always have an x^2 term, and often an x term and a number term, and generally have two different solutions.

Solution by factors

Consider the equation $a \times b = 0$, where a and b are numbers. The product $a \times b$ can only be zero if either a or b (or both) is equal to zero. Can you think of other possible pairs of numbers which multiply together to give zero?

Example 1

Solve the equation $x^2 + x - 12 = 0$

Factorising, $(x - 3)(x + 4) = 0$
either $x - 3 = 0$ or $x + 4 = 0$
$\qquad\qquad x = 3 \qquad\qquad x = -4$

Example 2

Solve the equation $6x^2 + x - 2 = 0$

Factorising, $(2x - 1)(3x + 2) = 0$
either $2x - 1 = 0$ or $3x + 2 = 0$
$\qquad\qquad 2x = 1 \qquad\qquad 3x = -2$
$\qquad\qquad x = \tfrac{1}{2} \qquad\qquad x = -\tfrac{2}{3}$

Exercise 27

Solve the following equations:

1. $x^2 + 7x + 12 = 0$

2. $x^2 + 7x + 10 = 0$

3. $x^2 + 2x - 15 = 0$

4. $x^2 + x - 6 = 0$

5. $x^2 - 8x + 12 = 0$

6. $x^2 + 10x + 21 = 0$

7. $x^2 - 5x + 6 = 0$

8. $x^2 - 4x - 5 = 0$

9. $x^2 + 5x - 14 = 0$

10. $2x^2 - 3x - 2 = 0$

11. $3x^2 + 10x - 8 = 0$

12. $2x^2 + 7x - 15 = 0$

13. $6x^2 - 13x + 6 = 0$

14. $4x^2 - 29x + 7 = 0$

15. $10x^2 - x - 3 = 0$

16. $y^2 - 15y + 56 = 0$

17. $12y^2 - 16y + 5 = 0$

18. $y^2 + 2y - 63 = 0$

19. $x^2 + 2x + 1 = 0$

20. $x^2 - 6x + 9 = 0$

21. $x^2 + 10x + 25 = 0$

22. $x^2 - 14x + 49 = 0$

23. $6a^2 - a - 1 = 0$

24. $4a^2 - 3a - 10 = 0$

25. $z^2 - 8z - 65 = 0$

26. $6x^2 + 17x - 3 = 0$

27. $10k^2 + 19k - 2 = 0$

28. $y^2 - 2y + 1 = 0$

29. $36x^2 + x - 2 = 0$

30. $20x^2 - 7x - 3 = 0$

Example 1

Solve the equation $x^2 - 7x = 0$

Factorising, $x(x - 7) = 0$

either $x = 0$ or $x - 7 = 0$

$x = 7$

The solutions are $x = 0$ and $x = 7$.

Example 2

Solve the equation $4x^2 - 9 = 0$

(a) Factorising, $(2x - 3)(2x + 3) = 0$

either $2x - 3 = 0$ or $2x + 3 = 0$

$2x = 3$ $2x = -3$

$x = \frac{3}{2}$ $x = -\frac{3}{2}$

(b) Alternative method

$4x^2 - 9 = 0$

$4x^2 = 9$

$x^2 = \frac{9}{4}$

$x = +\frac{3}{2}$ or $-\frac{3}{2}$.

> **Hint**
> You must give both the
> solutions. A common error is to
> only give the positive square
> root.

Exercise 28

Solve the following equations:

1. $x^2 - 3x = 0$

2. $x^2 + 7x = 0$

3. $2x^2 - 2x = 0$

4. $3x^2 - x = 0$

5. $x^2 - 16 = 0$

6. $x^2 - 49 = 0$

7. $4x^2 - 1 = 0$

8. $9x^2 - 4 = 0$

9. $6y^2 + 9y = 0$

10. $6a^2 - 9a = 0$

11. $10x^2 - 55x = 0$

12. $16x^2 - 1 = 0$

13. $y^2 - \frac{1}{4} = 0$

14. $56x^2 - 35x = 0$

15. $36x^2 - 3x = 0$

16. $x^2 = 6x$

17. $x^2 = 11x$

18. $2x^2 = 3x$

19. $x^2 = x$

20. $4x = x^2$

21. $3x - x^2 = 0$

22. $4x^2 = 1$

23. $9x^2 = 16$

24. $x^2 = 9$

25. $12x = 5x^2$

26. $1 - 9x^2 = 0$

27. $x^2 = \frac{x}{4}$

28. $2x^2 = \frac{x}{3}$

29. $4x^2 = \frac{1}{4}$

30. $\frac{x}{5} - x^2 = 0$

Solution by formula

The solutions of the quadratic equation $ax^2 + bx + c = 0$
are given by the formula

$$x = \frac{-b \pm \sqrt{(b^2 - 4ac)}}{2a}$$

Use this formula only after trying (and failing) to factorise.

Example

Solve the equation $2x^2 - 3x - 4 = 0$.

In this case $a = 2$, $b = -3$, $c = -4$.

$$x = \frac{-(-3) \pm \sqrt{[(-3)^2 - (4 \times 2 \times -4)]}}{2 \times 2}$$

$$x = \frac{3 \pm \sqrt{[9 + 32]}}{4} = \frac{3 \pm \sqrt{41}}{4} = \frac{3 \pm 6 \cdot 403}{4}$$

either $x = \dfrac{3 + 6 \cdot 403}{4}$ or $x = \dfrac{3 - 6 \cdot 403}{4} = \dfrac{-3 \cdot 403}{4}$

$\qquad\qquad = 2 \cdot 35$ (2 decimal places) $= -0 \cdot 85$ (2 decimal places).

Exercise 29

Solve the following, giving answers to two decimal places where necessary:

1. $2x^2 + 11x + 5 = 0$ **2.** $3x^2 + 11x + 6 = 0$ **3.** $6x^2 + 7x + 2 = 0$

4. $3x^2 - 10x + 3 = 0$ **5.** $5x^2 - 7x + 2 = 0$ **6.** $6x^2 - 11x + 3 = 0$

7. $2x^2 + 6x + 3 = 0$ **8.** $x^2 + 4x + 1 = 0$ **9.** $5x^2 - 5x + 1 = 0$

10. $x^2 - 7x + 2 = 0$ **11.** $2x^2 + 5x - 1 = 0$ **12.** $3x^2 + x - 3 = 0$

13. $3x^2 + 8x - 6 = 0$ **14.** $3x^2 - 7x - 20 = 0$ **15.** $2x^2 - 7x - 15 = 0$

16. $x^2 - 3x - 2 = 0$ **17.** $2x^2 + 6x - 1 = 0$ **18.** $6x^2 - 11x - 7 = 0$

19. $3x^2 + 25x + 8 = 0$ **20.** $3y^2 - 2y - 5 = 0$ **21.** $2y^2 - 5y + 1 = 0$

22. $\frac{1}{2}y^2 + 3y + 1 = 0$ **23.** $2 - x - 6x^2 = 0$ **24.** $3 + 4x - 2x^2 = 0$

25. $1 - 5x - 2x^2 = 0$ **26.** $3x^2 - 1 + 4x = 0$ **27.** $5x - x^2 + 2 = 0$

28. $24x^2 - 22x - 35 = 0$ **29.** $36x^2 - 17x - 35 = 0$ **30.** $20x^2 + 17x - 63 = 0$

31. $x^2 + 2 \cdot 5x - 6 = 0$ **32.** $0 \cdot 3y^2 + 0 \cdot 4y - 1 \cdot 5 - 0$ **33.** $10 - x - 3x^2 = 0$

34. $x^2 + 3 \cdot 3x - 0 \cdot 7 = 0$ **35.** $12 - 5x^2 - 11x = 0$ **36.** $5x - 2x^2 + 187 = 0$

The solution to a problem can involve an equation which does not at
first appear to be quadratic. The terms in the equation may need to be
rearranged as shown on the next page.

Example

Solve:
$$2x(x-1) = (x+1)^2 - 5$$
$$2x^2 - 2x = x^2 + 2x + 1 - 5$$
$$2x^2 - 2x - x^2 - 2x - 1 + 5 = 0$$
$$x^2 - 4x + 4 = 0$$
$$(x-2)(x-2) = 0$$
$$x = 2$$

In this example the quadratic has a repeated solution of $x = 2$.

Exercise 30

Solve the following, giving answers to two decimal places where necessary:

1. $x^2 = 6 - x$

2. $x(x+10) = -21$

3. $3x + 2 = 2x^2$

4. $x^2 + 4 = 5x$

5. $6x(x+1) = 5 - x$

6. $(2x)^2 = x(x-14) - 5$

7. $(x-3)^2 = 10$

8. $(x+1)^2 - 10 = 2x(x-2)$

9. $(2x-1)^2 = (x-1)^2 + 8$

10. $3x(x+2) - x(x-2) + 6 = 0$

11. $x = \dfrac{15}{x} - 22$

12. $x + 5 = \dfrac{14}{x}$

13. $4x + \dfrac{7}{x} = 29$

14. $10x = 1 + \dfrac{3}{x}$

15. $2x^2 = 7x$

16. $16 = \dfrac{1}{x^2}$

17. $2x + 2 = \dfrac{7}{x} - 1$

18. $\dfrac{2}{x} + \dfrac{2}{x+1} = 3$

19. $\dfrac{3}{x-1} + \dfrac{3}{x+1} = 4$

20. $\dfrac{2}{x-2} + \dfrac{4}{x+1} = 3$

21. One of the solutions published by Cardan in 1545 for the solution of cubic equations is given below. For an equation in the form $x^3 + px = q$

$$x = \sqrt[3]{\left[\sqrt{\left(\frac{p}{3}\right)^3 + \left(\frac{q}{2}\right)^2} + \frac{q}{2}\right]} - \sqrt[3]{\left[\sqrt{\left(\frac{p}{3}\right)^3 + \left(\frac{q}{2}\right)^2} - \frac{q}{2}\right]}$$

Use the formula to solve the following equations, giving answers to 4 s.f. where necessary.

(a) $x^3 + 7x = -8$

(b) $x^3 + 6x = 4$

(c) $x^3 + 3x = 2$

(d) $x^3 + 9x - 2 = 0$

Solution by completing the square

Look at the function $f(x) = x^2 + 6x$
Completing the square, this becomes $f(x) = (x+3)^2 - 9$

Functions and their notation are explained in more detail on page 280.

This is done as follows:

3 is half of 6 and gives $6x$.

Having added 3 to the square term, 9 needs to be subtracted from the expression to cancel the $+9$ obtained.

Here are some more examples.

(a) $x^2 - 12x = (x - 6)^2 - 36$

(b) $x^2 + 3x = (x + \frac{3}{2})^2 - \frac{9}{4}$

(c) $x^2 + 6x + 1 = (x + 3)^2 - 9 + 1$
$$= (x + 3)^2 - 8$$

(d) $x^2 - 10x - 17 = (x - 5)^2 - 25 - 17$
$$= (x - 5)^2 - 42$$

(e) $2x^2 - 12x + 7 = 2[x^2 - 6x + \frac{7}{2}]$
$$= 2[(x - 3)^2 - 9 + \frac{7}{2}]$$
$$= 2[(x - 3)^2 - \frac{11}{2}]$$

Example 1

Solve the quadratic equation $x^2 - 6x + 7 = 0$ by completing the square.
$$(x - 3)^2 - 9 + 7 = 0$$
$$(x - 3)^2 \qquad = 2$$
$$\therefore \quad x - 3 = +\sqrt{2} \quad \text{or} \ -\sqrt{2}$$
$$x = 3 + \sqrt{2} \ \text{or} \ 3 - \sqrt{2}$$
So, $\quad x = 4 \cdot 41 \qquad$ or $1 \cdot 59$ to 2 d.p.

Example 2

Given $f(x) = x^2 - 8x + 18$, show that $f(x) \geq 2$ for all values of x.

Completing the square, $f(x) = (x - 4)^2 - 16 + 18$
$$f(x) = (x - 4)^2 + 2.$$

Now $(x - 4)^2$ is always greater than or equal to zero because it is 'something squared'.
$$\therefore \quad f(x) \geq 2$$

Exercise 31

In Questions **1** to **10**, complete the square for each expression by writing each one in the form $(x + a)^2 + b$ where a and b can be positive or negative.

1. $x^2 + 8x$ **2.** $x^2 - 12x$ **3.** $x^2 + x$

4. $x^2 + 4x + 1$ **5.** $x^2 - 6x + 9$ **6.** $x^2 + 2x - 15$

7. $2x^2 + 16x + 5$ **8.** $2x^2 - 10x$ **9.** $6 + 4x - x^2$

10. $3 - 2x - x^2$

11. Solve these equations by completing the square

(a) $x^2 + 4x - 3 = 0$ (b) $x^2 - 3x - 2 = 0$ (c) $x^2 + 12x = 1$

12. Try to solve the equation $x^2 + 6x + 10 = 0$, by completing the square. Explain why you can find no solutions.

13. Given $f(x) = x^2 + 6x + 12$, show that $f(x) \geq 3$ for all values of x.

14. Given $g(x) = x^2 - 7x + \frac{1}{4}$, show that the least possible value of $g(x)$ is -12.

15. If $f(x) = x^2 + 4x + 7$ find

(a) the smallest possible value of $f(x)$

(b) the value of x for which this smallest value occurs

(c) the greatest possible value of $\dfrac{1}{(x^2 + 4x + 7)}$

2.11 Problems solved by quadratic equations

Example 1

The area of rectangle A is 16 cm² greater than the area of rectangle B. Find the height of rectangle A.

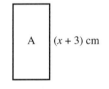

7 cm

A $(x + 3)$ cm

B $(x - 1)$ cm

$(x + 2)$ cm

Area of rectangle A = $7(x + 3)$

Area of rectangle B = $(x + 2)(x - 1)$

We are given $(x + 2)(x - 1) + 16 = 7(x + 3)$

Solve this equation
$$x^2 + 2x - x - 2 + 16 = 7x + 21$$
$$x^2 + x + 14 = 7x + 21$$
$$x^2 - 6x - 7 = 0$$
$$(x - 7)(x + 1) = 0$$
$$x = 7 \quad (x \text{ cannot be negative})$$
The height of rectangle A, $x + 3$, is 10 cm.

Example 2

A man bought a certain number of golf balls for $20. If each ball had cost 20 cents less, he could have bought five more for the same money. How many golf balls did he buy?

Let the number of balls bought be x.

Cost of each ball $= \dfrac{2000}{x}$ cents

If five more balls had been bought

Cost of each ball now $= \dfrac{2000}{(x+5)}$ cents

The new price is 20 cents less than the original price.

$\therefore \quad \dfrac{2000}{x} - \dfrac{2000}{(x+5)} = 20$

(multiply by x)

$x \cdot \dfrac{2000}{x} - x \cdot \dfrac{2000}{(x+5)} = 20x$

(multiply by $(x+5)$)

$2000(x+5) - x\dfrac{2000}{(x+5)}(x+5) = 20x(x+5)$

$$2000x + 10\,000 - 2000x = 20x^2 + 100x$$
$$20x^2 + 100x - 10\,000 = 0$$
$$x^2 + 5x - 500 = 0$$
$$(x - 20)(x + 25) = 0$$
$$x = 20$$
$$\text{or} \quad x = -25$$

We discard $x = -25$ as meaningless.
The number of balls bought $= 20$.

Exercise 32

Solve by forming a quadratic equation:

Questions 4, 6 and 7 use Pythagoras' Theorem. For more information on Pythagoras' Theorem see page 129.

1. Two numbers, which differ by 3, have a product of 88. Find them.

2. The product of two consecutive odd numbers is 143. Find the numbers. (Hint: If the first odd number is x, what is the next odd number?)

3. The length of a rectangle exceeds the width by 7 cm. If the area is $60\,\text{cm}^2$, find the length of the rectangle.

4. The length of a rectangle exceeds the width by 2 cm. If the diagonal is 10 cm long, find the width of the rectangle.

5. The area of the rectangle exceeds the area of the square by $24\,\text{m}^2$. Find x.

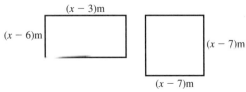

6. The perimeter of a rectangle is 68 cm. If the diagonal is 26 cm, find the dimensions of the rectangle.

7. Sang Jae walks a certain distance due North and then the same distance plus a further 7 km due East. If the final distance from the starting point is 17 km, find the distances he walks North and East.

8. A farmer makes a profit of x cents on each of the $(x + 5)$ eggs her hen lays. If her total profit was 84 cents, find the number of eggs the hen lays.

9. Sirak buys x eggs at $(x - 8)$ cents each and $(x - 2)$ bread rolls at $(x - 3)$ cents each. If the total bill is $1·75, how many eggs does he buy?

10. A number exceeds four times its reciprocal by 3. Find the number.

11. Two numbers differ by 3. The sum of their reciprocals is $\frac{7}{10}$; find the numbers.

12. A cyclist travels 40 km at a speed of x km/h. Find the time taken in terms of x. Find the time taken when his speed is reduced by 2 km/h. If the difference between the times is 1 hour, find the original speed x.

13. An increase of speed of 4 km/h on a journey of 32 km reduces the time taken by 4 hours. Find the original speed.

14. A train normally travels 240 km at a certain speed. One day, due to bad weather, the train's speed is reduced by 20 km/h so that the journey takes two hours longer. Find the normal speed.

15. The speed of a sparrow is x km/h in still air. When the wind is blowing at 1 km/h, the sparrow takes 5 hours to fly 12 kilometres to her nest and 12 kilometres back again. She goes out directly into the wind and returns with the wind behind her. Find her speed in still air.

16. An aircraft flies a certain distance on a bearing of 135° and then twice the distance on a bearing of 225°. Its distance from the starting point is then 350 km. Find the length of the first part of the journey.

> A **bearing** is a clockwise angle measured from North. For more information about bearings see page 200.

17. In Figure 1, ABCD is a rectangle with AB = 12 cm and BC = 7 cm. AK = BL = CM = DN = x cm. If the area of KLMN is 54 cm², find x.

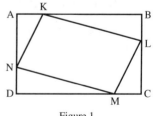

Figure 1

18. In Figure 1, AB = 14 cm, BC = 11 cm and AK = BL = CM = DN = x cm. If the area of KLMN is now 97 cm², find x.

19. The numerator of a fraction is 1 less than the denominator. When both numerator and denominator are increased by 2, the fraction is increased by $\frac{1}{12}$. Find the original fraction.

20. The perimeters of a square and a rectangle are equal. One side of the rectangle is 11 cm and the area of the square is $4\,\mathrm{cm}^2$ more than the area of the rectangle. Find the side of the square.

Revision exercise 2A

1. Solve the equations:
  (a) $x + 4 = 3x + 9$
  (b) $9 - 3a = 1$
  (c) $y^2 + 5y = 0$
  (d) $x^2 - 4 = 0$
  (e) $3x^2 + 7x - 40 = 0$

Questions 13 and 20 use Pythagoras' Theorem. See page 129.

2. Given $a = 3$, $b = 4$ and $c = -2$, evaluate:
  (a) $2a^2 - b$
  (b) $a(b - c)$
  (c) $2b^2 - c^2$

3. Factorise completely:
  (a) $4x^2 - y^2$
  (b) $2x^2 + 8x + 6$
  (c) $6m + 4n - 9km - 6kn$
  (d) $2x^2 - 5x - 3$

4. Solve the simultaneous equations:
  (a) $3x + 2y = 5$
     $2x - y = 8$
  (b) $2m - n = 6$
     $2m + 3n = -6$
  (c) $3x - 4y = 19$
     $x + 6y = 10$
  (d) $3x - 7y = 11$
     $2x - 3y = 4$

5. Given that $x = 4$, $y = 3$, $z = -2$, evaluate:
  (a) $2x(y + z)$
  (b) $(xy)^2 - z^2$
  (c) $x^2 + y^2 + z^2$
  (d) $(x + y)(x - z)$
  (e) $\sqrt{[x(1 - 4z)]}$
  (f) $\frac{xy}{z}$

6. (a) Simplify $3(2x - 5) - 2(2x + 3)$.
  (b) Factorise $2a - 3b - 4xa + 6xb$.
  (c) Solve the equation $\dfrac{x - 11}{2} - \dfrac{x - 3}{5} = 2$.

7. Solve the equations:
  (a) $5 - 7x = 4 - 6x$
  (b) $\dfrac{7}{x} = \dfrac{2}{3}$
  (c) $2x^2 - 7x = 0$
  (d) $x^2 + 5x + 6 = 0$
  (e) $\dfrac{1}{x} + \dfrac{1}{4} = \dfrac{1}{3}$

8. Factorise completely:
  (a) $z^3 - 16z$
  (b) $x^2y^2 + x^2 + y^2 + 1$
  (c) $2x^2 + 11x + 12$

9. Find the value of $\dfrac{2x - 3y}{5x + 2y}$ when $x = 2a$ and
   $y = -a.$

10. Solve the simultaneous equations:
    (a) $7c + 3d = 29$ (b) $2x - 3y = 7$
        $5c - 4d = 33$ $2y - 3x = -8$
    (c) $5x = 3(1 - y)$ (d) $5s + 3t = 16$
        $3x + 2y + 1 = 0$ $11s + 7t = 34$

11. Solve the equations:
    (a) $4(y + 1) = \dfrac{3}{1 - y}$
    (b) $4(2x - 1) - 3(1 - x) = 0$
    (c) $\dfrac{x + 3}{x} = 2$
    (d) $x^2 = 5x$

12. Solve the following, giving your answers correct to two decimal
    places.
    (a) $2x^2 - 3x - 1 = 0$ (b) $x^2 - x - 1 = 0$
    (c) $3x^2 + 2x - 4 = 0$ (d) $x + 3 = \dfrac{7}{x}$

13. Find x by forming a suitable equation.   (a) 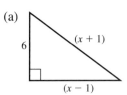   (b)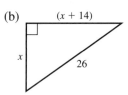

14. Given that $m = -2$, $n = 4$, evaluate:
    (a) $5m + 3n$ (b) $5 + 2m - m^2$
    (c) $m^2 + 2n^2$ (d) $(2m + n)(2m - n)$
    (e) $(n - m)^2$ (f) $n - mn - 2m^2$

15. A car travels for x hours at a speed of $(x + 2)\,$km/h. If the distance
    travelled is 15 km, write down an equation for x and solve it to find
    the speed of the car.

16. ABCD is a rectangle, where $AB = x\,$cm and BC is 1·5 cm less
    than AB.

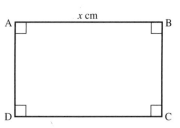

    If the area of the rectangle is 52 cm^2, form an equation in x and
    solve it to find the dimensions of the rectangle.

17. Solve the equations:
    (a) $(2x + 1)^2 = (x + 5)^2$
    (b) $\dfrac{x + 2}{2} - \dfrac{x - 1}{3} = \dfrac{x}{4}$
    (c) $x^2 - 7x + 5 = 0$, giving the answers correct to two decimal places.

18. Solve the equation:
    $$\dfrac{x}{x + 1} - \dfrac{x + 1}{3x - 1} = \dfrac{1}{4}$$

19. Given that $a + b = 2$ and that $a^2 + b^2 = 6$, prove that $2ab = -2$. Find also the value of $(a - b)^2$.

20. The sides of a right-angled triangle have lengths $(x - 3)$ cm, $(x + 11)$ cm and $2x$ cm, where $2x$ is the hypotenuse. Find x.

21. A jar contains 50 coins, all either 2 cents or 5 cents. The total value of the coins is $1·87. How many 2 cents coins are there?

22. Pat bought 45 stamps, some for 10c and some for 18c. If he spent $6·66 altogether, how many 10c stamps did he buy?

23. When each edge of a cube is decreased by 1 cm, its volume is decreased by 91 cm^3. Find the length of a side of the original cube.

24. One solution of the equation $2x^2 - 7x + k = 0$ is $x = -\frac{1}{2}$. Find the value of k.

Examination exercise 2B

1. (a) $\dfrac{2}{3} + \dfrac{5}{6} = \dfrac{x}{2}$.

    Find the value of x.   [1]

    (b) $\dfrac{5}{3} \div \dfrac{3}{y} = \dfrac{40}{9}$.

    Find the value of y.   [1]

    Cambridge IGCSE Mathematics 0580
    Paper 2 Q2 November 2006

2. Find the coordinates of the point of intersection of the straight lines
    $2x + 3y = 11$,
    $3x - 5y = -12$.   [3]

    Cambridge IGCSE Mathematics 0580
    Paper 22 Q16 June 2008

3. Solve the equations

(a) $\dfrac{2x}{3} - 9 = 0$,   (b) $x^2 - 3x - 4 = 0$.   [2] [2]

Cambridge IGCSE Mathematics 0580
Paper 2 Q14 November 2007

4. Solve the simultaneous equations
$0.4x + 2y = 10$,
$0.3x + 5y = 18$.   [3]

Cambridge IGCSE Mathematics 0580
Paper 2 Q12 June 2006

5. $x^2 + 4x - 8$ can be written in the form $(x + p)^2 + q$.
Find the value of p and q.   [3]

Cambridge IGCSE Mathematics 0580
Paper 2 Q9 November 2007

6.

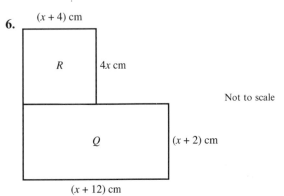

(a) (i) Write down an expression for the area of rectangle R.   [1]
    (ii) Show that the total area of rectangles R and Q is
          $5x^2 + 30x + 24$ square centimetres.   [1]
(b) The total area of rectangles R and Q is 64 cm^2.
    Calculate the value of x correct to 1 decimal place.   [4]

Cambridge IGCSE Mathematics 0580
Paper 2 Q20 November 2006

7. (a) (i) Factorise $x^2 - x - 20$.   [2]
       (ii) Solve the equation $x^2 - x - 20 = 0$.   [1]
   (b) Solve the equation $3x^2 - 2x - 2 = 0$.
       Show all your working and give your answers correct to
       2 decimal places.   [4]
   (c) $y = m^2 - 4n^2$.
        (i) Factorise $m^2 - 4n^2$.   [1]
       (ii) Find the value of y when $m = 4.4$ and $n = 2.8$.   [1]
      (iii) $m = 2x + 3$ and $n = x - 1$.
             Find y in terms of x, in its simplest form.   [2]
       (iv) Make n the subject of the formula $y = m^2 - 4n^2$.   [3]
   (d) (i) $m^4 - 16n^4$ can be written as $(m^2 - kn^2)(m^2 + kn^2)$.
            Write down the value of k.   [1]
       (ii) Factorise completely $m^4n - 16n^5$.   [2]

Cambridge IGCSE Mathematics 0580
Paper 4 Q2 June 2008

8. (a)   In triangle ABC, the line BD is perpendicular to AC.
$AD = (x + 6)$ cm, $DC = (x + 2)$ cm and the height
$BD = (x + 1)$ cm.
The area of triangle ABC is 40 cm^2.
  (i) Show that $x^2 + 5x - 36 = 0$.   [3]
 (ii) Solve the equation $x^2 + 5x - 36 = 0$.   [2]
(iii) Calculate the length of BC.   [2]

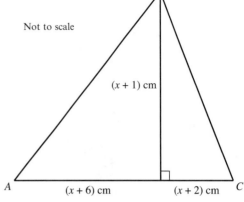

Not to scale

B

$(x + 1)$ cm

A $(x + 6)$ cm $(x + 2)$ cm C

(b) Amira takes 9 hours 25 minutes to complete a long walk.
  (i) Show that the time of 9 hours 25 minutes can be written as
      $\frac{113}{12}$ hours.   [1]
 (ii) She walks $(3y + 2)$ kilometres at 3km/h and then a further
      $(y + 4)$ kilometres at 2km/h.

      Show that the total time taken is $\frac{9y+16}{6}$ hours.   [2]

(iii) Solve the equation $\frac{9y+16}{6} = \frac{113}{12}$.   [2]

(iv) Calculate Amira's average speed, in kilometres
     per hour, for the whole walk.   [3]

Cambridge IGCSE Mathematics 0580
Paper 4 Q6 June 2009

3 MENSURATION

Archimedes of Samos (287–212 B.C.) studied at Alexandria as a young man. One of the first to apply scientific thinking to everyday problems, he was a practical man of common sense. He gave proofs for finding the area, the volume and the centre of gravity of circles, spheres, conics and spirals. By drawing polygons with many sides, he arrived at a value of π between $3\frac{10}{71}$ and $3\frac{10}{70}$. He was killed in the siege of Syracuse at the age of 75.

31 Carry out calculations involving the perimeter and area of a rectangle and triangle, the circumference and area of a circle, the area of a parallelogram and trapezium, the volume of a cuboid, prism and cylinder, and the surface area of a cylinder.
Solve problems involving the arc length and sector area of a circle, the surface area and volume of a sphere, pyramid and cone.

3.1 Area

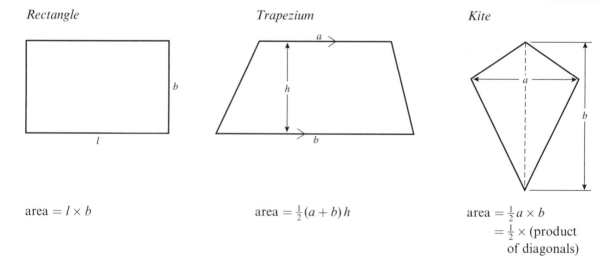

Rectangle

area $= l \times b$

Trapezium

area $= \frac{1}{2}(a + b)h$

Kite

area $= \frac{1}{2}a \times b$
$= \frac{1}{2} \times$ (product of diagonals)

Exercise 1

For questions **1** to **7**, find the area of each shape. Decide which information to use: you may not need all of it.

1.
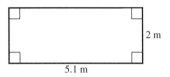

2.
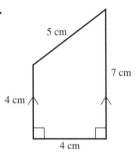

3.
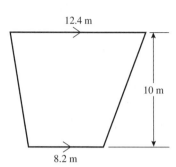

4.
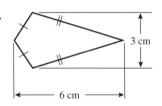

5.
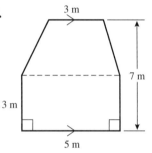

6.
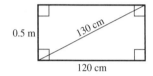

7.
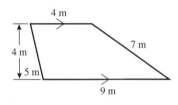

8. Find the area shaded.
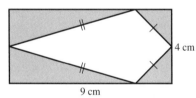

9. Find the area shaded.
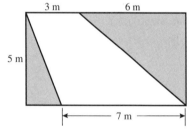

10. A rectangle has an area of $117 \, \text{m}^2$ and a width of 9 m. Find its length.

11. A trapezium of area $105 \, \text{cm}^2$ has parallel sides of length 5 cm and 9 cm. How far apart are the parallel sides?

12. A kite of area $252 \, \text{m}^2$ has one diagonal of length 9 m. Find the length of the other diagonal.

13. A kite of area $40 \, \text{m}^2$ has one diagonal 2 m longer than the other. Find the lengths of the diagonals.

14. A trapezium of area $140 \, \text{cm}^2$ has parallel sides 10 cm apart and one of these sides is 16 cm long. Find the length of the other parallel side.

15. A floor 5 m by 20 m is covered by square tiles of side 20 cm. How many tiles are needed?

16. On squared paper draw the triangle with vertices at (1, 1), (5, 3), (3, 5). Find the area of the triangle.

17. Draw the quadrilateral with vertices at (1, 1), (6, 2), (5, 5), (3, 6). Find the area of the quadrilateral.

18. A square wall is covered with square tiles. There are 85 tiles altogether along the two diagonals. How many tiles are there on the whole wall?

19. On squared paper draw a 7×7 square. Divide it up into nine smaller squares.

20. A rectangular field, 400 m long, has an area of 6 hectares. Calculate the perimeter of the field [1 hectare $= 10\,000\,\text{m}^2$].

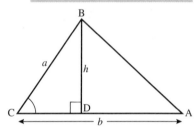

Triangle
area $= \frac{1}{2} \times b \times h$

In triangle BCD, $\sin C = \dfrac{h}{a}$

$\therefore \qquad\qquad h = a \sin C$

$\therefore \quad$ area of triangle $= \frac{1}{2} \times b \times a \sin C$

This formula is useful when *two sides* and the *included* angle are known.

Example

Find the area of the triangle shown.

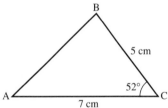

You can find out more about $\sin \theta$ in Trigonometry on page 193.

Area $= \frac{1}{2} ab \sin C$
$\qquad = \frac{1}{2} \times 5 \times 7 \times \sin 52°$
$\qquad = 13{\cdot}8\,\text{cm}^2$ (1 d.p.)

Parallelogram

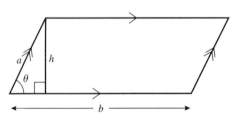

area $= b \times h$
area $= ba \sin \theta$

Exercise 2

In questions **1** to **12** find the area of $\triangle ABC$ where $AB = c$, $AC = b$ and
$BC = a$. (Sketch the triangle in each case.) You will need some basic
trigonometry (see page 192).

1. $a = 7\,\text{cm}$, $b = 14\,\text{cm}$, $\widehat{C} = 80°$.   **2.** $b = 11\,\text{cm}$, $a = 9\,\text{cm}$, $\widehat{C} = 35°$.

3. $c = 12\,\text{m}$, $b = 12\,\text{m}$, $\widehat{A} = 67\cdot2°$.   **4.** $a = 5\,\text{cm}$, $c = 6\,\text{cm}$, $\widehat{B} = 11\cdot8°$.

5. $b = 4\cdot2\,\text{cm}$, $a = 10\,\text{cm}$, $\widehat{C} = 120°$.   **6.** $a = 5\,\text{cm}$, $c = 8\,\text{cm}$, $\widehat{B} = 142°$.

7. $b = 3\cdot2\,\text{cm}$, $c = 1\cdot8\,\text{cm}$, $\widehat{B} = 10°$, $\widehat{C} = 65°$.   **8.** $a = 7\,\text{m}$, $b = 14\,\text{m}$, $\widehat{A} = 32°$, $\widehat{B} = 100°$.

9. $a = b = c = 12\,\text{m}$.   **10.** $a = c = 8\,\text{m}$, $\widehat{B} = 72°$.

11. $b = c = 10\,\text{cm}$, $\widehat{B} = 32°$.   **12.** $a = b = c = 0\cdot8\,\text{m}$.

In questions **13** to **20**, find the area of each shape.

13.

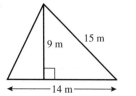

14.

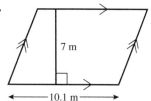

15.

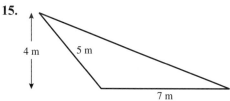

16.

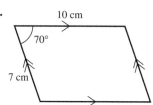

17.

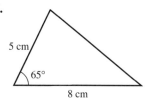

18.

19. Find the area shaded.

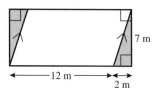

20.

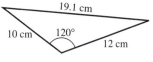

21. Find the area of a parallelogram ABCD with AB = 7 m,
AD = 20 m and $\widehat{BAD} = 62°$.

22. Find the area of a parallelogram ABCD with AD = 7 m,
CD = 11 m and $\widehat{BAD} = 65°$.

23. In the diagram if $AE = \frac{1}{3}AB$, find the area shaded.

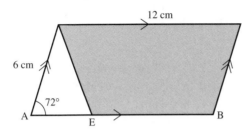

24. The area of an equilateral triangle ABC is $50\,\text{cm}^2$. Find AB.

25. The area of a triangle ABC is $64\,\text{cm}^2$. Given $AB = 11\,\text{cm}$ and $BC = 15\,\text{cm}$, find $A\widehat{B}C$.

26. The area of a triangle XYZ is $11\,\text{m}^2$. Given $YZ = 7\,\text{m}$ and $X\widehat{Y}Z = 130°$, find XY.

27. Find the length of a side of an equilateral triangle of area $10{\cdot}2\,\text{m}^2$.

28. A rhombus has an area of $40\,\text{cm}^2$ and adjacent angles of $50°$ and $130°$. Find the length of a side of the rhombus.

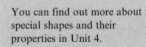

You can find out more about special shapes and their properties in Unit 4.

29. A regular hexagon is circumscribed by a circle of radius $3\,\text{cm}$ with centre O.

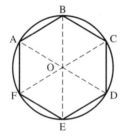

(a) What is angle EOD?
(b) Find the area of triangle EOD and hence find the area of the hexagon ABCDEF.

30. Hexagonal tiles of side $20\,\text{cm}$ are used to tile a room which measures $6{\cdot}25\,\text{m}$ by $4{\cdot}85\,\text{m}$. Assuming we complete the edges by cutting up tiles, how many tiles are needed?

31. Find the area of a regular pentagon of side $8\,\text{cm}$.

32. The diagram shows a part of the perimeter of a regular polygon with n sides

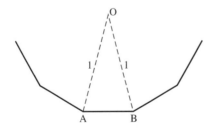

The centre of the polygon is at O and OA = OB = 1 unit.
(a) What is the angle AOB in terms of n?
(b) Work out an expression in terms of n for the area of the polygon.
(c) Find the area of polygons where $n = 6, 10, 300, 1000, 10\,000$. What do you notice?

33. The area of a regular pentagon is $600\,\text{cm}^2$. Calculate the length of one side of the pentagon.

3.2 The circle

For any circle, the ratio $\left(\dfrac{\text{circumference}}{\text{diameter}}\right)$ is equal to π.

The value of π is usually taken to be 3·14, but this is not an exact value. Through the centuries, mathematicians have been trying to obtain a better value for π.

For example, in the third century A.D., the Chinese mathematician Liu Hui obtained the value 3·14159 by considering a regular polygon having 3072 sides! Ludolph van Ceulen (1540–1610) worked even harder to produce a value correct to 35 significant figures. He was so proud of his work that he had this value of π engraved on his tombstone.

Electronic computers are now able to calculate the value of π to many thousands of figures, but its value is still not exact. It was shown in 1761 that π is an *irrational number* which, like $\sqrt{2}$ or $\sqrt{3}$ cannot be expressed exactly as a fraction.

The first fifteen significant figures of π can be remembered from the number of letters in each word of the following sentence.

How I need a drink, cherryade of course, after the silly lectures involving Italian kangaroos.

There remain a lot of unanswered questions concerning π, and many mathematicians today are still working on them.

The following formulae should be memorised.

$$\text{circumference} = \pi d$$
$$= 2\pi r$$
$$\text{area} = \pi r^2$$

Example

Find the circumference and area of a circle of diameter 8 cm. (Take $\pi = 3 \cdot 142$.)

$$\text{Circumference} = \pi d$$
$$= 3 \cdot 142 \times 8$$
$$= 25 \cdot 1 \text{ cm (1 d.p.)}$$

$$\text{Area} = \pi r^2$$
$$= 3 \cdot 142 \times 4^2$$
$$= 50 \cdot 3 \text{ cm}^2 \text{ (1 d.p.)}$$

Exercise 3

For each shape find (a) the perimeter, (b) the area. All lengths are in cm. Use the π button on a calculator or take $\pi = 3 \cdot 142$. All the arcs are either semi-circles or quarter circles.

1.
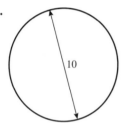

2.

3.

4.

5.
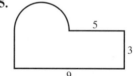

6.

7.

8.

9.

10.
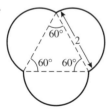

11.
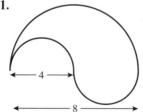

12.
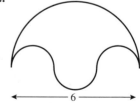

Example 1

A circle has a circumference of 20 m. Find the radius of the circle.

Let the radius of the circle by r m.

$$\text{Circumference} = 2\pi r$$
$$\therefore \quad 2\pi r = 20$$
$$\therefore \quad r = \frac{20}{2\pi}$$
$$r = 3{\cdot}18$$

The radius of the circle is $3{\cdot}18$ m (3 s.f.).

Example 2

A circle has an area of 45 cm². Find the radius of the circle.

Let the radius of the circle by r cm.

$$\pi r^2 = 45$$
$$r^2 = \frac{45}{\pi}$$
$$r = \sqrt{\left(\frac{45}{\pi}\right)} = 3{\cdot}78 \text{ (3 s.f.)}$$

The radius of the circle is $3{\cdot}78$ cm.

Exercise 4

Use the π button on a calculator and give answers to 3 s.f.

1. A circle has an area of 15 cm². Find its radius.

2. A circle has a circumference of 190 m. Find its radius.

3. Find the radius of a circle of area 22 km².

4. Find the radius of a circle of circumference 58·6 cm.

5. A circle has an area of 16 mm². Find its circumference.

6. A circle has a circumference of 2500 km. Find its area.

7. A circle of radius 5 cm is inscribed inside a square as shown. Find the area shaded.

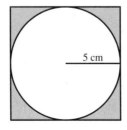

5 cm

8. A circular pond of radius 6 m is surrounded
   by a path of width 1 m.
   (a) Find the area of the path.
   (b) The path is resurfaced with astroturf
       which is bought in packs each
       containing enough to cover an area of
       7 m². How many packs are required?

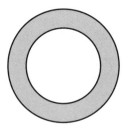

9. Discs of radius 4 cm are cut from a rectangular plastic sheet of
   length 84 cm and width 24 cm.
   (a) How many complete discs can be cut out?
   Find:
   (b) the total area of the discs cut
   (c) the area of the sheet wasted.

10. The tyre of a car wheel has an outer diameter of 30 cm. How many
    times will the wheel rotate on a journey of 5 km?

11. A golf ball of diameter 1·68 inches rolls a distance of 4 m in a
    straight line. How many times does the ball rotate completely?
    (1 inch = 2·54 cm)

12. 100 yards of cotton is wound without stretching onto a reel of
    diameter 3 cm. How many times does the reel rotate?
    (1 yard = 0·914 m. Ignore the thickness of the cotton.)

13. A rectangular metal plate has a length of 65 cm and a width of
    35 cm. It is melted down and recast into circular discs of the same
    thickness. How many complete discs can be formed if
    (a) the radius of each disc is 3 cm?
    (b) the radius of each disc is 10 cm?

14. Calculate the radius of a circle whose area is equal to the sum of the
    areas of three circles of radii 2 cm, 3 cm and 4 cm respectively.

15. The diameter of a circle is given as 10 cm, correct to the nearest cm.
    Calculate:
    (a) the maximum possible circumference
    (b) the minimum possible area of the circle consistent with this
        data.

16. A square is inscribed in a circle of radius
    7 cm. Find:
    (a) the area of the square
    (b) the area shaded.

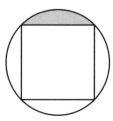

17. An archery target has three concentric regions. The diameters of the regions are in the ratio 1 : 2 : 3. Find the ratio of their areas.

18. The governor of a prison has 100 m of wire fencing. What area can he enclose if he makes a circular compound?

19. The semi-circle and the isosceles triangle have the same base AB and the same area. Find the angle x.

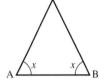

20. Lakmini decided to measure the circumference of the Earth using a very long tape measure. She held the tape measure 1 m from the surface of the (perfectly spherical) Earth all the way round. When she had finished her friend said that her measurement gave too large an answer and suggested taking off 6 m. Was her friend correct? [Take the radius of the Earth to be 6400 km (if you need it).]

21. The large circle has a radius of 10 cm. Find the radius of the largest circle which will fit in the middle.

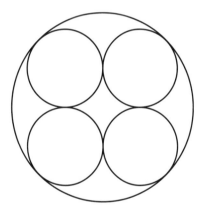

3.3 Arc length and sector area

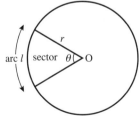

Arc length, $l = \dfrac{\theta}{360} \times 2\pi r$

We take a fraction of the whole circumference depending on the angle at the centre of the circle.

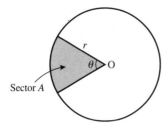

Sector area, $A = \dfrac{\theta}{360} \times \pi r^2$

We take a fraction of the whole area depending on the angle at the centre of the circle.

Example 1

Find the length of an arc which subtends an angle of $140°$ at the centre of a circle of radius $12\,cm$.

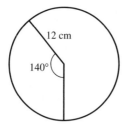

Arc length $= \dfrac{140}{360} \times 2 \times \pi \times 12$

$= \dfrac{28}{3}\pi$

$= 29\tfrac{1}{3}\,cm$

Example 2

A sector of a circle of radius $10\,cm$ has an area of $25\,cm^2$. Find the angle at the centre of the circle.

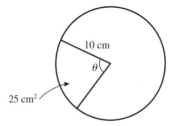

Let the angle at the centre of the circle be θ.

$\dfrac{\theta}{360} \times \pi \times 10^2 = 25$

$$\therefore \qquad \theta = \frac{25 \times 360}{\pi \times 100}$$

$$\theta = 28{\cdot}6° \text{ (3 s.f.)}$$

The angle at the centre of the circle is $28{\cdot}6°$.

Exercise 5

[Use the π button on a calculator unless told otherwise.]

1. Arc AB subtends an angle θ at the centre of circle radius r.

Find the arc length and sector area when:
(a) $r = 4\,\text{cm}$, $\theta = 30°$
(b) $r = 10\,\text{cm}$, $\theta = 45°$
(c) $r = 2\,\text{cm}$, $\theta = 235°$.

In questions **2** and **3** find the total area of the shape.

2.

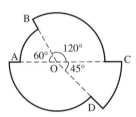

OA = 2 cm, OB = 3 cm, OC = 5 cm, OD = 3 cm.

3.

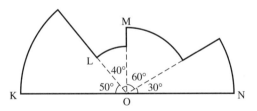

ON = 6 cm, OM = 3 cm, OL = 2 cm, OK = 6 cm.

4. Find the shaded areas.

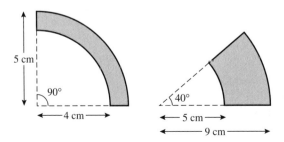

5. In the diagram the arc length is l and the sector area is A.

(a) Find θ, when $r = 5$ cm and $l = 7 \cdot 5$ cm.
(b) Find θ, when $r = 2$ m and $A = 2$ m^2.
(c) Find r, when $\theta = 55°$ and $l = 6$ cm.

6. The length of the minor arc AB of a circle, centre O, is 2π cm and the length of the major arc is 22π cm. Find:
(a) the radius of the circle
(b) the acute angle AOB.

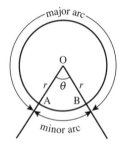

7. The lengths of the minor and major arcs of a circle are $5 \cdot 2$ cm and $19 \cdot 8$ respectively. Find:
(a) the radius of the circle
(b) the angle subtended at the centre by the minor arc.

8. A wheel of radius 10 cm is turning at a rate of 5 revolutions per minute. Calculate:
(a) the angle through which the wheel turns in 1 second
(b) the distance moved by a point on the rim in 2 seconds.

9. The length of an arc of a circle is 12 cm. The corresponding sector area is 108 cm^2. Find:
(a) the radius of the circle
(b) the angle subtended at the centre of the circle by the arc.

10. The length of an arc of a circle is $7 \cdot 5$ cm. The corresponding sector area is $37 \cdot 5$ cm^2. Find:
(a) the radius of the circle
(b) the angle subtended at the centre of the circle by the arc.

11. In the diagram the arc length is l and the sector area is A.

(a) Find l, when $\theta = 72°$ and $A = 15$ cm^2.
(b) Find l, when $\theta = 135°$ and $A = 162$ m^2.
(c) Find A, when $l = 11$ cm and $r = 5 \cdot 2$ cm.

12. A long time ago Dulani found an island shaped like a triangle with three straight shores of length 3 km, 4 km and 5 km. He said nobody could come within 1 km of his shore. What was the area of his exclusion zone?

3.4 Chord of a circle

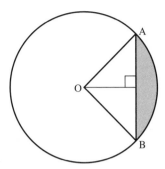

The line AB is a chord. The area of a circle cut off by a chord is called a *segment*. In the diagram the *minor* segment is shaded and the *major* segment is unshaded.

(a) The line from the centre of a circle to the mid-point of a chord *bisects* the chord at *right angles*.
(b) The line from the centre of a circle to the mid-point of a chord bisects the angle subtended by the chord at the centre of the circle.

Example

XY is a chord of length 12 cm of a circle of radius 10 cm, centre O. Calculate:
(a) the angle XOY
(b) the area of the minor segment cut off by the chord XY.

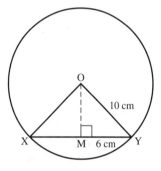

Let the mid-point of XY be M.

$$\therefore \qquad MY = 6\,\text{cm}$$
$$\sin M\hat{O}Y = \frac{6}{10}$$
$$\therefore \qquad M\hat{O}Y = 36\cdot87°$$
$$\therefore \qquad X\hat{O}Y = 2 \times 36\cdot87$$
$$= 73\cdot74°$$

area of minor segment = area of sector XOY − area of △XOY

$$\text{area of sector XOY} = \frac{73\cdot74}{360} \times \pi \times 10^2$$
$$= 64\cdot32\,\text{cm}^2.$$
$$\text{area of △XOY} = \tfrac{1}{2} \times 10 \times 10 \times \sin 73\cdot74°$$
$$= 48\cdot00\,\text{cm}^2$$

$$\therefore \quad \text{Area of minor segment} = 64\cdot32 - 48\cdot00$$
$$= 16\cdot3\,\text{cm}^2 \ (3\ \text{s.f.})$$

> You can find out more about trigonometry in Unit 6 on page 192.

Exercise 6

Use the π button on a calculator. You will need basic trigonometry (page 192).

1. The chord AB subtends an angle of 130° at the centre O. The radius of the circle is 8 cm. Find:
   (a) the length of AB
   (b) the area of sector OAB
   (c) the area of triangle OAB
   (d) the area of the minor segment (shown shaded).

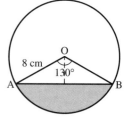

2. Find the shaded area when:
   (a) $r = 6$ cm, $\theta = 70°$
   (b) $r = 14$ cm, $\theta = 104°$
   (c) $r = 5$ cm, $\theta = 80°$

3. Find θ and hence the shaded area when:
   (a) AB $= 10$ cm, $r = 10$ cm
   (b) AB $= 8$ cm, $r = 5$ cm

4. How far is a chord of length 8 cm from the centre of a circle of radius 5 cm?

5. How far is a chord of length 9 cm from the centre of a circle of radius 6 cm?

6. The diagram shows the cross-section of a cylindrical pipe with water lying in the bottom.
   (a) If the maximum depth of the water is 2 cm and the radius of the pipe is 7 cm, find the area shaded.
   (b) What is the *volume* of water in a length of 30 cm?

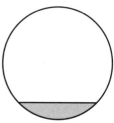

7. An equilateral triangle is inscribed in a circle of radius 10 cm. Find:
   (a) the area of the triangle
   (b) the area shaded.

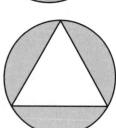

8. An equilateral triangle is inscribed in a circle of radius 18·8 cm. Find:
   (a) the area of the triangle
   (b) the area of the three segments surrounding the triangle.

9. A regular hexagon is circumscribed by a circle of radius 6 cm. Find the area shaded.

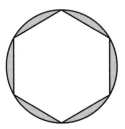

10. A regular octagon is circumscribed by a circle of radius r cm. Find the area enclosed between the circle and the octagon. (Give the answer in terms of r.)

11. Find the radius of the circle:
(a) when $\theta = 90°$, $A = 20$ cm^2
(b) when $\theta = 30°$, $A = 35$ cm^2
(c) when $\theta = 150°$, $A = 114$ cm^2

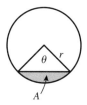

12. (Harder) The diagram shows a regular pentagon of side 10 cm with a star inside. Calculate the area of the star.

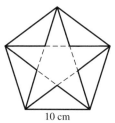

10 cm

3.5 Volume

Prism

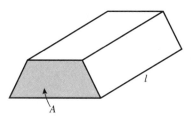

A prism is an object with the same cross-section throughout its length.

Volume of prism = (area of cross-section) × length
$$= A \times l$$

A *cuboid* is a prism whose six faces are all rectangles. A cube is a special case of a cuboid in which all six faces are squares.

Cylinder

radius = r
height = h

A cylinder is a prism whose cross-section is a circle.

Volume of cylinder = (area of cross-section) × length
Volume = $\pi r^2 h$

Example

Calculate the height of a cylinder of volume 500 cm^3 and base radius 8 cm. Let the height of the cylinder be h cm.

$$\pi r^2 h = 500$$
$$3{\cdot}142 \times 8^2 \times h = 500$$
$$h = \frac{500}{3{\cdot}142 \times 64}$$
$$h = 2{\cdot}49 \text{ (3 s.f.)}$$

The height of the cylinder is 2·49 cm.

Exercise 7

1. Calculate the volume of the prisms. All lengths are in cm.

(a)

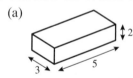

(b)

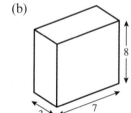

(c)

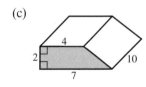

(d)

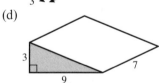

(e)

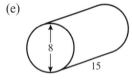

(f)

2. Calculate the volume of the following cylinders:
  (a) $r = 4$ cm, $h = 10$ cm
  (b) $r = 11$ m, $h = 2$ m
  (c) $r = 2{\cdot}1$ cm, $h = 0{\cdot}9$ cm

3. Find the height of a cylinder of volume $200\,\text{cm}^3$ and radius $4\,\text{cm}$.

4. Find the length of a cylinder of volume 2 litres and radius $10\,\text{cm}$.

5. Find the radius of a cylinder of volume $45\,\text{cm}^3$ and length $4\,\text{cm}$.

6. A prism has volume $100\,\text{cm}^3$ and length $8\,\text{cm}$. If the cross-section is an equilateral triangle, find the length of a side of the triangle.

7. When 3 litres of oil is removed from an upright cylindrical can the level falls by $10\,\text{cm}$. Find the radius of the can.

8. A solid cylinder of radius $4\,\text{cm}$ and length $8\,\text{cm}$ is melted down and recast into a solid cube. Find the side of the cube.

9. A solid rectangular block of copper $5\,\text{cm}$ by $4\,\text{cm}$ by $2\,\text{cm}$ is drawn out to make a cylindrical wire of diameter $2\,\text{mm}$. Calculate the length of the wire.

10. Water flows through a circular pipe of internal diameter $3\,\text{cm}$ at a speed of $10\,\text{cm/s}$. If the pipe is full, how much water flows from the pipe in one minute? (Answer in litres.)

11. Water flows from a hose-pipe of internal diameter $1\,\text{cm}$ at a rate of 5 litres per minute. At what speed is the water flowing through the pipe?

12. A cylindrical metal pipe has external diameter of $6\,\text{cm}$ and internal diameter of $4\,\text{cm}$. Calculate the volume of metal in a pipe of length $1\,\text{m}$. If $1\,\text{cm}^3$ of the metal has a mass of $8\,\text{g}$, find the mass of the pipe.

13. For two cylinders A and B, the ratio of lengths is $3:1$ and the ratio of diameters is $1:2$. Calculate the ratio of their volumes.

14. A machine makes boxes which are either perfect cylinders of diameter and length $4\,\text{cm}$, or perfect cubes of side $5\,\text{cm}$. Which boxes have the greater volume, and by how much? (Take $\pi = 3$)

15. Natalia decided to build a garage and began by calculating the number of bricks required. The garage was to be $6\,\text{m}$ by $4\,\text{m}$ and $2{\cdot}5\,\text{m}$ in height. Each brick measures $22\,\text{cm}$ by $10\,\text{cm}$ by $7\,\text{cm}$. Natalia estimated that she would need about $40\,000$ bricks. Is this a reasonable estimate?

16. A cylindrical can of internal radius $20\,\text{cm}$ stands upright on a flat surface. It contains water to a depth of $20\,\text{cm}$. Calculate the rise in the level of the water when a brick of volume $1500\,\text{cm}^3$ is immersed in the water.

17. A cylindrical tin of height $15\,\text{cm}$ and radius $4\,\text{cm}$ is filled with sand from a rectangular box. How many times can the tin be filled if the dimensions of the box are $50\,\text{cm}$ by $40\,\text{cm}$ by $20\,\text{cm}$?

18. Rain which falls onto a flat rectangular surface of length 6 m and width 4 m is collected in a cylinder of internal radius 20 cm. What is the depth of water in the cylinder after a storm in which 1 cm of rain fell?

Pyramid

Volume $= \frac{1}{3}$(base area) × height.

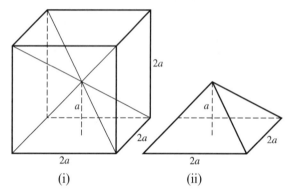

(i)   (ii)

Figure (i) shows a cube of side $2a$ broken down into six pyramids of height a as shown in figure (ii).

If the volume of each pyramid is V,

then $6V = 2a \times 2a \times 2a$

$V = \frac{1}{6} \times (2a)^2 \times 2a$

so $V = \frac{1}{3} \times (2a)^2 \times a$

$V = \frac{1}{3}$(base area) × height

Cone

Volume $= \frac{1}{3}\pi r^2 h$
(note the similarity with the pyramid)

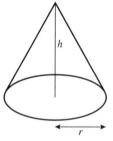

Sphere

Volume $= \frac{4}{3}\pi r^3$

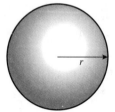

Example 1

A pyramid has a square base of side 5 m and vertical height 4 m. Find its volume.

$$\text{Volume of pyramid} = \tfrac{1}{3}(5 \times 5) \times 4$$
$$= 33\tfrac{1}{3} \text{ m}^3$$

Example 2

Calculate the radius of a sphere of volume 500 cm^3.

Let the radius of the sphere be r cm

$$\tfrac{4}{3}\pi r^3 = 500$$
$$r^3 = \frac{3 \times 500}{4\pi}$$
$$r = \sqrt[3]{\left(\frac{3 \times 500}{4\pi}\right)} = 4 \cdot 92 \text{ (3 s.f.)}$$

The radius of the sphere is 4·92 cm.

Exercise 8

Find the volumes of the following objects:

1. cone: height = 5 cm, radius = 2 cm

2. sphere: radius = 5 cm

3. sphere: radius = 10 cm

4. cone: height = 6 cm, radius = 4 cm

5. sphere: diameter = 8 cm

6. cone: height = x cm, radius = $2x$ cm

7. sphere: radius = 0·1 m

8. cone: height = $\dfrac{1}{\pi}$ cm, radius = 3 cm

9. pyramid: rectangular base 7 cm by 8 cm; height = 5 cm

10. pyramid: square base of side 4 m, height = 9 m

11. pyramid: equilateral triangular base of side = 8 cm, height = 10 cm

12. Find the volume of a hemisphere of radius 5 cm.

13. A cone is attached to a hemisphere of radius 4 cm. If the total height of the object is 10 cm, find its volume.

14. A toy consists of a cylinder of diameter 6 cm 'sandwiched' between a hemisphere and a cone of the same diameter. If the cone is of height 8 cm and the cylinder is of height 10 cm, find the total volume of the toy.

15. Find the height of a pyramid of volume 20 m³ and base area 12 m².

16. Find the radius of a sphere of volume 60 cm³.

17. Find the height of a cone of volume 2·5 litre and radius 10 cm.

18. Six square-based pyramids fit exactly onto the six faces of a cube of side 4 cm. If the volume of the object formed is 256 cm³, find the height of each of the pyramids.

19. A solid metal cube of side 6 cm is recast into a solid sphere. Find the radius of the sphere.

20. A hollow spherical vessel has internal and external radii of 6 cm and 6·4 cm respectively. Calculate the mass of the vessel if it is made of metal of density 10 g/cm³.

21. Water is flowing into an inverted cone, of diameter and height 30 cm, at a rate of 4 litres per minute. How long, in seconds, will it take to fill the cone?

22. A solid metal sphere is recast into many smaller spheres. Calculate the number of the smaller spheres if the initial and final radii are as follows:
(a) initial radius = 10 cm, final radius = 2 cm
(b) initial radius = 7 cm, final radius = $\frac{1}{2}$ cm
(c) initial radius = 1 m, final radius = $\frac{1}{3}$ cm.

23. A spherical ball is immersed in water contained in a vertical cylinder.
Assuming the water covers the ball, calculate the rise in the water level if:
(a) sphere radius = 3 cm, cylinder radius = 10 cm
(b) sphere radius = 2 cm, cylinder radius = 5 cm.

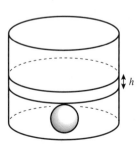

24. A spherical ball is immersed in water contained in a vertical cylinder. The rise in water level is measured in order to calculate the radius of the spherical ball. Calculate the radius of the ball in the following cases:
    (a) cylinder of radius 10 cm, water level rises 4 cm
    (b) cylinder of radius 100 cm, water level rises 8 cm.

25. One corner of a solid cube of side 8 cm is removed by cutting through the mid-points of three adjacent sides. Calculate the volume of the piece removed.

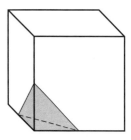

26. The cylindrical end of a pencil is sharpened to produce a perfect cone at the end with no overall loss of length. If the diameter of the pencil is 1 cm, and the cone is of length 2 cm, calculate the volume of the shavings.

27. Metal spheres of radius 2 cm are packed into a rectangular box of internal dimensions 16 cm × 8 cm × 8 cm. When 16 spheres are packed the box is filled with a preservative liquid. Find the volume of this liquid.

28. The diagram shows the cross-section of an inverted cone of height MC = 12 cm. If AB = 6 cm and XY = 2 cm, use similar triangles to find the length NC.
    (You can find out about similar triangles on page 135.)

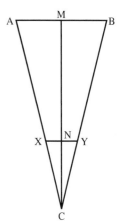

29. An inverted cone of height 10 cm and base radius 6·4 cm contains water to a depth of 5 cm, measured from the vertex. Calculate the volume of water in the cone.

30. An inverted cone of height 15 cm and base radius 4 cm contains water to a depth of 10 cm. Calculate the volume of water in the cone.

31. An inverted cone of height 12 cm and base radius 6 cm contains 20 cm^3 of water. Calculate the depth of water in the cone, measured from the vertex.

32. A frustum is a cone with 'the end chopped off'. A bucket in the shape of a frustum as shown has diameters of 10 cm and 4 cm at its ends and a depth of 3 cm. Calculate the volume of the bucket.

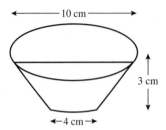

33. Find the volume of a frustum with end diameters of 60 cm and 20 cm and a depth of 40 cm.

34. The diagram shows a sector of a circle of radius 10 cm.
(a) Find, as a multiple of π, the arc length of the sector.
The straight edges are brought together to make a cone. Calculate:
(b) the radius of the base of the cone,
(c) the vertical height of the cone.

35. Calculate the volume of a regular octahedron whose edges are all 10 cm.

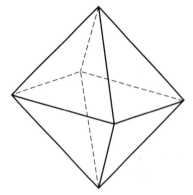

36. A sphere passes through the eight corners of a cube of side 10 cm. Find the volume of the sphere.

37. (Harder) Find the volume of a regular tetrahedron of side 20 cm.

38. Find the volume of a regular tetrahedron of side 35 cm.

3.6 Surface area

We are concerned here with the surface areas of the *curved* parts of
cylinders, spheres and cones. The areas of the plane faces are easier to find.

(a) Cylinder
   Curved surface area $= 2\pi rh$

(b) Sphere
   Surface area $= 4\pi r^2$

(c) Cone
   Curved surface area $= \pi rl$
   where l is the slant height.

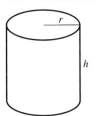

Example

Find the *total* surface area of a solid cone of radius 4 cm and vertical
height 3 cm.

Let the slant height of the cone be l cm.

$l^2 = 3^2 + 4^2$ (Pythagoras' theorem)
$l = 5$

Curved surface area $= \pi \times 4 \times 5$
$= 20\pi \text{ cm}^2$
Area of end face $= \pi \times 4^2 = 16\pi \text{ cm}^2$
$\therefore$ Total surface area $= 20\pi + 16\pi$
$= 36\pi \text{ cm}^2$
$= 113 \text{ cm}^2$ to 3 s.f.

Exercise 9

Use the π button on a calculator unless otherwise instructed.

1. Copy the table and find the quantities marked *.
   (Leave π in your answers.)

solid object	radius	vertical height	curved surface area	total surface area
(a) sphere	3 cm		*	
(b) cylinder	4 cm	5 cm		*
(c) cone	6 cm	8 cm	*	
(d) cylinder	0·7 m	1 m		*
(e) sphere	10 m		*	
(f) cone	5 cm	12 cm	*	
(g) cylinder	6 mm	10 mm		*
(h) cone	2·1 cm	4·4 cm	*	
(i) sphere	0·01 m		*	
(j) hemisphere	7 cm		*	*

2. Find the radius of a sphere of surface area $34\,\text{cm}^2$.

3. Find the slant height of a cone of curved surface area $20\,\text{cm}^2$ and radius $3\,\text{cm}$.

4. Find the height of a solid cylinder of radius $1\,\text{cm}$ and *total* surface area $28\,\text{cm}^2$.

5. Copy the table and find the quantities marked *. (Take $\pi = 3$)

object	radius	vertical height	curved surface area	total surface area
(a) cylinder	4 cm	*	$72\,\text{cm}^2$	
(b) sphere	*		$192\,\text{cm}^2$	
(c) cone	4 cm	*	$60\,\text{cm}^2$	
(d) sphere	*		$0{\cdot}48\,\text{m}^2$	
(e) cylinder	5 cm	*		$330\,\text{cm}^2$
(f) cone	6 cm	*		$225\,\text{cm}^2$
(g) cylinder	2 m	*		$108\,\text{m}^2$

6. A solid wooden cylinder of height $8\,\text{cm}$ and radius $3\,\text{cm}$ is cut in two along a vertical axis of symmetry. Calculate the total surface area of the two pieces.

7. A tin of paint covers a surface area of $60\,\text{m}^2$ and costs \$4·50. Find the cost of painting the outside surface of a hemispherical dome of radius $50\,\text{m}$. (Just the curved part.)

8. A solid cylinder of height $10\,\text{cm}$ and radius $4\,\text{cm}$ is to be plated with material costing \$11 per cm^2. Find the cost of the plating.

9. Find the volume of a sphere of surface area $100\,\text{cm}^2$.

10. Find the surface area of a sphere of volume $28\,\text{cm}^3$.

11. Calculate the total surface area of the combined cone/cylinder/hemisphere.

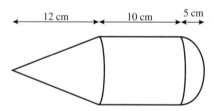

12. A man wants to spray the entire surface of the Earth (including the oceans) with a new weed killer. If it takes him 10 seconds to spray $1\,\text{m}^2$, how long will it take to spray the whole world? (Radius of the Earth $= 6370\,\text{km}$; ignore leap years)

13. An inverted cone of vertical height $12\,\text{cm}$ and base radius $9\,\text{cm}$ contains water to a depth of $4\,\text{cm}$. Find the area of the interior surface of the cone not in contact with the water.

14. A circular piece of paper of radius 20 cm is cut in half and each half is made into a hollow cone by joining the straight edges. Find the slant height and base radius of each cone.

15. A golf ball has a diameter of 4·1 cm and the surface has 150 dimples of radius 2 mm. Calculate the total surface area which is exposed to the surroundings. (Assume the 'dimples' are hemispherical.)

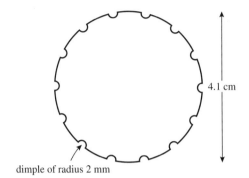

dimple of radius 2 mm

16. A cone of radius 3 cm and slant height 6 cm is cut into four identical pieces. Calculate the total surface area of the four pieces.

Revision exercise 3A

1. Find the area of the following shapes:

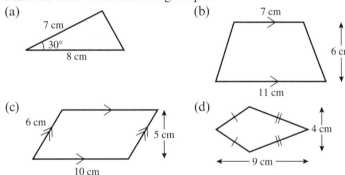

(a) 7 cm 30° 8 cm

(b) 7 cm 6 cm 11 cm

(c) 6 cm 5 cm 10 cm

(d) 4 cm 9 cm

2. (a) A circle has radius 9 m. Find its circumference and area.
(b) A circle has circumference 34 cm. Find its diameter.
(c) A circle has area 50 cm². Find its radius.

3. A target consists of concentric circles of radii 3 cm and 9 cm.
(a) Find the area of A, in terms of π.
(b) Find the ratio $\dfrac{\text{area of B}}{\text{area of A}}$

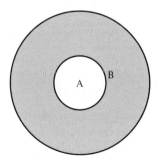

4. In Figure 1 a circle of radius 4 cm is inscribed in a square. In Figure 2 a square is inscribed in a circle of radius 4 cm.
Calculate the shaded area in each diagram.

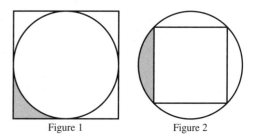

Figure 1   Figure 2

5. Given that OA = 10 cm and AÔB = 70° (where O is the centre of the circle), calculate:
(a) the arc length AB
(b) the area of minor sector AOB.

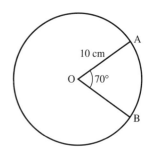

6. The points X and Y lie on the circumference of a circle, of centre O and radius 8 cm, where XÔY = 80°. Calculate:
(a) the length of the minor arc XY
(b) the length of the chord XY
(c) the area of sector XOY
(d) the area of triangle XOY
(e) the area of the minor segment of the circle cut off by XY.

7. Given that ON = 10 cm and minor arc MN = 18 cm, calculate the angle MÔN (shown as $x°$).

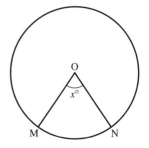

8. A cylinder of radius 8 cm has a volume of 2 litres. Calculate the height of the cylinder.

9. Calculate:
(a) the volume of a sphere of radius 6 cm
(b) the radius of a sphere whose volume is 800 cm^3.

10. A sphere of radius 5 cm is melted down and made into a solid cube. Find the length of a side of the cube.

11. The curved surface area of a solid circular cylinder of height 8 cm is 100 cm^2. Calculate the volume of the cylinder.

12. A cone has base radius 5 cm and vertical height 10 cm, correct to the nearest cm. Calculate the maximum and minimum possible volumes of the cone, consistent with this data.

13. Calculate the radius of a hemispherical solid whose total surface area is $48\pi\,\text{cm}^2$.

14. Calculate:
   (a) the area of an equilateral triangle of side 6 cm.
   (b) the area of a regular hexagon of side 6 cm.
   (c) the volume of a regular hexagonal prism of length 10 cm, where the side of the hexagon is 12 cm.

15. Ten spheres of radius 1 cm are immersed in liquid contained in a vertical cylinder of radius 6 cm. Calculate the rise in the level of the liquid in the cylinder.

16. A cube of side 10 cm is melted down and made into ten identical spheres. Calculate the surface area of one of the spheres.

17. The square has sides of length 3 cm and the arcs have centres at the corners. Find the shaded area.

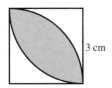

3 cm

18. A copper pipe has external diameter 18 mm and thickness 2 mm. The density of copper is $9\,\text{g/cm}^3$ and the price of copper is \$150 per tonne. What is the cost of the copper in a length of 5 m of this pipe?

19. Twenty-seven small wooden cubes fit exactly inside a cubical box without a lid. How many of the cubes are touching the sides or the bottom of the box?

20. In the diagram the area of the smaller square is $10\,\text{cm}^2$.
Find the area of the larger square.

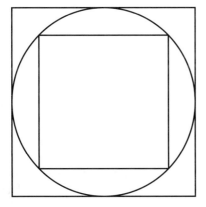

Examination exercise 3B

1. A spacecraft made 58 376 orbits of the Earth and travelled a
   distance of $2 \cdot 656 \times 10^9$ kilometres.
   (a) Calculate the distance travelled in 1 orbit correct to the nearest
       kilometre. [2]
   (b) The orbit of the spacecraft is a circle.
       Calculate the radius of the orbit. [2]

Cambridge IGCSE Mathematics 0580
Paper 21 Q14 November 2008

2. A large conference table is made from four rectangular sections and
   four corner sections.
   Each rectangular section is 4 m long and 1.2 m wide.
   Each corner section is a quarter circle, radius 1.2 m.

Not to scale

Each person sitting at the conference table requires one metre of its
outside perimeter.
Calculate the greatest number of people who can sit around the
outside of the table.
Show all your working. [3]

Cambridge IGCSE Mathematics 0580
Paper 2 Q11 November 2005

3. In triangle ABC, $AB = 6$ cm, $AC = 8$ cm, and $BC = 12$ cm. Angle
   $ACB = 26 \cdot 4°$.
   Calculate the area of the triangle ABC. [2]

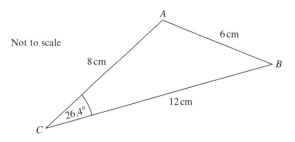

Cambridge IGCSE Mathematics 0580
Paper 2 Q5 June 2006

4.

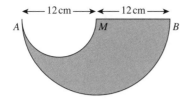

The shape above is made by removing a small semi-circle from a
large semi-circle.
$AM = MB = 12$ cm   [3]
Calculate the area of the shape.   **Cambridge IGCSE Mathematics 0580**
Paper 2 Q10 November 2007

5.

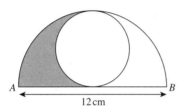

Not to scale

The largest possible circle is drawn inside a semi-circle, as shown in
the diagram.
The distance AB is 12 centimetres.
(a) Find the shaded area.   [4]
(b) Find the perimeter of the shaded area.   [2]
Cambridge IGCSE Mathematics 0580
Paper 2 Q23 June 2007

6.

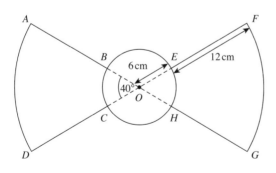

Not to scale

The diagram shows part of a fan.
OFG and OAD are sectors, centre O, with radius 18 cm and sector
angle $40°$.
B, C, H and E lie on a circle, centre O and radius 6 cm.   [4]
Calculate the shaded area.   **Cambridge IGCSE Mathematics 0580**
Paper 22 Q19 June 2009

7.

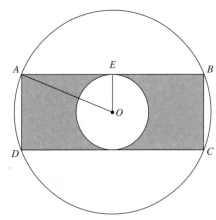

Not to scale

A, B, C and D lie on a circle, centre O, radius 8 cm.
AB and CD are tangents to a circle, centre O, radius 4 cm.
$ABCD$ is a rectangle.
(a) Calculate the distance AE [2]
(b) Calculate the shaded area.   [3]

Cambridge IGCSE Mathematics 0580
Paper 2 Q21 November 2005

8.

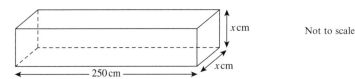

Not to scale

A solid metal bar is in the shape of a cuboid of length of 250 cm.
The cross-section is a square of side x cm.
The volume of the cuboid is 4840 cm^3.
(a) Show that $x = 4.4$.   [2]
(b) The mass of 1 cm^3 of the metal is 8.8 grams.
    Calculate the mass of the whole metal bar in kilograms.   [2]
(c) A box, in the shape of a cuboid measures 250 cm by 88 cm by
    h cm.
    120 of the metal bars fit exactly in the box.
    Calculate the value of h.   [2]
(d) One metal bar, of volume 4840 cm^3, is melted down to make
    4200 identical small spheres.
    All the metal is used.
    (i) Calculate the radius of each sphere. Show that your answer
        rounds to 0.65 cm, correct to 2 decimal places.   [4]
        [The volume, V, of a sphere, radius r, is given by $V = \frac{4}{3}\pi r^3$.]
    (ii) Calculate the surface area of each sphere, using 0.65 cm for
         the radius.   [1]
         [The surface area, A, of a sphere, radius r, is given by $A = 4\pi r^2$.]
    (iii) Calculate the total surface area of all 4200 spheres as a
          percentage of the surface area of the metal bar.   [4]

Cambridge IGCSE Mathematics 0580
Paper 4 Q7 June 2009

9.

Not to scale

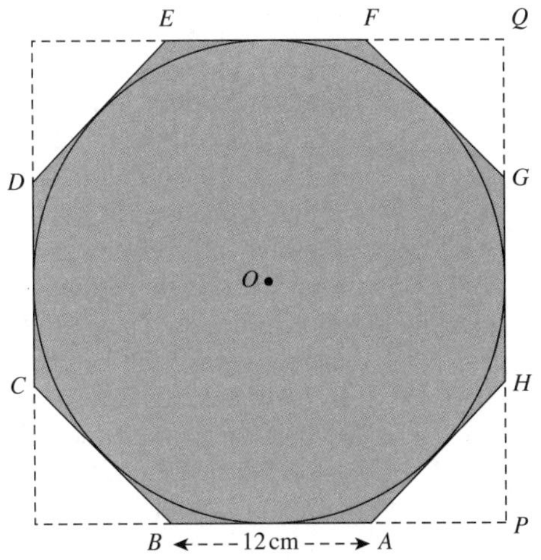

A circle, centre O, touches all the sides of the regular octagon
ABCDEFGH shaded in the diagram.
The sides of the octagon are of length 12 cm.
BA and *GH* are extended to meet at *P*. *HG* and *EF* are extended to
meet at *Q*.

(a) (i) Show that angle *BAH* is 135°.   [2]
   (ii) Show that angle *APH* is 90°.   [1]

(b) Calculate
   (i) the length of *PH*,   [2]
   (ii) the length of *PQ*,   [2]
   (iii) the area of triangle *APH*,   [2]
   (iv) the area of the octagon.   [3]

(c) Calculate
   (i) the radius of the circle,   [2]
   (ii) the area of the circle as a percentage of the area of the
       octagon.   [3]

Cambridge IGCSE Mathematics 0580
Paper 4 Q5 June 2008

4 GEOMETRY

Pythagoras (569–500 B.C.) was one of the first of the great mathematical names in Greek antiquity. He settled in southern Italy and formed a mysterious brotherhood with his students who were bound by an oath not to reveal the secrets of numbers and who exercised great influence. They laid the foundations of arithmetic through geometry but failed to resolve the concept of irrational numbers. The work of these and others was brought together by Euclid at Alexandria in a book called 'The Elements' which was still studied in English schools as recently as 1900.

26 Use and interpret geometrical terms, including similarity and congruence; use the relationships between areas, volumes and surface areas of similar figures; use and interpret vocabulary of shapes and simple solid figures including nets

27 Measure lines and angles; construct a triangle given three sides; construct angle bisectors and perpendicular bisectors

28 Recognise rotational and line symmetry

29 Calculate unknown angles using geometrical properties, including irregular polygons and circle theorems

30 Use loci in two dimensions

32 Apply Pythagoras' theorem

4.1 Fundamental results

You should already be familiar with the following results. They are used later in this section and are quoted here for reference.

- The angles on a straight line add up to 180°:

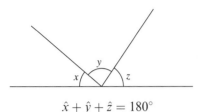

$$\hat{x} + \hat{y} + \hat{z} = 180°$$

- The angles at a point add up to 360°:

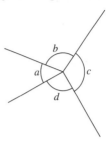

$$\hat{a} + \hat{b} + \hat{c} + \hat{d} = 360°$$

- Vertically opposite angles are equal:

- The angle sum of a triangle is 180°.
- An isosceles triangle has 2 sides and 2 angles the same:

- The angle sum of a quadrilateral is 360°.
- An equilateral triangle has 3 sides and 3 angles the same:

Exercise 1

Find the angles marked with letters. (AB is always a straight line.)

1.

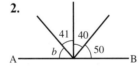

2.

3.

4.

5.

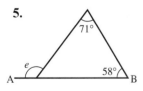

6.

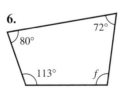

7.

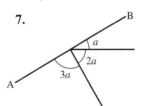

8.

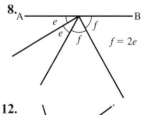

$f = 2e$

9.

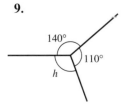

10.

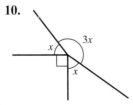

11.

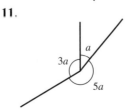

12.

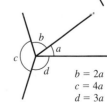

$b = 2a$
$c = 4a$
$d = 3a$

13.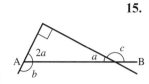

14.

15.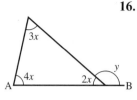

16.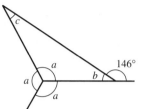

17.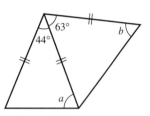

18.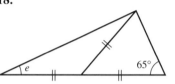

19.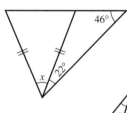

20.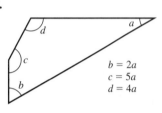

21.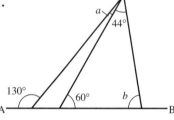

22.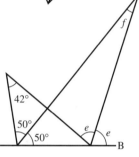

23. Calculate the largest angle of a triangle in which one angle is eight times each of the others.

24. In $\triangle ABC$, $\widehat{A}$ is a right angle and D is a point on AC such that BD bisects $\widehat{B}$. If $\widehat{BDC} = 100°$, calculate $\widehat{C}$.

25. WXYZ is a quadrilateral in which $\widehat{W} = 108°$, $\widehat{X} = 88°$, $\widehat{Y} = 57°$ and $\widehat{WXZ} = 31°$. Calculate $\widehat{WZX}$ and $\widehat{XZY}$.

26. In quadrilateral ABCD, AB produced is perpendicular to DC produced. If $\widehat{A} = 44°$ and $\widehat{C} = 148°$, calculate $\widehat{D}$ and $\widehat{B}$.

27. Triangles ABD, CBD and ADC are all isosceles. Find the angle x.

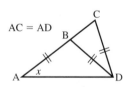

AC = AD

Polygons

(i) The exterior angles of a polygon add up to 360°
$(\hat{a} + \hat{b} + \hat{c} + \hat{d} + \hat{e} = 360°)$.

(ii) The sum of the interior angles of a polygon is $(n - 2) \times 180°$ where n is the number of sides of the polygon.
This result is investigated in question **3** in the next exercise.

(iii) A regular polygon has equal sides and equal angles.

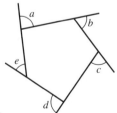

> **Remember**
> **pent**agon = 5 sides
> **hex**agon = 6 sides
> **oct**agon = 8 sides
> **deca**gon = 10 sides.

Example

Find the angles marked with letters.

The sum of the interior angles $= (n - 2) \times 180°$

where n is the number of sides of the polygon.

In this case $n = 6$.

$$\therefore \quad 110 + 120 + 94 + 114 + 2t = 4 \times 180$$
$$438 + 2t = 720$$
$$2t = 282$$
$$t = 141°$$

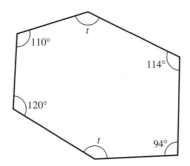

Exercise 2

1. Find angles a and b for the regular pentagon.

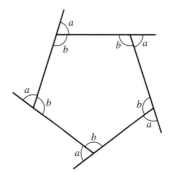

2. Find x and y.

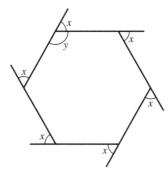

3. Consider the pentagon below which has been divided into three triangles.

$$\widehat{A} = a + f + g, \ \widehat{B} = b, \ \widehat{C} = c + d, \ \widehat{D} = e + i, \ \widehat{E} = h$$

Now $a + b + c = d + e + f = g + h + i = 180°$

$$\therefore \quad \widehat{A} + \widehat{B} + \widehat{C} + \widehat{D} + \widehat{E} = a + b + c + d + e$$
$$+ f + g + h + i$$
$$= 3 \times 180°$$
$$= 6 \times 90°$$

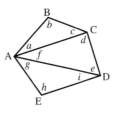

Draw further polygons and make a table of results.

Number of sides n	5	6	7	8 ...
Sum of interior angles	$3 \times 180°$			

What is the sum of the interior angles for a polygon with n sides?

4. Find *a*.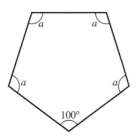

5. Find *m*.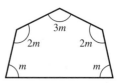

6. Find *a*.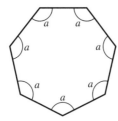

7. Calculate the number of sides of a regular polygon whose interior angles are each 156°.

8. Calculate the number of sides of a regular polygon whose interior angles are each 150°.

9. Calculate the number of sides of a regular polygon whose exterior angles are each 40°.

10. In a regular polygon each interior angle is 140° greater than each exterior angle. Calculate the number of sides of the polygon.

11. In a regular polygon each interior angle is 120° greater than each exterior angle. Calculate the number of sides of the polygon.

12. Two sides of a regular pentagon are produced to form angle *x*. What is *x*?

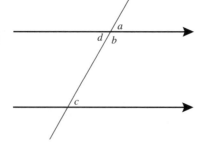

Parallel lines

(i) $\hat{a} = \hat{c}$ (corresponding angles)
(ii) $\hat{c} = \hat{d}$ (alternate angles)
(iii) $\hat{b} + \hat{c} = 180°$ (allied angles)

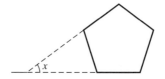

Remember: 'The acute angles (angles less than 90°) are the same and the obtuse angles (angles between 90° and 180°) are the same.'

Exercise 3

In questions **1** to **9** find the angles marked with letters.

1.

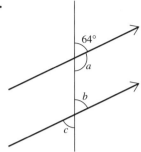

2.

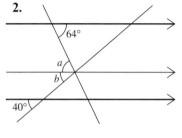

3.

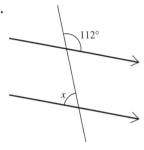

4.

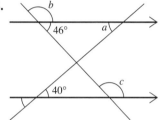

5.

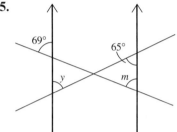

6.

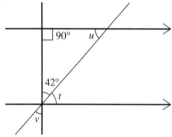

7.

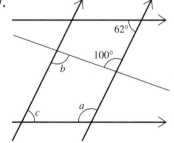

8.

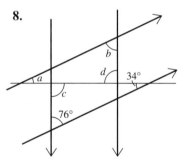

9.

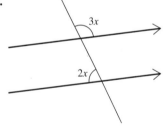

4.2 Pythagoras' theorem

In a right-angled triangle the square on
the hypotenuse is equal to the sum of the
squares on the other two sides.

$$a^2 + b^2 = c^2$$

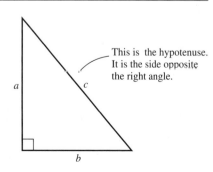

This is the hypotenuse.
It is the side opposite
the right angle.

Example

Find the side marked d.

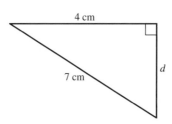

$d^2 + 4^2 = 7^2$
$\quad d^2 = 49 - 16$
$\quad\quad d = \sqrt{33} = 5 \cdot 74 \, \text{cm} \ (3 \text{ s.f.})$

The *converse* is also true:
'If the square on one side of a triangle is equal to the sum of the squares on the other two sides, then the triangle is right-angled.'

Exercise 4

In questions **1** to **10**, find x. All the lengths are in cm.

1.
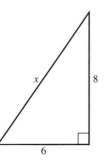

2.
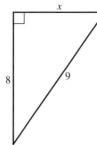

3.
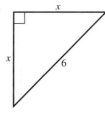

4.
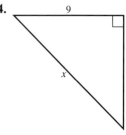

5.
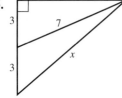

6.
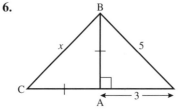

7.
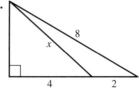

8.
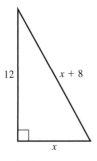

9.
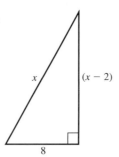

10.
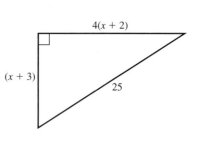

11. Find the length of a diagonal of a rectangle of length 9 cm and width 4 cm.

12. A square has diagonals of length 10 cm. Find the sides of the square.

13. A 4 m ladder rests against a vertical wall with its foot 2 m from the wall. How far up the wall does the ladder reach?

14. A ship sails 20 km due North and then 35 km due East. How far is it from its starting point?

15. Find the length of a diagonal of a rectangular box of length 12 cm, width 5 cm and height 4 cm.

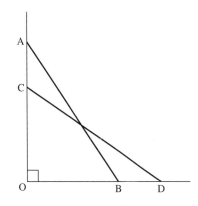

16. Find the length of a diagonal of a rectangular room of length 5 m, width 3 m and height 2·5 m.

17. Find the height of a rectangular box of length 8 cm, width 6 cm where the length of a diagonal is 11 cm.

18. An aircraft flies equal distances South-East and then South-West to finish 120 km due South of its starting-point. How long is each part of its journey?

19. The diagonal of a rectangle exceeds the length by 2 cm. If the width of the rectangle is 10 cm, find the length.

20. A cone has base radius 5 cm and *slant* height 11 cm. Find its vertical height.

21. It is possible to find the sides of a right-angled triangle, with lengths which are whole numbers, by substituting different values of x into the expressions:
(a) $2x^2 + 2x + 1$ (b) $2x^2 + 2x$ (c) $2x + 1$
((a) represents the hypotenuse, (b) and (c) the other two sides.)
(i) Find the sides of the triangles when $x = 1, 2, 3, 4$ and 5.
(ii) Confirm that $(2x + 1)^2 + (2x^2 + 2x)^2 = (2x^2 + 2x + 1)^2$

22. The diagram represents the starting position (AB) and the finishing position (CD) of a ladder as it slips. The ladder is leaning against a vertical wall.

Given: AC $= x$, OC $=$ 4AC, BD $=$ 2AC and OB $=$ 5 m. Form an equation in x, find x and hence find the length of the ladder.

23. A thin wire of length 18 cm is bent into the shape shown. Calculate the length from A to B.

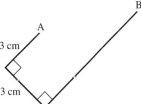

24. An aircraft is vertically above a point which is 10 km West and 15 km North of a control tower. If the aircraft is 4000 m above the ground, how far is it from the control tower?

4.3 Symmetry

Line symmetry

The letter A has one line of symmetry, shown dotted.

Rotational symmetry

The shape may be turned about O into three identical positions. It has rotational symmetry of order 3.

Quadrilaterals

1. *Square*
   All sides are equal, all angles 90°, opposite sides parallel; diagonals bisect at right angles. Four lines of symmetry. Rotational symmetry of order of 4.

2. *Rectangle*
   Opposite sides parallel and equal, all angles 90°, diagonals bisect each other. Two lines of symmetry. Rotational symmetry of order 2.

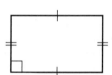

3. *Parallelogram*
   Opposite sides parallel and equal, opposite angles equal, diagonals bisect each other (but not equal). No lines of symmetry. Rotational symmetry of order 2.

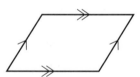

4. *Rhombus*
   A parallelogram with all sides equal, diagonals bisect each other at right angles and bisect angles. Two lines of symmetry. Rotational symmetry of order 2.

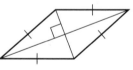

5. *Trapezium*
   One pair of sides is parallel. No line or rotational symmetry.

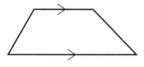

6. *Kite*
   Two pairs of adjacent sides equal, diagonals meet at right angles bisecting one of them. One line of symmetry. No rotational symmetry.

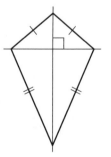

Exercise 5

1. For each shape state:
(a) the number of lines of symmetry (b) the order of rotational symmetry.

(a)

(b)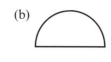

(c)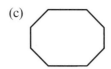

(d)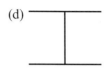

(e)

(f)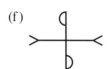

(g)

(h)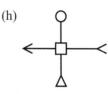

(i)

(j)

(k)

(l)

(m)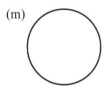

(n)

2. Add one line to each of the diagrams below so that the resulting figure has rotational symmetry but not line symmetry.

(a)

(b)

3. Draw a hexagon with just two lines of symmetry.

4. For each of the following shapes, find:
(a) the number of lines of symmetry
(b) the order of rotational symmetry.

square; rectangle; parallelogram; rhombus; trapezium; kite; equilateral triangle; regular hexagon.

In questions **5** to **15**, begin by drawing a diagram.

5. In a rectangle KLMN, $\widehat{\text{LNM}} = 34°$. Calculate:
(a) $\widehat{\text{KLN}}$ (b) $\widehat{\text{KML}}$

6. In a trapezium ABCD; $\widehat{\text{ABD}} = 35°$, $\widehat{\text{BAD}} = 110°$ and AB is parallel to DC. Calculate:
(a) $\widehat{\text{ADB}}$ (b) $\widehat{\text{BDC}}$

7. In a parallelogram WXYZ, $W\widehat{X}Y = 72°$, $Z\widehat{W}Y = 80°$. Calculate:
   (a) $W\widehat{Z}Y$ (b) $X\widehat{W}Z$ (c) $W\widehat{Y}Z$

8. In a kite ABCD, AB = AD; BC = CD; $C\widehat{A}D = 40°$ and $C\widehat{B}D = 60°$. Calculate:
   (a) $B\widehat{A}C$ (b) $B\widehat{C}A$ (c) $A\widehat{D}C$

9. In a rhombus ABCD, $A\widehat{B}C = 64°$. Calculate:
   (a) $B\widehat{C}D$ (b) $A\widehat{D}B$ (c) $B\widehat{A}C$

10. In a rectangle WXYZ, M is the mid-point of WX and $Z\widehat{M}Y = 70°$. Calculate:
    (a) $M\widehat{Z}Y$ (b) $Y\widehat{M}X$

11. In a trapezium ABCD, AB is parallel to DC, AB = AD, BD = DC and $B\widehat{A}D = 128°$. Find:
    (a) $A\widehat{B}D$ (b) $B\widehat{D}C$ (c) $B\widehat{C}D$

12. In a parallelogram KLMN, KL = KM and $K\widehat{M}L = 64°$. Find:
    (a) $M\widehat{K}L$ (b) $K\widehat{N}M$ (c) $L\widehat{M}N$

13. In a kite PQRS with PQ = PS and RQ = RS, $Q\widehat{R}S = 40°$ and $Q\widehat{P}S = 100°$. Find:
    (a) $Q\widehat{S}R$ (b) $P\widehat{S}Q$ (c) $P\widehat{Q}R$

14. In a rhombus PQRS, $R\widehat{P}Q = 54°$. Find:
    (a) $P\widehat{R}Q$ (b) $P\widehat{S}R$ (c) $R\widehat{Q}S$

15. In a kite PQRS, $R\widehat{P}S = 2P\widehat{R}S$, PQ = QS = PS and QR = RS. Find:
    (a) $Q\widehat{P}S$ (b) $P\widehat{R}S$ (c) $Q\widehat{S}R$ (d) $P\widehat{Q}R$

Planes of symmetry

A **plane of symmetry** divides a 3-D shape into two congruent shapes. One shape must be a mirror image of the other shape.

The diagrams show two planes of symmetry of a cube.

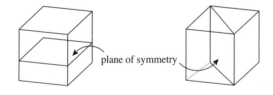

plane of symmetry

Exercise 6

1. How many planes of symmetry does this cuboid have?

2. How many planes of symmetry do these shapes have?

(a) 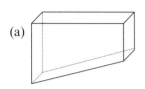 (b) 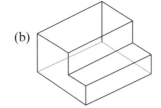 (c)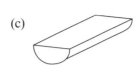

3. (a) Draw a diagram of a cube like the one on the previous page and draw a different plane of symmetry.
(b) How many planes of symmetry does a cube have?

4. Draw a pyramid with a square base so that the point of the pyramid is vertically above the centre of the square base. Show any planes of symmetry by shading.

5. The diagrams show the plan view and the side view of an object.

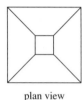

plan view   side view

> The plan view is the view looking down on the object.

> The side view is the view from one side.

How many planes of symmetry does this object have?

6. (a) How many planes of symmetry does a cylinder have?
(b) Describe the symmetry, if any, of a cone.

4.4 Similarity

Two triangles are similar if they have the same angles. For other shapes, not only must corresponding angles be equal, but also corresponding sides must be in the same proportion.
The two rectangles A and B are *not* similar even though they have the same angles.

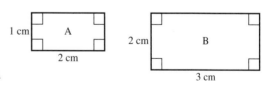

Example

In the triangles ABC and XYZ

$$\widehat{A} = \widehat{X} \text{ and } \widehat{B} = \widehat{Y}$$

so the triangles are similar. ($\widehat{C}$ must be equal to $\widehat{Z}$.)

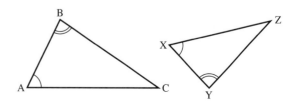

We have $\dfrac{BC}{YZ} = \dfrac{AC}{XZ} = \dfrac{AB}{XY}$

Exercise 7

Find the sides marked with letters in questions **1** to **11**; all lengths are given in centimetres.

1.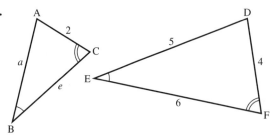

2.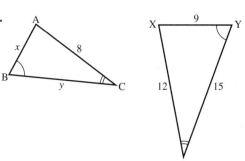

3.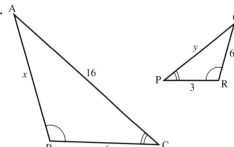

4.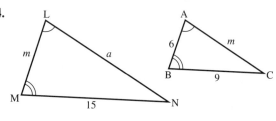

5.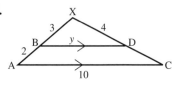

6.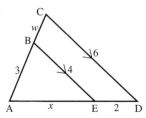

7.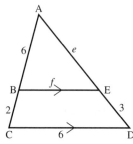

8.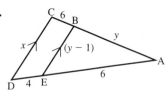

9. $B\widehat{A}C = D\widehat{B}C$

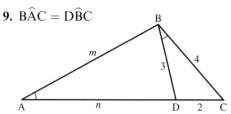

10.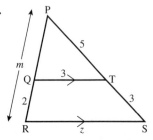

11.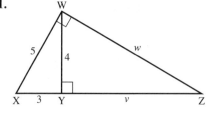

12. The drawing shows a rectangular picture 16 cm × 8 cm surrounded by a border of width 4 cm. Are the two rectangles similar?

13. The diagonals of a trapezium ABCD intersect at O. AB is parallel to DC, AB = 3 cm and DC = 6 cm. If CO = 4 cm and OB = 3 cm, find AO and DO.

14. A tree of height 4 m casts a shadow of length 6·5 m. Find the height of a house casting a shadow 26 m long.

15. Which of the following *must* be similar to each other?
(a) two equilateral triangles
(b) two rectangles
(c) two isosceles triangles
(d) two squares
(e) two regular pentagons
(f) two kites
(g) two rhombuses
(h) two circles

16. In the diagram $A\widehat{B}C = A\widehat{D}B = 90°$, AD = p and DC = q.
(a) Use similar triangles to show that $x^2 = pz$.
(b) Find a similar expression for y^2.
(c) Add the expressions for x^2 and y^2 and hence prove Pythagoras' theorem.

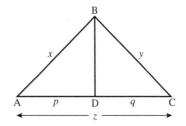

17. In a triangle ABC, a line is drawn parallel to BC to meet AB at D and AC at E. DC and BE meet at X. Prove that:
(a) the triangles ADE and ABC are similar
(b) the triangles DXE and BXC are similar
(c) $\dfrac{AD}{AB} = \dfrac{EX}{XB}$

18. From the rectangle ABCD a square is cut off to leave rectangle BCEF.
Rectangle BCEF is similar to ABCD. Find x and hence state the ratio of the sides of rectangle ABCD. ABCD is called the Golden Rectangle and is an important shape in architecture.

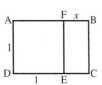

Congruence

Two plane figures are congruent if one fits exactly on the other. They must be the same size and the same shape.

Exercise 8

1. Identify pairs of congruent shapes below.

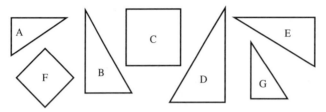

2. Triangle LMN is isosceles with LM = LN; X and Y are points on LM, LN respectively such that LX = LY. Prove that triangles LMY and LNX are congruent.

3. ABCD is a quadrilateral and a line through A parallel to BC meets DC at X. If $\hat{D} = \hat{C}$, prove that $\triangle$ADX is isosceles.

4. In the diagram, N lies on a side of the square ABCD, AM and LC are perpendicular to DN. Prove that:
(a) $A\hat{D}N = L\hat{C}D$ (b) AM = LD

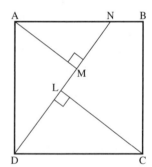

5. Points L and M on the side YZ of a triangle XYZ are drawn so that L is between Y and M. Given that XY = XZ and $Y\hat{X}L = M\hat{X}Z$, prove that YL = MZ.

6. Squares AMNB and AOPC are drawn on the sides of triangle ABC, so that they lie outside the triangle. Prove that MC = OB.

7. In the diagram, $L\hat{M}N = O\hat{N}M = 90°$. P is the mid-point of MN, MN = 2ML and MN = NO. Prove that:
(a) the triangles MNL and NOP are congruent
(b) $O\hat{P}N = L\hat{N}O$
(c) $L\hat{Q}O = 90°$

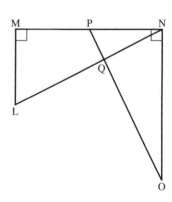

8. PQRS is a parallelogram in which the bisectors of the angles P and Q meet at X. Prove that the angle PXQ is a right angle.

Areas of similar shapes

The two rectangles are similar, the ratio of corresponding sides being k.

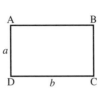
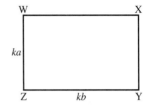

area of $ABCD = ab$
area of $WXYZ = ka \times kb = k^2ab$

$$\therefore \quad \frac{\text{area } WXYZ}{\text{area } ABCD} = \frac{k^2ab}{ab} = k^2$$

This illustrates an important general rule for all similar shapes:

If two figures are similar and the ratio of corresponding sides is k, then the ratio of their areas is k^2.

Note: k is sometimes called the *linear scale factor*.

> This result also applies for the surface areas of similar three-dimensional objects.

Example 1

XY is parallel to BC.

$$\frac{AB}{AX} = \frac{3}{2}$$

If the area of $\triangle AXY = 4 \, \text{cm}^2$, find the area of $\triangle ABC$.

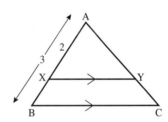

The triangles ABC and AXY are similar.

Ratio of corresponding sides $(k) = \frac{3}{2}$

$\therefore \qquad\qquad$ Ratio of areas $(k^2) = \frac{9}{4}$

$\therefore$ Area of $\triangle ABC = \frac{9}{4} \times$ (area of $\triangle AXY$)

$$= \frac{9}{4} \times (4) = 9 \, \text{cm}^2$$

Example 2

Two similar triangles have areas of $18 \, \text{cm}^2$ and $32 \, \text{cm}^2$ respectively. If the base of the smaller triangle is $6 \, \text{cm}$, find the base of the larger triangle.

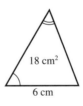
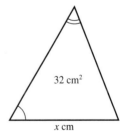

Ratio of areas $(k^2) = \dfrac{32}{18} = \dfrac{16}{9}$

$\therefore$ Ratio of corresponding sides $(k) = \sqrt{\left(\dfrac{16}{9}\right)}$

$$= \frac{4}{3}$$

$\therefore$ Base of larger triangle $= 6 \times \dfrac{4}{3} = 8 \, \text{cm}$

Exercise 9

In this exercise a number written inside a figure represents the area of
the shape in cm². Numbers on the outside give linear dimensions in cm.
In questions **1** to **6** find the unknown area *A*. In each case the shapes
are similar.

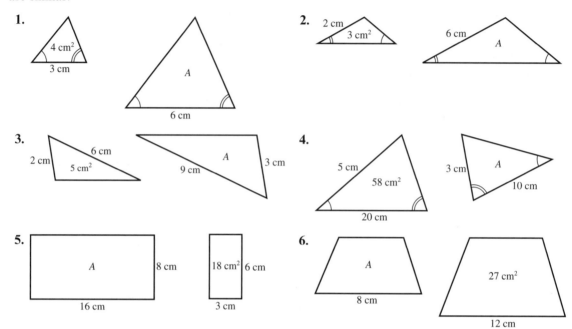

In questions **7** to **12**, find the lengths marked for each pair of similar shapes.

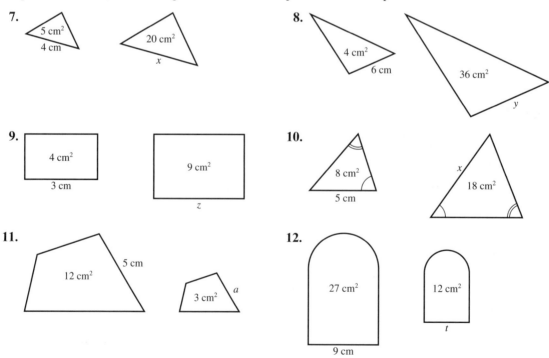

In questions **13** to **16** you have a pair of similar three-dimensional objects. Find the surface area indicated.

13.

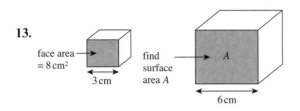

14.

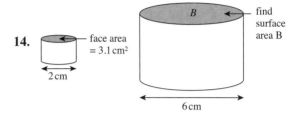

15.

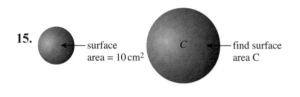

The radius of the large sphere is twice the radius of the small sphere.

16.

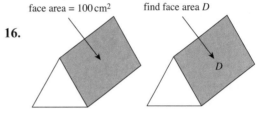

The length of large solid is 1·5 times the length of the small solid.

17. Given: AD = 3 cm, AB = 5 cm and area of △ADE = 6 cm².
Find:
(a) area of △ABC   (b) area of DECB

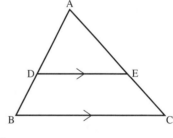

18. Given: XY = 5 cm, MY = 2 cm and area of △MYN = 4 cm².
Find:
(a) area of △XYZ   (b) area of MNZX

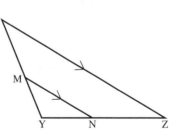

19. Given XY = 2 cm, BC = 3 cm and area of XYCB = 10 cm², find the area of △AXY.

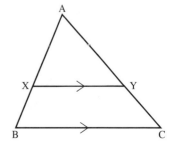

20. Given $KP = 3$ cm, area of $\triangle KOP = 2$ cm^2 and area of OPML $= 16$ cm^2, find the length of PM.

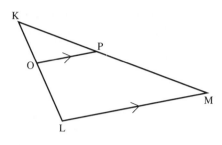

21. The triangles ABC and EBD are similar (AC and DE are *not* parallel).
If $AB = 8$ cm, $BE = 4$ cm and the area of $\triangle DBE = 6$ cm^2, find the area of $\triangle ABC$.

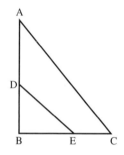

22. Given: $AZ = 3$ cm, $ZC = 2$ cm, $MC = 5$ cm, $BM = 3$ cm. Find:
(a) XY
(b) YZ
(c) the ratio of areas AXY : AYZ
(d) the ratio of areas AXY : ABM

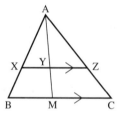

23. A floor is covered by 600 tiles which are 10 cm by 10 cm. How many 20 cm by 20 cm tiles are needed to cover the same floor?

24. A wall is covered by 160 tiles which are 15 cm by 15 cm. How many 10 cm by 10 cm tiles are needed to cover the same wall?

25. When potatoes are peeled do you lose more peel when you use big potatoes or small potatoes?

Volumes of similar objects

When solid objects are similar, one is an accurate enlargement of the other.
If two objects are similar and the ratio of corresponding sides is k, then the ratio of their volumes is k^3.

A line has one dimension, and the scale factor is used once.

An area has two dimensions, and the scale factor is used twice.

A volume has three dimensions, and the scale factor is used three times.

Example 1

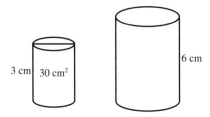

Two similar cylinders have heights of 3 cm and 6 cm respectively. If the volume of the smaller cylinder is 30 cm^3, find the volume of the larger cylinder.

If linear scale factor $= k$, then ratio of heights $(k) = \frac{6}{3} = 2$

$\therefore$ ratio of volumes $(k^3) = 2^3$

$\qquad\qquad\qquad\qquad = 8$

and volume of larger cylinder $= 8 \times 30$

$\qquad\qquad\qquad\qquad\qquad = 240 \, \text{cm}^3$

Example 2

Two similar spheres made of the same material have masses of 32 kg and 108 kg respectively. If the radius of the larger sphere is 9 cm, find the radius of the smaller sphere.

We may take the ratio of masses to be the same as the ratio of volumes.

ratio of volumes $(k^3) = \dfrac{32}{108}$

$\qquad\qquad\qquad\quad = \dfrac{8}{27}$

ratio of corresponding lengths $(k) = \sqrt[3]{\left(\dfrac{8}{27}\right)}$

$\qquad\qquad\qquad\qquad\qquad\qquad = \dfrac{2}{3}$

$\therefore$ Radius of smaller sphere $= \dfrac{2}{3} \times 9$

$\qquad\qquad\qquad\qquad\qquad\quad = 6 \, \text{cm}$

Exercise 10

In this exercise, the objects are similar and a number written inside a figure represents the volume of the object in cm^3.
Numbers on the outside give linear dimensions in cm. In questions **1** to **8**, find the unknown volume V.

1.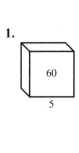

2.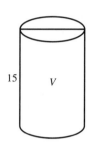

3.

4. radius = 1.2 cm　　radius = 12 cm

5.

6.

7.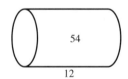

8.

In questions **9** to **14**, find the lengths marked by a letter.

9.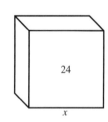

10.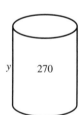

11.

12.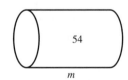

13.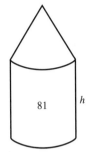

14.

15. Two similar jugs have heights of 4 cm and 6 cm respectively. If the capacity of the smaller jug is 50 cm³, find the capacity of the larger jug.

16. Two similar cylindrical tins have base radii of 6 cm and 8 cm respectively. If the capacity of the larger tin is 252 cm³, find the capacity of the small tin.

17. Two solid metal spheres have masses of 5 kg and 135 kg respectively. If the radius of the smaller one is 4 cm, find the radius of the larger one.

18. Two similar cones have surface areas in the ratio 4 : 9. Find the ratio of:
(a) their lengths,   (b) their volumes.

19. The area of the bases of two similar glasses are in the ratio 4 : 25. Find the ratio of their volumes.

20. Two similar solids have volumes V_1 and V_2 and corresponding sides of length x_1 and x_2. State the ratio $V_1 : V_2$ in terms of x_1 and x_2.

21. Two solid spheres have surface areas of 5 cm² and 45 cm² respectively and the mass of the smaller sphere is 2 kg. Find the mass of the larger sphere.

22. The masses of two similar objects are 24 kg and 81 kg respectively. If the surface area of the larger object is 540 cm², find the surface area of the smaller object.

23. A cylindrical can has a circumference of 40 cm and a capacity of 4·8 litres. Find the capacity of a similar cylinder of circumference 50 cm.

24. A container has a surface area of 5000 cm² and a capacity of 12·8 litres. Find the surface area of a similar container which has a capacity of 5·4 litres.

4.5 Circle theorems

(a) The angle subtended at the centre of a circle is twice the angle subtended at the circumference.

$$\hat{AOB} = 2 \times \hat{ACB}$$

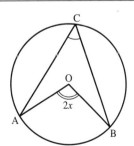

Proof:
Draw the straight line COD.
Let $A\widehat{C}O = y$ and $B\widehat{C}O = z$.
In triangle AOC,

$$AO = OC \qquad \text{(radii)}$$
$$\therefore \quad O\widehat{C}A = O\widehat{A}C \quad \text{(isosceles triangle)}$$
$$\therefore \quad C\widehat{O}A = 180 - 2y \text{ (angle sum of triangle)}$$
$$\therefore \quad A\widehat{O}D = 2y \qquad \text{(angles on a straight line)}$$

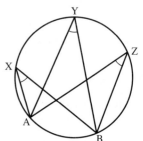

Similarly from triangle COB, we find

$$D\widehat{O}B = 2z$$

Now $A\widehat{C}B = y + z$

and $A\widehat{O}B = 2y + 2z$

$\therefore \quad A\widehat{O}B = 2 \times A\widehat{C}B$ as required.

(b) Angles subtended by an arc in the same segment of a circle are equal.

$$A\widehat{X}B = A\widehat{Y}B = A\widehat{Z}B$$

Example 1

Given $A\widehat{B}O = 50°$, find $B\widehat{C}A$.
Triangle OBA is isosceles (OA = OB).

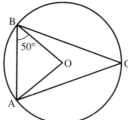

$\therefore \quad O\widehat{A}B = 50°$
$\therefore \quad B\widehat{O}A = 80°$ (angle sum of a triangle)
$\therefore \quad B\widehat{C}A = 40°$ (angle at the circumference)

Example 2

Given $B\widehat{D}C = 62°$ and $D\widehat{C}A = 44°$, find $B\widehat{A}C$ and $A\widehat{B}D$.
$B\widehat{D}C = B\widehat{A}C$ (both subtended by arc BC)
$\therefore \quad B\widehat{A}C = 62°$

$D\widehat{C}A = A\widehat{B}D$ (both subtended by arc DA)
$\therefore \quad A\widehat{B}D = 44°$

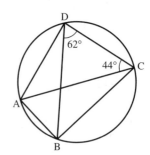

Exercise 11

Find the angles marked with letters. A line passes through the centre
only when point O is shown.

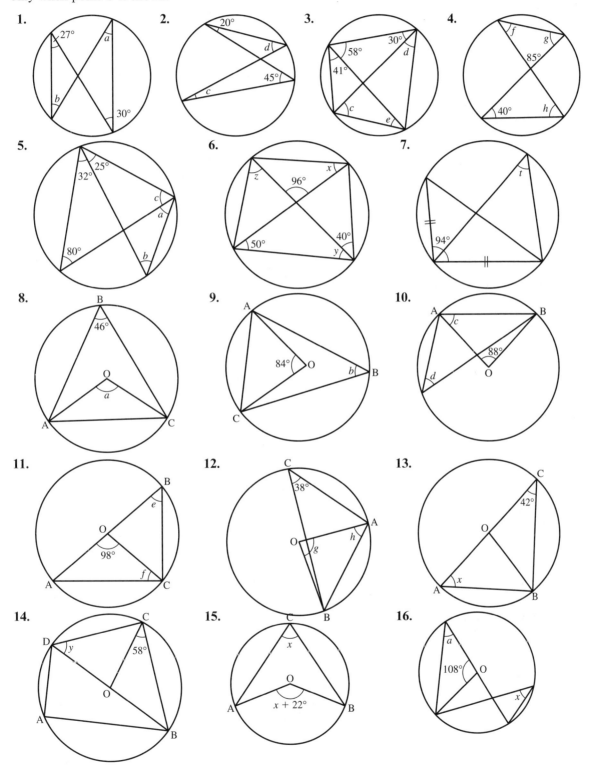

- ABCD is a cyclic quadrilateral. The corners touch the circle.

(c) The opposite angles in a cyclic quadrilateral add up to 180°
   (the angles are supplementary).

$$\widehat{A} + \widehat{C} = 180°$$
$$\widehat{B} + \widehat{D} = 180°$$

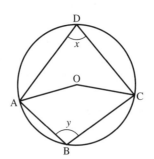

Proof:
Draw radii OA and OC.
Let $A\widehat{D}C = x$ and $A\widehat{B}C = y$.

$\quad$ AÔC obtuse $= 2x$ $\quad$ (angle at the centre)
$\quad$ AÔC reflex $= 2y$ $\quad$ (angle at the centre)
∴ $\quad\quad 2x + 2y = 360°$ (angles at a point)
∴ $\quad\quad\quad x + y = 180°$ as required

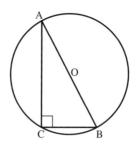

(d) The angle in a semi-circle is a right angle.

In the diagram, AB is a diameter.

$A\widehat{C}B = 90°$.

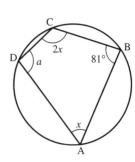

Example 1

Find a and x.
$\quad a = 180° - 81°$ (opposite angles of a cyclic quadrilateral)
∴ $\quad a = 99°$

$x + 2x = 180°$ (opposite angles of a cyclic quadrilateral)
$\quad\quad 3x = 180°$
$\quad\quad\ x = 60°$

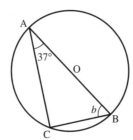

Example 2

Find b.
$\quad A\widehat{C}B = 90°$ (angle in a semi-circle)
∴ $\quad\quad b = 180° - (90 + 37)°$
$\quad\quad\quad = 53°$

Exercise 12

Find the angles marked with a letter.

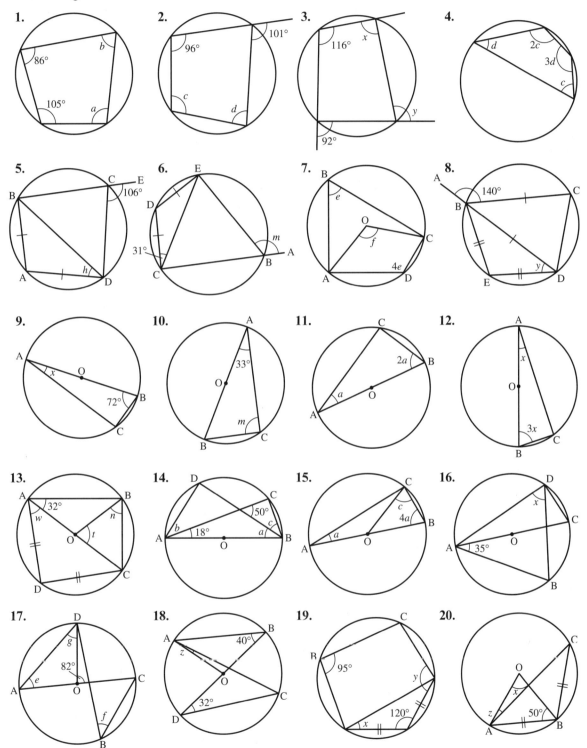

Tangents to circles

(a) The angle between a tangent and the radius drawn to the point of contact is 90°.

$$A\widehat{B}O = 90°$$

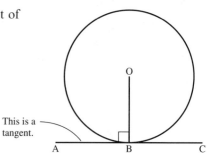

This is a tangent.

(b) From any point outside a circle just two tangents to the circle may be drawn and they are of equal length.

$$TA = TB$$

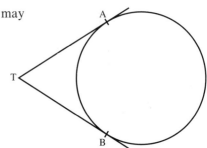

Example

TA and TB are tangents to the circle, centre O.
Given $A\widehat{T}B = 50°$, find
(a) $A\widehat{B}T$
(b) $O\widehat{B}A$

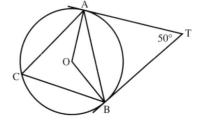

(a) $\triangle TBA$ is isosceles $(TA = TB)$
   $\therefore\quad A\widehat{B}T = \frac{1}{2}(180 - 50) = 65°$
(b) $\quad O\widehat{B}T = 90°$ (tangent and radius)
   $\therefore\quad O\widehat{B}A = 90 - 65$
   $\qquad\qquad = 25°$

Exercise 13

For questions **1** to **12**, find the angles marked with a letter.

1.

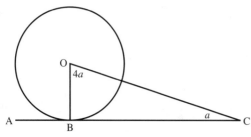

2.

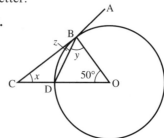

3.

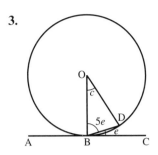

4.

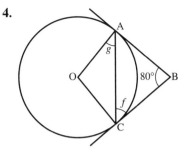

5.

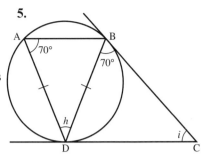

6.

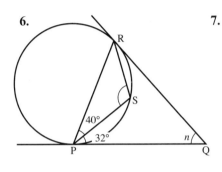

7.

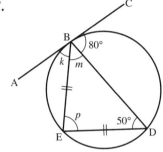

8.

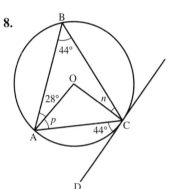

9. Find x, y and z.

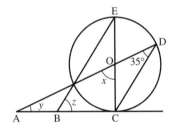

4.6 Constructions and loci

When the word 'construct' is used, the diagram should be drawn using equipment such as compasses, a ruler, a protractor etc.

Three basic constructions are shown below.

(a) Perpendicular bisector of a line joining two points

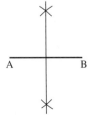

(b) Bisector of an angle

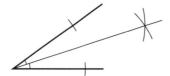

(c) 60° angle construction

Exercise 14

1. Construct a triangle ABC in which AB = 8 cm, AC = 6 cm and BC = 5 cm. Measure the angle $A\widehat{C}B$.

2. Construct a triangle PQR in which PQ = 10 cm, PR = 7 cm and RQ = 6 cm. Measure the angle $R\widehat{P}Q$.

3. Construct an equilateral triangle of side 7 cm.

4. Draw a line AB of length 10 cm. Construct the perpendicular bisector of AB.

5. Draw two lines AB and AC of length 8 cm, where $B\widehat{A}C$ is approximately 40°. Construct the line which bisects $B\widehat{A}C$.

6. Draw a line AB of length 12 cm and draw a point X approximately 6 cm above the middle of the line. Construct the line through X which is perpendicular to AB.

7. Construct an equilateral triangle ABC of side 9 cm. Construct a line through A to meet BC at 90° at the point D. Measure the length AD.

8. Construct the triangles shown and measure the length x.

(a)

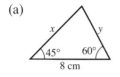

(b)

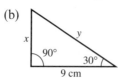

(c)

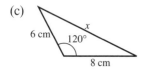

(d)

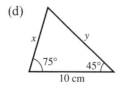

9. Construct a parallelogram WXYZ in which WX = 10 cm, WZ = 6 cm and $X\widehat{W}Z = 60°$. By construction, find the point A that lies on ZY and is equidistant from lines WZ and WX. Measure the length WA.

10. (a) Draw a line OX = 10 cm and construct an angle XOY = 60°.
    (b) Bisect the angle XOY and mark a point A on the bisector so that OA = 7 cm.
    (c) Construct a circle with centre A to touch OX and OY and measure the radius of the circle.

11. (a) Construct a triangle PQR with PQ = 8 cm, PR = 12 cm and $P\widehat{Q}R = 90°$.
    (b) Construct the bisector of $Q\widehat{P}R$.
    (c) Construct the perpendicular bisector of PR and mark the point X where this line meets the bisector of $Q\widehat{P}R$.
    (d) Measure the length PX.

12. (a) Construct a triangle ABC in which AB = 8 cm, AC = 6 cm and BC = 9 cm.
    (b) Construct the bisector of $B\widehat{A}C$.
    (c) Construct the line through C perpendicular to CA and mark the point X where this line meets the bisector of $B\widehat{A}C$.
    (d) Measure the lengths CX and AX.

The locus of a point

The locus of a point is the path which it describes as it moves.

Example

Draw a line AB of length 8 cm.
Construct the locus of a point P which moves so that $B\widehat{A}P = 90°$.

Construct the perpendicular at A.
This line is the locus of P.
These are the basic loci you will come across:
1. Given distance from a given point. Locus is a circle.
2. Given distance from a straight line. Locus is a parallel line.
3. Equidistant from two given points. Locus is the perpendicular bisector of the line joining the two points.
4. Equidistant from two intersecting lines. Locus is the angle bisector of the two lines.

P is anywhere on this line

Exercise 15

1. Draw a line XY of length 10 cm. Construct the locus of a point which is equidistant from X and Y.

2. Draw two lines AB and AC of length 8 cm, where $B\widehat{A}C$ is approximately 70°. Construct the locus of a point which is equidistant from the lines AB and AC.

3. Draw a circle, centre O, of radius 5 cm and draw a radius OA. Construct the locus of a point P which moves so that $O\widehat{A}P = 90°$.

4. Draw a line AB of length 10 cm and construct the circle with diameter AB. Indicate the locus of a point P which moves so that $A\widehat{P}B = 90°$.

5. (a) Describe in words the locus of M, the tip of the minute hand of a clock as the time changes from 3 o'clock to 4 o'clock.
   (b) Sketch the locus of H, the tip of the hour hand, as the time changes from 3 o'clock to 4 o'clock.
   (c) Describe the locus of the tip of the second hand as the time goes from 3 o'clock to 4 o'clock.

6. A detective has put a radio transmitter on a suspect's car. From the strength of the signals received at points R and P she knows that the car is
   (a) not more than 40 km from R, and
   (b) not more than 20 km from P.

   Make a scale drawing [1 cm ≡ 10 km] and show the possible positions of the suspect's car.

R •

50 km

P •

7. A treasure is buried in the rectangular garden shown. The treasure is: (a) within 4 m of A and (b) more than 3 m from the line AD. Draw a plan of the garden and shade the points where the treasure could be.

8. A goat is tied to one corner on the outside of a barn. The diagram shows a plan view. Sketch two plan views of the barn and show the locus of points where the goat can graze if
   (a) the rope is 4 m long,
   (b) the rope is 7 m long.

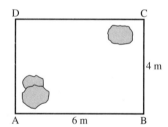

9. Draw a line AB of length 10 cm. With AB as base draw a triangle ABP so that the *area* of the triangle is 30 cm². Describe the locus of P if P moves so that the area of the triangle ABP is always 30 cm².

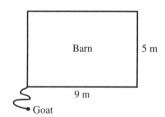

10. As the second hand of a clock goes through a vertical position, a money spider starts walking from C along the hand. After one minute the spider is at the top of the clock T. Describe the locus of the spider.

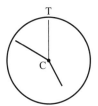

11. Sketch a side view of the locus of the valve on a bicycle wheel as the bicycle goes past in a straight line.

4.7 Nets

If the cube below was made of cardboard, and you cut along some of the edges and laid it out flat, you would have the *net* of the cube.

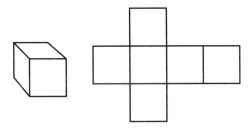

Exercise 16

1. Which of the nets below can be used to make a cube?

(a)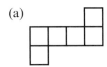

(b)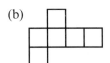

(c)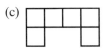

(d)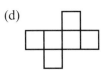

2. The diagram shows the net of a closed rectangular box. All lengths are in cm.
   (a) Find the lengths a, x, y.
   (b) Calculate the volume of the box.

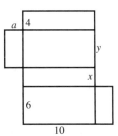

3. The diagram shows the net of a pyramid. The base is shaded. The lengths are in cm.

(a) Find the lengths a, b, c, d.
(b) Find the volume of the pyramid.

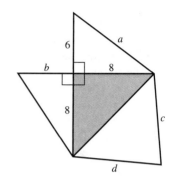

4. The diagram shows the net of a prism.
   (a) Find the area of one of the triangular faces (shown shaded).
   (b) Find the volume of the prism.

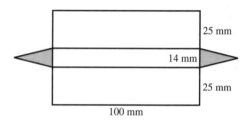

5. This is the net of a square-based pyramid.
   What are the lengths a, b, c, x, y?

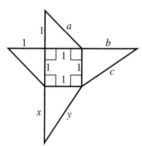

6. Sketch nets for the following:
   (a) Closed rectangular box 7 cm × 9 cm × 5 cm.
   (b) Closed cylinder: length 10 cm, radius 6 cm.
   (c) Prism of length 12 cm, cross-section an equilateral triangle of side 4 cm.

Revision exercise 4A

1. ABCD is a parallelogram and AE bisects angle A. Prove that DE = BC.

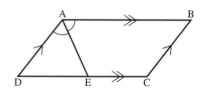

2. In a triangle PQR, $P\widehat{Q}R = 50°$ and point X lies on PQ such QX = XR. Calculate $Q\widehat{X}R$.

3. (a) ABCDEF is a regular hexagon. Calculate $F\widehat{D}E$.
   (b) ABCDEFGH is a regular eight-sided polygon. Calculate $A\widehat{G}H$.
   (c) Each interior angle of a regular polygon measures 150°. How many sides has the polygon?

4. In the quadrilateral PQRS, PQ = QS = QR,
PS is parallel to QR and QR̂S = 70°. Calculate:
(a) RQ̂S
(b) PQ̂S

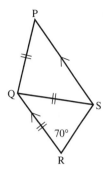

5. Find x.

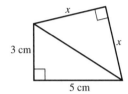

6. In the triangle ABC, AB = 7 cm, BC = 8 cm and AB̂C = 90°. Point
P lies inside the triangle such that BP = PC = 5 cm. Calculate:
(a) the perpendicular distance from P to BC
(b) the length AP.

7. In triangle PQR the bisector of PQ̂R meets PR at S and the point T
lies on PQ such that ST is parallel to RQ.
(a) Prove that QT = TS.
(b) Prove that the triangles PTS and PQR are similar.
(c) Given that PT = 5 cm and TQ = 2 cm, calculate the length of QR.

8. In the quadrilateral ABCD, AB is parallel to DC and DÂB = DB̂C.
(a) Prove that the triangles ABD and DBC are similar.
(b) If AB = 4 cm and DC = 9 cm, calculate the length of BD.

9. A rectangle 11 cm by 6 cm is similar to a rectangle 2 cm by x cm.
Find the two possible values of x.

10. In the diagram, triangles ABC and EBD are similar but DE is *not*
parallel to AC. Given that AD = 5 cm, DB = 3 cm and BE = 4 cm,
calculate the length of BC.

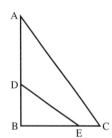

11. The radii of two spheres are in the ratio 2 : 5. The volume of the
smaller sphere is 16 cm³. Calculate the volume of the larger sphere.

12. The surface areas of two similar jugs are 50 cm² and 450 cm²
respectively.
(a) If the height of the larger jug is 10 cm, find the height of the smaller jug.
(b) If the volume of the smaller jug is 60 cm³, find the volume of the
larger jug.

13. A car is an enlargement of a model, the scale factor being 10.
    (a) If the windscreen of the model has an area of $100\,\text{cm}^2$, find the area of the windscreen on the actual car (answer in m^2).
    (b) If the capacity of the boot of the car is $1\,\text{m}^3$, find the capacity of the boot on the model (answer in cm^3).

14. Find the angles marked with letters. (O is the centre of the circle.)

    (a)
    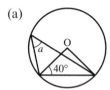

    (b)
    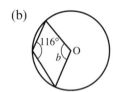

    (c)
    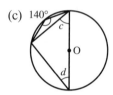

    (d)
    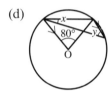

15. ABCD is a cyclic quadrilateral in which
    AB = BC and $A\hat{B}C = 70°$.
    AD produced meets BC produced at the point P, where $A\hat{P}B = 30°$.
    Calculate:
    (a) $A\hat{D}B$ (b) $A\hat{B}D$

16. Using ruler and compasses only:
    (a) Construct the triangle ABC in which AB = 7 cm, BC = 5 cm and AC = 6 cm.
    (b) Construct the circle which passes through A, B and C and measure the radius of this circle.

17. Construct:
    (a) the triangle XYZ in which XY = 10 cm, YZ = 11 cm and XZ = 9 cm.
    (b) the locus of points, inside the triangle, which are equidistant from the lines XZ and YZ.
    (c) the locus of points which are equidistant from Y and Z.
    (d) the circle which touches YZ at its mid-point and also touches XZ.

Examination exercise 4B

1.

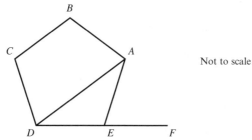

Not to scale

ABCDE is a regular pentagon.
DEF is a straight line.
Calculate
(a) angle *AEF*,   [2]
(b) angle *DAE*.   [1]

Cambridge IGCSE Mathematics 0580
Paper 2 Q17 November 2005

2. A square *ABCD*, of side 8 cm, has another square, *PQRS*, drawn
inside it.
P, *Q*, *R* and *S* are at the mid-points of each side of the square
ABCD, as shown in the diagram.

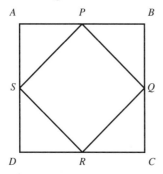

Not to scale

(a) Calculate the length of *PQ*.   [2]
(b) Calculate the area of the square *PQRS*.   [1]

Cambridge IGCSE Mathematics 0580
Paper 2 Q6 June 2005

3. Two similar vases have heights which are in the ratio 3 : 2.
(a) The volume of the larger vase is 1080 cm^3.
    Calculate the volume of the smaller vase.   [2]
(b) The surface area of the smaller vase is 252 cm^2.
    Calculate the surface area of the larger vase.   [2]

Cambridge IGCSE Mathematics 0580
Paper 21 Q18 June 2009

4. A company makes two models of television.
   Model A has a rectangular screen that measures 44 cm by 32 cm.
   Model B has a larger screen with these measurements increased in the ratio 5 : 4.
   (a) Work out the measurements of the larger screen. [2]
   (b) Find the **fraction** $\dfrac{\text{model A screen area}}{\text{model B screen area}}$ in its simplest form. [1]

   Cambridge IGCSE Mathematics 0580
   Paper 2 Q14 June 2006

5. A cylindrical glass has a radius of 4 centimetres and a height of 6 centimetres.
   A large cylindrical jar full of water is a similar shape to the glass.
   The glass can be filled with water from the jar exactly 216 times.
   Work out the radius and height of the jar. [3]

   Cambridge IGCSE Mathematics 0580
   Paper 22 Q10 June 2008

6.

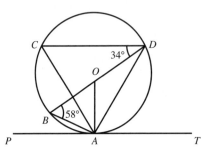

Not to scale

   A, B, C and D lie on the circle, centre O.
   BD is a diameter and PAT is the tangent at A.
   Angle $ABD = 58°$ and angle $CDB = 34°$.
   Find
   (a) angle ACD, [1]
   (b) angle ADB, [1]
   (c) angle DAT, [1]
   (d) angle CAO. [2]

   Cambridge IGCSE Mathematics 0580
   Paper 22 Q22 June 2009

7.

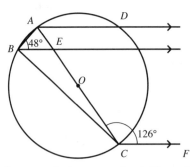

Not to scale

   A, B, C and D lie on a circle centre O. AC is a diameter of the circle.
   AD, BE and CF are parallel lines. Angle $ABE = 48°$ and angle
   $ACF = 126°$.

Find
(a) angle *DAE*, [1]
(b) angle *EBC*, [1]
(c) angle *BAE*. [1]

Cambridge IGCSE Mathematics 0580
Paper 2 Q15 November 2005

8.

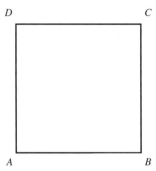

ABCD is a square.
It is rotated through 90° clockwise about *B*.
Draw accurately the locus of the point *D*. [2]

Cambridge IGCSE Mathematics 0580
Paper 22 Q5 November 2008

9. Using a straight edge and compasses only, draw the locus of all
points inside the quadrilateral *ABCD* which are equidistant from
the lines *AC* and *BD*.

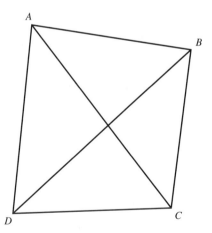

Show clearly all your construction arcs, [4]

Cambridge IGCSE Mathematics 0580
Paper 22 Q16 June 2009

5 ALGEBRA 2

Girolamo Cardan (1501–1576) was a colourful character who became Professor of Mathematics at Milan. As well as being a distinguished academic, he was an astrologer, a physician, a gambler and a heretic, yet he received a pension from the Pope. His mathematical genius enabled him to open up the general theory of cubic and quartic equations, although a method for solving cubic equations which he claimed as his own was pirated from Niccolo Tartaglia.

10	Express direct and inverse variation in algebraic terms
20	Construct and transform more complicated formulae and equations
21	Manipulate algebraic fractions
23	Use and interpret indices, including fractional indices
24	Solve simple linear inequalities
25	Represent inequalities graphically and solve simple linear programming problems

5.1 Algebraic fractions

Simplifying fractions

Example

Simplify: (a) $\dfrac{32}{56}$ (b) $\dfrac{3a}{5a^2}$ (c) $\dfrac{3y + y^2}{6y}$

(a) $\dfrac{32}{56} = \dfrac{\cancel{8} \times 4}{\cancel{8} \times 7} = \dfrac{4}{7}$

(b) $\dfrac{3a}{5a^2} = \dfrac{3 \times \cancel{a}}{5 \times a \times \cancel{a}} = \dfrac{3}{5a}$

(c) $\dfrac{y(3 + y)}{6y} = \dfrac{3 + y}{6}$

Exercise 1

Simplify as far as possible, where you can.

1. $\dfrac{25}{35}$

2. $\dfrac{84}{96}$

3. $\dfrac{5y^2}{y}$

4. $\dfrac{y}{2y}$

5. $\dfrac{8x^2}{2x^2}$

6. $\dfrac{2x}{4y}$

7. $\dfrac{6y}{3y}$

8. $\dfrac{5ab}{10b}$

9. $\dfrac{8ab^2}{12ab}$

10. $\dfrac{7a^2b}{35ab^2}$

11. $\dfrac{(2a)^2}{4a}$

12. $\dfrac{7yx}{8xy}$

13. $\dfrac{5x + 2x^2}{3x}$

14. $\dfrac{9x + 3}{3x}$

15. $\dfrac{25 + 7}{25}$

16. $\dfrac{4a + 5a^2}{5a}$

17. $\dfrac{3x}{4x - x^2}$

18. $\dfrac{5ab}{15a + 10a^2}$

19. $\dfrac{5x + 4}{8x}$

20. $\dfrac{12x + 6}{6y}$

21. $\dfrac{5x + 10y}{15xy}$

22. $\dfrac{18a - 3ab}{6a^2}$

23. $\dfrac{4ab + 8a^2}{2ab}$

24. $\dfrac{(2x)^2 - 8x}{4x}$

Example

Simplify:

(a) $\dfrac{x^2 + x - 6}{x^2 + 2x - 3} = \dfrac{(x - 2)\cancel{(x + 3)}}{\cancel{(x + 3)}(x - 1)} = \dfrac{x - 2}{x - 1}$

(b) $\dfrac{x^2 + 3x - 10}{x^2 - 4} = \dfrac{\cancel{(x - 2)}(x + 5)}{\cancel{(x - 2)}(x + 2)} = \dfrac{x + 5}{x + 2}$

(c) $\dfrac{3x^2 - 9x}{x^2 - 4x + 3} = \dfrac{3x\cancel{(x - 3)}}{(x - 1)\cancel{(x - 3)}} = \dfrac{3x}{x - 1}$

Exercise 2

Simplify as far as possible:

1. $\dfrac{x^2 + 2x}{x^2 - 3x}$

2. $\dfrac{x^2 - 3x}{x^2 - 2x - 3}$

3. $\dfrac{x^2 + 4x}{2x^2 - 10x}$

4. $\dfrac{x^2 + 6x + 5}{x^2 - x - 2}$

5. $\dfrac{x^2 - 4x - 21}{x^2 - 5x - 14}$

6. $\dfrac{x^2 + 7x + 10}{x^2 - 4}$

7. $\dfrac{x^2 + x - 2}{x^2 - x}$

8. $\dfrac{3x^2 - 6x}{x^2 + 3x - 10}$

9. $\dfrac{6x^2 - 2x}{12x^2 - 4x}$

10. $\dfrac{3x^2 + 15x}{x^2 - 25}$

11. $\dfrac{12x^2 - 20x}{4x^2}$

12. $\dfrac{x^2 + x - 6}{x^2 + 2x - 3}$

Addition and subtraction of algebraic fractions

Example

Write as a single fraction:

(a) $\dfrac{2}{3} + \dfrac{3}{4}$

(b) $\dfrac{2}{x} + \dfrac{3}{y}$

Compare these two workings line for line:

(a) $\dfrac{2}{3} + \dfrac{3}{4}$; the L.C.M. of 3 and 4 is 12.

$$\therefore \quad \frac{2}{3} + \frac{3}{4} = \frac{8}{12} + \frac{9}{12}$$

$$= \frac{17}{12}$$

(b) $\dfrac{2}{x} + \dfrac{3}{y}$; the L.C.M. of x and y is xy.

$$\therefore \quad \frac{2}{x} + \frac{3}{y} = \frac{2y}{xy} + \frac{3x}{xy}$$

$$= \frac{2y + 3x}{xy}$$

Exercise 3

Simplify the following:

1. $\dfrac{2}{5} + \dfrac{1}{5}$

2. $\dfrac{2x}{5} + \dfrac{x}{5}$

3. $\dfrac{2}{x} + \dfrac{1}{x}$

4. $\dfrac{1}{7} + \dfrac{3}{7}$

5. $\dfrac{x}{7} + \dfrac{3x}{7}$

6. $\dfrac{1}{7x} + \dfrac{3}{7x}$

7. $\dfrac{5}{8} + \dfrac{1}{4}$

8. $\dfrac{5x}{8} + \dfrac{x}{4}$

9. $\dfrac{5}{8x} + \dfrac{1}{4x}$

10. $\dfrac{2}{3} + \dfrac{1}{6}$

11. $\dfrac{2x}{3} + \dfrac{x}{6}$

12. $\dfrac{2}{3x} + \dfrac{1}{6x}$

13. $\dfrac{3}{4} + \dfrac{2}{5}$

14. $\dfrac{3x}{4} + \dfrac{2x}{5}$

15. $\dfrac{3}{4x} + \dfrac{2}{5x}$

16. $\dfrac{3}{4} - \dfrac{2}{3}$

17. $\dfrac{3x}{4} - \dfrac{2x}{3}$

18. $\dfrac{3}{4x} - \dfrac{2}{3x}$

19. $\dfrac{x}{2} + \dfrac{x+1}{3}$

20. $\dfrac{x-1}{3} + \dfrac{x+2}{4}$

21. $\dfrac{2x-1}{5} + \dfrac{x+3}{2}$

22. $\dfrac{x+1}{3} - \dfrac{2x+1}{4}$

23. $\dfrac{x-3}{3} - \dfrac{x-2}{5}$

24. $\dfrac{2x+1}{7} - \dfrac{x+2}{2}$

25. $\dfrac{1}{x} + \dfrac{2}{x+1}$

26. $\dfrac{3}{x-2} + \dfrac{4}{x}$

27. $\dfrac{5}{x-2} + \dfrac{3}{x+3}$

28. $\dfrac{7}{x+1} - \dfrac{3}{x+2}$

29. $\dfrac{2}{x+3} - \dfrac{5}{x-1}$

30. $\dfrac{3}{x-2} - \dfrac{4}{x+1}$

5.2 Changing the subject of a formula

The operations involved in solving ordinary linear equations are exactly the same as the operations required in changing the subject of a formula.

Example 1

(a) Solve the equation $3x + 1 = 12$.
(b) Make x the subject of the formula $Mx + B = A$.

(a) $3x + 1 = 12$
$$3x = 12 - 1$$
$$x = \frac{12 - 1}{3} = \frac{11}{3}$$

(b) $Mx + B = A$
$$Mx = A - B$$
$$x = \frac{A - B}{M}$$

Example 2

(a) Solve the equation $3(y - 2) = 5$.
(b) Make y the subject of the formula $x(y - a) = e$.

(a) $3(y - 2) = 5$
$$3y - 6 = 5$$
$$3y = 11$$
$$y = \frac{11}{3}$$

(b) $x(y - a) = e$
$$xy - xa = e$$
$$xy = e + xa$$
$$y = \frac{e + xa}{x}$$

Exercise 4

Make x the subject of the following:

1. $2x = 5$	**2.** $7x = 21$	**3.** $Ax = B$
4. $Nx = T$	**5.** $Mx = K$	**6.** $xy = 4$
7. $Bx = C$	**8.** $4x = D$	**9.** $9x = T + N$
10. $Ax = B - R$	**11.** $Cx = R + T$	**12.** $Lx = N - R^2$
13. $R - S^2 = Nx$	**14.** $x + 5 = 7$	**15.** $x + 10 = 3$
16. $x + A = T$	**17.** $x + B - S$	**18.** $N = x + D$
19. $M = x + B$	**20.** $L = x + D^2$	**21.** $N^2 + x = T$
22. $L + x = N + M$	**23.** $Z + x = R - S$	**24.** $x - 5 = 2$
25. $x - R = A$	**26.** $x - A = E$	**27.** $F = x - B$
28. $F^2 = x - B^2$	**29.** $x - D = A + B$	**30.** $x - E = A^2$

Make y the subject of the following:

31. $L = y - B$
32. $N = y - T$
33. $3y + 1 = 7$
34. $2y - 4 = 5$
35. $Ay + C = N$
36. $By + D = L$
37. $Dy + E = F$
38. $Ny - F = H$
39. $Yy - Z = T$
40. $Ry - L = B$
41. $Vy + m = Q$
42. $ty - m = n + a$
43. $qy + n = s - t$
44. $ny - s^2 = t$
45. $V^2 y + b = c$
46. $r = ny - 6$
47. $s = my + d$
48. $t = my - b$
49. $j = my + c$
50. $2(y + 1) = 6$
51. $3(y - 1) = 5$
52. $A(y + B) = C$
53. $D(y + E) = F$
54. $h(y + n) = a$
55. $b(y - d) = q$
56. $n = r(y + t)$
57. $t(y - 4) = b$
58. $z = S(y + t)$
59. $s = v(y - d)$
60. $g = m(y + n)$

Example 1

(a) Solve the equation $\dfrac{3a + 1}{2} = 4$.

(b) Make a the subject of the formula $\dfrac{na + b}{m} = n$.

(a) $\dfrac{3a + 1}{2} = 4$

$3a + 1 = 8$

$3a = 7$

$a = \dfrac{7}{3}$

(b) $\dfrac{na + b}{m} = n$

$na + b = mn$

$na = mn - b$

$a = \dfrac{mn - b}{n}$

Example 2

Make a the subject of the formula

$$x - na = y$$

Make the 'a' term positive

$$x = y + na$$
$$x - y = na$$
$$\dfrac{x - y}{n} = a$$

Exercise 5

Make a the subject.

1. $\dfrac{a}{4} = 3$
2. $\dfrac{a}{5} = 2$
3. $\dfrac{a}{D} = B$

4. $\dfrac{a}{B} = T$
5. $\dfrac{a}{N} = R$
6. $b = \dfrac{a}{m}$

7. $\dfrac{a - 2}{4} = 6$
8. $\dfrac{a - A}{B} = T$
9. $\dfrac{a - D}{N} = A$

10. $\dfrac{a + Q}{N} = B^2$
11. $g = \dfrac{a - r}{e}$
12. $\dfrac{2a + 1}{5} = 2$

13. $\dfrac{Aa + B}{C} = D$
14. $\dfrac{na + m}{p} = q$
15. $\dfrac{ra - t}{S} = v$

16. $\dfrac{za - m}{q} = t$ **17.** $\dfrac{m + Aa}{b} = c$ **18.** $A = \dfrac{Ba + D}{E}$

19. $n = \dfrac{ea - f}{h}$ **20.** $q = \dfrac{ga + b}{r}$ **21.** $6 - a = 2$

22. $7 - a = 9$ **23.** $5 = 7 - a$ **24.** $A - a = B$

25. $C - a = E$ **26.** $D - a = H$ **27.** $n - a = m$

28. $t = q - a$ **29.** $b = s - a$ **30.** $v = r - a$

31. $t = m - a$ **32.** $5 - 2a = 1$ **33.** $T - Xa = B$

34. $M - Na = Q$ **35.** $V - Ma = T$ **36.** $L = N - Ra$

37. $r = v^2 - ra$ **38.** $t^2 = w - na$ **39.** $n - qa = 2$

40. $\dfrac{3 - 4a}{2} = 1$ **41.** $\dfrac{5 - 7a}{3} = 2$ **42.** $\dfrac{B - Aa}{D} = E$

43. $\dfrac{D - Ea}{N} = B$ **44.** $\dfrac{h - fa}{b} = x$ **45.** $\dfrac{v^2 - ha}{C} = d$

46. $\dfrac{M(a + B)}{N} = T$ **47.** $\dfrac{f(Na - e)}{m} = B$ **48.** $\dfrac{T(M - a)}{E} = F$

49. $\dfrac{y(x - a)}{z} = t$ **50.** $\dfrac{k^2(m - a)}{x} = x$

Example 1

(a) Solve the equation $\dfrac{4}{z} = 7$.

(b) Make z the subject of the formula $\dfrac{n}{z} = k$.

(a) $\dfrac{4}{z} = 7$ (b) $\dfrac{n}{z} = k$

$\quad 4 = 7z$ $\quad n = kz$

$\quad \dfrac{4}{7} = z$ $\quad \dfrac{n}{k} = z$

Example 2

Make t the subject of the formula $\dfrac{x}{t} + m = a$.

$\qquad \dfrac{x}{t} = a - m$

$\qquad x = (a - m)t$

$\dfrac{x}{(a - m)} = t$

Exercise 6

Make a the subject.

1. $\dfrac{7}{a} = 14$ **2.** $\dfrac{5}{a} = 3$ **3.** $\dfrac{B}{a} = C$ **4.** $\dfrac{T}{a} = X$

5. $\dfrac{M}{a} = B$ **6.** $m = \dfrac{n}{a}$ **7.** $t = \dfrac{v}{a}$ **8.** $\dfrac{n}{a} = \sin 20°$

9. $\dfrac{7}{a} = \cos 30°$ **10.** $\dfrac{B}{a} = x$ **11.** $\dfrac{5}{a} = \dfrac{3}{4}$ **12.** $\dfrac{N}{a} = \dfrac{B}{D}$

13. $\dfrac{H}{a} = \dfrac{N}{M}$ **14.** $\dfrac{t}{a} = \dfrac{b}{e}$ **15.** $\dfrac{v}{a} = \dfrac{m}{s}$ **16.** $\dfrac{t}{b} = \dfrac{m}{a}$

17. $\dfrac{5}{a+1} = 2$ **18.** $\dfrac{7}{a-1} = 3$ **19.** $\dfrac{B}{a+D} = C$ **20.** $\dfrac{Q}{a-C} = T$

21. $\dfrac{V}{a-T} = D$ **22.** $\dfrac{L}{Ma} = B$ **23.** $\dfrac{N}{Ba} = C$ **24.** $\dfrac{m}{ca} = d$

25. $t = \dfrac{b}{c-a}$ **26.** $x = \dfrac{z}{y-a}$

Make x the subject.

27. $\dfrac{2}{x} + 1 = 3$ **28.** $\dfrac{5}{x} - 2 = 4$ **29.** $\dfrac{A}{x} + B = C$ **30.** $\dfrac{V}{x} + G = H$

31. $\dfrac{r}{x} - t = n$ **32.** $q = \dfrac{b}{x} + d$ **33.** $t = \dfrac{m}{x} - n$ **34.** $h = d - \dfrac{b}{x}$

35. $C - \dfrac{d}{x} = e$ **36.** $r - \dfrac{m}{x} = e^2$ **37.** $t^2 = b - \dfrac{n}{x}$ **38.** $\dfrac{d}{x} + b = mn$

39. $\dfrac{M}{x+q} - N = 0$ **40.** $\dfrac{Y}{x-c} - T = 0$ **41.** $3M = M + \dfrac{N}{P+x}$ **42.** $A = \dfrac{B}{c+x} - 5A$

43. $\dfrac{K}{Mx} + B = C$ **44.** $\dfrac{z}{xy} - z = y$ **45.** $\dfrac{m^2}{x} - n = -p$ **46.** $t = w - \dfrac{q}{x}$

Example

Make x the subject of the formulae.

(a) $\sqrt{(x^2 + A)} = B$

$\quad x^2 + A = B^2$ (square both sides)

$\quad\quad x^2 = B^2 - A$

$\quad\quad\ x = \pm\sqrt{(B^2 - A)}$

(b) $(Ax - B)^2 = M$

$\quad Ax - B = \pm\sqrt{M}$ (square root both sides)

$\quad\quad Ax = B \pm \sqrt{M}$

$\quad\quad\ x = \dfrac{B \pm \sqrt{M}}{A}$

(c) $\sqrt{(R - x)} = T$

$\quad R - x = T^2$

$\quad\quad R = T^2 + x$

$\quad R - T^2 = x$

Exercise 7

Make x the subject.

1. $\sqrt{x} = 2$ **2.** $\sqrt{(x+1)} = 5$ **3.** $\sqrt{(x-2)} = 3$ **4.** $\sqrt{(x+a)} = B$

5. $\sqrt{(x+C)} = D$ **6.** $\sqrt{(x-E)} = H$ **7.** $\sqrt{(ax+b)} = c$ **8.** $\sqrt{(x-m)} = a$

9. $b = \sqrt{(gx-t)}$ **10.** $r = \sqrt{(b-x)}$ **11.** $\sqrt{(d-x)} = t$ **12.** $b = \sqrt{(x-d)}$

13. $c = \sqrt{(n-x)}$ **14.** $f = \sqrt{(b-x)}$ **15.** $g = \sqrt{(c-x)}$ **16.** $\sqrt{(M - Nx)} = P$

17. $\sqrt{(Ax+B)} = \sqrt{D}$ **18.** $\sqrt{(x-D)} = A^2$ **19.** $x^2 = g$ **20.** $x^2 + 1 = 17$

21. $x^2 = B$

22. $x^2 + A = B$

23. $x^2 - A = M$

24. $b = a + x^2$

25. $C - x^2 = m$

26. $n = d - x^2$

27. $mx^2 = n$

28. $b = ax^2$

Make k the subject.

29. $\dfrac{kz}{a} = t$

30. $ak^2 - t = m$

31. $n = a - k^2$

32. $\sqrt{(k^2 - 4)} = 6$

33. $\sqrt{(k^2 - A)} = B$

34. $\sqrt{(k^2 + y)} = x$

35. $t = \sqrt{(m + k^2)}$

36. $2\sqrt{(k + 1)} = 6$

37. $A\sqrt{(k + B)} = M$

38. $\sqrt{\left(\dfrac{M}{k}\right)} = N$

39. $\sqrt{\left(\dfrac{N}{k}\right)} = B$

40. $\sqrt{(a - k)} = b$

41. $\sqrt{(a^2 - k^2)} = t$

42. $\sqrt{(m - k^2)} = x$

43. $2\pi\sqrt{(k + t)} = 4$

44. $A\sqrt{(k + 1)} = B$

45. $\sqrt{(ak^2 - b)} = C$

46. $a\sqrt{(k^2 - x)} = b$

47. $k^2 + b = x^2$

48. $\dfrac{k^2}{a} + b = c$

49. $\sqrt{(c^2 - ak)} = b$

50. $\dfrac{m}{k^2} = a + b$

Example

Make x the subject of the formulae.

(a) $Ax - B = Cx + D$
$\quad Ax - Cx = D + B$
$\quad x(A - C) = D + B$ (factorise)
$$x = \frac{D + B}{A - C}$$

(b) $x + a = \dfrac{x + b}{c}$
$\quad c(x + a) = x + b$
$\quad cx + ca = x + b$
$\quad cx - x = b - ca$
$\quad x(c - 1) = b - ca$ (factorise)
$$x = \frac{b - ca}{c - 1}$$

Exercise 8

Make y the subject.

1. $5(y - 1) = 2(y + 3)$

2. $7(y - 3) = 4(3 - y)$

3. $Ny + B = D - Ny$

4. $My - D = E - 2My$

5. $ay + b = 3b + by$

6. $my - c = e - ny$

7. $xy + 4 = 7 - ky$

8. $Ry + D = Ty + C$

9. $ay - x = z + by$

10. $m(y + a) = n(y + b)$

11. $x(y - b) = y + d$

12. $\dfrac{a - y}{a + y} = b$

13. $\dfrac{1 - y}{1 + y} = \dfrac{c}{d}$

14. $\dfrac{M - y}{M + y} = \dfrac{a}{b}$

15. $m(y + n) = n(n - y)$

16. $y + m = \dfrac{2y - 5}{m}$

17. $y \quad n = \dfrac{y + 2}{n}$

18. $y + b = \dfrac{ay + e}{b}$

19. $\dfrac{ay + x}{x} = 4 - y$

20. $c - dy = e - ay$

21. $y(a - c) = by + d$

22. $y(m + n) = a(y + b)$

23. $t - ay = s - by$

24. $\dfrac{y + x}{y - x} = 3$

25. $\dfrac{v-y}{v+y} = \dfrac{1}{2}$

26. $y(b-a) = a(y+b+c)$

27. $\sqrt{\left(\dfrac{y+x}{y-x}\right)} = 2$

28. $\sqrt{\left(\dfrac{z+y}{z-y}\right)} = \dfrac{1}{3}$

29. $\sqrt{\left[\dfrac{m(y+n)}{y}\right]} = p$

30. $n-y = \dfrac{4y-n}{m}$

Example

Make w the subject of the formula $\sqrt{\left(\dfrac{w}{w+a}\right)} = c$.

Squaring both sides, $\dfrac{w}{w+a} = c^2$

Multiplying by $(w+a)$, $w = c^2(w+a)$

$$w = c^2 w + c^2 a$$
$$w - c^2 w = c^2 a$$
$$w(1 - c^2) = c^2 a$$
$$w = \dfrac{c^2 a}{1 - c^2}$$

Exercise 9

Make the letter in square brackets the subject.

1. $ax + by + c = 0$ [x]

2. $\sqrt{\{a(y^2 - b)\}} = e$ [y]

3. $\dfrac{\sqrt{(k-m)}}{n} = \dfrac{1}{m}$ [k]

4. $a - bz = z + b$ [z]

5. $\dfrac{x+y}{x-y} = 2$ [x]

6. $\sqrt{\left(\dfrac{a}{z} - c\right)} = e$ [z]

7. $lm + mn + a = 0$ [n]

8. $t = 2\pi\sqrt{\left(\dfrac{d}{g}\right)}$ [d]

9. $t = 2\pi\sqrt{\left(\dfrac{d}{g}\right)}$ [g]

10. $\sqrt{(x^2 + a)} = 2x$ [x]

11. $\sqrt{\left\{\dfrac{b(m^2 + a)}{e}\right\}} = t$ [m]

12. $\sqrt{\left(\dfrac{x+1}{x}\right)} = a$ [x]

13. $a + b - mx = 0$ [m]

14. $\sqrt{(a^2 + b^2)} = x^2$ [a]

15. $\dfrac{a}{k} + b = \dfrac{c}{k}$ [k]

16. $a - y = \dfrac{b+y}{a}$ [y]

17. $G = 4\pi\sqrt{(x^2 + T^2)}$ [x]

18. $M(ax + by + c) = 0$ [y]

19. $x = \sqrt{\left(\dfrac{y-1}{y+1}\right)}$ [y]

20. $a\sqrt{\left(\dfrac{x^2 - n}{m}\right)} = \dfrac{a^2}{b}$ [x]

21. $\dfrac{M}{N} + E = \dfrac{P}{N}$ [N]

22. $\dfrac{Q}{P-x} = R$ [x]

23. $\sqrt{(z - ax)} = t$ [a]

24. $e + \sqrt{(x+f)} = g$ [x]

25. $\dfrac{m(ny - e^2)}{p} + n = 5n$ [y]

5.3 Variation

Direct variation

There are several ways of expressing a relationship between two
quantities x and y. Here are some examples.

> x varies as y
> x varies directly as y
> x is proportional to y

These three all mean the same and they are written in symbols as
follows.

> $x \propto y$

The '$\propto$' sign can always be replaced by '$= k$' where k is a constant:

> $x = ky$

Suppose $x = 3$ when $y = 12$;
then $3 = k \times 12$
and $k = \frac{1}{4}$
We can then write $x = \frac{1}{4}y$, and this allows us to find the value of x for
any value of y and vice versa.

Example 1

y varies as z, and $y = 2$ when $z = 5$; find
(a) the value of y when $z = 6$
(b) the value of z when $y = 5$

Because $y \propto z$, then $y = kz$ where k is a constant.

$$y = 2 \text{ when } z = 5$$
$$2 = k \times 5$$
$$k = \frac{2}{5}$$

So $y = \frac{2}{5}z$

(a) When $z = 6$, $y = \frac{2}{5} \times 6 = 2\frac{2}{5}$
(b) When $y = 5$, $5 = \frac{2}{5}z$

$$z = \frac{25}{2} = 12\frac{1}{2}$$

Example 2

The value V of a diamond is proportional to the square of its mass M.
If a diamond with a mass of 10 grams is worth \$200, find:
(a) the value of a diamond with a mass of 30 grams
(b) the mass of a diamond worth \$5000.

$$V \propto M^2$$

or $V = kM^2$ where k is a constant.

$$V = 200 \text{ when } M = 10$$
$\therefore$ $200 = k \times 10^2$
$$k = 2$$
So $V = 2M^2$

(a) When $M = 30$,
$$V = 2 \times 30^2 = 2 \times 900$$
$$V = \$1800$$
So a diamond with a mass of 30 grams is worth \$1800.

(b) When $V = 5000$,
$$5000 = 2 \times M^2$$
$$M^2 = \frac{5000}{2} = 2500$$
$$M = \sqrt{2500} = 50$$
So a diamond of value \$5000 has a mass of 50 grams.

Exercise 10

1. Rewrite the statement connecting each pair of variables using a constant k instead of '$\propto$'.

(a) $S \propto e$ (b) $v \propto t$ (c) $x \propto z^2$

(d) $y \propto \sqrt{x}$ (e) $T \propto \sqrt{L}$ (f) $C \propto r$

(g) $A \propto r^2$ (h) $V \propto r^3$

2. y varies as t. If $y = 6$ when $t = 4$, calculate:

(a) the value of y, when $t = 6$

(b) the value of t, when $y = 4$.

3. z is proportional to m. If $z = 20$ when $m = 4$, calculate:

(a) the value of z, when $m = 7$

(b) the value of m, when $z = 55$.

4. A varies directly as r^2. If $A = 12$, when $r = 2$, calculate:

(a) the value of A, when $r = 5$

(b) the value of r, when $A = 48$.

5. Given that $z \propto x$, copy and complete the table.

x	1	3		$5\frac{1}{2}$
z	4		16	

6. Given that $V \propto r^3$, copy and complete the table.

r	1	2		$1\frac{1}{2}$
V	4		256	

7. Given that $w \propto \sqrt{h}$, copy and complete the table.

h	4	9		$2\frac{1}{4}$
w	6		15	

8. s is proportional to $(v - 1)^2$. If $s = 8$, when $v = 3$, calculate:

(a) the value of s, when $v = 4$

(b) the value of v, when $s = 2$.

9. m varies as $(d + 3)$. If $m = 28$ when $d = 1$, calculate:
   (a) the value of m, when $d = 3$
   (b) the value of d, when $m = 49$.

10. The pressure of the water P at any point below the surface of the sea varies as the depth of the point below the surface d. If the pressure is 200 newtons/cm^2 at a depth of 3 m, calculate the pressure at a depth of 5 m.

11. The distance d through which a stone falls from rest is proportional to the square of the time taken t. If the stone falls 45 m in 3 seconds, how far will it fall in 6 seconds? How long will it take to fall 20 m?

12. The energy E stored in an elastic band varies as the square of the extension x. When the elastic is extended by 3 cm, the energy stored is 243 joules. What is the energy stored when the extension is 5 cm? What is the extension when the stored energy is 36 joules?

13. In the first few days of its life, the length of an earthworm l is thought to be proportional to the square root of the number of hours n which have elapsed since its birth. If a worm is 2 cm long after 1 hour, how long will it be after 4 hours? How long will it take to grow to a length of 14 cm?

14. The number of eggs which a goose lays in a week varies as the cube root of the average number of hours of sleep she has. When she has 8 hours sleep, she lays 4 eggs. How long does she sleep when she lays 5 eggs?

15. The resistance to motion of a car is proportional to the square of the speed of the car. If the resistance is 4000 newtons at a speed of 20 m/s, what is the resistance at a speed of 30 m/s?
    At what speed is the resistance 6250 newtons?

16. A road research organisation recently claimed that the damage to road surfaces was proportional to the fourth power of the axle load. The axle load of a 44-tonne HGV is about 15 times that of a car. Calculate the ratio of the damage to road surfaces made by a 44-tonne HGV and a car.

Inverse variation

There are several ways of expressing an inverse relationship between two variables,

   x varies inversely as y
   x is inversely proportional to y.

We write $x \propto \dfrac{1}{y}$ for both statements and proceed using the method outlined in the previous section.

Example

z is inversely proportional to t^2 and $z = 4$ when $t = 1$. Calculate:
(a) z when $t = 2$
(b) t when $z = 16$.

We have $z \propto \dfrac{1}{t^2}$

or $z = k \times \dfrac{1}{t^2}$ (k is a constant)

$z = 4$ when $t = 1$,

$\therefore \quad 4 = k \left(\dfrac{1}{1^2} \right)$

so $k = 4$

$\therefore \quad z = 4 \times \dfrac{1}{t^2}$

(a) when $t = 2$, $z = 4 \times \dfrac{1}{2^2} = 1$

(b) when $z = 16$, $16 = 4 \times \dfrac{1}{t^2}$

$$16t^2 = 4$$
$$t^2 = \tfrac{1}{4}$$
$$t = \pm \tfrac{1}{2}$$

Exercise 11

1. Rewrite the statements connecting the variables using a constant of variation, k.

(a) $x \propto \dfrac{1}{y}$ (b) $s \propto \dfrac{1}{t^2}$ (c) $t \propto \dfrac{1}{\sqrt{q}}$

(d) m varies inversely as w
(e) z is inversely proportional to t^2.

2. b varies inversely as e. If $b = 6$ when $e = 2$, calculate:
(a) the value of b when $e = 12$
(b) the value of e when $b = 3$.

3. q varies inversely as r. If $q = 5$ when $r = 2$, calculate:
(a) the value of q when $r = 4$
(b) the value of r when $q = 20$.

4. x is inversely proportional to y^2. If $x = 4$ when $y = 3$, calculate:
(a) the value of x when $y = 1$
(b) the value of y when $x = 2\frac{1}{4}$.

5. R varies inversely as v^2. If $R = 120$ when $v = 1$, calculate:
(a) the value of R when $v = 10$
(b) the value of v when $R = 30$.

6. T is inversely proportional to x^2. If $T = 36$ when $x = 2$, calculate:
(a) the value of T when $x = 3$
(b) the value of x when $T = 1 \cdot 44$.

7. p is inversely proportional to $\sqrt{y}$. If $p = 1\cdot2$ when $y = 100$, calculate:
(a) the value of p when $y = 4$
(b) the value of y when $p = 3$.

8. y varies inversely as z. If $y = \frac{1}{8}$ when $z = 4$, calculate:
(a) the value of y when $z = 1$
(b) the value of z when $y = 10$.

9. Given that $z \propto \dfrac{1}{y}$, copy and complete the table:

y	2	4		$\frac{1}{4}$
z	8		16	

10. Given that $v \propto \dfrac{1}{t^2}$, copy and complete the table:

t	2	5		10
v	25		$\frac{1}{4}$	

11. Given that $r \propto \dfrac{1}{\sqrt{x}}$, copy and complete the table:

x	1	4		
r	12		$\frac{3}{4}$	2

12. e varies inversely as $(y - 2)$. If $e = 12$ when $y = 4$, find
(a) e when $y = 6$ (b) y when $e = \frac{1}{2}$.

13. M is inversely proportional to the square of l.
If $M = 9$ when $l = 2$, find:
(a) M when $l = 10$ (b) l when $M = 1$.

14. Given $z = \dfrac{k}{x^n}$, find k and n, then copy and complete the table.

x	1	2	4	
z	100	$12\frac{1}{2}$		$\frac{1}{10}$

15. Given $y = \dfrac{k}{\sqrt[n]{v}}$, find k and n, then copy and complete the table

v	1	4	36	
y	12	6		$\frac{3}{25}$

16. The volume V of a given mass of gas varies inversely as the pressure P. When $V = 2\,\text{m}^3$, $P = 500\,\text{N/m}^2$. Find the volume when the pressure is $400\,\text{N/m}^2$. Find the pressure when the volume is $5\,\text{m}^3$.

17. The number of hours N required to dig a certain hole is inversely proportional to the number of men available x. When 6 men are digging, the hole takes 4 hours. Find the time taken when 8 men are available. If it takes $\frac{1}{2}$ hour to dig the hole, how many men are there?

18. The life expectancy L of a rat varies inversely as the square of the density d of poison distributed around his home. When the density of poison is $1\,\text{g/m}^2$ the life expectancy is 50 days. How long will he survive if the density of poison is:
    (a) $5\,\text{g/m}^2$?   (b) $\frac{1}{2}\,\text{g/m}^2$?

19. The force of attraction F between two magnets varies inversely as the square of the distance d between them. When the magnets are 2 cm apart, the force of attraction is 18 newtons. How far apart are they if the attractive force is 2 newtons?

5.4 Indices

Rules of indices

1. $a^n \times a^m = a^{n+m}$ e.g. $7^2 \times 7^4 = 7^6$
2. $a^n \div a^m = a^{n-m}$ e.g. $6^6 \div 6^2 = 6^4$
3. $(a^n)^m = a^{nm}$ e.g. $(3^2)^5 = 3^{10}$

Also, $a^{-n} = \dfrac{1}{a^n}$ e.g. $5^{-2} = \dfrac{1}{5^2}$

$a^{\frac{1}{n}}$ means 'the nth root of a'   e.g. $9^{\frac{1}{2}} = \sqrt[2]{9}$

$a^{\frac{m}{n}}$ means 'the nth root of a raised to the power m'
$$\text{e.g. } 4^{\frac{3}{2}} = (\sqrt{4})^3 = 8$$

Example

Simplify:
(a) $x^7 \times x^{13}$ (b) $x^3 \div x^7$
(c) $(x^4)^3$ (d) $(3x^2)^3$
(e) $(2x^{-1})^2 \div x^{-5}$ (f) $3y^2 \times 4y^3$

(a) $x^7 \times x^{13} = x^{7+13} = x^{20}$

(b) $x^3 \div x^7 = x^{3-7} = x^{-4} = \dfrac{1}{x^4}$

(c) $(x^4)^3 = x^{12}$

(d) $(3x^2)^3 = 3^3 \times (x^2)^3 = 27x^6$

(e) $(2x^{-1})^2 \div x^{-5} = 4x^{-2} \div x^{-5}$
$$= 4x^{(-2--5)}$$
$$= 4x^3$$

(f) $3y^2 \times 4y^3 = 12y^5$

Exercise 12

Express in index form:

1. $3 \times 3 \times 3 \times 3$

2. $4 \times 4 \times 5 \times 5 \times 5$

3. $3 \times 7 \times 7 \times 7$

4. $2 \times 2 \times 2 \times 7$

5. $\dfrac{1}{10 \times 10 \times 10}$

6. $\dfrac{1}{2 \times 2 \times 3 \times 3 \times 3}$

7. $\sqrt{15}$

8. $\sqrt[3]{3}$

9. $\sqrt[5]{10}$

10. $(\sqrt{5})^3$

Simplify:

11. $x^3 \times x^4$

12. $y^6 \times y^7$

13. $z^7 \div z^3$

14. $z^{50} \times z^{50}$

15. $m^3 \div m^2$

16. $e^{-3} \times e^{-2}$

17. $y^{-2} \times y^4$

18. $w^4 \div w^{-2}$

19. $y^{\frac{1}{2}} \times y^{\frac{1}{2}}$

20. $(x^2)^5$

21. $x^{-2} \div x^{-2}$

22. $w^{-3} \times w^{-2}$

23. $w^{-7} \times w^2$

24. $x^3 \div x^{-4}$

25. $(a^2)^4$

26. $(k^{\frac{1}{2}})^6$

27. $e^{-4} \times e^4$

28. $x^{-1} \times x^{30}$

29. $(y^4)^{\frac{1}{2}}$

30. $(x^{-3})^{-2}$

31. $z^2 \div z^{-2}$

32. $t^{-3} \div t$

33. $(2x^3)^2$

34. $(4y^5)^2$

35. $2x^2 \times 3x^2$

36. $5y^3 \times 2y^2$

37. $5a^3 \times 3a$

38. $(2a)^3$

39. $3x^3 \div x^3$

40. $8y^3 \div 2y$

41. $10y^2 \div 4y$

42. $8a \times 4a^3$

43. $(2x)^2 \times (3x)^3$

44. $4z^4 \times z^{-7}$

45. $6x^{-2} \div 3x^2$

46. $5y^3 \div 2y^{-2}$

47. $(x^2)^{\frac{1}{2}} \div (x^{\frac{1}{3}})^3$

48. $7w^{-2} \times 3w^{-1}$

49. $(2n)^4 \div 8n^0$

50. $4x^{\frac{3}{2}} \div 2x^{\frac{1}{2}}$

Example

Evaluate:

(a) $9^{\frac{1}{2}}$

(b) 5^{-1}

(c) $4^{-\frac{1}{2}}$

(d) $25^{\frac{3}{2}}$

(e) $(5^{\frac{1}{2}})^3 \times 5^{\frac{1}{2}}$

(f) 7^0

$\qquad$ (a) $9^{\frac{1}{2}} = \sqrt{9} = 3$

$\qquad$ (b) $5^{-1} = \frac{1}{5}$

$\qquad$ (c) $4^{-\frac{1}{2}} = \dfrac{1}{4^{\frac{1}{2}}} = \dfrac{1}{\sqrt{4}} = \dfrac{1}{2}$

$\qquad$ (d) $25^{\frac{3}{2}} = (\sqrt{25})^3 = 5^3 = 125$

$\qquad$ (e) $(5^{\frac{1}{2}})^3 \times 5^{\frac{1}{2}} = 5^{\frac{3}{2}} \times 5^{\frac{1}{2}} = 5^2$

$\qquad\qquad\qquad\qquad = 25$

$\qquad$ (f) $7^0 = 1 \left[\text{consider} \dfrac{7^3}{7^3} = 7^{3-3} = 7^0 = 1\right]$

> **Remember**
> $a^0 = 1$ for any non-zero value of a.

Exercise 13

Evaluate the following:

1. $3^2 \times 3$

2. 100^0

3. 3^{-2}

4. $(5^{-1})^{-2}$

5. $4^{\frac{1}{2}}$

6. $16^{\frac{1}{2}}$

7. $81^{\frac{1}{2}}$

8. $8^{\frac{1}{3}}$

9. $9^{\frac{3}{2}}$

10. $27^{\frac{1}{3}}$

11. $9^{-\frac{1}{2}}$

12. $8^{-\frac{1}{3}}$

13. $1^{\frac{5}{2}}$

14. $25^{-\frac{1}{2}}$

15. $1000^{\frac{1}{3}}$

16. $2^{-2} \times 2^5$

17. $2^4 \div 2^{-1}$

18. $8^{\frac{2}{3}}$

19. $27^{-\frac{2}{3}}$

20. $4^{-\frac{3}{2}}$

21. $36^{\frac{1}{2}} \times 27^{\frac{1}{3}}$ **22.** $10\,000^{\frac{1}{4}}$ **23.** $100^{\frac{3}{2}}$ **24.** $(100^{\frac{1}{2}})^{-3}$

25. $(9^{\frac{1}{2}})^{-2}$ **26.** $(-16 \cdot 371)^{0}$ **27.** $81^{\frac{1}{4}} \div 16^{\frac{1}{4}}$ **28.** $(5^{-4})^{\frac{1}{2}}$

29. $1000^{-\frac{1}{3}}$ **30.** $(4^{-\frac{1}{2}})^{2}$ **31.** $8^{-\frac{2}{3}}$ **32.** $100^{\frac{5}{2}}$

33. $1^{\frac{4}{3}}$ **34.** 2^{-5} **35.** $(0 \cdot 01)^{\frac{1}{2}}$ **36.** $(0 \cdot 04)^{\frac{1}{2}}$

37. $(2 \cdot 25)^{\frac{1}{2}}$ **38.** $(7 \cdot 63)^{0}$ **39.** $3^{5} \times 3^{-3}$ **40.** $(3\frac{3}{8})^{\frac{1}{3}}$

41. $(11\frac{1}{9})^{-\frac{1}{2}}$ **42.** $(\frac{1}{8})^{-2}$ **43.** $(\frac{1}{1000})^{\frac{2}{3}}$ **44.** $(\frac{9}{25})^{-\frac{1}{2}}$

45. $(10^{-6})^{\frac{1}{3}}$ **46.** $7^{2} \div (7^{\frac{1}{2}})^{4}$ **47.** $(0 \cdot 0001)^{-\frac{1}{2}}$ **48.** $\dfrac{9^{\frac{1}{2}}}{4^{-\frac{1}{2}}}$

49. $\dfrac{25^{\frac{3}{2}} \times 4^{\frac{1}{2}}}{9^{-\frac{1}{2}}}$ **50.** $(-\frac{1}{7})^{2} \div (-\frac{1}{7})^{3}$

Example

Simplify:

(a) $(2a)^{3} \div (9a^{2})^{\frac{1}{2}}$ (b) $(3ac^{2})^{3} \times 2a^{-2}$ (c) $(2x)^{2} \div 2x^{2}$

$$(a)\ (2a)^{3} \div (9a^{2})^{\frac{1}{2}} = 8a^{3} \div 3a$$
$$= \tfrac{8}{3}a^{2}$$
$$(b)\ (3ac^{2})^{3} \times 2a^{-2} = 27a^{3}c^{6} \times 2a^{-2}$$
$$= 54ac^{6}$$
$$(c)\ (2x)^{2} \div 2x^{2} = 4x^{2} \div 2x^{2}$$
$$= 2$$

Exercise 14

Rewrite without brackets:

1. $(5x^{2})^{2}$ **2.** $(7y^{3})^{2}$ **3.** $(10ab)^{2}$ **4.** $(2xy^{2})^{2}$

5. $(4x^{2})^{\frac{1}{2}}$ **6.** $(9y)^{-1}$ **7.** $(x^{-2})^{-1}$ **8.** $(2x^{-2})^{-1}$

9. $(5x^{2}y)^{0}$ **10.** $(\frac{1}{2}x)^{-1}$ **11.** $(3x)^{2} \times (2x)^{2}$ **12.** $(5y)^{2} \div y$

13. $(2x^{\frac{1}{2}})^{4}$ **14.** $(3y^{\frac{1}{3}})^{3}$ **15.** $(5x^{0})^{2}$ **16.** $[(5x)^{0}]^{2}$

17. $(7y^{0})^{2}$ **18.** $[(7y)^{0}]^{2}$ **19.** $(2x^{2}y)^{3}$ **20.** $(10xy^{3})^{2}$

Simplify the following:

21. $(3x^{-1})^{2} \div 6x^{-3}$ **22.** $(4x)^{\frac{1}{2}} \div x^{\frac{3}{2}}$ **23.** $x^{2}y^{2} \times xy^{3}$ **24.** $4xy \times 3x^{2}y$

25. $10x^{-1}y^{3} \times xy$ **26.** $(3x)^{2} \times (\frac{1}{9}x^{2})^{\frac{1}{2}}$ **27.** $z^{3}yx \times x^{2}yz$ **28.** $(2x)^{-2} \times 4x^{3}$

29. $(3y)^{-1} \div (9y^{2})^{-1}$ **30.** $(xy)^{0} \times (9x)^{\frac{3}{2}}$ **31.** $(x^{2}y)(2xy)(5y^{3})$ **32.** $(4x^{\frac{1}{2}}) \times (8x^{\frac{3}{2}})$

33. $5x^{-3} \div 2x^{-5}$ **34.** $[(3x^{-1})^{-2}]^{-1}$ **35.** $(2a)^{-2} \times 8a^{4}$ **36.** $(abc^{2})^{3}$

37. Write in the form 2^{p} (e.g. $4 = 2^{2}$):
   (a) 32   (b) 128   (c) 64   (d) 1

38. Write in the form 3^{q}:
   (a) $\frac{1}{27}$ (b) $\frac{1}{81}$ (c) $\frac{1}{3}$ (d) $9 \times \frac{1}{81}$

Evaluate, with $x = 16$ and $y = 8$.

39. $2x^{\frac{1}{2}} \times y^{\frac{1}{3}}$ **40.** $x^{\frac{1}{4}} \times y^{-1}$ **41.** $(y^{2})^{\frac{1}{6}} \div (9x)^{\frac{1}{2}}$ **42.** $(x^{2}y^{3})^{0}$

43. $x + y^{-1}$ **44.** $x^{-\frac{1}{2}} + y^{-1}$ **45.** $y^{\frac{1}{3}} \div x^{\frac{3}{4}}$ **46.** $(1000y)^{\frac{1}{3}} \times x^{-\frac{5}{2}}$

47. $(x^{\frac{1}{4}} + y^{-1}) \div x^{\frac{1}{4}}$ **48.** $x^{\frac{1}{2}} - y^{\frac{2}{3}}$ **49.** $(x^{\frac{3}{4}}y)^{-\frac{1}{3}}$ **50.** $\left(\dfrac{x}{y}\right)^{-2}$

Solve the equations for x.

51. $2^x = 8$

52. $3^x = 81$

53. $5^x = \frac{1}{5}$

54. $10^x = \frac{1}{100}$

55. $3^{-x} = \frac{1}{27}$

56. $4^x = 64$

57. $6^{-x} = \frac{1}{6}$

58. $100\,000^x = 10$

59. $12^x = 1$

60. $10^x = 0.0001$

61. $2^x + 3^x = 13$

62. $(\frac{1}{2})^x = 32$

63. $5^{2x} = 25$

64. $1\,000\,000^{3x} = 10$

65. These two are more difficult. Use a calculator to find solutions correct to three significant figures.

(a) $x^x = 100$

(b) $x^x = 10\,000$

5.5 Inequalities

$x < 4$ means 'x is *less than* 4'

$y > 7$ means 'y is *greater than* 7'

$z \leqslant 10$ means 'z is *less than or equal to* 10'

$t \geqslant -3$ means 't is *greater than or equal to* -3'

Solving inequalities

We follow the same procedure used for solving equations except that when we multiply or divide by a *negative* number the inequality is *reversed*.

e.g. $4 > -2$

but multiplying by -2,

$\quad -8 < 4$

Example

Solve the inequalities:

(a) $2x - 1 > 5$

$\quad 2x > 5 + 1$

$\quad x > \dfrac{6}{2}$

$\quad x > 3$

(b) $5 - 3x \leqslant 1$

$\quad 5 \leqslant 1 + 3x$

$\quad 5 - 1 \leqslant 3x$

$\quad \dfrac{4}{3} \leqslant x$

Exercise 15

Introduce one of the symbols $<$, $>$ or $=$ between each pair of numbers.

1. $-2, 1$

2. $(-2)^2, 1$

3. $\frac{1}{4}, \frac{1}{5}$

4. $0.2, \frac{1}{5}$

5. $10^2, 2^{10}$

6. $\frac{1}{4}, 0.4$

7. $40\%, 0.4$

8. $(-1)^2, (-\frac{1}{2})^2$

9. $5^2, 2^5$

10. $3\frac{1}{3}, \sqrt{10}$

11. $\pi^2, 10$

12. $-\frac{1}{3}, -\frac{1}{2}$

13. $2^{-1}, 3^{-1}$

14. $50\%, \frac{1}{5}$

15. $1\%, 100^{-1}$

State whether the following are true or false:

16. $0 \cdot 7^2 > \frac{1}{2}$ **17.** $10^3 = 30$ **18.** $\frac{1}{8} > 12\%$

19. $(0 \cdot 1)^3 = 0 \cdot 0001$ **20.** $(-\frac{1}{5})^0 = -1$ **21.** $\dfrac{1}{5^2} > \dfrac{1}{2^5}$

22. $(0 \cdot 2)^3 < (0 \cdot 3)^2$ **23.** $\frac{6}{7} > \frac{7}{8}$ **24.** $0 \cdot 1^2 > 0 \cdot 1$

Solve the following inequalities:

25. $x - 3 > 10$ **26.** $x + 1 < 0$ **27.** $5 > x - 7$ **28.** $2x + 1 \leqslant 6$
29. $3x - 4 > 5$ **30.** $10 \leqslant 2x - 6$ **31.** $5x < x + 1$ **32.** $2x \geqslant x - 3$
33. $4 + x < -4$ **34.** $3x + 1 < 2x + 5$ **35.** $2(x + 1) > x - 7$ **36.** $7 < 15 - x$
37. $9 > 12 - x$ **38.** $4 - 2x \leqslant 2$ **39.** $3(x - 1) < 2(1 - x)$ **40.** $7 - 3x < 0$

The number line

The inequality $x < 4$ is represented on the number line as

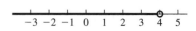

$x \geqslant -2$ is shown as

In the first case, 4 is *not* included so we have ∘.
In the second case, -2 *is* included so we have •.

$-1 \leqslant x < 3$ is shown as

Exercise 16

For questions **1** to **25**, solve each inequality and show the result on a
number line.

1. $2x + 1 > 11$ **2.** $3x - 4 \leqslant 5$ **3.** $2 < x - 4$
4. $6 \geqslant 10 - x$ **5.** $8 < 9 - x$ **6.** $8x - 1 < 5x - 10$
7. $2x > 0$ **8.** $1 < 3x - 11$ **9.** $4 - x > 6 - 2x$

10. $\dfrac{x}{3} < -1$ **11.** $1 < x < 4$ **12.** $-2 \leqslant x \leqslant 5$

13. $1 \leqslant x < 6$ **14.** $0 \leqslant 2x < 10$ **15.** $-3 \leqslant 3x \leqslant 21$

16. $1 < 5x < 10$ **17.** $\dfrac{x}{4} > 20$ **18.** $3x - 1 > x + 19$

19. $7(x + 2) < 3x + 4$ **20.** $1 < 2x + 1 < 9$ **21.** $10 \leqslant 2x \leqslant x + 9$
22. $x < 3x + 2 < 2x + 6$ **23.** $10 \leqslant 2x - 1 \leqslant x + 5$ **24.** $x < 3x - 1 < 2x + 7$
25. $x - 10 < 2(x - 1) < x$
(Hint: in questions **20** to **25**, solve the two inequalities separately.)

For questions **26** to **35**, find the solutions, subject to the given
condition.

26. $3a + 1 < 20$; a is a positive integer
27. $b - 1 \geqslant 6$; b is a prime number less than 20
28. $2e - 3 < 21$; e is a positive even number

29. $1 < z < 50$; z is a square number
30. $0 < 3x < 40$; x is divisible by 5
31. $2x > -10$; x is a negative integer
32. $x + 1 < 2x < x + 13$; x is an integer
33. $x^2 < 100$; x is a positive square number
34. $0 \leqslant 2z - 3 \leqslant z + 8$; z is a prime number
35. $\dfrac{a}{2} + 10 > a$; a is a positive even number
36. State the smallest integer n for which $4n > 19$.
37. Find an integer value of x such that $2x - 7 < 8 < 3x - 11$.
38. Find an integer value of y such that $3y - 4 < 12 < 4y - 5$.
39. Find any value of z such that $9 < z + 5 < 10$.
40. Find any value of p such that $9 < 2p + 1 < 11$.
41. Find a simple fraction q such that $\frac{4}{9} < q < \frac{5}{9}$.
42. Find an integer value of a such that $a - 3 \leqslant 11 \leqslant 2a + 10$.
43. State the largest prime number z for which $3z < 66$.
44. Find a simple fraction r such that $\frac{1}{3} < r < \frac{2}{3}$.
45. Find the largest prime number p such that $p^2 < 400$.
46. Illustrate on a number line the solution set of each pair of simultaneous inequalities:
    (a) $x < 6$; $-3 \leqslant x \leqslant 8$ (b) $x > -2$; $-4 < x < 2$
    (c) $2x + 1 \leqslant 5$; $-12 \leqslant 3x - 3$ (d) $3x - 2 < 19$; $2x \geqslant -6$
47. Find the integer n such that $n < \sqrt{300} < n + 1$.

Graphical display

It is useful to represent inequalities on a graph, particularly where two variables are involved.

> Drawing accurate graphs is explained in Unit 7.

Example

Draw a sketch graph and leave unshaded the area which represents the set of points that satisfy each of these inequalities:
(a) $x > 2$ (b) $1 \leqslant y \leqslant 5$ (c) $x + y \leqslant 8$

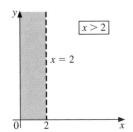   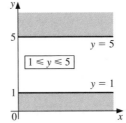   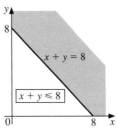

In each graph, the unwanted region is shaded so that the region representing the set of points is left clearly visible.

In (a), the line $x = 2$ is shown as a broken line to indicate that the points on the line are *not* included.

In (b) and (c), the lines $y = 1$, $y = 5$ and $x + y = 8$ are shown as solid lines because points on the line *are* included in the solution set.

An inequality can thus be regarded as a set of points, for example, the unshaded region in (c) may be described as

$$\{(x, y): x + y \leqslant 8\}$$

i.e. the set of points (x, y) such that $x + y \leqslant 8$.

Exercise 17

In questions **1** to **9** describe the region left unshaded.

1.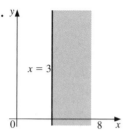

2.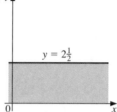

3.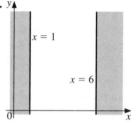

4.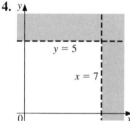

5.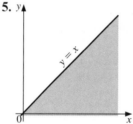

6.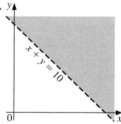

7.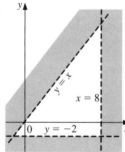

8.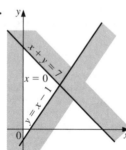

9.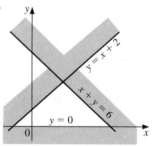

For questions **10** to **27**, draw a sketch graph similar to those above and indicate the set of points which satisfy the inequalities by shading the unwanted regions.

10. $2 \leqslant x \leqslant 7$

11. $0 \leqslant y \leqslant 3\frac{1}{2}$

12. $-2 < x < 2$

13. $x < 6$ and $y \leqslant 4$

14. $0 < x < 5$ and $y < 3$

15. $1 \leqslant x \leqslant 6$ and $2 \leqslant y \leqslant 8$

16. $-3 < x < 0$ and $-4 < y < 2$

17. $y \leqslant x$

18. $x + y < 5$

19. $y > x + 2$ and $y < 7$

20. $x \geqslant 0$ and $y \geqslant 0$ and $x + y \leqslant 7$

21. $x \geqslant 0$ and $x + y < 10$ and $y > x$

22. $8 \geqslant y \geqslant 0$ and $x + y > 3$

23. $x + 2y < 10$ and $x \geqslant 0$ and $y \geqslant 0$

24. $3x + 2y \leqslant 18$ and $x \geqslant 0$ and $y \geqslant 0$

25. $x \geqslant 0$, $y \geqslant x - 2$, $x + y \leqslant 10$

26. $3x + 5y \leqslant 30$ and $y > \dfrac{x}{2}$

27. $y \geqslant \dfrac{x}{2}$, $y \leqslant 2x$ and $x + y \leqslant 8$

5.6 Linear programming

In most linear programming problems, there are two stages:
1. to interpret the information given as a series of simultaneous inequalities and display them graphically.
2. to investigate some characteristic of the points in the unshaded solution set.

Example

A shopkeeper buys two types of cat food for his shop: Bruno at 40c a tin and Blaze at 60c a tin. He has $15 available and decides to buy at least 30 tins altogether. He also decides that at least one third of the tins should be Blaze. He buys x tins of Bruno and y tins of Blaze.

(a) Write down three inequalities which correspond to the above conditions.
(b) Illustrate these inequalities on a graph.
(c) He makes a profit of 10c a tin on Bruno and a profit of 20c a tin on Blaze. Assuming he can sell all his stock, find how many tins of each type he should buy to maximise his profit and find that profit.

(a) Cost $40x + 60y \leqslant 1500$ $2x + 3y \leqslant 75 \ldots$[line A on graph]
   Total number $x + y \geqslant 30 \ldots$[line B on graph]

   At least one third Blaze $\dfrac{y}{x} \geqslant \dfrac{1}{2}$ $2y \geqslant x \ldots$[line C on graph]

(b) The graph below shows these three equations.

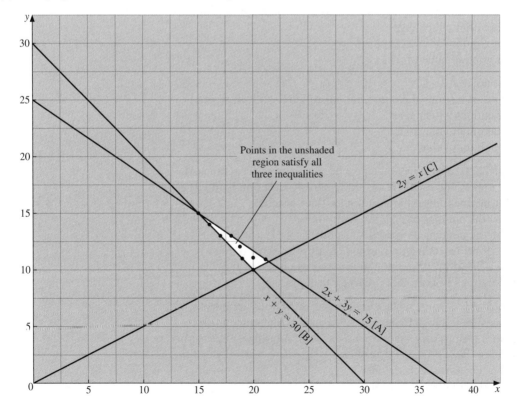

(c) The table below shows the points on the graph in the unshaded region together with the corresponding figure for the profit.

The points marked * will clearly not provide a maximum profit.

x	15	16	17*	18	19	19*	20*	20*	21
y	15	14	13	13	12	11	11	10	11

profit	150	160		180	190				210
	$+300$	$+280$		$+260$	$+240$				$+220$
	$\overline{450c}$	$\overline{440c}$		$\overline{440c}$	$\overline{430c}$				$\overline{430c}$

Conclusion: he should buy 15 tins of Bruno and 15 tins of Blaze. His maximised profit is then 450c.

Exercise 18

For questions **1** to **3**, draw an accurate graph to represent the inequalities listed, using shading to show the unwanted regions.

1. $x + y \leqslant 11$; $y \geqslant 3$; $y \leqslant x$.
Find the point having whole number coordinates and satisfying these inequalities which gives:
(a) the maximum value of $x + 4y$
(b) the minimum value of $3x + y$

2. $3x + 2y > 24$; $x + y < 12$; $y < \frac{1}{2}x$; $y > 1$.
Find the point having whole number coordinates and satisfying these inequalities which gives:
(a) the maximum value of $2x + 3y$
(b) the minimum value of $x + y$

3. $3x + 2y \leqslant 60$; $x + 2y \leqslant 30$; $x \geqslant 10$; $y \geqslant 0$.
Find the point having whole number coordinates and satisfying these inequalities which gives:
(a) the maximum value of $2x + y$
(b) the maximum value of xy

4. Kojo is given $1·20 to buy some peaches and apples. Peaches cost 20c each, apples 10c each. He is told to buy at least 6 individual fruits, but he must not buy more apples than peaches.
  Let x be the number of peaches Kojo buys.
  Let y be the number of apples Kojo buys.
(a) Write down three inequalities which must be satisfied.
(b) Draw a linear programming graph and use it to list the combinations of fruit that are open to Kojo.

5. Ulla is told to buy some melons and oranges. Melons are 50c each and oranges 25c each, and she has $2 to spend. She must not buy more than 2 melons and she must buy at least 4 oranges. She is also told to buy at least 6 fruits altogether.
  Let x be the number of melons.
  Let y be the number of oranges.

(a) Write down four inequalities which must be satisfied.

(b) Draw a graph and use it to list the combinations of fruit that are open to Ulla.

6. A chef is going to make some fruit cakes and sponge cakes. He has plenty of all ingredients except for flour and sugar. He has only 2000 g of flour and 1200 g of sugar.

A fruit cake uses 500 g of flour and 100 g of sugar.

A sponge cake uses 200 g of flour and 200 g of sugar.

He wishes to make *more than* 4 cakes altogether.

Let the number of fruit cakes be x.

Let the number of sponge cakes be y.

(a) Write down three inequalities which must be satisfied.

(b) Draw a graph and use it to list the possible combinations of fruit cakes and sponge cakes which he can make.

7. Kwame has a spare time job spraying cars and vans. Vans take 2 hours each and cars take 1 hour each. He has 14 hours available per week. He has an agreement with one firm to do 2 of their vans every week. Apart from that he has no fixed work.

Kwame's permission to use his back garden contains the clause that he must do at least twice as many cars as vans.

Let x be the number of vans sprayed each week.

Let y be the number of cars sprayed each week.

(a) Write down three inequalities which must be satisfied.

(b) Draw a graph and use it to list the possible combinations of vehicles which Kwame can spray each week.

8. The manager of a football team has \$100 000 to spend on buying new players. He can buy defenders at \$6000 each or forwards at \$8000 each. There must be at least 6 of each sort. To cover for injuries he must buy at least 13 players altogether. Let x represent the number of defenders he buys and y the number of forwards.

(a) In what ways can he buy players?

(b) If the wages are \$10 000 per week for each defender and \$20 000 per week for each forward, what is the combination of players which has the lowest wage bill?

9. A tennis-playing golfer has €15 to spend on golf balls (x) costing €1 each and tennis balls (y) costing 60c each. He must buy at least 16 altogether and he must buy *more* golf balls than tennis balls.

(a) What is the greatest number of balls he can buy?

(b) After using them, he can sell golf balls for 10c each and tennis balls for 20c each. What is his maximum possible income from sales?

10. A travel agent has to fly 1000 people and 35 000 kg of baggage from Hong Kong to Shanghai. Two types of aircraft are available: A which takes 100 people and 2000 kg of baggage, or B which takes 60 people and 3000 kg of baggage. He can use no more than 16 aircraft altogether. Write down three inequalities which must be satisfied if he uses x of A and y of B.

(a) What is the smallest number of aircraft he could use?
(b) If the hire charge for each aircraft A is $10 000 and for each aircraft B is $12 000, find the cheapest option available to him.
(c) If the hire charges are altered so that each A costs $10 000 and each B costs $20 000, find the cheapest option now available to him.

11. A farmer has to transport 20 people and 32 sheep to a market. He can use either Fiats (x) which take 2 people and 1 sheep, or Rolls Royces (y) which take 2 people and 4 sheep.
He must not use more than 15 cars altogether.
(a) What is the lowest total numbers of cars he could use?
(b) If it costs £10 to hire each Fiat and £30 for each Rolls Royce, what is the *cheapest* solution?

12. Ahmed wishes to buy up to 20 notebooks for his shop. He can buy either type A for $1.50 each or type B for $3 each. He has a total of $45 he can spend. He must have at least 6 of each type in stock.
(a) If he buys x of type A and y of type B, write down four inequalities which must be satisfied and represent the information on a graph.
(b) If he makes a profit of 40c on each of type A and $1 on each of type B, how many of each should he buy for maximum profit?
(c) If the profit is 80c on each of type A and $1 on each of type B, how many of each should he buy now?

13. A farmer needs to buy up to 25 cows for a new herd. He can buy either brown cows (x) at $50 each or black cows ($y$) at $80 each and he can spend a total of no more than $1600. He must have at least 9 of each type.
On selling the cows he makes a profit of $50 on each brown cow and $60 on each black cow. How many of each sort should he buy for maximum profit?

14. The manager of a car park allows $10 \, m^2$ of parking space for each car and $30 \, m^2$ for each lorry. The total space available is $300 \, m^2$. He decides that the maximum number of vehicles at any time must not exceed 20 and he also insists that there must be at least as many cars as lorries. If the number of cars is x and the number of lorries is y, write down three inequalities which must be satisfied.
(a) If the parking charge is $1 for a car and $5 for a lorry, find how many vehicles of each kind he should admit to maximise his income.
(b) If the charges are changed to $2 for a car and $3 for a lorry, find how many of each kind he would be advised to admit.

Revision exercise 5A

1. Express the following as single fractions:

(a) $\dfrac{x}{4} + \dfrac{x}{5}$

(b) $\dfrac{1}{2x} + \dfrac{2}{3x}$

(c) $\dfrac{x+2}{2} + \dfrac{x-4}{3}$

(d) $\dfrac{7}{x-1} - \dfrac{2}{x+3}$

2. (a) Factorise $x^2 - 4$

(b) Simplify $\dfrac{3x - 6}{x^2 - 4}$

3. Given that $s - 3t = rt$, express:
(a) s in terms of r and t (b) r in terms of s and t
(c) t in terms of s and r.

4. (a) Given that $x - z = 5y$, express z in terms of x and y.
(b) Given that $mk + 3m = 11$, express m in terms of k.
(c) For the formula $T = C\sqrt{z}$, express z in terms of T and C.

5. It is given that $y = \dfrac{k}{x}$ and that $1 \leqslant x \leqslant 10$.
(a) If the smallest possible value of y is 5, find the value of the constant k.
(b) Find the largest possible value of y.

6. Given that y varies as x^2 and that $y = 36$ when $x = 3$, find:
(a) the value of y when $x = 2$ (b) the value of x when $y = 64$.

7. (a) Evaluate: (i) $9^{\frac{1}{2}}$ (ii) $8^{\frac{2}{3}}$ (iii) $16^{-\frac{1}{2}}$
(b) Find x, given that
    (i) $3^x = 81$ (ii) $7^x = 1$.

8. List the integer values of x which satisfy.
(a) $2x - 1 < 20 < 3x - 5$ (b) $5 < 3x + 1 < 17$.

9. Given that $t = k\sqrt{(x + 5)}$, express x in terms of t and k.

10. Given that $z = \dfrac{3y + 2}{y - 1}$, express y in terms of z.

11. Given that $y = \dfrac{k}{k + w}$
(a) Find the value of y when $k = \frac{1}{2}$ and $w = \frac{1}{3}$
(b) Express w in terms of y and k.

12. On a suitable sketch graph, identify clearly the region A defined by
$x > 0$, $x + y \leqslant 8$ and $y \geqslant x$.

13. Without using a calculator, calculate the value of:
(a) $9^{-\frac{1}{2}} + (\frac{1}{8})^{\frac{1}{3}} + (-3)^0$ (b) $(1000)^{-\frac{1}{3}} - (0\cdot1)^2$

14. It is given that $10^x = 3$ and $10^y = 7$. What is the value of 10^{x+y}?

15. Make x the subject of the following formulae:

(a) $x + a = \dfrac{2x - 5}{a}$ (b) $cz + ax + b = 0$ (c) $a = \sqrt{\left(\dfrac{x+1}{x-1}\right)}$

16. Write the following as single fractions:

(a) $\dfrac{3}{x} + \dfrac{1}{2x}$ (b) $\dfrac{3}{a-2} + \dfrac{1}{a^2 - 4}$ (c) $\dfrac{3}{x(x+1)} - \dfrac{2}{x(x-2)}$

17. p varies jointly as the square of t and inversely as s. Given that $p = 5$ when $t = 1$ and $s = 2$, find a formula for p in terms of t and s.

18. A positive integer r is such that $pr^2 = 168$, where p lies between 3 and 5. List the possible values of r.

19. The shaded region A is formed by the lines $y = 2$, $y = 3x$ and $x + y = 6$. Write down the three inequalities which define A.

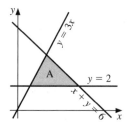

20. The shaded region B is formed by the lines $x = 0$, $y = x - 2$ and $x + y = 7$.

Write down the four inequalities which define B.

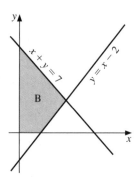

21. In the diagram, the solution set $-1 \leqslant x < 2$ is shown on a number line.

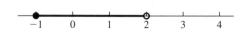

Illustrate, on similar diagrams, the solution set of the following pairs of simultaneous inequalities.
(a) $x > 2$, $x \leqslant 7$ (b) $4 + x \geqslant 2$, $x + 4 < 10$
(c) $2x + 1 \geqslant 3$, $x - 3 \leqslant 3$.

22. In a laboratory we start with 2 cells in a dish. The number of cells in the dish doubles every 30 minutes.
(a) How many cells are in the dish after four hours?
(b) After what time are there 2^{13} cells in the dish?
(c) After $10\frac{1}{2}$ hours there are 2^{22} cells in the dish and an experimental fluid is added which eliminates half of the cells. How many cells are left?

Examination exercise 5B

1. Write as a fraction in its simplest form

$$\frac{x-3}{4} + \frac{4}{x-3}.$$

[3]
Cambridge IGCSE Mathematics 0580
Paper 2 Q10 June 2007

2. Write as a single fraction in its simplest form

$$\frac{4}{2x+3} - \frac{2}{x-3}.$$

[3]
Cambridge IGCSE Mathematics 0580
Paper 21 Q11 November 2008

3. $(0.8)^{\frac{1}{2}}, \qquad 0.8, \qquad \sqrt{0.8}, \qquad (0.8)^{-1}, \qquad (0.8)^2.$
From the numbers above, write down
(a) the smallest,   [1]
(b) the largest.   [1]
Cambridge IGCSE Mathematics 0580
Paper 2 Q6 November 2005

4. Simplify $(27x^3)^{\frac{2}{3}}$.

[2]
Cambridge IGCSE Mathematics 0580
Paper 21 Q8 June 2008

5. Find the value of n in each of the following statements.
(a) $32^n = 1$ [1]
(b) $32^n = 2$ [1]
(c) $32^n = 8$ [1]
Cambridge IGCSE Mathematics 0580
Paper 2 Q7 November 2006

6. (a) Simplify $(27x^6)^{\frac{1}{3}}$.   [2]

(b) $(512)^{-\frac{2}{3}} = 2^p$. Find p.   [2]
Cambridge IGCSE Mathematics 0580
Paper 2 Q21 November 2007

7. Write $\dfrac{1}{c} + \dfrac{1}{d} - \dfrac{c-d}{cd}$ as a single fraction in its

simplest form.   [3]
Cambridge IGCSE Mathematics 0580
Paper 21 Q10 June 2009

8. Solve the inequality

$$\frac{2x-5}{8} > \frac{x+4}{3}.$$

[3]
Cambridge IGCSE Mathematics 0580
Paper 22 Q13 June 2008

9. Rearrange the formula to make y the subject

$$x + \frac{\sqrt{y}}{8} = 1.$$

[3]

Cambridge IGCSE Mathematics 0580
Paper 22 Q9 June 2009

10. (a) Factorise $ax^2 + bx^2$. [1]
(b) Make x the subject of the formula [2]
$ax^2 + bx^2 - d^2 = p^2$.

Cambridge IGCSE Mathematics 0580
Paper 22 Q8 November 2008

11. The quantity p varies inversely as the square of $(q + 2)$.
$p = 5$ when $q = 3$.
Find p when $q = 8$.

[3]

Cambridge IGCSE Mathematics 0580
Paper 21 Q13 November 2008

12. A spray can is used to paint a wall.
The thickness of the paint on the wall is t. The distance of the spray
can from the wall is d.
t is inversely proportional to the square of d.
$t = 0.2$ when $d = 8$. [3]
Find t when $d = 10$.

Cambridge IGCSE Mathematics 0580
Paper 21 Q13 June 2009

13. The length, y, of a solid is inversely proportional to the square of its
height, x.
(a) Write down a general equation for x and y.
Show that when $x = 5$ and $y = 4.8$ the equation becomes
$x^2 y = 120$. [2]
(b) Find y when $x = 2$. [1]
(c) Find x when $y = 10$. [2]
(d) Find x when $y = x$. [2]
(e) Describe exactly what happens to y when x is doubled. [2]
(f) Describe exactly what happens to x when y is decreased by 36%. [2]
(g) Make x the subject of the formula $x^2 y = 120$. [2]

Cambridge IGCSE Mathematics 0580
Paper 4 Q5 June 2006

14. Answer the whole of this question on a sheet of graph paper.
Tiago does some work during the school holidays.
In one week he spends x hours cleaning cars and y hours repairing
cycles.
The time he spends repairing cycles is at least equal to the time he
spends cleaning cars.
This can be written as $y \geqslant x$.

He spends no more than 12 hours working.
He spends at least 4 hours cleaning cars.

(a) Write down two more inequalities in x and/or y to show this
   information. [3]
(b) Draw x and y axes from 0 to 12, using a scale of 1 cm to
   represent 1 unit on each axis. [1]
(c) Draw three lines to show the three inequalities. Shade the
   unwanted regions. [5]
(d) Tiago receives $3 each hour for cleaning cars and $1.50 each
   hour for repairing cycles.
   (i) What is the least amount he could receive? [2]
   (ii) What is the largest amount he could receive? [2]

Cambridge IGCSE Mathematics 0580
Paper 4 Q9 November 2006

15. (a) The surface area, A, of a cylinder, radius r and height h, is given
   by the formula.

   $A = 2\pi rh + 2\pi r^2$.

   (i) Calculate the surface area of a cylinder of radius 5 cm and
      height 9 cm. [2]
   (ii) Make h the subject of the formula. [2]
   (iii) A cylinder has a radius of 6 cm and a surface area of
      377 cm². Calculate the height of this cylinder. [2]
   (iv) A cylinder has a surface area of 1200 cm² and its radius and
      height are equal. Calculate the radius. [3]

   (b) (i) On Monday a shop receives $60.30 by selling bottles of
      water at 45 cents each. How many bottles are sold? [1]
   (ii) On Tuesday the shop receives x **cents** by selling bottles of
      water at 45 cents each. In terms of x, how many bottles are
      sold? [1]
   (iii) On Wednesday the shop receives $(x - 75)$ **cents** by selling
      bottles of water at 48 cents each. In terms of x, how many
      bottles are sold? [1]
   (iv) The number of bottles sold on Tuesday was 7 more than the
      number of bottles sold on Wednesday.
      Write down an equation in x and solve your equation. [4]

Cambridge IGCSE Mathematics 0580
Paper 4 Q8 November 2006

6 TRIGONOMETRY

Leonard Euler (1707–1783) was born near Basel in Switzerland but moved to St Petersburg in Russia and later to Berlin. He had an amazing facility for figures but delighted in speculating in the realms of pure intellect. In trigonometry he introduced the use of small letters for the sides and capitals for the angles of a triangle. He also wrote r, R and s for the radius of the inscribed and of the circumscribed circles and the semi-perimeter, giving the beautiful formula $4rRs = abc$.

32 Interpret and use three-figure bearings; apply the sine, cosine and tangent ratios for acute angles; solve trigonometrical problems in two dimensions involving angles of elevation and depression; extend sine and cosine values to angles between 90° and 180°; solve problems using the sine and cosine rules; solve simple trigonometrical problems in three dimensions

6.1 Right-angled triangles

The side opposite the right angle is called the hypotenuse (we will use H). It is the longest side.

The side opposite the marked angle of 35° is called the opposite (we will use O).

The other side is called the adjacent (we will use A).

Consider two triangles, one of which is an enlargement of the other.

It is clear that the *ratio* $\dfrac{O}{H}$ will be the same in

both triangles.

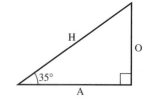

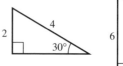

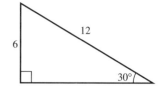

Sine, cosine and tangent

Three important functions are defined as follows:

$$\sin x = \frac{O}{H}$$

$$\cos x = \frac{A}{H}$$

$$\tan x = \frac{O}{A}$$

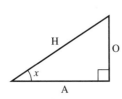

It is important to get the letters in the right order. Some people find a simple sentence helpful when the first letters of each word describe sine, cosine or tangent and Hypotenuse, Opposite and Adjacent. An example is:

Silly Old Harry Caught A Herring Trawling Off Afghanistan.

e.g. $S\,O\,H : \sin = \dfrac{O}{H}$

For any angle x the values for $\sin x$, $\cos x$ and $\tan x$ can be found using either a calculator or tables.

Exercise 1

1. Draw a circle of radius 10 cm and construct a tangent to touch the circle at T.
   Draw OA, OB and OC where $A\widehat{O}T = 20°$
   $$B\widehat{O}T = 40°$$
   $$C\widehat{O}T = 50°$$
   Measure the length AT and compare it with the value for tan 20° given on a calculator or in tables. Repeat for BT, CT and for other angles of your own choice.

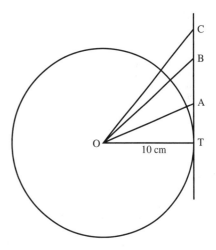

Finding the length of a side

Example 1

Find the side marked x.

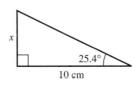

(a) Label the sides of the triangle H, O, A (in brackets).

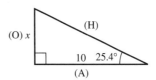

(b) In this example, we know nothing about H so we need the function involving O and A.

$$\tan 25{\cdot}4° = \frac{O}{A} = \frac{x}{10}$$

(c) Find $\tan 25{\cdot}4°$ from tables.

$$0{\cdot}4748 = \frac{x}{10}$$

(d) Solve for x.
$$x = 10 \times 0{\cdot}4748 = 4{\cdot}748$$
$$x = 4{\cdot}75 \text{ cm (3 significant figures)}$$

Example 2

Find the side marked z.

(a) Label H, O, A.

(b) $\sin 31{\cdot}3° = \dfrac{O}{H} = \dfrac{7{\cdot}4}{z}$

(c) Multiply by z.
$$z \times (\sin 31{\cdot}3°) = 7{\cdot}4$$
$$z = \frac{7{\cdot}4}{\sin 31{\cdot}3}$$

(d) On a calculator, press the keys as follows:

| 7.4 | ÷ | 31·3 | sin | = |

$$z = 14{\cdot}2 \text{ cm (to 3 s.f.)}$$

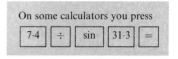

On some calculators you press

| 7·4 | ÷ | sin | 31·3 | = |

Exercise 2

In questions **1** to **22** all lengths are in centimetres. Find the sides marked with letters. Give your answers to three significant figures.

1.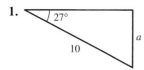

2.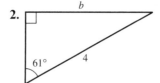

3.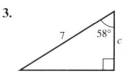

4.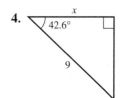

5.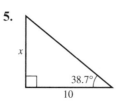

6.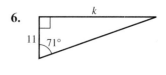

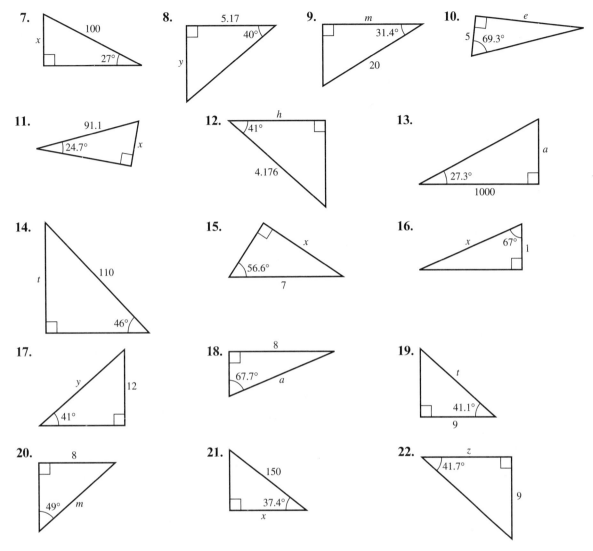

In questions **23** to **34**, the triangle has a right angle at the middle letter.

23. In $\triangle ABC$, $\widehat{C} = 40°$, $BC = 4\,cm$. Find AB.

24. In $\triangle DEF$, $\widehat{F} = 35 \cdot 3°$, $DF = 7\,cm$. Find ED.

25. In $\triangle GHI$, $\widehat{I} = 70°$, $GI = 12\,m$. Find HI.

26. In $\triangle JKL$, $\widehat{L} = 55°$, $KL = 8 \cdot 21\,m$. Find JK.

27. In $\triangle MNO$, $\widehat{M} = 42 \cdot 6°$, $MO = 14\,cm$. Find ON.

28. In $\triangle PQR$, $\widehat{P} = 28°$, $PQ = 5 \cdot 071\,m$. Find PR.

29. In $\triangle STU$, $\widehat{S} = 39°$, $TU = 6\,cm$. Find SU.

30. In $\triangle VWX$, $\widehat{X} = 17°$, $WV = 30 \cdot 7\,m$. Find WX.

31. In $\triangle ABC$, $\widehat{A} = 14 \cdot 3°$, $BC = 14\,m$. Find AC.

32. In $\triangle KLM$, $\widehat{K} = 72 \cdot 8°$, $KL = 5 \cdot 04\,cm$. Find LM.

33. In $\triangle PQR$, $\widehat{R} = 31 \cdot 7°$, $QR = 0 \cdot 81\,cm$. Find PR.

34. In $\triangle XYZ$, $\widehat{X} = 81 \cdot 07°$, $YZ = 52 \cdot 6\,m$. Find XY.

Example

Find the length marked x.

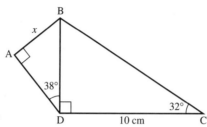

(a) Find BD from triangle BDC.

$$\tan 32° = \frac{BD}{10}$$

∴ $\qquad$ BD $= 10 \times \tan 32°$ $\qquad\qquad\qquad$ ...[1]

(b) Now find x from triangle ABD.

$$\sin 38° = \frac{x}{BD}$$

∴ $\qquad x = BD \times \sin 38°$
$\qquad x = 10 \times \tan 32° \times \sin 38°$ (from [1])
$\qquad x = 3·85$ cm (to 3 s.f.)

Notice that BD was *not* calculated in [1].
It is better to do all the multiplications at one time.

Exercise 3

In questions **1** to **10**, find each side marked with a letter.
All lengths are in centimetres.

1.

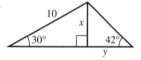

2.

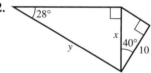

3.

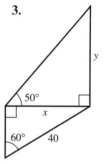

4.

5.

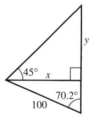

6.

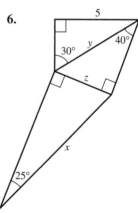

7.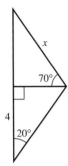

8.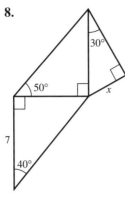

9.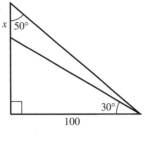

10.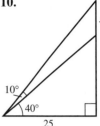

11. $\widehat{BAD} = \widehat{ACD} = 90°$
$\widehat{CAD} = 35°$
$\widehat{BDA} = 41°$
$AD = 20\,cm$
Calculate:
(a) AB
(b) DC
(c) BD

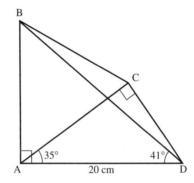

12. $\widehat{ABD} = \widehat{ADC} = 90°$
$\widehat{CAD} = 31°$
$\widehat{BDA} = 43°$
$AD = 10\,cm$
Calculate:
(a) AB
(b) CD
(c) DB

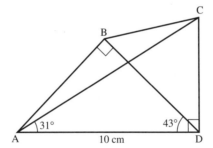

Finding an unknown angle

Example

Find the angle marked m.

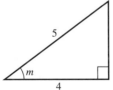

(a) Label the sides of the triangle H, O, Λ in
   relation to angle m.

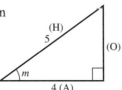

(b) In this example, we do not know 'O' so we need the cosine.

$$\cos m = \left(\frac{A}{H}\right) = \frac{4}{5}$$

(c) Change $\frac{4}{5}$ to a decimal: $\frac{4}{5} = 0\cdot 8$

(d) $\cos m = 0\cdot 8$

Find angle m from the cosine table: $m = 36\cdot 9°$

Note: On a calculator, angles can be found as follows:

If $\cos m = \frac{4}{5}$

(a) Press $\boxed{4}$ $\boxed{\div}$ $\boxed{5}$ $\boxed{=}$

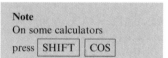

Note
On some calculators

press $\boxed{\text{SHIFT}}$ $\boxed{\text{COS}}$

(b) Press $\boxed{\text{INV}}$ and then $\boxed{\text{COS}}$

This will give the angle as $36\cdot 86989765°$. We require the angle to 1 place of decimals so $m = 36\cdot 9°$.

Exercise 4

In questions **1** to **15**, find the angle marked with a letter. All lengths are in cm.

1.

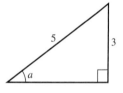

2.

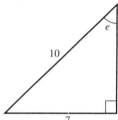

3.

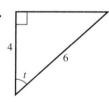

4.

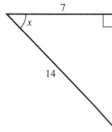

5.

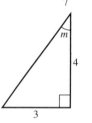

6.

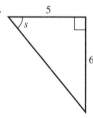

7.

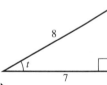

8.

9.

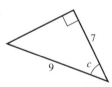

10.

11.

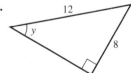

12.

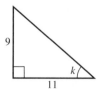

13.

14.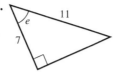

15.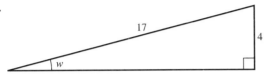

In questions **16** to **20**, the triangle has a right angle at the middle letter.

16. In $\triangle$ABC, BC = 4, AC = 7. Find $\widehat{A}$.

17. In $\triangle$DEF, EF = 5, DF = 10. Find $\widehat{F}$.

18. In $\triangle$GHI, GH = 9, HI = 10. Find $\widehat{I}$.

19. In $\triangle$JKL, JL = 5, KL = 3. Find $\widehat{J}$.

20. In $\triangle$MNO, MN = 4, NO = 5. Find $\widehat{M}$.

In questions **21** to **26**, find the angle x.

21.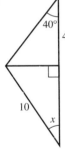

22.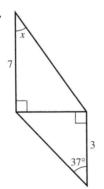

23.

24.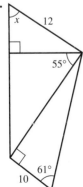

25.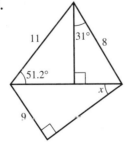

26.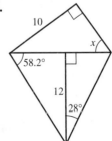

Bearings

A bearing is an angle measured clockwise from North.
It is given using three digits.
In the diagram:
the bearing of B from A is 052°
the bearing of A from B is 232°.

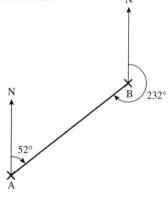

Example

A ship sails 22 km from A on a bearing of 042°, and a further 30 km on a bearing of 090° to arrive at B. What is the distance and bearing of B from A?

(a) Draw a clear diagram and label extra points as shown.

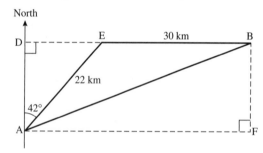

(b) Find DE and AD.

   (i) $\sin 42° = \dfrac{DE}{22}$

      $\therefore$ $DE = 22 \times \sin 42° = 14\text{·}72\,\text{km}$

   (ii) $\cos 42° = \dfrac{AD}{22}$

      $\therefore$ $AD = 22 \times \cos 42° = 16\text{·}35\,\text{km}$

(c) Using triangle ABF,
   $AB^2 = AF^2 + BF^2$ (Pythagoras' theorem)
   and $AF = DE + EB$
          $AF = 14\text{·}72 + 30 = 44\text{·}72\,\text{km}$
   and $BF = AD = 16\text{·}35\,\text{km}$

   $\therefore$ $AB^2 = 44\text{·}72^2 + 16\text{·}35^2$
           $= 2267\text{·}2$

      $AB = 47\text{·}6\,\text{km}$ (to 3 s.f.)

(d) The bearing of B from A is given by the angle DAB.
But $D\hat{A}B = A\hat{B}F$.

$$\tan A\hat{B}F = \frac{AF}{BF} = \frac{44\cdot72}{16\cdot35}$$
$$= 2\cdot7352$$

$\therefore \qquad A\hat{B}F = 69\cdot9°$

B is 47·6 km from A on a bearing of 069·9°.

Exercise 5

In this exercise, start by drawing a clear diagram.

1. A ladder of length 6 m leans against a vertical wall so that the base of the ladder is 2 m from the wall. Calculate the angle between the ladder and the wall.

2. A ladder of length 8 m rests against a wall so that the angle between the ladder and the wall is 31°. How far is the base of the ladder from the wall?

3. A ship sails 35 km on a bearing of 042°.
   (a) How far north has it travelled?
   (b) How far east has it travelled?

4. A ship sails 200 km on a bearing of 243·7°.
   (a) How far south has it travelled?
   (b) How far west has it travelled?

5. Find TR if PR = 10 m and QT = 7 m.

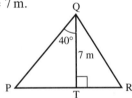

6. Find d.

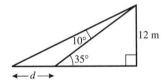

7. An aircraft flies 400 km from a point O on a bearing of 025° and then 700 km on a bearing of 080° to arrive at B.
   (a) How far north of O is B?
   (b) How far east of O is B?
   (c) Find the distance and bearing of B from O.

8. An aircraft flies 500 km on a bearing of 100° and then 600 km on a bearing of 160°.
   Find the distance and bearing of the finishing point from the starting point.

For questions **9** to **12**, plot the points for each question on a sketch graph with *x*- and *y*-axes drawn to the same scale.

9. For the points A(5, 0) and B(7, 3), calculate the angle between AB and the *x*-axis.

10. For the points C(0, 2) and D(5, 9), calculate the angle between CD and the *y*-axis.

11. For the points A(3, 0), B(5, 2) and C(7, −2), calculate the angle BAC.

12. For the points P(2, 5), Q(5, 1) and R(0, −3), calculate the angle PQR.

13. From the top of a tower of height 75 m, a man sees two goats, both due west of him. If the angles of depression of the two goats are 10° and 17°, calculate the distance between them.

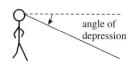

angle of depression

14. An isosceles triangle has sides of length 8 cm, 8 cm and 5 cm. Find the angle between the two equal sides.

15. The angles of an isosceles triangle are 66°, 66° and 48°. If the shortest side of the triangle is 8·4 cm, find the length of one of the two equal sides.

16. A chord of length 12 cm subtends an angle of 78·2° at the centre of a circle. Find the radius of the circle.

17. Find the acute angle between the diagonals of a rectangle whose sides are 5 cm and 7 cm.

18. A kite flying at a height of 55 m is attached to a string which makes an angle of 55° with the horizontal. What is the length of the string?

19. A boy is flying a kite from a string of length 150 m. If the string is taut and makes an angle of 67° with the horizontal, what is the height of the kite?

20. A rocket flies 10 km vertically, then 20 km at an angle of 15° to the vertical and finally 60 km at an angle of 26° to the vertical. Calculate the vertical height of the rocket at the end of the third stage.

21. Find *x*, given
AD = BC = 6 m.

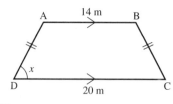

22. Find *x*.

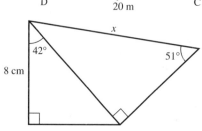

23. Ants can hear each other up to a range of 2 m. An ant at A, 1 m from a wall sees her friend at B about to be eaten by a spider. If the angle of elevation of B from A is 62°, will the spider have a meal or not? (Assume B escapes if he hears A calling.)

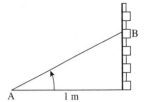

24. A hedgehog wishes to cross a road without being run over. He observes the angle of elevation of a lamp post on the other side of the road to be 27° from the edge of the road and 15° from a point 10 m back from the road. How wide is the road? If he can run at 1 m/s, how long will he take to cross?
If cars are travelling at 20 m/s, how far apart must they be if he is to survive?

25. From a point 10 m from a vertical wall, the angles of elevation of the bottom and the top of a statue of Sir Isaac Newton, set in the wall, are 40° and 52°. Calculate the height of the statue.

6.2 Scale drawing

On a scale drawing you must always state the scale you use.

Exercise 6

Make a scale drawing and then answer the questions.

1. A field has four sides as shown below:

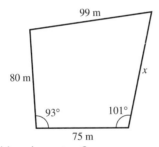

How long is the side x in metres?

2. Two ships leave a port at the same time. The first ship sails at 38 knots on a bearing of 042° and the second ship sails at 25 knots on a bearing of 315°. How far apart are the ships two hours later?
[1 knot is a speed of 1 nautical mile per hour.]

3. Two radar stations A and B are 80 km apart and B is due east of A. One aircraft is on a bearing of 030° from A and 346° from B. A second aircraft is on a bearing of 325° from A and 293° from B. How far apart are the two aircraft?

4. A ship sails 95 km on a bearing of 140°, then a further 102 km on a bearing of 260° and then returns directly to its starting point. Find the length and bearing of the return journey.

5. A control tower observes the flight of an aircraft.
   At 09:23 the aircraft is 580 km away on a bearing of 043°.
   At 09:25 the aircraft is 360 km away on a bearing of 016°.
   What is the speed and the course of the aircraft?
   [Use a scale of 1 cm to 50 km.]

6. Make a scale drawing of the diagram and find
   the length of CD in km.

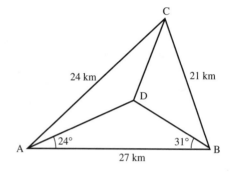

6.3 Three-dimensional problems

Always draw a large, clear diagram. It is often helpful to redraw the
triangle which contains the length or angle to be found.

Example

A rectangular box with top WXYZ and base
ABCD has AB = 6 cm, BC = 8 cm and WA = 3 cm.
Calculate:
(a) the length of AC
(b) the angle between WC and AC.

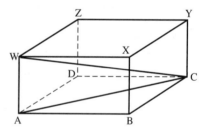

(a) Redraw triangle ABC.

   $AC^2 = 6^2 + 8^2 = 100$
   $AC = 10$ cm

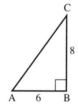

(b) Redraw triangle WAC.

   Let $W\widehat{C}A = \theta$
   $\tan \theta = \frac{3}{10}$
   $\theta = 16.7°$
   The angle between WC and AC is 16·7°.

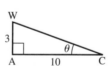

Exercise 7

1. In the rectangular box shown, find:
   (a) AC
   (b) AR
   (c) the angle between AC and AR.

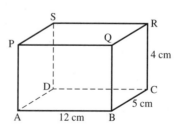

2. A vertical pole BP stands at one corner of a horizontal rectangular field as shown.

If AB = 10 m, AD = 5 m and the angle of elevation of P from A is 22°, calculate:

(a) the height of the pole

(b) the angle of elevation of P from C

(c) the length of a diagonal of the rectangle ABCD

(d) the angle of elevation of P from D.

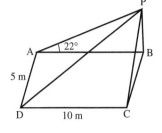

3. In the cube shown, find:

(a) BD

(b) AS

(c) BS

(d) the angle SBD

(e) the angle ASB

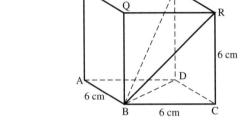

4. In the cuboid shown, find:

(a) WY

(b) DY

(c) WD

(d) the angle WDY

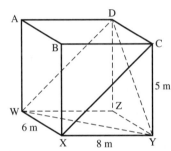

5. In the square-based pyramid, V is vertically above the middle of the base, AB = 10 cm and VC = 20 cm. Find:

(a) AC

(b) the height of the pyramid

(c) the angle between VC and the base ABCD

(d) the angle AVB

(e) the angle AVC

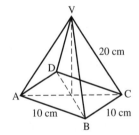

6. In the wedge shown, PQRS is perpendicular to ABRQ; PQRS and ABRQ are rectangles with AB = QR = 6 m, BR = 4 m, RS = 2 m. Find:

(a) BS   (b) AS

(c) angle BSR   (d) angle ASR

(e) angle PAS

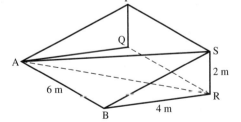

7. The edges of a box are 4 cm, 6 cm and 8 cm. Find the length of a diagonal and the angle it makes with the diagonal on the largest face.

8. In the diagram A, B and O are points in a horizontal plane and P is vertically above O, where OP = h m.

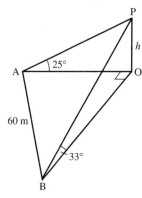

A is due West of O, B is due South of O and AB = 60 m. The angle of elevation of P from A is 25° and the angle of elevation of P from B is 33°.

(a) Find the length AO in terms of h.

(b) Find the length BO in terms of h.

(c) Find the value of h.

9. The angle of elevation of the top of a tower is 38° from a point A due south of it. The angle of elevation of the top of the tower from another point B, due east of the tower is 29°. Find the height of the tower if the distance AB is 50 m.

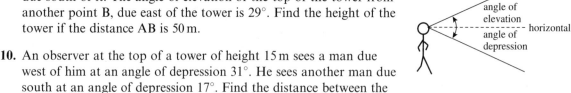

10. An observer at the top of a tower of height 15 m sees a man due west of him at an angle of depression 31°. He sees another man due south at an angle of depression 17°. Find the distance between the men.

11. The angle of elevation of the top of a tower is 27° from a point A due east of it. The angle of elevation of the top of the tower is 11° from another point B due south of the tower. Find the height of the tower if the distance AB is 40 m.

12. The figure shows a triangular pyramid on a horizontal base ABC, V is vertically above B where VB = 10 cm, $A\widehat{B}C = 90°$ and AB = BC = 15 cm. Point M is the mid-point of AC.

Calculate the size of angle VMB.

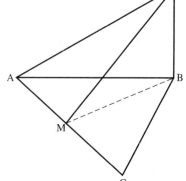

6.4 Sine, cosine, tangent for any angle

So far we have used sine, cosine and tangent only in right-angled triangles. For angles greater than 90°, we will see that there is a close connection between trigonometric ratios and circles.

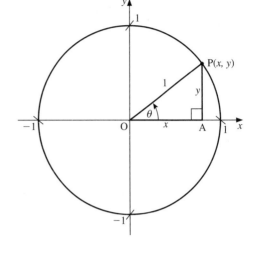

The circle on the right is of radius 1 unit with centre (0, 0). A point P with coordinates (x, y) moves round the circumference of the circle. The angle that OP makes with the positive x-axis as it turns in an anticlockwise direction is θ.

In triangle OAP, $\cos \theta = \dfrac{x}{1}$ and $\sin \theta = \dfrac{y}{1}$

The x-coordinate of P is $\cos \theta$.
The y-coordinate of P is $\sin \theta$.
This idea is used to define the cosine and the sine of any angle, including angles greater than 90°.
Here are two angles that are greater than 90°.

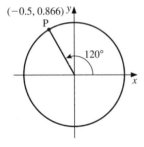

$$\cos 120° = -0.5$$
$$\sin 120° = 0.866$$

Note: Only sine and cosine of angles to 180° will be assessed in the IGCSE examination.

A graphics calculator can be used to show the graph of $y = \sin x$ for any range of angles. The graph on the right shows $y = \sin x$ for x from 0° to 180°. The curve above the x-axis has symmetry about $x = 90°$.

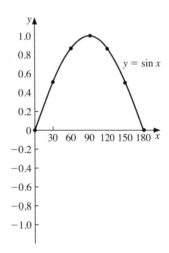

Note:
$\sin 150° = \sin 30°$ and $\cos 150° = -\cos 30°$
$\sin 110° = \sin 70°$ $\cos 110° = -\cos 70°$
$\sin 163° = \sin 17°$ $\cos 163° = -\cos 17°$

or $\sin x = \sin (180° - x)$
or $\cos x = -\cos (180° - x)$

These two results are particularly important for use with obtuse angles ($90° < x < 180°$) in Sections 6.5 and 6.6 when applying the sine formula or the cosine formula.

Exercise 8

1. (a) Use a calculator to find the cosine of all the angles 0°, 30°, 60°, 90°, 120°, ..., 180°.
   (b) Draw a graph of $y = \cos x$ for $0 \leqslant x \leqslant 180°$. Use a scale of 1 cm to 30° on the x-axis and 5 cm to 1 unit on the y-axis.

2. Draw the graph of $y = \sin x$, using the same angles and scales as in question **1**.

In questions **3** to **11** do not use a calculator. Use the symmetry of the graphs $y = \sin x$ and $y = \cos x$.
Angles are given to the nearest degree.

3. If $\sin 18° = 0.309$, give another angle whose sine is 0.309.

4. If $\sin 27° = 0.454$, give another angle whose sine is 0.454.

5. Give another angle which has the same sine as:
   (a) 40°   (b) 70°   (c) 130°

6. If $\cos 70° = 0.342$, give an angle whose cosine is -0.342.

7. If $\cos 45° = 0.707$, give an angle whose cosine is -0.707.

8. If $\sin 20° = 0.342$, what other angle has a sine of 0.342?

9. If $\sin 98° = 0.990$, give another angle whose sine is 0.990.

10. Find two values for x, between 0° and 180°, if $\sin x = 0.848$. Give each angle to the nearest degree.

11. If $\sin x = 0.35$, find two solutions for x between 0° and 180°.

12. Find *two* solutions of the equation
    $(\cos x)^2 = \dfrac{1}{4}$ for x between 0° and 180°.

13. Draw the graph of $y = 2\sin x + 1$ for $0 \leqslant x \leqslant 180°$, taking 1 cm to 10° for x and 5 cm to 1 unit for y. Find approximate solutions to the equations:
    (a) $2\sin x + 1 = 2.3$
    (b) $\dfrac{1}{(2\sin x + 1)} = 0.5$

14. Draw the graph of $y = 2\sin x + \cos x$ for $0 \leqslant x \leqslant 180°$, taking 1 cm to 10° for x and 5 cm to 1 unit for y.
    (a) Solve approximately the equations:
        (i) $2\sin x + \cos x = 1.5$
        (ii) $2\sin x + \cos x = 0$
    (b) Estimate the maximum value of y.
    (c) Find the value of x at which the maximum occurs.

6.5 The sine rule

The sine rule enables us to calculate sides and angles in some triangles where there is not a right angle.

In $\triangle$ABC, we use the convention that
  a is the side opposite $\widehat{A}$
  b is the side opposite $\widehat{B}$, etc.

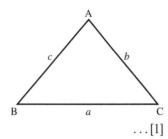

Either $\dfrac{a}{\sin A} = \dfrac{b}{\sin B} = \dfrac{c}{\sin C}$ ... [1]

or $\dfrac{\sin A}{a} = \dfrac{\sin B}{b} = \dfrac{\sin C}{c}$ ... [2]

Use [1] when finding a *side*,
and [2] when finding an *angle*.

Example 1
Find c.

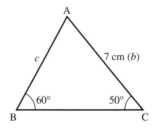

$$\dfrac{c}{\sin C} = \dfrac{b}{\sin B}$$

$$\dfrac{c}{\sin 50°} = \dfrac{7}{\sin 60°}$$

$$c = \dfrac{7 \times \sin 50°}{\sin 60°} = 6 \cdot 19 \, \text{cm} \quad (3 \text{ s.f.})$$

Although we cannot have an angle of more than 90° in a right-angled triangle, it is still useful to define sine, cosine and tangent for these angles.
For an obtuse angle x,
we have $\sin x = \sin(180 - x)$

Examples $\sin 130° = \sin 50°$
        $\sin 170° = \sin 10°$
        $\sin 116° = \sin 64°$

Most people simply use a calculator when finding the sine of an obtuse angle.

Example 2

Find $\widehat{B}$.

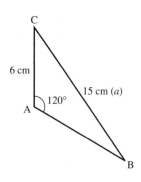

$$\frac{\sin B}{b} = \frac{\sin A}{a}$$

$$\frac{\sin B}{6} = \frac{\sin 120°}{15} \quad (\sin 120° = \sin 60°)$$

$$\sin B = \frac{6 \times \sin 60°}{15}$$

$$\sin B = 0{\cdot}346$$

$$\widehat{B} = 20{\cdot}3°$$

Exercise 9

For questions **1** to **6**, find each side marked with a letter.
Give answers to 3 s.f.

1.

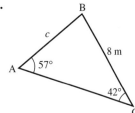

2.

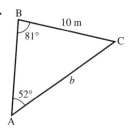

3.

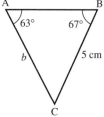

4.

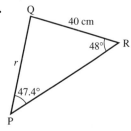

5.

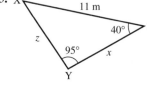

6.

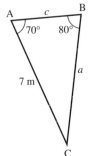

7. In $\triangle ABC$, $\widehat{A} = 61°$, $\widehat{B} = 47°$, $AC = 7{\cdot}2$ cm. Find BC.

8. In $\triangle XYZ$, $\widehat{Z} = 32°$, $\widehat{Y} = 78°$, $XY = 5{\cdot}4$ cm. Find XZ.

9. In $\triangle PQR$, $\widehat{Q} = 100°$, $\widehat{R} = 21°$, $PQ = 3{\cdot}1$ cm. Find PR.

10. In $\triangle LMN$, $\widehat{L} = 21°$, $\widehat{N} = 30°$, $MN = 7$ cm. Find LN.

In questions **11** to **18**, find each angle marked *. All lengths are in centimetres.

11.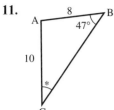

12.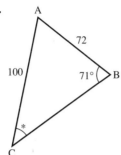

13.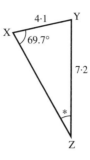

14.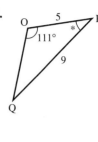

15.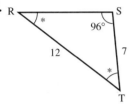

16.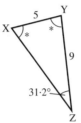

17.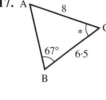

18.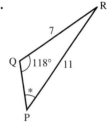

19. In $\triangle ABC$, $\widehat{A} = 62°$, $BC = 8$, $AB = 7$.
Find $\widehat{C}$.

20. In $\triangle XYZ$, $\widehat{Y} = 97·3°$, $XZ = 22$, $XY = 14$.
Find $\widehat{Z}$.

21. In $\triangle DEF$, $\widehat{D} = 58°$, $EF = 7·2$, $DE = 5·4$.
Find $\widehat{F}$.

22. In $\triangle LMN$, $\widehat{M} = 127·1°$, $LN = 11·2$, $LM = 7·3$. Find $\widehat{L}$.

6.6 The cosine rule

We use the cosine rule when we have either
(a) two sides and the included angle or
(b) all three sides.

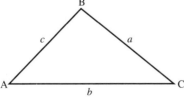

There are two forms.

1. To find the length of a side.
$$a^2 = b^2 + c^2 - (2bc \cos A)$$
$$\text{or} \quad b^2 = c^2 + a^2 - (2ac \cos B)$$
$$\text{or} \quad c^2 = a^2 + b^2 - (2ab \cos C)$$

2. To find an angle when given all three sides.

$$\cos A = \frac{b^2 + c^2 - a^2}{2bc}$$

or $$\cos B = \frac{a^2 + c^2 - b^2}{2ac}$$

or $$\cos C = \frac{a^2 + b^2 - c^2}{2ab}$$

For an obtuse angle x we have $\cos x = -\cos(180 - x)$

Examples $\cos 120° = -\cos 60°$
              $\cos 142° = -\cos 38°$

Example 1

Find b.

$b^2 = a^2 + c^2 - (2ac \cos B)$
$b^2 = 8^2 + 5^2 - (2 \times 8 \times 5 \times \cos 112°)$
$b^2 = 64 + 25 - [80 \times (-0.3746)]$
$b^2 = 64 + 25 + 29.968$

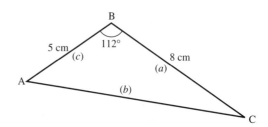

(Notice the change of sign for the obtuse angle.)

$b = \sqrt{(118.968)} = 10.9\,\text{cm}$ (to 3 s.f.)

Example 2

Find angle C.

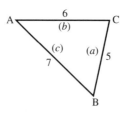

$$\cos C = \frac{a^2 + b^2 - c^2}{2ab}$$

$$\cos C = \frac{5^2 + 6^2 - 7^2}{2 \times 5 \times 6} = \frac{12}{60} = 0.200$$

$$\hat{C} = 78.5°$$

Exercise 10

Find the sides marked *. All lengths are in centimetres.

1.

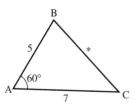

2.

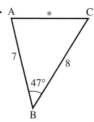

3.

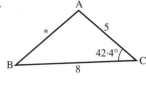

4.

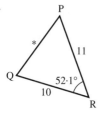

5.

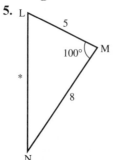

6.

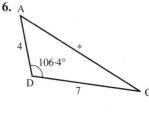

7. In △ABC, AB = 4 cm, AC = 7 cm, Â = 57°. Find BC.

8. In △XYZ, XY = 3 cm, YZ = 3 cm, Ŷ = 90°. Find XZ.

9. In △LMN, LM = 5·3 cm, MN = 7·9 cm, M̂ = 127°. Find LN.

10. △PQR, Q̂ = 117°, PQ = 80 cm, QR = 100 cm. Find PR.

In questions **11** to **16**, find each angle marked *.

11.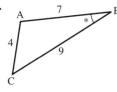

12.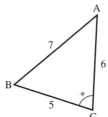

13.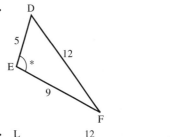

14.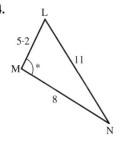

15.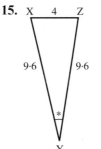

16.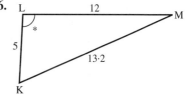

17. In △ABC, $a = 4\cdot3$, $b = 7\cdot2$, $c = 9$. Find Ĉ.

18. In △DEF, $d = 30$, $e = 50$, $f = 70$. Find Ê.

19. In △PQR, $p = 8$, $q = 14$, $r = 7$. Find Q̂.

20. In △LMN, $l = 7$, $m = 5$, $n = 4$. Find N̂.

21. In △XYZ, $x = 5\cdot3$, $y = 6\cdot7$, $z = 6\cdot14$. Find Ẑ.

22. In △ABC, $a = 4\cdot1$, $c = 6\cdot3$, B̂ = 112·2°. Find b.

23. In △PQR, $r = 0\cdot72$, $p = 1\cdot14$, Q̂ = 94·6°. Find q.

24. In △LMN, $n = 7\cdot206$, $l = 6\cdot3$, L̂ = 51·2°, N̂ = 63°. Find m.

Example

A ship sails from a port P a distance of 7 km on a bearing of 306° and then a further 11 km on a bearing of 070° to arrive at X. Calculate the distance from P to X.

$PX^2 = 7^2 + 11^2 - (2 \times 7 \times 11 \times \cos 56°)$
$= 49 + 121 - (86\cdot12)$

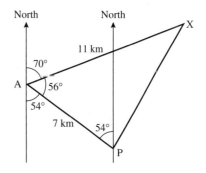

$PX^2 = 83{\cdot}88$
$PX = 9{\cdot}16\,km$ (to 3 s.f.)

The distance from P to X is $9{\cdot}16\,km$.

Exercise 11

Start each question by drawing a large, clear diagram.

1. In triangle PQR, $\widehat{Q} = 72°$, $\widehat{R} = 32°$ and $PR = 12\,cm$. Find PQ.

2. In triangle LMN, $\widehat{M} = 84°$, $LM = 7\,m$ and $MN = 9\,m$. Find LN.

3. A destroyer D and a cruiser C leave port P at the same time. The destroyer sails 25 km on a bearing 040° and the cruiser sails 30 km on a bearing of 320°. How far apart are the ships?

4. Two honeybees A and B leave the hive H at the same time; A flies 27 m due South and B flies 9 m on a bearing of 111°. How far apart are they?

5. Find all the angles of a triangle in which the sides are in the ratio 5 : 6 : 8.

6. A golfer hits his ball B a distance of 170 m towards a hole H which measures 195 m from the tee T to the green. If his shot is directed 10° away from the true line to the hole, find the distance between his ball and the hole.

7. From A, B lies 11 km away on a bearing of 041° and C lies 8 km away on a bearing of 341°. Find:
   (a) the distance between B and C
   (b) the bearing of B from C.

8. From a lighthouse L an aircraft carrier A is 15 km away on a bearing of 112° and a submarine S is 26 km away on a bearing of 200°. Find:
   (a) the distance between A and S
   (b) the bearing of A from S.

9. If the line BCD is horizontal find:
   (a) AE
   (b) EÂC

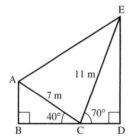

   (c) the angle of elevation of E from A.

10. An aircraft flies from its base 200 km on a bearing 162°, then 350 km on a bearing 260°, and then returns directly to base. Calculate the length and bearing of the return journey.

11. Town Y is 9 km due North of town Z. Town X is 8 km from Y, 5 km from Z and somewhere to the west of the line YZ.
(a) Draw triangle XYZ and find angle YZX.
(b) During an earthquake, town X moves due South until it is due West of Z. Find how far it has moved.

12. Calculate WX, given YZ = 15 m.

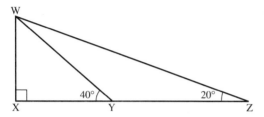

13. A golfer hits her ball a distance of 127 m so that it finishes 31 m from the hole. If the length of the hole is 150 m, calculate the angle between the line of her shot and the direct line to the hole.

Revision exercise 6A

1. Calculate the side or angle marked with a letter.

(a)

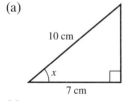

(b)

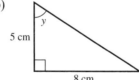

(c)

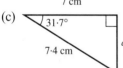

(d)

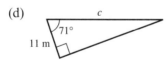

2. Given that x is an acute angle and that

$$3 \tan x - 2 = 4 \cos 35 \cdot 3°$$

calculate:
(a) $\tan x$
(b) the value of x in degrees correct to 1 d.p.

3. In the triangle XYZ, XY = 14 cm, XZ = 17 cm and angle YXZ = 25°. A is the foot of the perpendicular from Y to XZ.
Calculate:
(a) the length XA   (b) the length YA
(c) the angle ZYA

4. Calculate the length of AB.

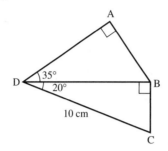

5. (a) A lies on a bearing of 040° from B.
    Calculate the bearing of B from A.
   (b) The bearing of X from Y is 115°.
    Calculate the bearing of Y from X.

6. Given BD = 1 m, calculate the length AC.

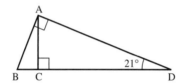

7. In the triangle PQR, angle PQR = 90° and angle RPQ = 31°. The length of PQ is 11 cm.
   Calculate:
   (a) the length of QR
   (b) the length of PR
   (c) the length of the perpendicular from Q to PR.

8. $B\widehat{A}D = D\widehat{C}A = 90°$, $C\widehat{A}D = 32\cdot4°$, $B\widehat{D}A = 41°$ and AD = 100 cm.
   Calculate:
   (a) the length of AB
   (b) the length of DC
   (c) the length of BD.

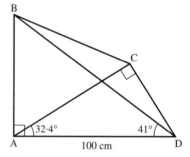

9. An observer at the top of a tower of height 20 m sees a man due East of him at an angle of depression of 27°. He sees another man due South of him at an angle of depression of 30°. Find the distance between the men on the ground.

10. The figure shows a cube of side 10 cm.
Calculate:
(a) the length of AC
(b) the angle YAC
(c) the angle ZBD.

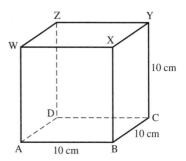

11. The diagram shows a rectangular block.
AY = 12 cm, AB = 8 cm, BC = 6 cm.
Calculate:
(a) the length YC
(b) the angle YÂZ

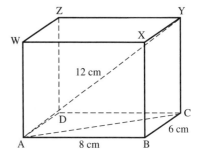

12. VABCD is a pyramid in which the base ABCD is a square of side
8 cm; V is vertically above the centre of the square and
VA = VB = VC = VD = 10 cm.
Calculate:
(a) the length AC
(b) the height of V above the base
(c) the angle VĈA.

Questions **13** to **18** may be answered either by scale drawing or by using
the sine and cosine rules.

13. Two lighthouses A and B are 25 km apart and A is due West of B.
A submarine S is on a bearing of 137° from A and on a bearing of
170° from B. Find the distance of S from A and the distance of S
from B.

14. In triangle PQR, PQ = 7 cm, PR = 8 cm and QR = 9 cm. Find
angle QPR.

15. In triangle XYZ, XY = 8 m, $\hat{X} = 57°$ and $\hat{Z} = 50°$. Find the lengths
YZ and XZ.

16. In triangle ABC, $\hat{A} = 22°$ and $\hat{C} = 44°$.
Find the ratio $\dfrac{BC}{AB}$.

17. Given cos AĈB = 0·6, AC = 4 cm, BC = 5 cm and
CD = 7 cm, find the length of AB and AD.

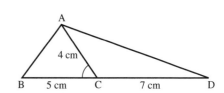

18. Find the smallest angle in a triangle whose sides are of
length $3x$, $4x$ and $6x$.

Examination exercise 6B

1.

A shop has a wheelchair ramp to its entrance from the pavement.
The ramp is 3.17 metres long and is inclined at 5° to the horizontal.
Calculate the height, h metres, of the entrance above the pavement.
Show all your working.   [2]

Cambridge IGCSE Mathematics 0580
Paper 2 Q2 June 2005

2.

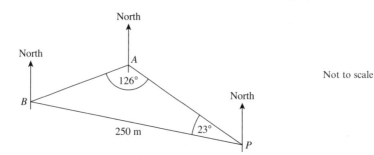

The diagram shows three straight horizontal roads in a town,
connecting points P, A, and B.
$PB = 250$ m, angle $APB = 23°$ and angle $BAP = 126°$.
(a) Calculate the length of the road AB.   [3]
(b) The bearing of A from P is 303°.

   Find the bearing of
   (i) B from P,   [1]
   (ii) A from B.   [2]

Cambridge IGCSE Mathematics 0580
Paper 4 Q4 June 2009

3. A plane flies from Auckland (*A*) to Gisborne (*G*) on a bearing of 115°.
The plane then flies on to Wellington (*W*). Angle *AGW* = 63°.

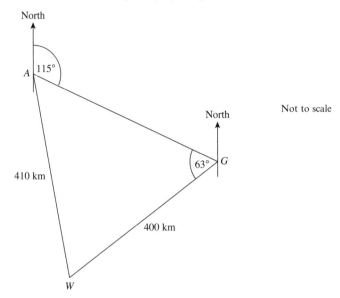

Not to scale

(a) Calculate the bearing of Wellington from Gisborne. [2]
(b) The distance from Wellington to Gisborne is 400 kilometres.
The distance from Auckland to Wellington is 410 kilometres.

Calculate the bearing of Wellington from Auckland. [4]

Cambridge IGCSE Mathematics 0580
Paper 2 Q20 June 2005

4.

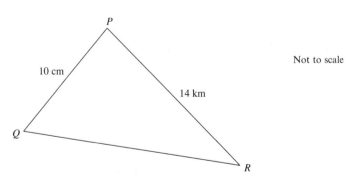

Not to scale

In triangle *PQR*, angle *QPR* is acute, *PQ* = 10 cm and *PR* = 14 cm.
(a) The area of triangle *PQR* is 48 cm^2.
Calculate angle *QPR* and show that it rounds to 43.3°, correct
to 1 decimal place.
You must show all your working. [3]
(b) Calculate the length of the side *QR*. [4]

Cambridge IGCSE Mathematics 0580
Paper 4 Q3 June 2009

5.

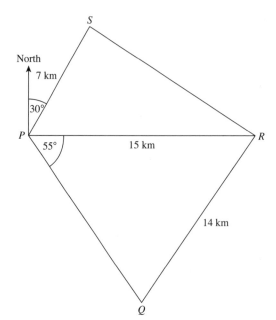

The quadrilateral *PQRS* shows the boundary of a forest.
A straight 15 kilometre road goes due East from *P* to *R*.

(a) The bearing of *S* from *P* is 030° and *PS* = 7 km.
    (i) Write down the size of angle *SPR*.   [1]
    (ii) Calculate the length of *RS*.   [4]

(b) Angle *RPQ* = 55° and *QR* = 14 km.
    (i) Write down the bearing of *Q* from *P*.   [1]
    (ii) Calculate the acute angle *PQR*.   [3]
    (iii) Calculate the length of *PQ*.   [3]

(c) Calculate the area of the forest, correct to the nearest square
    kilometre.   [4]

Cambridge IGCSE Mathematics 0580
Paper 4 Q3 November 2005

6. sin $x°$ = 0.707107 and $0 \leqslant x \leqslant 180$.

Find the two values of *x*.   [2]

Cambridge IGCSE Mathematics 0580
Paper 22 Q6 November 2008

7.

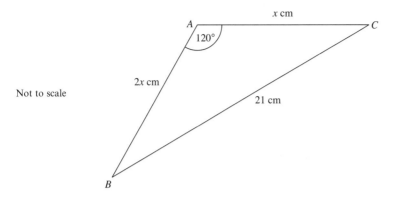

x cm

A

120°

2*x* cm

21 cm

B

C

Not to scale

In triangle *ABC*, *AB* = 2*x* cm, *AC* = *x* cm, *BC* = 21 cm and angle
BAC = 120°.
Calculate the value of *x*.   [3]

Cambridge IGCSE Mathematics 0580
Paper 21 Q11 June 2008

8.

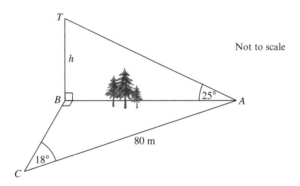

T

h

Not to scale

B

25°

A

80 m

18°

C

Mahmoud is working out the height, *h* metres, of a tower *BT* which
stands on level ground.
He measures the angle *TAB* as 25°.
He cannot measure the distance *AB* and so he walks 80 m from *A*
to *C*, where angle *ACB* = 18° and angle *ABC* = 90°.

Calculate
(a) the distance *AB*,   [2]
(b) the height of the tower, *BT*.   [2]

Cambridge IGCSE Mathematics 0580
Paper 21 Q15 June 2009

9.

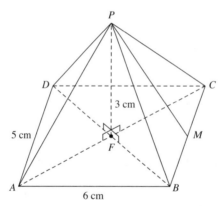

Not to scale

ABCD, *BEFC* and *AEFD* are all rectangles.
ABCD is horizontal, *BEFC* is vertical and *AEFD* represents a hillside.
AF is a path on the hillside.
AD = 800 m, *DC* = 600 m and *CF* = 200 m.
(a) Calculate the angle that the path *AF* makes with *ABCD*.   [5]
(b) In the diagram *D* is due south of *C*.
    Jasmine walks down the path from *F* to *A* in bad weather. She
    cannot see the path ahead.
    The compass bearing she must use is the bearing of *A* from *C*.
    Calculate this bearing.   [3]

Cambridge IGCSE Mathematics 0580
Paper 21 Q21 June 2008

10.

Not to scale

The diagram shows a pyramid on a rectangular base *ABCD*, with
AB = 6 cm and *AD* = 5 cm.
The diagonals *AC* and *BD* intersect at *F*.
The vertical height *FP* = 3 cm.
(a) How many planes of symmetry does the pyramid have?   [1]
(b) Calculate the volume of the pyramid.
    [The volume of a pyramid is $\frac{1}{3}$ × area of base × height.]   [2]
(c) The mid-point of *BC* is *M*.
    Calculate the angle between *PM* and the base.   [2]
(d) Calculate the angle between *PB* and the base.   [4]
(e) Calculate the length of *PB*.   [2]

Cambridge IGCSE Mathematics 0580
Paper 4 Q6 November 2005

7 GRAPHS

Rene Descartes (1596–1650) was one of the greatest philosophers of his time. Strangely his restless mind only found peace and quiet as a soldier and he apparently discovered the idea of 'cartesian' geometry in a dream before the battle of Prague. The word 'cartesian' is derived from his name and his work formed the link between geometry and algebra which inevitably led to the discovery of calculus. He finally settled in Holland for ten years, but later moved to Sweden where he soon died of pneumonia.

17   Apply rate of change to distance–time and speed–time graphs

18   Construct tables of values and draw graphs for functions of the form ax^n; estimate gradients of curves by drawing tangents

19   Interpret and obtain the equation of a straight-line graph in the form $y = mx + c$; calculate the gradient of a straight line from the coordinates of two points on it

7.1 Drawing accurate graphs

Example

Draw the graph of $y = 2x - 3$ for values of x from -2 to $+4$.

(a) The coordinates of points on the line are calculated in a table.

x	-2	-1	0	1	2	3	4
$2x$	-4	-2	0	2	4	6	8
-3	-3	-3	-3	-3	-3	-3	-3
y	-7	-5	3	1	1	3	5

(b) Draw and label axes using suitable scales.
(c) Plot the points and draw a pencil line through them. Label the line with its equation.

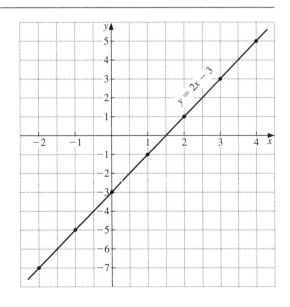

Exercise 1

Draw the following graphs, using a scale of 2 cm to 1 unit on the x-axis and 1 cm to 1 unit on the y-axis.

1. $y = 2x + 1$ for $-3 \leqslant x \leqslant 3$

3. $y = 2x - 1$ for $-3 \leqslant x \leqslant 3$

5. $y = 10 - 2x$ for $-2 \leqslant x \leqslant 4$

7. $y = 3(x - 2)$ for $-3 \leqslant x \leqslant 3$

9. $v = 2t - 3$ for $-2 \leqslant t \leqslant 4$

2. $y = 3x - 4$ for $-3 \leqslant x \leqslant 3$

4. $y = 8 - x$ for $-2 \leqslant x \leqslant 4$

6. $y = \dfrac{x + 5}{2}$ for $-3 \leqslant x \leqslant 3$

8. $y = \frac{1}{2}x + 4$ for $-3 \leqslant x \leqslant 3$

10. $z = 12 - 3t$ for $-2 \leqslant t \leqslant 4$

In each question from **11** to **16**, draw the graphs on the same page and hence find the coordinates of the vertices of the polygon formed. Give the answers as accurately as your graph will allow.

11. (a) $y = x$ (b) $y = 8 - 4x$ (c) $y = 4x$
Take $-1 \leqslant x \leqslant 3$ and $-4 \leqslant y \leqslant 14$.

12. (a) $y = 2x + 1$ (b) $y = 4x - 8$ (c) $y = 1$
Take $0 \leqslant x \leqslant 5$ and $-8 \leqslant y \leqslant 12$.

13. (a) $y = 3x$ (b) $y = 5 - x$ (c) $y = x - 4$
Take $-2 \leqslant x \leqslant 5$ and $-9 \leqslant y \leqslant 8$.

14. (a) $y = -x$ (b) $y = 3x + 6$ (c) $y = 8$ (d) $x = 3\frac{1}{2}$
Take $-2 \leqslant x \leqslant 5$ and $-6 \leqslant y \leqslant 10$.

15. (a) $y = \frac{1}{2}(x - 8)$ (b) $2x + y = 6$ (c) $y = 4(x + 1)$
Take $-3 \leqslant x \leqslant 4$ and $-7 \leqslant y \leqslant 7$.

16. (a) $y = 2x + 7$ (b) $3x + y = 10$ (c) $y = x$ (d) $2y + x = 4$
Take $-2 \leqslant x \leqslant 4$ and $0 \leqslant y \leqslant 13$.

17. The equation connecting the annual distance travelled M km, of a certain car and the annual running cost, $\$C$ is $C = \dfrac{M}{20} + 200$.
Draw the graph for $0 \leqslant M \leqslant 10\,000$ using scales of 1 cm for 1000 km for M and 2 cm for $100 for C.
From the graph find:
(a) the cost when the annual distance travelled is 7200 km,
(b) the annual mileage corresponding to a cost of $320.

18. The equation relating the cooking time t hours and the mass m kg for a cake is $t = \dfrac{3m + 1}{4}$.

Draw the graph for $0 \leqslant m \leqslant 5$. From the graph find:
(a) the mass of a cake requiring a cooking time of 2·8 hours,
(b) the cooking time for a cake with a mass of 4·1 kg.

19. Some drivers try to estimate their annual cost of repairs $c in relation to their average speed of driving s km/h using the equation $c = 6s + 50$. Draw the graph for $0 \leqslant s \leqslant 160$. From the graph find:
    (a) the estimated repair bill for a man who drives at an average speed of 65 km/h,
    (b) the average speed at which a motorist drives if his annual repair bill is $300,
    (c) the annual saving for a man who, on returning from a holiday, reduces his average speed of driving from 100 km/h to 65 km/h.

20. The value of a car $v is related to the number of km n which it has travelled by the equation

$$v = 4500 - \frac{n}{20}.$$

Draw the graph for $0 \leqslant n \leqslant 90\,000$. From the graph find:
(a) the value of a car which has travelled 3700 km,
(b) the number of km travelled by a car valued at $3200.

7.2 Gradients

The gradient of a straight line is a measure of how steep it is.

Example 1

Find the gradient of the line joining the points A (1, 2) and B (6, 5).

$$\text{gradient of AB} = \frac{BC}{AC} = \frac{3}{5}$$

It is possible to use the formula

$$\text{gradient} = \frac{\text{difference in } y\text{-coordinates}}{\text{difference in } x\text{-coordinates}}.$$

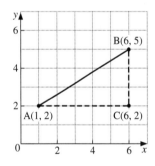

Example 2

Find the gradient of the line joining the points D (1, 5) and E (5, 2).

$$\text{gradient of DE} = \frac{5 - 2}{1 - 5} = \frac{3}{-4} = -\frac{3}{4}$$

Note:
(a) Lines which slope upward to the right have a *positive* gradient.
(b) Lines which slope downward to the right have a *negative* gradient.

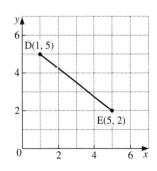

Example 3

(a) Find the length of the line segment drawn from P(2, 1) to
Q(6, 4).

(b) Find the coordinates of the mid-point of the line segment PQ.

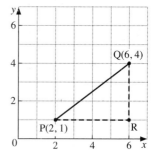

    (a)   Draw triangle PQR. PR $=4$ units and QR $=3$ units.
        By Pythagoras' theorem $PQ^2 = 4^2 + 3^2 = 25$
                    PQ $= 5$ units

    (b) Add the x-coordinates of P and Q and then divide by 2.
        Add the y-coordinates of P and Q and then divide by 2.

The mid-point has coordinates $\left[\dfrac{(2+6)}{2}, \dfrac{(1+4)}{2} \right]$

i.e. $(4, 2\frac{1}{2})$

Exercise 2

Calculate the gradient of the line joining the following pairs of points.

1. (3, 1)(5, 4)

2. (1, 1)(3, 5)

3. (3, 0)(4, 3)

4. (−1, 3)(1, 6)

5. (−2, −1)(0, 0)

6. (7, 5)(1, 6)

7. (2, −3)(1, 4)

8. (0, −2)(−2, 0)

9. $(\frac{1}{2}, 1)(\frac{3}{4}, 2)$

10. $(-\frac{1}{2}, 1)(0, -1)$

11. (3·1, 2)(3·2, 2·5)

12. (−7, 10)(0, 0)

13. $(\frac{1}{3}, 1)(\frac{1}{2}, 2)$

14. (3, 4)(−2, 4)

15. (2, 5)(1·3, 5)

16. (2, 3)(2, 7)

17. (−1, 4)(−1, 7.2)

18. (2·3, −2·2)(1·8, 1·8)

19. (0·75, 0)(0·375, −2)

20. (17·6, 1) (1·4, 1)

21. (a, b)(c, d)

22. (m, n)(a, −b)

23. (2a, f)(a, −f)

24. (2k, −k)(k, 3k)

25. (m, 3n)(−3m, 3n)

26. $\left(\dfrac{c}{2}, -d \right) \left(\dfrac{c}{4}, \dfrac{d}{2} \right)$

In questions **27** and **28**, find the gradient of each straight line.

27.

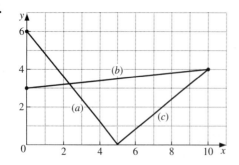

28.

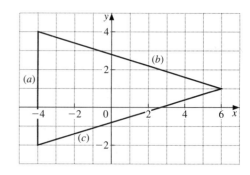

29. Find the value of a if the line joining the points $(3a, 4)$ and $(a, -3)$ has a gradient of 1.

30. (a) Write down the gradient of the line joining the points $(2m, n)$ and $(3, -4)$,
   (b) Find the value of n if the line is parallel to the x-axis,
   (c) Find the value of m if the line is parallel to the y-axis.

31. (a) Draw a pair of x and y axes and plot the points A(1, 2) and B(7, 6).
   (b) Find the length of the line segment AB. Give your answer correct to one decimal place.
   (c) Find the coordinates of the mid-point of AB.

32. (a) On a graph plot the points P(1, 4), Q(4, 8), R(5, 1).
   (b) Determine whether or not the triangle PQR is isosceles.
   (c) Find the coordinates of the mid-point of PR.

7.3 The form $y = mx + c$

When the equation of a straight line is written in the form $y = mx + c$, the gradient of the line is m and the intercept on the y-axis is c.

Example 1

Draw the line $y = 2x + 3$ on a *sketch* graph.

The word 'sketch' implies that we do not plot a series of points but simply show the position and slope of the line.

The line $y = 2x + 3$ has a gradient of 2 and cuts the y-axis at (0, 3).

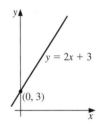

Example 2

Draw the line $x + 2y - 6 = 0$ on a sketch graph.

(a) Rearrange the equation to make y the subject.
$$x + 2y - 6 = 0$$
$$2y = -x + 6$$
$$y = -\tfrac{1}{2}x + 3$$

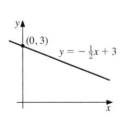

(b) The line has a gradient of $-\tfrac{1}{2}$ and cuts the y-axis at (0, 3).

Exercise 3

In questions **1** to **20**, find the gradient of the line and the intercept on the y-axis. Hence draw a small sketch graph of each line.

1. $y = x + 3$	**2.** $y = x - 2$	**3.** $y = 2x + 1$	**4.** $y = 2x - 5$
5. $y = 3x + 4$	**6.** $y = \tfrac{1}{2}x + 6$	**7.** $y = 3x - 2$	**8.** $y = 2x$
9. $y = \tfrac{1}{4}x - 4$	**10.** $y = -x + 3$	**11.** $y = 6 - 2x$	**12.** $y = 2 - x$
13. $y + 2x = 3$	**14.** $3x + y + 4 = 0$	**15.** $2y - x = 6$	**16.** $3y + x - 9 = 0$
17. $4x - y = 5$	**18.** $3x - 2y = 8$	**19.** $10x - y = 0$	**20.** $y - 4 = 0$

Finding the equation of a line

Example

Find the equation of the straight line which passes through (1, 3) and (3, 7).

(a) Let the equation of the line take the form $y = mx + c$.

The gradient, $m = \dfrac{7 - 3}{3 - 1} = 2$

so we may write the equation as

$y = 2x + c$ ...[1]

(b) Since the line passes through (1, 3), substitute 3 for y and 1 for x in [1].

$\therefore\quad 3 = 2 \times 1 + c$

$\qquad 1 = c$

The equation of the line is $y = 2x + 1$.

Exercise 4

In questions **1** to **11** find the equation of the line which:

1. Passes through (0, 7) at a gradient of 3

2. Passes through (0, −9) at a gradient of 2

3. Passes through (0, 5) at a gradient of −1

4. Passes through (2, 3) at a gradient of 2

5. Passes through (2, 11) at a gradient of 3

6. Passes through (4, 3) at a gradient of −1

7. Passes through (6, 0) at a gradient of $\frac{1}{2}$

8. Passes through (2, 1) and (4, 5)

9. Passes through (5, 4) and (6, 7)

10. Passes through (0, 5) and (3, 2)

11. Passes through (3, −3) and (9, −1)

Exercise 5

1. Find the equations of the lines A and B.

2. Find the equations of the lines C and D.

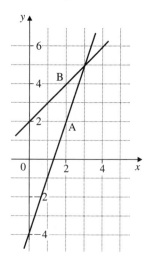

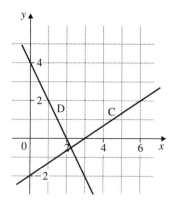

3. Look at the graph.

   (a) Find the equation of the line which is parallel to line A and which passes through the point (0, 5).

   (b) Find the equation of the line which is parallel to line B and which passes through the point (0, 3).

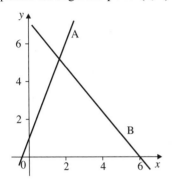

4. Look at the graphs in Question **1**.

   (a) Find the equation of the line which is parallel to line A and which passes through the point (0, 1).

   (b) Find the equation of the line which is parallel to line B and which passes through the point (0, −2).

7.4 Plotting curves

Example

Draw the graph of the function
$y = 2x^2 + x - 6$, for $-3 \leqslant x \leqslant 3$.

(a)

x	−3	−2	−1	0	1	2	3
$2x^2$	18	8	2	0	2	8	18
x	−3	−2	−1	0	1	2	3
−6	−6	−6	−6	−6	−6	−6	−6
y	9	0	−5	−6	−3	−4	15

(b) Draw and label axes using suitable scales.

(c) Plot the points and draw a smooth curve through them with a pencil.

(d) Check any points which interrupt the smoothness of the curve.

(e) Label the curve with its equation.

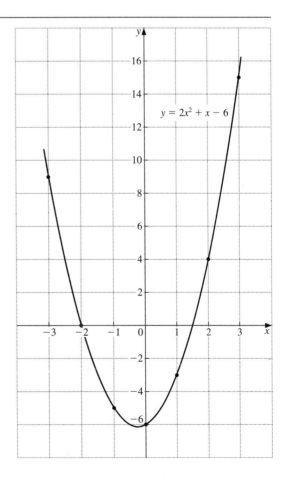

Sometimes the function notation $f(x)$ is used. $f(x)$ means 'a function of x'. If, for example, $f(x) = x^2 + 2x$ then the graph of $y = f(x)$ is simply the graph of $y = x^2 + 2x$.

> Functions are explained in greater detail on page 280.

> An alternative way of writing $f(x) = x^2 + 2x$ is $f : x \longrightarrow x^2 + 2x$

To find the value of y when $x = 1$ we obtain,
$$f(1) = 1^2 + 2 \times 1 = 3, \quad \text{so} \quad y = 3$$

Similarly,
$$f(2) = 2^2 + 2 \times 2 = 8$$
$$f(3) = 3^2 + 2 \times 3 = 15, \quad \text{and so on.}$$

Exercise 6

Draw the graphs of the following functions using a scale of 2 cm for 1 unit on the x-axis and 1 cm for 1 unit on the y-axis.

1. $y = x^2 + 2x$, for $-3 \leqslant x \leqslant 3$

2. $y = x^2 + 4x$, for $-3 \leqslant x \leqslant 3$

3. $y = x^2 - 3x$, for $-3 \leqslant x \leqslant 3$

4. $y = x^2 + 2$, for $-3 \leqslant x \leqslant 3$

5. $y = x^2 - 7$, for $-3 \leqslant x \leqslant 3$

6. $y = x^2 + x - 2$, for $-3 \leqslant x \leqslant 3$

7. $y = x^2 + 3x - 9$, for $-4 \leqslant x \leqslant 3$

8. $y = x^2 - 3x - 4$, for $-2 \leqslant x \leqslant 4$

9. $y = x^2 - 5x + 7$, for $0 \leqslant x \leqslant 6$

10. $y = 2x^2 - 6x$, for $-1 \leqslant x \leqslant 5$

11. $y = 2x^2 + 3x - 6$, for $-4 \leqslant x \leqslant 2$

12. $y = 3x^2 - 6x + 5$, for $-1 \leqslant x \leqslant 3$

13. $y = 2 + x - x^2$, for $-3 \leqslant x \leqslant 3$

14. $f(x) = 1 - 3x - x^2$, for $-5 \leqslant x \leqslant 2$

15. $f(x) = 3 + 3x - x^2$, for $-2 \leqslant x \leqslant 5$

16. $f(x) = 7 - 3x - 2x^2$, for $-3 \leqslant x \leqslant 3$

17. $f(x) = 6 + x - 2x^2$, for $-3 \leqslant x \leqslant 3$

18. $f : x \rightarrow 8 + 2x - 3x^2$, for $-2 \leqslant x \leqslant 3$

19. $f : x \rightarrow x(x - 4)$, for $-1 \leqslant x \leqslant 6$

20. $f : x \rightarrow (x + 1)(2x - 5)$, for $-3 \leqslant x \leqslant 3$.

Example

Draw the graph of $y = \dfrac{12}{x} + x - 6$, for $1 \leqslant x \leqslant 8$.

Use the graph to find approximate values for:

(a) the minimum value of $\dfrac{12}{x} + x - 6$

(b) the value of $\dfrac{12}{x} + x - 6$, when $x = 2 \cdot 25$

(c) the gradient of the tangent to the curve drawn at the point where $x = 5$.

Here is the table of values:

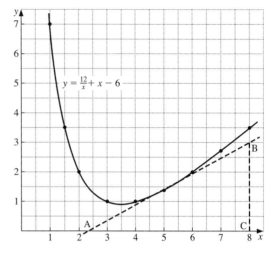

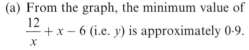

x	1	2	3	4	5	6	7	8	1·5
$\dfrac{12}{x}$	12	6	4	3	2·4	2	1·71	1·5	8
x	1	2	3	4	5	6	7	8	1·5
-6	-6	-6	-6	-6	-6	-6	-6	-6	-6
y	7	2	1	1	1·4	2	2·71	3·5	3·5

Notice that an 'extra' value of y has been calculated at $x = 1\cdot5$ because of the large difference between the y-values at $x = 1$ and $x = 2$.

(a) From the graph, the minimum value of $\dfrac{12}{x} + x - 6$ (i.e. y) is approximately $0\cdot9$.

(b) At $x = 2\cdot25$, y is approximately $1\cdot6$.

(c) The tangent AB is drawn to touch the curve at $x = 5$

The gradient of $AB = \dfrac{BC}{AC}$

$$\text{gradient} = \frac{3}{8 - 2\cdot4} = \frac{3}{5\cdot6} \approx 0\cdot54$$

It is difficult to obtain an accurate value for the gradient of a tangent so the above result is more realistically 'approximately $0\cdot5$'.

> The gradient is the 'rate of change of y with respect to x'. On a velocity–time graph the gradient is the *acceleration* at that point.

Exercise 7

Draw the following curves. The scales given are for one unit of x and y.

1. $y = x^2$, for $0 \leqslant x \leqslant 6$.
  (Scales: 2 cm for x, $\frac{1}{2}$ cm for y)
  Find:
  (a) the gradient of the tangent to the curve at $x = 2$,
  (b) the gradient of the tangent to the curve at $x = 4$,
  (c) the y-value at $x = 3\cdot25$.

2. $y = x^2 - 3x$, for $-2 \leqslant x \leqslant 5$.
  (Scales: 2 cm for x, 1 cm for y)
  Find:
  (a) the gradient of the tangent to the curve at $x = 3$,
  (b) the gradient of the tangent to the curve at $x = -1$,
  (c) the value of x where the gradient of the curve is zero.

3. $y = 5 + 3x - x^2$, for $-2 \leqslant x \leqslant 5$.
  (Scales: 2 cm for x, 1 cm for y)
  Find:
  (a) the maximum value of the function $5 + 3x - x^2$,
  (b) the gradient of the tangent to the curve at $x = 2\cdot5$,
  (c) the two values of x for which $y = 2$.

4. $y = \dfrac{12}{x}$, for $1 \leqslant x \leqslant 10$.

(Scales: 1 cm for x and y)

5. $y = \dfrac{9}{x}$, for $1 \leqslant x \leqslant 10$.

(Scales: 1 cm for x and y)

6. $y = \dfrac{12}{x+1}$, for $0 \leqslant x \leqslant 8$.)

(Scales: 2 cm for x, 1 cm for y

7. $y = \dfrac{8}{x-4}$, for $-4 \leqslant x \leqslant 3 \cdot 5$.

(Scales: 2 cm for x, 1 cm for y)

8. $y = \dfrac{15}{3-x}$, for $-4 \leqslant x \leqslant 2$.

(Scales: 2 cm for x, 1 cm for y)

9. $y = \dfrac{x}{x+4}$, for $-3 \cdot 5 \leqslant x \leqslant 4$.

(Scales: 2 cm for x and y)

10. $y = \dfrac{3x}{5-x}$, for $-3 \leqslant x \leqslant 4$.

(Scales: 2 cm for x, 1 cm for y)

11. $y = \dfrac{x+8}{x+1}$, for $0 \leqslant x \leqslant 8$.

(Scales: 2 cm for x and y)

12. $y = \dfrac{x-3}{x+2}$, for $-1 \leqslant x \leqslant 6$.

(Scales: 2 cm for x and y)

13. $y = \dfrac{10}{x} + x$, for $1 \leqslant x \leqslant 7$.

(Scales: 2 cm for x, 1 cm for y)

14. $y = \dfrac{12}{x} - x$, for $1 \leqslant x \leqslant 7$.

(Scales: 2 cm for x, 1 cm for y)

15. $y = \dfrac{15}{x} + x - 7$, for $1 \leqslant x \leqslant 7$.

(Scales: 1 cm for x and y)

Find: (a) the minimum value of y,

(b) the y value when $x = 5 \cdot 5$.

16. $y = x^3 - 2x^2$, for $0 \leqslant x \leqslant 4$.

(Scales: 2 cm for x, $\frac{1}{2}$ cm for y)

Find: (a) the y value at $x = 2 \cdot 5$,

(b) the x value at $y = 15$.

17. $y = \frac{1}{10}(x^3 + 2x + 20)$, for $-3 \leqslant x \leqslant 3$.

(Scales: 2 cm for x and y)

Find:

(a) the x value where $x^3 + 2x + 20 = 0$,

(b) the gradient of the tangent to the curve at $x = 2$.

18. Copy and complete the table for the function $y = 7 - 5x - 2x^2$, giving values of y correct to one decimal place.

x	-4	$-3 \cdot 5$	-3	$-2 \cdot 5$	-2	$-1 \cdot 5$
7	7	7		7		7
$-5x$	20	17·5		12·5		7·5
$-2x^2$	-32	$-24·5$		$-12·5$		$-4·5$
y	5	0		7		10

x	-1	$-0 \cdot 5$	0	$0 \cdot 5$	1	$1 \cdot 5$	2
7		7		7		7	
$-5x$		2·5		$-2·5$		$-7·5$	
$-2x^2$		$-0·5$		$-0·5$		$-4·5$	
y		9		4		-5	

Draw the graph, using a scale of 2 cm for x and 1 cm for y. Find:
(a) the gradient of the tangent to the curve at $x = -2.5$,
(b) the maximum value of y,
(c) the value of x at which this maximum value occurs.

19. Draw the graph of $y = \dfrac{x}{x^2 + 1}$, for $-6 \leqslant x \leqslant 6$.

    (Scales: 1 cm for x, 10 cm for y)

20. Draw the graph of $E = \dfrac{5000}{x} + 3x$ for

    $10 \leqslant x \leqslant 80$. (Scales: 1 cm to 5 units for x and 1 cm to 25 units for E)
    From the graph find:
    (a) the minimum value of E,
    (b) the value of x corresponding to this minimum value,
    (c) the range of values of x for which E is less than 275.

Sketch graphs

You need to recognise and be able to sketch the graphs of:

$$y = \frac{a}{x}$$ 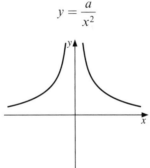 $$y = \frac{a}{x^2}$$ $$y = a^x$$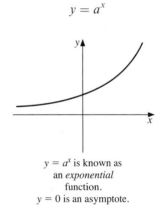

It discontinues
at $x = 0$.
$x = 0$ is an asymptote.

It discontinues
at $x = 0$.
$x = 0$ is an asymptote.

$y = a^x$ is known as
an *exponential*
function.
$y = 0$ is an asymptote.

Exercise 8

1. A rectangle has a perimeter of 14 cm and length x cm. Show that
   the width of the rectangle is $(7 - x)$ cm and hence that the area A of
   the rectangle is given by the formula $A = x(7 - x)$. Draw the graph,
   plotting x on the horizontal axis with a scale of 2 cm to 1 unit, and
   A on the vertical axis with a scale of 1 cm to 1 unit. Take x from 0
   to 7. From the graph find:
   (a) the area of the rectangle when $x = 2.25$ cm,
   (b) the dimensions of the rectangle when its area is 9 cm²,
   (c) the maximum area of the rectangle,
   (d) the length and width of the rectangle corresponding to the
       maximum area,
   (e) what shape of rectangle has the largest area.

2. A farmer has 60 m of wire fencing which he uses to make a rectangular pen for his sheep. He uses a stone wall as one side of the pen so the wire is used for only 3 sides of the pen.

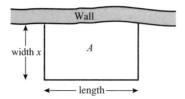

If the width of the pen is x m, what is the length (in terms of x)?
What is the area A of the pen?
Draw a graph with area A on the vertical axis and the width x on the horizontal axis. Take values of x from 0 to 30.
What dimensions should the pen have if the farmer wants to enclose the largest possible area?

3. A ball is thrown in the air so that t seconds after it is thrown, its height h metres above its starting point is given by the function $h = 25t - 5t^2$. Draw the graph of the function for $0 \leqslant t \leqslant 6$, plotting t on the horizontal axis with a scale of 2 cm to 1 second, and h on the vertical axis with a scale of 2 cm for 10 metres. Use the graph to find:
(a) the time when the ball is at its greatest height,
(b) the greatest height reached by the ball,
(c) the interval of time during which the ball is at a height of more than 30 m.

4. The velocity v m/s of a rocket t seconds after launching is given by the equation $v = 54t - 2t^3$. Draw a graph, plotting t on the horizontal axis with a scale of 2 cm to 1 second, and v on the vertical axis with a scale of 1 cm for 10 m/s. Take values of t from 0 to 5.
Use the graph to find:
(a) the maximum velocity reached,
(b) the time taken to accelerate to a velocity of 70 m/s,
(c) the interval of time during which the missile is travelling at more than 100 m/s.

5. Draw the graph of $y = 2^x$, for $-4 \leqslant x \leqslant 4$.
(Scales: 2 cm for x, 1 cm for y)

6. Draw the graph of $y = 3^x$, for $-3 \leqslant x \leqslant 3$.
(Scales: 2 cm for x, $\frac{1}{2}$ cm for y)
Find the gradient of the tangent to the curve at $x = 1$.

7. Consider the equation $y = \dfrac{1}{x}$.

When $x = \dfrac{1}{2}$, $y = \dfrac{1}{\frac{1}{2}} = 2$.

When $x = \dfrac{1}{100}$, $y = \dfrac{1}{\frac{1}{100}} = 100$.

As the denominator of the fraction $\dfrac{1}{x}$ gets smaller, the

answer gets larger. An 'infinitely small' denominator gives an
'infinitely large' answer.

We write $\dfrac{1}{0} \to \infty$. '$\dfrac{1}{0}$ tends to an infinitely large number.'

Draw the graph of $y = \dfrac{1}{x}$ for $x = -4, -3, -2,$

$-1, -0.5, -0.25, 0.5, 1, 2, 3, 4$

(Scales: 2 cm for x and y)

8. Draw the graph of $y = x + \dfrac{1}{x}$ for $x = -4, -3,$

$-2, -1, -0.5, -0.25, 0.25, 0.5, 1, 2, 3, 4$

(Scales: 2 cm for x and y)

9. Draw the graph of $y = x + \dfrac{1}{x^2}$ for $x = -4,$

$-3, -2, -1, -0.5, -0.25, 0.25, 0.5, 1, 2, 3, 4$

(Scales: 2 cm for x, 1 cm for y)

10. This sketch shows a water tank with a square base. It is 1·5 m high,
and the length of the base is x metres.

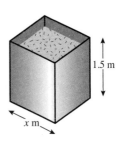

(a) Explain why the volume of the tank is given by the formula
$V = 1.5x^2$

(b) Complete the table to show the volume for various values of x.

x	0·1	0·2	0·3	0·4	0·5	0·6	0·7	0·8
V	0·02	0·06	0·14	0·24		0·54	0·74	0·96

x	0·9	1·0	1·1	1·2	1·3	1·4	1·5
V	1·2		1·82		2·54		3·38

(c) Draw the graph of $V = 1.5x^2$ for values of x from 0 to 1·5.

(d) What value of x will give a volume of 3 m^3?

(e) A guest house needs a tank with a volume at least 2 m^3. To fit
the tank into the loft, it must not be more than 1·3 m wide.
Write down the range of values for x which will satisfy these
conditions.

7.5 Interpreting graphs

Exercise 9

1. The graph shows how to convert miles
   into kilometres.

   (a) Use the graph to find approximately how many
       kilometres are the same as:
       (i) 25 miles   (ii) 15 miles
       (iii) 45 miles   (iv) 5 miles

   (b) Use the graph to find approximately how many
       miles are the same as:
       (i) 64 km   (ii) 56 km
       (iii) 16 km   (iv) 32 km

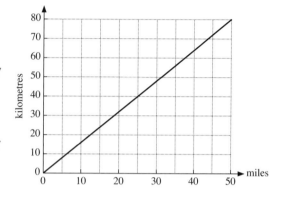

2. The graph shows how to convert
   pounds into euros.

   (a) Use the graph to find approximately
       how many euros are the same as:
       (i) £20   (ii) £80   (iii) £50

   (b) Use the graph to find approximately
       how many pounds are the same as:
       (i) €56   (ii) €84   (iii) €140

   (c) Tim spends €154 on clothes in Paris.
       How many pounds has he spent?

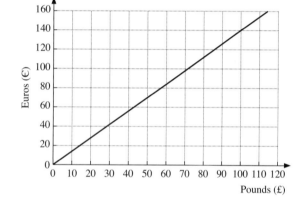

3. A company hires out vans at a basic charge of
   $35 plus a charge of 20c per km travelled. Copy
   and complete the table where x is the number of
   km travelled and C is the total cost in dollars.

x	0	50	100	150	200	250	300
C	35			65			95

   Draw a graph of C against x, using scales
   of 2 cm for 50 km on the x-axis and
   1 cm for $10 on the C-axis.
   (a) Use the graph to find the number of
       miles travelled when the total cost was $71.
   (b) What is the formula connecting C and x?

4. A car travels along a motorway and the amount of petrol in its tank is monitored as shown on the graph.

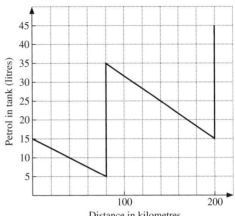

(a) How much petrol was bought at the first stop?

(b) What was the petrol consumption in km per litre:
   (i) before the first stop,
   (ii) between the two stops?

(c) What was the average petrol consumption over the 200 km?

After it leaves the second service station the car is stuck in slow traffic for the next 20 km. Its petrol consumption is reduced to 4 km per litre. After that, the road clears and the car travels a further 75 km during which time the consumption is 7·5 km/litre. Draw the graph above and extend it to show the next 95 km. How much petrol is in the tank at the end of the journey?

5. A firm makes a profit of P thousand dollars from producing x thousand tiles.
Corresponding values of P and x are given below

x	0	0·5	1·0	1·5	2·0	2·5	3·0
P	−1·0	0·75	2·0	2·75	3·0	2·75	2·0

Using a scale of 4 cm to one unit on each axis, draw the graph of P against x. [Plot x on the horizontal axis.] Use your graph to find:
(a) the number of tiles the firm should produce in order to make the maximum profit,

(b) the minimum number of tiles that should be produced to cover the cost of production,

(c) the range of values of x for which the profit is more than $2850.

6. A small firm increases its monthly expenditure on advertising and records its monthly income from sales.

Month	1	2	3	4	5	6	7
Expenditure ($)	100	200	300	400	500	600	700
Income ($)	280	450	560	630	680	720	740

Draw a graph to display this information.
(a) Is it wise to spend $100 per month on advertising?
(b) Is it wise to spend $700 per month?
(c) What is the most sensible level of expenditure on advertising?

7.6 Graphical solution of equations

Accurately drawn graphs enable us to find approximate solutions to a
wide range of equations, many of which are impossible to solve exactly
by 'conventional' methods.

Example 1

Draw the graph of the function

$$y = 2x^2 - x - 3$$

for $-2 \leqslant x \leqslant 3$. Use the graph to find approximate solutions to the
following equations.

(a) $2x^2 - x - 3 = 6$
(b) $2x^2 - x = x + 5$

The table of values for $y = 2x^2 - x - 3$ is found. Note the 'extra' value
at $x = \frac{1}{2}$.

x	-2	-1	0	1	2	3	$\frac{1}{2}$
$2x^2$	8	2	0	2	8	18	$\frac{1}{2}$
$-x$	2	1	0	-1	-2	-3	$-\frac{1}{2}$
-3	-3	-3	-3	-3	-3	-3	-3
y	7	0	-3	-2	3	12	-3

The graph drawn from this table is opposite.

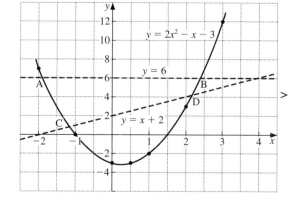

(a) To solve the equation $2x^2 - x - 3 = 6$, the
line $y = 6$ is drawn. At the points of
intersection (A and B), y simultaneously
equals both 6 and $(2x^2 - x - 3)$.
So we may write

$$2x^2 - x - 3 = 6$$

The solutions are the x-values of the points
A and B.
    i.e. $x = -1 \cdot 9$ and $x = 2 \cdot 4$ approx.

(b) To solve the equation $2x^2 - x = x + 5$, we rearrange the
equation to obtain the function $(2x^2 - x - 3)$ on the left-hand
side. In this case, subtract 3 from both sides.

$$2x^2 - x - 3 = x + 5 - 3$$
$$2x^2 - x - 3 = x + 2$$

If we now draw the line $y = x + 2$, the solutions of the equation
are given by the x-values of C and D, the points of intersection.
    i.e. $x = -1 \cdot 2$ and $x = 2 \cdot 2$ approx.

It is important to rearrange the equation to be solved so that the
function already plotted is on one side.

Example 2

Assuming that the graph of $y = x^2 - 3x + 1$ has been drawn, find the equation of the line which should be drawn to solve the equation:

$$x^2 - 4x + 3 = 0$$

Rearrange $x^2 - 4x + 3 = 0$ in order to obtain $(x^2 - 3x + 1)$ on the left-hand side.

$$x^2 - 4x + 3 = 0$$
add x $x^2 - 3x + 3 = x$
subtract 2 $x^2 - 3x + 1 = x - 2$

Therefore draw the line $y = x - 2$ to solve the equation.

Exercise 10

1. In the diagram, the graphs of $y = x^2 - 2x - 3$, $y = -2$ and $y = x$ have been drawn.
   Use the graphs to find approximate solutions to the following equations:
   (a) $x^2 - 2x - 3 = -2$
   (b) $x^2 - 2x - 3 = x$
   (c) $x^2 - 2x - 3 = 0$
   (d) $x^2 - 2x - 1 = 0$

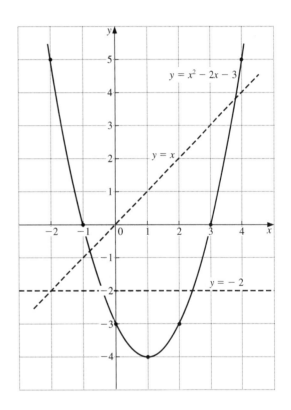

In questions **2** to **4**, use a scale of 2 cm to 1 unit for x and 1 cm to 1 unit for y.

2. Draw the graphs of the functions $y = x^2 - 2x$ and $y = x + 1$ for $-1 \leqslant x \leqslant 4$. Hence find approximate solutions of the equation $x^2 - 2x = x + 1$.

3. Draw the graphs of the functions $y = x^2 - 3x + 5$ and $y = x + 3$ for $-1 \leqslant x \leqslant 5$. Hence find approximate solutions of the equation $x^2 - 3x + 5 = x + 3$.

4. Draw the graphs of the functions $y = 6x - x^2$ and $y = 2x + 1$ for $0 \leqslant x \leqslant 5$. Hence find approximate solutions of the equation $6x - x^2 = 2x + 1$.

In questions **5** to **9**, do *not* draw any graphs.

5. Assuming the graph of $y = x^2 - 5x$ has been drawn, find the equation of the line which should be drawn to solve the equations:

(a) $x^2 - 5x = 3$ (b) $x^2 - 5x = -2$
(c) $x^2 - 5x = x + 4$ (d) $x^2 - 6x = 0$
(e) $x^2 - 5x - 6 = 0$

6. Assuming the graph of $y = x^2 + x + 1$ has been drawn, find the equation of the line which should be drawn to solve the equations:

(a) $x^2 + x + 1 = 6$ (b) $x^2 + x + 1 = 0$
(c) $x^2 + x - 3 = 0$ (d) $x^2 - x + 1 = 0$
(e) $x^2 - x - 3 = 0$

7. Assuming the graph of $y = 6x - x^2$ has been drawn, find the equation of the line which should be drawn to solve the equations:

(a) $4 + 6x - x^2 = 0$ (b) $4x - x^2 = 0$
(c) $2 + 5x - x^2 = 0$ (d) $x^2 - 6x = 3$
(e) $x^2 - 6x = -2$

8. Assuming the graph of $y = x + \dfrac{4}{x}$ has been drawn, find the equation of the line which should be drawn to solve the equations:

(a) $x + \dfrac{4}{x} - 5 = 0$ (b) $\dfrac{4}{x} - x = 0$

(c) $x + \dfrac{4}{x} = 0{\cdot}2$ (d) $2x + \dfrac{4}{x} - 3 = 0$

(e) $x^2 + 4 = 3x$

9. Assuming the graph of $y = x^2 - 8x - 7$ has been drawn, find the equation of the line which should be drawn to solve the equations:

(a) $x = 8 + \dfrac{7}{x}$ (b) $2x^2 = 16x + 9$

(c) $x^2 = 7$ (d) $x = \dfrac{4}{x - 8}$

(e) $2x - 5 = \dfrac{14}{x}$.

For questions **10** to **14**, use scales of 2 cm to 1 unit for x and 1 cm to 1 unit for y.

10. Draw the graph of $y = x^2 - 2x + 2$ for $-2 \leqslant x \leqslant 4$. By drawing other graphs, solve the equations:

(a) $x^2 - 2x + 2 = 8$

(b) $x^2 - 2x + 2 = 5 - x$

(c) $x^2 - 2x - 5 = 0$

11. Draw the graph of $y = x^2 - 7x$ for $0 \leqslant x \leqslant 7$. Draw suitable straight lines to solve the equations:

(a) $x^2 - 7x + 9 = 0$

(b) $x^2 - 5x + 1 = 0$

12. Draw the graph of $y = x^2 + 4x + 5$ for $-6 \leqslant x \leqslant 1$. Draw suitable straight lines to find approximate solutions of the equations:

(a) $x^2 + 3x - 1 = 0$ (b) $x^2 + 5x + 2 = 0$

13. Draw the graph of $y = 2x^2 + 3x - 9$ for $-3 \leqslant x \leqslant 2$. Draw suitable straight lines to find approximate solutions of the equations:

(a) $2x^2 + 3x - 4 = 0$

(b) $2x^2 + 2x - 9 = 1$

14. Draw the graph of $y = 2 + 3x - 2x^2$ for $-2 \leqslant x \leqslant 4$.

(a) Draw suitable straight lines to find approximate solutions of the equations:

(i) $2 + 4x - 2x^2 = 0$

(ii) $2x^2 - 3x - 2 = 0$

(b) Find the range of values of x for which $2 + 3x - 2x^2 \geqslant -5$.

15. Draw the graph of $y = \dfrac{18}{x}$ for $1 \leqslant x \leqslant 10$, using scales of 1 cm to one unit on both axes. Use the graph to solve approximately:

(a) $\dfrac{18}{x} = x + 2$ (b) $\dfrac{18}{x} + x = 10$

(c) $x^2 = 18$

16. Draw the graph of $y = \frac{1}{2}x^2 - 6$ for $-4 \leqslant x \leqslant 4$, taking 2 cm to 1 unit on each axis.

(a) Use your graph to solve approximately the equation $\frac{1}{2}x^2 - 6 = 1$.

(b) Using tables or a calculator confirm that your solutions are approximately $\pm\sqrt{14}$ and explain why this is so.

(c) Use your graph to find the square roots of 8.

17. Draw the graph of $y = 6 - 2x - \frac{1}{2}x^3$ for $x = \pm 2, \pm 1\frac{1}{2}, \pm 1, \pm \frac{1}{2}, 0$.
Take 4 cm to 1 unit for x and 1 cm to 1 unit for y.
Use your graph to find approximate solutions of the equations:

(a) $\frac{1}{2}x^3 + 2x - 6 = 0$ (b) $x - \frac{1}{2}x^3 = 0$

Using tables confirm that two of the solutions to the equation in part (b) are $\pm\sqrt{2}$ and explain why this is so.

18. Draw the graph of $y = x + \dfrac{12}{x} - 5$ for $x = 1, 1\frac{1}{2}, 2, 3, 4, 5, 6, 7, 8$,

taking 2 cm to 1 unit on each axis.

(a) From your graph find the range of values of x for which $x + \dfrac{12}{x} \leqslant 9$

(b) Find an approximate solution of the equation $2x - \dfrac{12}{x} - 12 = 0$.

19. Draw the graph of $y = 2^x$ for $-4 \leqslant x \leqslant 4$, taking 2 cm to one unit for x and 1 cm to one unit for y. Find approximate solutions to the equations:

(a) $2^x = 6$ (b) $2^x = 3x$ (c) $x2^x = 1$
Find also the approximate value of $2^{2.5}$.

20. Draw the graph of $y = \dfrac{1}{x}$ for $-4 \leqslant x \leqslant 4$

taking 2 cm to one unit on each axis. Find approximate solutions to the equations:

(a) $\dfrac{1}{x} = x + 1$

(b) $2x^2 - x - 1 = 0$

7.7 Distance–time graphs

When a distance–time graph is drawn the *gradient* of the graph gives the *speed* of the object.

From O to A : constant speed
     A to B : speed goes down to zero
     B to C : at rest
     C to D : accelerates
     D to E : constant speed (not as fast as O to A)

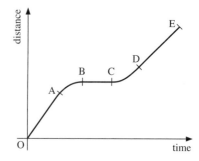

Exercise 11

1. The graph shows the journeys
   made by a van and a car starting
   at Baden, travelling to St Gallen
   and returning to Baden.
   (a) For how long was the van
       stationary during the journey?
   (b) At what time did the car
       first overtake the van?
   (c) At what speed was the van
       travelling between 09:30 and
       10:00?
   (d) What was the greatest speed
       attained by the car during the
       entire journey?
   (e) What was the average speed
       of the car over its entire journey?

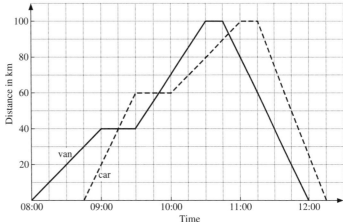

2. The graph shows the journeys of a
   bus and a car along the same road.
   The bus goes from Sofia to
   Rila and back to Sofia.
   The car goes from Rila to
   Sofia and back to Rila.

   (a) When did the bus and the car
       meet for the second time?
   (b) At what speed did the car
       travel from Rila to Sofia?
   (c) What was the average speed
       of the bus over its entire
       journey?
   (d) Approximately how far
       apart were the bus and the
       car at 09:45?
   (e) What was the greatest speed
       attained by the car during its
       entire journey?

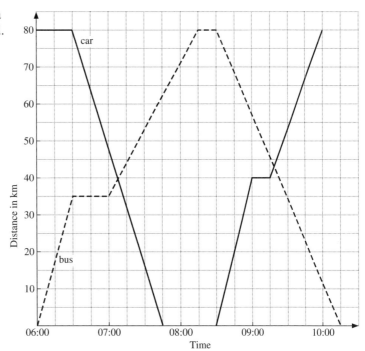

In questions 3, 4, 5 draw a travel graph to illustrate the journey
described. Draw axes with the same scales as in question 2.

3. Mrs Chuong leaves home at 08:00 and drives at a speed of 50 km/h.
   After $\frac{1}{2}$ hour she reduces her speed to 40 km/h and continues at this
   speed until 09:30. She stops from 09:30 until 10:00 and then returns
   home at a speed of 60 km/h.
   Use the graph to find the approximate time at which she arrives home.

4. Kemen leaves home at 09:00 and drives at a speed of 20 km/h. After $\frac{3}{4}$ hour he increases his speed to 45 km/h and continues at this speed until 10:45. He stops from 10:45 until 11:30 and then returns home at a speed of 50 km/h.
Draw a graph and use it to find the approximate time at which he arrives home.

5. At 10:00 Akram leaves home and cycles to his grandparents' house which is 70 km away. He cycles at a speed of 20 km/h until 11:15, at which time he stops for $\frac{1}{2}$ hour. He then completes the journey at a speed of 30 km/h. At 11:45 Akram's sister, Hameeda, leaves home and drives her car at 60 km/h. Hameeda also goes to her grandparents' house and uses the same road as Akram.
At approximately what time does Hameeda overtake Akram?

6. A boat can travel at a speed of 20 km/h in still water. The current in a river flows at 5 km/h so that downstream the boat travels at 25 km/h and upstream it travels at only 15 km/h.

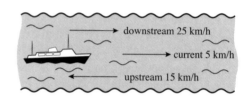

The boat has only enough fuel to last 3 hours.
The boat leaves its base and travels downstream.
Draw a distance–time graph and draw lines to indicate the outward and return journeys.
After what time must the boat turn round so that it can get back to base without running out of fuel?

7. The boat in question **6** sails in a river where the current is 10 km/h and it has fuel for four hours. At what time must the boat turn round this time if it is not to run out of fuel?

8. The graph shows the motion of three cars A, B and C along the same road.
Answer the following questions giving estimates where necessary.
   (a) Which car is in front after
       (i) 10 s,   (ii) 20 s?
   (b) When is B in the front?
   (c) When are B and C going at the same speed?
   (d) When are A and C going at the same speed?
   (e) Which car is going fastest after 5 s?
   (f) Which car starts slowly and then goes faster and faster?

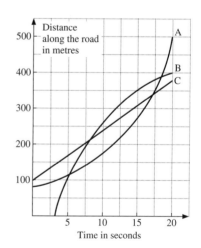

9. Three girls Hanna, Fateema and Carine took part in an egg and spoon race. Describe what happened, giving as many details as possible.

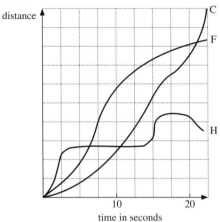

7.8 Speed–time graphs

The diagram is the speed–time graph of the first 30 seconds of a car journey. Two quantities are obtained from such graphs:
(a) acceleration = gradient of speed–time graph,
(b) distance travelled = area under graph.

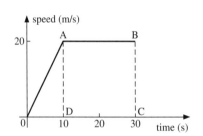

In this example,

(a) The gradient of line OA $= \frac{20}{10} = 2$

∴   The acceleration in the first 10 seconds is $2\,\text{m/s}^2$.

(b) The distance travelled in the first 30 seconds is given by the area of OAD plus the area of ABCD.

Distance $= (\frac{1}{2} \times 10 \times 20) + (20 \times 20)$
$\qquad\quad = 500\,\text{m}$

Exercise 12

On the graphs in this exercise speeds are in m/s and all times are in seconds.

1. Find:
   (a) the acceleration when $t = 4$,
   (b) the total distance travelled,
   (c) the average speed for the whole journey.

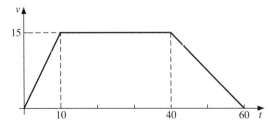

2. Find:
(a) the total distance travelled,
(b) the average speed for the whole journey,
(c) the distance travelled in the first 10 seconds,
(d) the acceleration when $t = 20$.

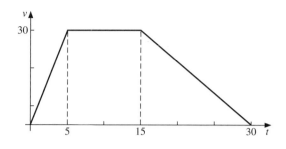

3 Find:
(a) the total distance travelled,
(b) the distance travelled in the first 40 seconds,
(c) the acceleration when $t = 15$.

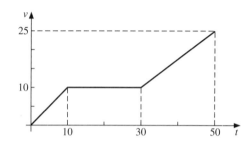

4. Find:
(a) V if the total distance travelled is 900 m,
(b) the distance travelled in the first 60 seconds.

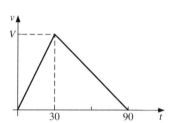

5. Find:
(a) T if the initial acceleration is $2\,\text{m/s}^2$,
(b) the total distance travelled,
(c) the average speed for the whole journey.

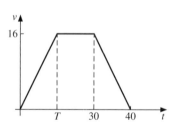

6. Given that the total distance travelled $= 810\,\text{m}$, find:
(a) the value of V,
(b) the rate of change of the speed when $t = 30$,
(c) the time taken to travel the first 420 m of the journey.

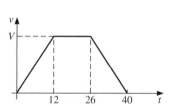

7. Given that the total distance travelled is $1{\cdot}5\,\text{km}$, find:
(a) the value of V,
(b) the rate of deceleration after 10 seconds.

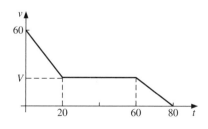

8. Given that the total distance travelled is 1·4 km, and the acceleration is 4 m/s² for the first T seconds, find:
   (a) the value of V,
   (b) the value of T.

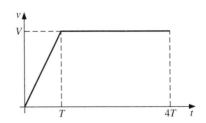

9. Given that the average speed for the whole journey is 37·5 m/s and that the deceleration between T and $2T$ is 2·5 m/s², find:
   (a) the value of V,
   (b) the value of T.

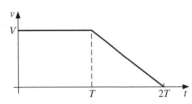

10. Given that the total distance travelled is 4 km and that the initial deceleration is 4 m/s², find:
    (a) the value of V,
    (b) the value of T.

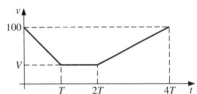

Exercise 13

Sketch a speed–time graph for each question.
All accelerations are taken to be uniform.

1. A car accelerated from 0 to 50 m/s in 9s. How far did it travel in this time?

2. A motor cycle accelerated from 10 m/s to 30 m/s in 6s. How far did it travel in this time?

3. A train slowed down from 50 km/h to 10 km/h in 2 minutes. How far did it travel in this time?

4. When taking off, an aircraft accelerates from 0 to 100 m/s in a distance of 500 m. How long did it take to travel this distance?

5. An earthworm accelerates from a speed of 0·01 m/s to 0·02 m/s over a distance of 0·9 m. How long did it take?

6. A car travelling at 60 km/h is stopped in 6 seconds. How far does it travel in this time? [Hint: Change 6 seconds into hours.]

7. A car accelerates from 15 km/h to 60 km/h in 3 seconds. How far does it travel in this time?

8. At lift off a rocket accelerates from 0 to 1000 km/h in just 10 s. How far does it travel in this time?

9. A coach accelerated from 0 to 60 km/h in 30 s. How many metres did it travel in this time?

10. Hamad was driving a car at 30 m/s when he saw an obstacle 45 m in front of him. It took a reaction time of 0·3 seconds before he could press the brakes and a further 2·5 seconds to stop the car. Did he hit the obstacle?

11. An aircraft is cruising at a speed of 200 m/s. When it lands it must be travelling at a speed of 50 m/s. In the air it can slow down at a rate of 0·2 m/s². On the ground it slows down at a rate of 2 m/s². Draw a velocity–time graph for the aircraft as it reduces it speed from 200 m/s to 50 m/s and then to 0 m/s.
How far does it travel in this time?

12. The speed of a train is measured at regular intervals of time from $t = 0$ to $t = 60$ s, as shown below.

t s	0	10	20	30	40	50	60
v m/s	0	10	16	19·7	22·2	23·8	24·7

Draw a speed–time graph to illustrate the motion. Plot t on the horizontal axis with a scale of 1 cm to 5 s and plot v on the vertical axis with a scale of 2 cm to 5 m/s.
Use the graph to estimate:
(a) the acceleration at $t = 10$,
(b) the distance travelled by the train from $t = 30$ to $t = 60$.

[An approximate value for the area under a curve can be found by splitting the area into several trapeziums.]

13. The speed of a car is measured at regular intervals of time from $t = 0$ to $t = 60$ s, as shown below.

t s	0	10	20	30	40	50	60
v m/s	0	1·3	3·2	6	10·1	16·5	30

Draw a speed–time graph using the same scales as in question 11.
Use the graph to estimate:
(a) the acceleration at $t = 30$.
(b) the distance travelled by the car from $t = 20$ to $t = 50$.

Revision exercise 7A

1. Find the equation of the straight line satisfied by the following points:

(a)
x	2	7	10
y	−5	0	3

(b)
x	1	2	3
y	7	9	11

(c)
x	1	2	3
y	8	6	4

(d)
x	3	4	5
y	2	$2\frac{1}{2}$	3

2. Find the gradient of the line joining each pair of points.
 (a) (3, 3)(5, 7)   (b) (3, −1)(7, 3)
 (c) (−1, 4)(1, −3)   (d) (2, 4)(−3, 4)
 (e) (0·5, −3)(0·4, −4)

3. Find the gradient and the intercept on the y-axis for the following
 lines. Draw a *sketch* graph of each line.
 (a) $y = 2x - 7$ (b) $y = 5 - 4x$
 (c) $2y = x + 8$ (d) $2y = 10 - x$
 (e) $y + 2x = 12$ (f) $2x + 3y = 24$

4. In the diagram, the equations of the lines are $y = 3x$, $y = 6$,
 $y = 10 - x$ and $y = \frac{1}{2}x - 3$.
 Find the equation corresponding to each line.

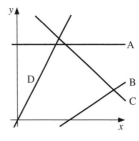

5. In the diagram, the equations of the lines are $2y = x - 8$,
 $2y + x = 8$, $4y = 3x - 16$ and $4y + 3x = 16$.
 Find the equation corresponding to each line.

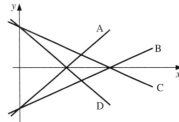

6. Find the equations of the lines which pass through the following
 pairs of points:
 (a) (2, 1)(4, 5)   (b) (0, 4)(−1, 1)
 (c) (2, 8)(−2, 12)   (d) (0, 7)(−3, 7)

7. The sketch represents a section of the curve $y = x^2 - 2x - 8$.
 Calculate:
 (a) the coordinates of A and of B,
 (b) the gradient of the line AB,
 (c) the equation of the straight line AB.

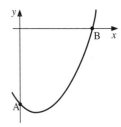

8. Find the area of the triangle formed by the intersection of the lines
 $y = x$, $x + y = 10$ and $x = 0$.

9. Draw the graph of $y = 7 - 3x - 2x^2$ for $-4 \leqslant x \leqslant 2$.
 Find the gradient of the tangent to the curve at the point where the
 curve cuts the y-axis.

10. Draw the graph of $y = \dfrac{4000}{x} + 3x$ for $10 \leqslant x \leqslant 80$. Find the
 minimum value of y.

11. Draw the graph of $y = \dfrac{1}{x} + 2^x$ for $x = \frac{1}{4}, \frac{1}{2}, \frac{3}{4}, 1, 1\frac{1}{2}, 2, 3$.

12. Assuming that the graph of $y = 4 - x^2$ has been drawn, find the equation of the straight line which should be drawn in order to solve the following equations:

(a) $4 - 3x - x^2 = 0$ (b) $\frac{1}{2}(4 - x^2) = 0$

(c) $x^2 - x + 7 = 0$ (d) $\dfrac{4}{x} - x = 5$

13. Draw the graph of $y = 5 - x^2$ for $-3 \leqslant x \leqslant 3$, taking 2 cm to one unit for x and 1 cm to one unit for y.
Use the graph to find:
(a) approximate solutions to the equation $4 - x - x^2 = 0$,
(b) the square roots of 5,
(c) the square roots of 7.

14. Draw the graph of $y = \dfrac{5}{x} + 2x - 3$, for $\frac{1}{2} \leqslant x \leqslant 7$, taking 2 cm to one unit for x and 1 cm to one unit for y.
Use the graph to find:
(a) approximate solutions to the equation $2x^2 - 10x + 7 = 0$,
(b) the range of values of x for which
$$\dfrac{5}{x} + 2x - 3 < 6.$$
(c) the minimum value of y.

15. Draw the graph of $y = 4^x$ for $-2 \leqslant x \leqslant 2$.
Use the graph to find:
(a) the approximate value of $4^{1.6}$,
(b) the approximate value of $4^{-\frac{1}{3}}$,
(c) the gradient of the curve at $x = 0$
(d) an approximate solution to the equation $4^x = 10$.

16. The diagram is the speed–time graph of a bus.
Calculate:
(a) the acceleration during the first 50 seconds,
(b) the total distance travelled,
(c) how long it takes before it is moving at 12 m/s for the first time.

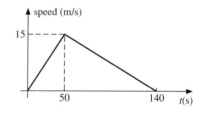

17. The diagram is the speed–time graph of a car.
Given that the total distance travelled is 2·4 km, calculate:
(a) the value of the maximum speed V,
(b) the distance travelled in the first 30 seconds of the motion.

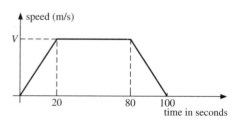

Examination exercise 7B

1.

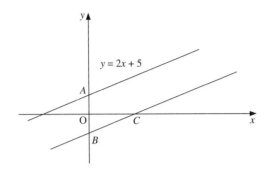

Not to scale

The distance AB is 11 units.

(a) Write down the equation of the line through B which is
   parallel to $y = 2x + 5$. [2]
(b) Find the coordinates of the point C where this line
   crosses the x-axis. [1]

Cambridge IGCSE Mathematics 0580
Paper 22 Q9 November 2008

2. The equation of a straight line can be written in the form
$3x + 2y - 8 = 0$.
(a) Rearrange this equation to make y the subject. [2]
(b) Write down the gradient of the line. [1]
(c) Write down the coordinates of the point where the line
   crosses the y-axis. [1]

Cambridge IGCSE Mathematics 0580
Paper 2 Q18 June 2007

3. A straight line passes through two points with coordinates
$(6,8)$ and $(0,5)$.
Work out the equation of the line. [3]

Cambridge IGCSE Mathematics 0580
Paper 21 Q9 June 2008

4. In an experiment, the number of bacteria, N, after x days, is
$N = 1000 \times 1.4^x$.

(a) Copy and complete the table.

x	0	1	2	3	4
N					

[2]

(b) Draw a graph to show this information.   [2]
(c) How many days does it take for the number of bacteria to reach 3000?
   Give your answer correct to 1 decimal place.   [1]

Cambridge IGCSE Mathematics 0580
Paper 21 Q16 November 2008

5.

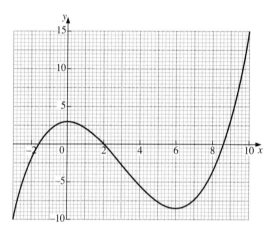

The diagram shows the accurate graph of $y = f(x)$.
(a) Use the graph to find
   (i) $f(0)$,   [1]
   (ii) $f(8)$,   [1]
(b) Use the graph to solve
   (i) $f(x) = 0$,   [2]
   (ii) $f(x) = 5$.   [1]
(c) k is an integer for which the equation $f(x) = k$ has exactly two solutions.
   Use the graph to find the two values of k.   [2]
(d) Write down the range of values of x for which the graph of $y = f(x)$ has a negative gradient.   [2]
(e) The equation $f(x) + x - 1 = 0$ can be solved by drawing a line on the grid.
   (i) Write down the equation of this line.   [1]
   (ii) How many solutions are there for $f(x) + x - 1 = 0$?   [1]

Cambridge IGCSE Mathematics 0580
Paper 4 Q4 June 2007

6. Answer the whole of this question on a sheet of graph paper.
The table shows some of the values of the function
$f(x) = x^2 - \frac{1}{x}, x \neq 0.$

x	-3	-2	-1	-0.5	-0.2	0.2	0.5	1	2	3
y	9.3	4.5	2.0	2.3	p	-5.0	-1.8	q	3.5	r

(a) Find the values of p, q and r correct to 1 decimal place.   [3]
(b) Using a scale of 2 cm to represent 1 unit on the x-axis and 1 cm
    to represent 1 unit on the y-axis, draw an x-axis for $-3 \leqslant x \leqslant 3$
    and a y-axis for $-6 \leqslant y \leqslant 10$.
    Draw the graph of $y = f(x)$ for $-3 \leqslant x \leqslant -0.2$ and
    $0.2 \leqslant x \leqslant 3$.   [6]
(c) (i) By drawing a suitable straight line, find the three values of x
        where $f(x) = -3x$.   [3]
    (ii) $x^2 - \frac{1}{x} = -3x$ can be written as $x^3 + ax^2 + b = 0$.
        Find the values of a and b.   [2]
(d) Draw a tangent to the graph of $y = f(x)$ at the point where
    $x = -2$.
    Use it to estimate the gradient of $y = f(x)$ when
    $x = -2$.   [3]

Cambridge IGCSE Mathematics 0580
Paper 4 Q3 November 2008

7.

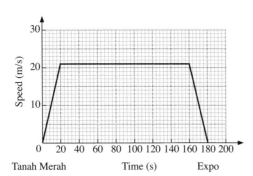

The graph shows the train journey between Tanah Merah and Expo
in Singapore.
Work out
(a) the acceleration of the train when it leaves Tanah Merah,   [2]
(b) the distance between Tanah Merah and Expo,   [3]
(c) the average speed of the train for the journey.   [1]

Cambridge IGCSE Mathematics 0580
Paper 21 Q18 November 2008

8. (a) A train completed a journey of 850 kilometres with an average speed of 80 kilometres per hour.

Calculate, giving **exact** answers, the time taken for this journey in

(i) hours,   [2]

(ii) hours, minutes and seconds.   [1]

(b) Another train took 10 hours 48 minutes to complete the same 850 km journey.

(i) It departed at 19:20.

At what time, on the next day, did this train complete the journey?   [1]

(ii) Calculate the average speed, in kilometres per hour, for the journey.   [2]

(c)

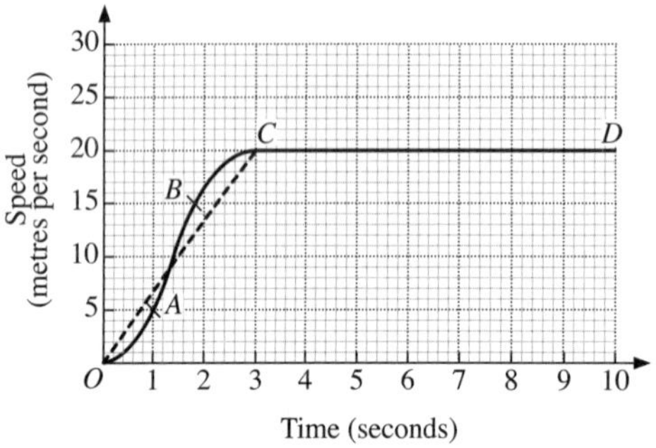

Time (seconds)

The **solid** line $OABCD$ on the grid shows the first 10 seconds of a car journey.

(i) Describe briefly what happens to the **speed** of the car between B and C.   [1]

(ii) Describe briefly what happens to the **acceleration** of the car between B and C.   [1]

(iii) Calculate the acceleration between A and B.   [2]

(iv) Using the **broken** straight line OC, estimate the total distance travelled by the car in the whole 10 seconds.   [3]

(v) Explain briefly why, in this case, using the broken line makes the answer to **part (iv)** a good estimate of the distance travelled.   [1]

(vi) Calculate the average speed of the car during the 10 seconds.

Give your answer in kilometres per hour.   [2]

Cambridge IGCSE Mathematics 0580
Paper 4 Q1 June 2006

9. **Answer the whole of this question on one sheet of graph paper.**

$$f(x) = 1 - \frac{1}{x^2}, x \neq 0.$$

(a)

x	-3	-2	-1	-0.5	-0.4	-0.3		0.3	0.4	0.5	1	2	3
$f(x)$	p	0.75	0	-3	-5.25	q		q	-5.25	-3	0	0.75	p

Find the values p and q. [2]

(b) (i) Draw an x-axis for $-3 \leqslant x \leqslant 3$ using 2 cm to represent
1 unit and a y-axis for $-11 \leqslant y \leqslant 2$ using 1 cm to represent
1 unit. [1]
   (ii) Draw the graph of $y = f(x)$ for $-3 \leqslant x \leqslant -0.3$ and for
$0.3 \leqslant x \leqslant 3$. [5]
(c) Write down an integer k such that $f(x) = k$ has no solutions. [1]
(d) **On the same grid,** draw the graph of $y = 2x - 5$ for
$-3 \leqslant x \leqslant 3$. [2]
(e) (i) Use your graphs to find solutions of the equation
$$1 - \frac{1}{x^2} = 2x - 5.$$ [3]
   (ii) Rearrange $1 - \frac{1}{x^2} = 2x - 5$ into the form $ax^3 + bx^2 + c = 0$,
where a, b and c are integers. [2]
(f) (i) Draw a tangent to the graph of $y = f(x)$ which is parallel to
the line $y = 2x - 5$. [1]
   (ii) Write down the equation of this tangent. [2]

Cambridge IGCSE Mathematics 0580
Paper 4 Q5 November 2005

8 SETS, VECTORS AND FUNCTIONS

Bertrand Russell (1872–1970) tried to reduce all mathematics to formal logic. He showed that the idea of a set of all sets which are not members of themselves leads to contradictions. He wrote to Gottlieb Frege just as he was putting the finishing touches to a book that represented his life's work, pointing out that Frege's work was invalidated. Russell's elder brother, the second Earl Russell, showed great foresight in 1903 by queueing overnight outside the vehicle licensing office in London to have his car registered as A1.

1 Use language, notation and Venn diagrams to describe sets

22 Use function notation to describe simple functions, and the notation $f^{-1}(x)$ to describe their inverses; form composite functions

35 Add and subtract vectors; multiply a vector by a scalar; calculate the magnitude of a vector; represent vectors by directed line segments; express given vectors in terms of two coplanar vectors; use position vectors

8.1 Sets

1. ∩ 'intersection'
A ∩ B is shaded.

2. ∪ 'union'
A ∪ B is shaded.

3. ⊂ 'is a subset of'
A ⊂ B
[B ⊄ A means 'B is *not* a subset of A']

4. ∈ 'is a member of'
    'belongs to'
      b ∈ X
    [e ∉ X means 'e is not a member of set X']

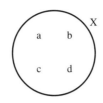

5. ℰ 'universal set'
    ℰ = {a, b, c, d, e}

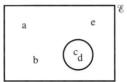

6. A′ 'complement of'
    'not in A'
  A′ is shaded
  $(A \cup A' = \mathscr{E})$

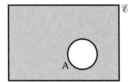

7. $n(A)$ 'the number of elements in set A'
    $n(A) = 3$

8. A = $\{x : x$ is an integer, $2 \leqslant x \leqslant 9\}$

A is the ⬚set of⬚ elements x ⬚such that⬚ x is an integer and

$2 \leqslant x \leqslant 9$.

The set A is {2, 3, 4, 5, 6, 7, 8, 9}.

9. ∅ or { } 'empty set'
    (Note: ∅ ⊂ A for any set A)

Exercise 1

1. In the Venn diagram,
    ℰ = {people in an hotel}
    T = {people who like toast}
    E = {people who like eggs}

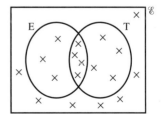

  (a) How many people like toast?
  (b) How many people like eggs but not toast?
  (c) How many people like toast and eggs?
  (d) How many people are in the hotel?
  (e) How many people like neither toast nor eggs?

2. In the Venn diagram,

$\mathscr{E} = \{\text{boys in Year 10}\}$

R = {members of the rugby team}

C = {members of the cricket team}

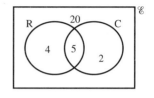

(a) How many are in the rugby team?

(b) How many are in both teams?

(c) How many are in the rugby team but not in the cricket team?

(d) How many are in neither team?

(e) How many are there in Year 10?

3. In the Venn diagram,

$\mathscr{E} = \{\text{cars in a street}\}$

B = {blue cars}

L = {cars with left-hand drive}

F = {cars with four doors}

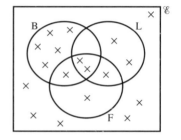

(a) How many cars are blue?

(b) How many blue cars have four doors?

(c) How many cars with left-hand drive have four doors?

(d) How many blue cars have left-hand drive?

(e) How many cars are in the street?

(f) How many blue cars with left-hand drive do not have four doors?

4. In the Venn diagram,

$\mathscr{E} = \{\text{houses in the street}\}$

C = {houses with central heating}

T = {houses with a colour T.V.}

G = {houses with a garden}

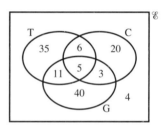

(a) How many houses have gardens?

(b) How many houses have a colour T.V. and central heating?

(c) How many houses have a colour T.V. and central heating and a garden?

(d) How many houses have a garden but not a T.V. or central heating?

(e) How many houses have a T.V. and a garden but not central heating?

(f) How many houses are there in the street?

5. In the Venn diagram,

$\mathscr{E} = \{\text{children in a mixed school}\}$

G = {girls in the school}

S = {children who can swim}

L = {children who are left-handed}

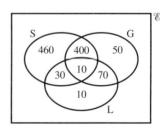

(a) How many left-handed children are there?

(b) How many girls cannot swim?

(c) How many boys can swim?

(d) How many girls are left-handed?

(e) How many boys are left-handed?

(f) How many left-handed girls can swim?

(g) How many boys are there in the school?

Example

$\mathscr{E} = \{1, 2, 3, \ldots, 12\}$, $A = \{2, 3, 4, 5, 6\}$ and $B = \{2, 4, 6, 8, 10\}$.

(a) $A \cup B = \{2, 3, 4, 5, 6, 8, 10\}$
(b) $A \cap B = \{2, 4, 6\}$
(c) $A' = \{1, 7, 8, 9, 10, 11, 12\}$
(d) $n(A \cup B) = 7$
(e) $B' \cap A = \{3, 5\}$

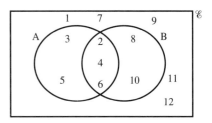

Exercise 2

In this exercise, be careful to use set notation only when the answer *is* a set.

1. If $M = \{1, 2, 3, 4, 5, 6, 7, 8\}$, $N = \{5, 7, 9, 11, 13\}$,

   find:
   (a) $M \cap N$ (b) $M \cup N$ (c) $n(N)$ (d) $n(M \cup N)$

   State whether true or false:
   (e) $5 \in M$ (f) $7 \in (M \cup N)$
   (g) $N \subset M$ (h) $\{5, 6, 7\} \subset M$

2. If $A = \{2, 3, 5, 7\}$, $B = \{1, 2, 3, \ldots, 9\}$,

   find:
   (a) $A \cap B$ (b) $A \cup B$ (c) $n(A \cap B)$ (d) $\{1, 4\} \cap A$

   State whether true or false:
   (e) $A \in B$ (f) $A \subset B$ (g) $9 \subset B$ (h) $3 \in (A \cap B)$

3. If $X = \{1, 2, 3, \ldots, 10\}$, $Y = \{2, 4, 6, \ldots, 20\}$ and
   $Z = \{x : x \text{ is an integer}, 15 \leqslant x \leqslant 25\}$,

   find:
   (a) $X \cap Y$ (b) $Y \cap Z$ (c) $X \cap Z$
   (d) $n(X \cup Y)$ (e) $n(Z)$ (f) $n(X \cup Z)$

   State whether true or false:
   (g) $5 \in Y$ (h) $20 \in X$
   (i) $n(X \cap Y) = 5$ (j) $\{15, 20, 25\} \subset Z$.

4. If $D = \{1, 3, 5\}$, $E = \{3, 4, 5\}$, $F = \{1, 5, 10\}$,

   find:
   (a) $D \cup E$ (b) $D \cap F$ (c) $n(E \cap F)$
   (d) $(D \cup E) \cap F$ (e) $(D \cap E) \cup F$ (f) $n(D \cup F)$

   State whether true or false:
   (g) $D \subset (E \cup F)$ (h) $3 \in (E \cap F)$ (i) $4 \notin (D \cap E)$

5. Find:
   (a) $n(E)$ (b) $n(F)$ (c) $E \cap F$
   (d) $E \cup F$ (e) $n(E \cup F)$ (f) $n(E \cap F)$

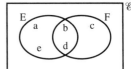

6. Find:

(a) $n(M \cap N)$ (b) $n(N)$ (c) $M \cup N$

(d) $M' \cap N$ (e) $N' \cap M$ (f) $(M \cap N)'$

(g) $M \cup N'$ (h) $N \cup M'$ (i) $M' \cup N'$

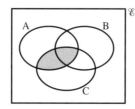

Example

On a Venn diagram, shade the regions:

(a) $A \cap C$ (b) $(B \cap C) \cap A'$

where A, B, C are intersecting sets.

(a) $A \cap C$

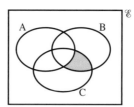

(b) $(B \cap C) \cap A'$

   [find $(B \cap C)$ first]

Exercise 3

1. Draw six diagrams similar to Figure 1 and shade the following sets:

(a) $A \cap B$ (b) $A \cup B$ (c) A'

(d) $A' \cap B$ (e) $B' \cap A$ (f) $(B \cup A)'$

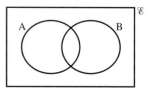

Figure 1

2. Draw four diagrams similar to Figure 2 and shade the following sets:

(a) $A \cap B$ (b) $A \cup B$ (c) $B' \cap A$ (d) $(B \cup A)'$

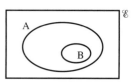

Figure 2

3. Draw four diagrams similar to Figure 3 and shade the following sets:

(a) $A \cup B$ (b) $A \cap B$ (c) $A \cap B'$ (d) $(B \cup A)'$

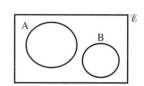

Figure 3

4. Draw eleven diagrams similar to Figure 4 and shade the following sets:

(a) $A \cap B$ (b) $A \cup C$ (c) $A \cap (B \cap C)$

(d) $(A \cup B) \cap C$ (e) $B \cap (A \cup C)$ (f) $A \cap B'$

(g) $A \cap (B \cup C)'$ (h) $(B \cup C) \cap A$ (i) $C' \cap (A \cap B)$

(j) $(A \cup C) \cup B'$ (k) $(A \cup C) \cap (B \cap C)$

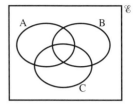

Figure 4

5. Draw nine diagrams similar to Figure 5 and shade the following sets:

(a) $(A \cup B) \cap C$ (b) $(A \cap B) \cup C$ (c) $(A \cup B) \cup C$

(d) $A \cap (B \cup C)$ (e) $A' \cap C$ (f) $C' \cap (A \cup B)$

(g) $(A \cap B) \cap C$ (h) $(A \cap C) \cup (B \cap C)$ (i) $(A \cup B \cup C)'$

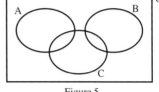

Figure 5

6. Copy each diagram and shade the region indicated.

(a)
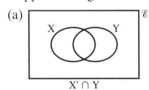
$X' \cap Y$

(b)
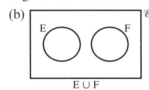
$E \cup F$

(c)

$A \cap B$

(d)
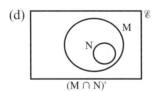
$(M \cap N)'$

7. Describe the shaded region.

(a)
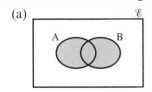

(b)
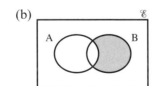

(c)
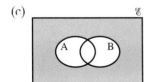

(d)
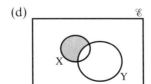

8.2 Logical problems

Example 1

In a form of 30 girls, 18 play netball and 14 play hockey, whilst 5 play neither. Find the number who play both netball and hockey.

Let $\quad \mathscr{E} = \{\text{girls in the form}\}$
$\qquad N = \{\text{girls who play netball}\}$
$\qquad H = \{\text{girls who play hockey}\}$

and $\quad x = $ the number of girls who play both netball and hockey

The number of girls in each portion of the universal set is shown in the Venn diagram.

Since $\qquad\qquad\qquad n(\mathscr{E}) = 30$
$$18 - x + x + 14 - x + 5 = 30$$
$$37 - x = 30$$
$$x = 7$$

$\therefore$ Seven girls play both netball and hockey.

Example 2

If $A = \{\text{sheep}\}$
$\quad B = \{\text{horses}\}$
$\quad C = \{\text{'intelligent' animals}\}$
$\quad D = \{\text{animals which make good pets}\}$

(a) Express the following sentences in set language:
   (i) No sheep are 'intelligent' animals.
   (ii) All horses make good pets.
   (iii) Some sheep make good pets.

(b) Interpret the following statements:
   (i) $B \subset C$
   (ii) $B \cup C = D$

(a)   (i) $A \cap C = \varnothing$
   (ii) $B \subset D$
   (iii) $A \cap D \neq \varnothing$

(b)   (i) All horses are intelligent animals.
   (ii) Animals which make good pets are either horses or 'intelligent' animals (or both).

Exercise 4

1. In the Venn diagram $n(A) = 10$, $n(B) = 13$, $n(A \cap B) = x$ and $n(A \cup B) = 18$.

   (a) Write in terms of x the number of elements in A but not in B.
   (b) Write in terms of x the number of elements in B but not in A.
   (c) Add together the number of elements in the three parts of the diagram to obtain the equation $10 - x + x + 13 - x = 18$.
   (d) Hence find the number of elements in both A and B.

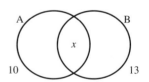

2. In the Venn diagram $n(A) = 21$, $n(B) = 17$, $n(A \cap B) = x$ and $n(A \cup B) = 29$.

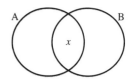

   (a) Write down in terms of x the number of elements in each part of the diagram.
   (b) Form an equation and hence find x.

3. The sets M and N intersect such that $n(M) = 31$, $n(N) = 18$ and $n(M \cup N) = 35$. How many elements are in both M and N?

4. The sets P and Q intersect such that $n(P) = 11$, $n(Q) = 29$ and $n(P \cup Q) = 37$. How many elements are in both P and Q?

5. The sets A and B intersect such that $n(A \cap B) = 7$, $n(A) = 20$ and $n(B) = 23$. Find $n(A \cup B)$.

6. Twenty boys all play either football or basketball (or both). If thirteen play football and ten play basketball, how many play both sports?

7. Of the 53 staff at a school, 36 drink tea, 18 drink coffee and 10 drink neither tea nor coffee. How many drink both tea and coffee?

8. Of the 32 pupils in a class, 18 play golf, 16 play the piano and 7 play both. How many play neither?

9. Of the pupils in a class, 15 can spell 'parallel', 14 can spell 'Pythagoras', 5 can spell both words and 4 can spell neither. How many pupils are there in the class?

10. In a school, students must take at least one of these subjects: Maths, Physics or Chemistry. In a group of 50 students, 7 take all three subjects, 9 take Physics and Chemistry only, 8 take Maths and Physics only and 5 take Maths and Chemistry only. Of these 50 students, x take Maths only, x take Physics only and $x + 3$ take Chemistry only. Draw a Venn diagram, find x, and hence find the number taking Maths.

11. All of 60 different vitamin pills contain at least one of the vitamins A, B and C. Twelve have A only, 7 have B only, and 11 have C only. If 6 have all three vitamins and there are x having A and B only, B and C only and A and C only, how many pills contain vitamin A?

12. The IGCSE results of the 30 members of a rugby squad were as follows: All 30 players passed at least two subjects, 18 players passed at least three subjects, and 3 players passed four subjects or more. Calculate:
   (a) how many passed exactly two subjects,
   (b) what fraction of the squad passed exactly three subjects.

13. In a group of 59 people, some are wearing hats, gloves or scarves (or a combination of these), 4 are wearing all three, 7 are wearing just a hat and gloves, 3 are wearing just gloves and a scarf and 9 are wearing just a hat and scarf. The number wearing only a hat or only gloves is x, and the number wearing only a scarf or none of the three items is $(x - 2)$. Find x and hence the number of people wearing a hat.

14. In a street of 150 houses, three different newspapers are delivered:
T, G and M. Of these, 40 receive T, 35 receive G, and 60 receive M;
7 receive T and G, 10 receive G and M and 4 receive T and M; 34
receive no paper at all. How many receive all three?
Note: If '7 receive T and G', this information does not mean 7
receive T and G *only*.

15. If S = {Serbian men}, G = {good footballers}, express the
following statements in words:
(a) G ⊂ S
(b) G ∩ S = ∅
(c) G ∩ S ≠ ∅
(Ignore the truth or otherwise of the statements.)

16. Given that $\mathscr{E}$ = {pupils in a school}, B = {boys},
H = {hockey players}, F = {football players}, express the following
in words:
(a) F ⊂ B   (b) H ⊂ B′   (c) F ∩ H ≠ ∅   (d) B ∩ H = ∅
Express in set notation:
(e) No boys play football.
(f) All pupils play either football or hockey.

17. If $\mathscr{E}$ = {living creatures}, S = {spiders}, F = {animals that fly},
T = {animals which taste nice}, express in set notation:
(a) No spiders taste nice.
(b) All animals that fly taste nice.
(c) Some spiders can fly.
Express in words:
(d) S ∪ F ∪ T = $\mathscr{E}$ (e) T ⊂ S

18. $\mathscr{E}$ = {tigers}, T = {tigers who believe in fairies},
X = {tigers who believe in Eskimos}, H = {tigers in hospital}.
Express in words:
(a) T ⊂ X   (b) T ∪ X = H   (c) H ∩ X = ∅
Express in set notation:
(d) All tigers in hospital believe in fairies.
(e) Some tigers believe in both fairies and Eskimos.

19. $\mathscr{E}$ = {school teachers}, P = {teachers called Peter},
B = {good bridge players}, W = {women teachers}.
Express in words:
(a) P ∩ B = ∅   (b) P ∪ B ∪ W = $\mathscr{E}$ (c) P ∩ W ≠ ∅
Express in set notation:
(d) Women teachers cannot play bridge well.
(e) All good bridge players are women called Peter.

8.3 Vectors

A vector quantity has both magnitude and direction. Problems involving forces, velocities and displacements are often made easier when vectors are used.

Addition of vectors

Vectors **a** and **b** represented by the line segments can be added using the parallelogram rule or the 'nose-to-tail' method.

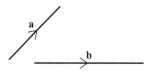

The 'tail' of vector **b** is joined to the 'nose' of vector **a**.

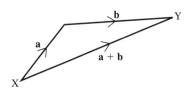

Alternatively the tail of **a** can be joined to the 'nose' of vector **b**.

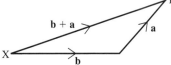

In both cases the vector $\overrightarrow{XY}$ has the same length and direction and therefore

$$\mathbf{a} + \mathbf{b} = \mathbf{b} + \mathbf{a}$$

Multiplication by a scalar

A scalar quantity has a magnitude but no direction (e.g. mass, volume, temperature). Ordinary numbers are scalars.
When vector **x** is multiplied by 2, the result is 2**x**.

When **x** is multiplied by −3 the result is −3**x**.

Note:
(1) The negative sign reverses the direction of the vector.
(2) The result of **a** − **b** is **a** + −**b**.
   i.e. Subtracting **b** is equivalent to adding the negative of **b**.

Example

The diagram shows vectors **a** and **b**.

Find $\overrightarrow{OP}$ and $\overrightarrow{OQ}$ such that

$\overrightarrow{OP} = 3\mathbf{a} + \mathbf{b}$

$\overrightarrow{OQ} = -2\mathbf{a} - 3\mathbf{b}$

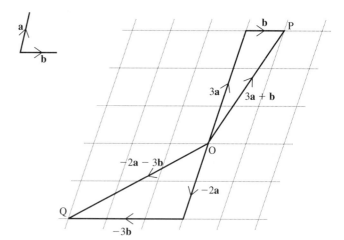

Exercise 5

In questions **1** to **26**, use the diagram below to describe the vectors given in terms of **c** and **d** where $\mathbf{c} = \overrightarrow{QN}$ and $\mathbf{d} = \overrightarrow{QR}$.

e.g. $\overrightarrow{QS} = 2\mathbf{d}$, $\overrightarrow{TD} = \mathbf{c} + \mathbf{d}$

1. $\overrightarrow{AB}$	**2.** $\overrightarrow{SG}$	**3.** $\overrightarrow{VK}$
4. $\overrightarrow{KH}$	**5.** $\overrightarrow{OT}$	**6.** $\overrightarrow{WJ}$
7. $\overrightarrow{FH}$	**8.** $\overrightarrow{FT}$	**9.** $\overrightarrow{KV}$
10. $\overrightarrow{NQ}$	**11.** $\overrightarrow{OM}$	**12.** $\overrightarrow{SD}$
13. $\overrightarrow{PI}$	**14.** $\overrightarrow{YG}$	**15.** $\overrightarrow{OI}$
16. $\overrightarrow{RE}$	**17.** $\overrightarrow{XM}$	**18.** $\overrightarrow{ZH}$
19. $\overrightarrow{MR}$	**20.** $\overrightarrow{KA}$	**21.** $\overrightarrow{RZ}$
22. $\overrightarrow{CR}$	**23.** $\overrightarrow{NV}$	**24.** $\overrightarrow{EV}$
25. $\overrightarrow{JS}$	**26.** $\overrightarrow{LE}$	

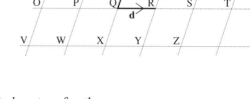

In questions **27** to **38**, use the same diagram to find vectors for the following in terms of the capital letters, starting from Q each time.

e.g. $3\mathbf{d} = \overrightarrow{QT}$, $\mathbf{c} + \mathbf{d} = \overrightarrow{QA}$.

27. $2\mathbf{c}$	**28.** $4\mathbf{d}$	**29.** $2\mathbf{c} + \mathbf{d}$	**30.** $2\mathbf{d} + \mathbf{c}$
31. $3\mathbf{d} + 2\mathbf{c}$	**32.** $2\mathbf{c} - \mathbf{d}$	**33.** $-\mathbf{c} + 2\mathbf{d}$	**34.** $\mathbf{c} - 2\mathbf{d}$
35. $2\mathbf{c} + 4\mathbf{d}$	**36.** $-\mathbf{c}$	**37.** $-\mathbf{c} - \mathbf{d}$	**38.** $2\mathbf{c} - 2\mathbf{d}$

In questions **39** to **43**, write each vector in terms of **a** and/or **b**.

39. (a) $\overrightarrow{BA}$ (b) $\overrightarrow{AC}$
    (c) $\overrightarrow{DB}$ (d) $\overrightarrow{AD}$

40. (a) $\overrightarrow{ZX}$ (b) $\overrightarrow{YW}$
    (c) $\overrightarrow{XY}$ (d) $\overrightarrow{XZ}$

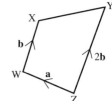

41. (a) $\overrightarrow{MK}$ (b) $\overrightarrow{NL}$ (c) $\overrightarrow{NK}$ (d) $\overrightarrow{KN}$

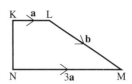

42. (a) $\overrightarrow{FE}$ (b) $\overrightarrow{BC}$ (c) $\overrightarrow{FC}$ (d) $\overrightarrow{DA}$

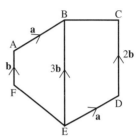

43. (a) $\overrightarrow{EC}$ (b) $\overrightarrow{BE}$ (c) $\overrightarrow{AE}$ (d) $\overrightarrow{EA}$

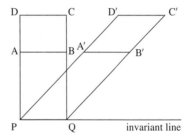

In questions **44** to **46**, write each vector in terms of **a**, **b** and **c**.

44. (a) $\overrightarrow{FC}$ (b) $\overrightarrow{GB}$ (c) $\overrightarrow{AB}$ (d) $\overrightarrow{HE}$ (e) $\overrightarrow{CA}$

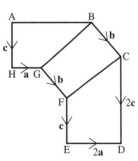

45. (a) $\overrightarrow{OF}$ (b) $\overrightarrow{OC}$ (c) $\overrightarrow{BC}$ (d) $\overrightarrow{EB}$ (e) $\overrightarrow{FB}$

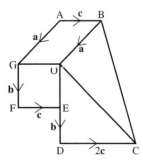

46. (a) $\overrightarrow{GD}$ (b) $\overrightarrow{GE}$ (c) $\overrightarrow{AD}$ (d) $\overrightarrow{AF}$ (e) $\overrightarrow{FE}$

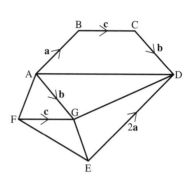

Example

Using Figure 1, express each of the following vectors in terms of **a** and/or **b**.

(a) $\overrightarrow{AP}$ (b) $\overrightarrow{AB}$ (c) $\overrightarrow{OQ}$ (d) $\overrightarrow{PO}$ (e) $\overrightarrow{PQ}$

(f) $\overrightarrow{PN}$ (g) $\overrightarrow{ON}$ (h) $\overrightarrow{AN}$ (i) $\overrightarrow{BP}$ (j) $\overrightarrow{QA}$

OA = AP
BQ = 3OB
N is the mid-point of PQ
$\overrightarrow{OA} = \mathbf{a}$, $\overrightarrow{OB} = \mathbf{b}$

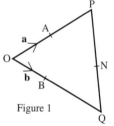

Figure 1

(a) $\overrightarrow{AP} = \mathbf{a}$

(b) $\overrightarrow{AB} = -\mathbf{a} + \mathbf{b}$

(c) $\overrightarrow{OQ} = 4\mathbf{b}$

(d) $\overrightarrow{PO} = -2\mathbf{a}$

(e) $\overrightarrow{PQ} = \overrightarrow{PO} + \overrightarrow{OQ}$
  $= -2\mathbf{a} + 4\mathbf{b}$

(f) $\overrightarrow{PN} = \frac{1}{2}\overrightarrow{PQ}$
  $= -\mathbf{a} + 2\mathbf{b}$

(g) $\overrightarrow{ON} = \overrightarrow{OP} + \overrightarrow{PN}$
  $= 2\mathbf{a} + (-\mathbf{a} + 2\mathbf{b})$
  $= \mathbf{a} + 2\mathbf{b}$

(h) $\overrightarrow{AN} = \overrightarrow{AP} + \overrightarrow{PN}$
  $= \mathbf{a} + (-\mathbf{a} + 2\mathbf{b})$
  $= 2\mathbf{b}$

(i) $\overrightarrow{BP} = \overrightarrow{BO} + \overrightarrow{OP}$
  $= -\mathbf{b} + 2\mathbf{a}$

(j) $\overrightarrow{QA} = \overrightarrow{QO} + \overrightarrow{OA}$
  $= -4\mathbf{b} + \mathbf{a}$

Exercise 6

In questions **1** to **6**, $\overrightarrow{OA} = \mathbf{a}$ and $\overrightarrow{OB} = \mathbf{b}$. Copy each diagram and use the information given to express the following vectors in terms of **a** and/or **b**.

(a) $\overrightarrow{AP}$ (b) $\overrightarrow{AB}$ (c) $\overrightarrow{OQ}$ (d) $\overrightarrow{PO}$ (e) $\overrightarrow{PQ}$

(f) $\overrightarrow{PN}$ (g) $\overrightarrow{ON}$ (h) $\overrightarrow{AN}$ (i) $\overrightarrow{BP}$ (j) $\overrightarrow{QA}$

1. A, B and N are mid-points of OP, OQ and PQ respectively.

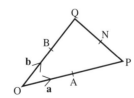

2. A and N are mid-points of OP and PQ and BQ = 2OB.

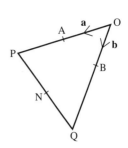

3. AP = 2OA, BQ = OB, PN = NQ.

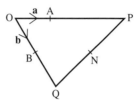

4. OA = 2AP, BQ = 3OB, PN = 2NQ.

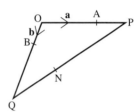

5. AP = 5OA, OB = 2BQ, NP = 2QN.

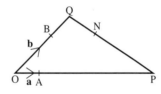

6. OA = $\frac{1}{5}$OP, OQ = 3OB, N is $\frac{1}{4}$ of the way along PQ.

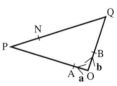

7. In $\triangle$XYZ, the mid-point of YZ is M.
   If $\overrightarrow{XY} = \mathbf{s}$ and $\overrightarrow{ZX} = \mathbf{t}$, find $\overrightarrow{XM}$ in terms of $\mathbf{s}$ and $\mathbf{t}$.

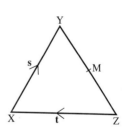

8. In $\triangle$AOB, AM : MB = 2 : 1. If $\overrightarrow{OA} = \mathbf{a}$ and $\overrightarrow{OB} = \mathbf{b}$, find $\overrightarrow{OM}$ in terms of $\mathbf{a}$ and $\mathbf{b}$.

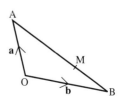

9. O is any point in the plane of the square ABCD.
The vectors $\overrightarrow{OA}$, $\overrightarrow{OB}$ and $\overrightarrow{OC}$ are **a**, **b** and **c** respectively. Find the vector $\overrightarrow{OD}$ in terms of **a**, **b** and **c**.

10. ABCDEF is a regular hexagon with $\overrightarrow{AB}$ representing the vector **m** $\overrightarrow{AF}$ representing the vector **n**. Find the vector representing $\overrightarrow{AD}$.

11. ABCDEF is a regular hexagon with centre O.
$\overrightarrow{FA} = \mathbf{a}$ and $\overrightarrow{FB} = \mathbf{b}$.

Express the following vectors in terms of **a** and/or **b**.

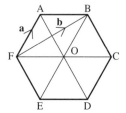

(a) $\overrightarrow{AB}$ (b) $\overrightarrow{FO}$ (c) $\overrightarrow{FC}$
(d) $\overrightarrow{BC}$ (e) $\overrightarrow{AO}$ (f) $\overrightarrow{FD}$

12. In the diagram, M is the mid-point of CD, BP:PM = 2:1, $\overrightarrow{AB} = \mathbf{x}$, $\overrightarrow{AC} = \mathbf{y}$ and $\overrightarrow{AD} = \mathbf{z}$.

Express the following vectors in terms of **x**, **y** and **z**.

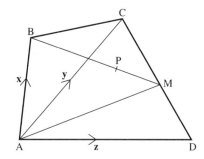

(a) $\overrightarrow{DC}$ (b) $\overrightarrow{DM}$ (c) $\overrightarrow{AM}$
(d) $\overrightarrow{BM}$ (e) $\overrightarrow{BP}$ (f) $\overrightarrow{AP}$

8.4 Column vectors

The vector $\overrightarrow{AB}$ may be written as a *column vector*.

$$AB = \begin{pmatrix} 5 \\ 3 \end{pmatrix}.$$

The top number is the horizontal component of $\overrightarrow{AB}$ (i.e. 5) and the bottom number is the vertical component (i.e. 3).

Similarly $\quad \overrightarrow{CD} = \begin{pmatrix} 4 \\ -2 \end{pmatrix}$

$$\overrightarrow{EF} = \begin{pmatrix} 0 \\ 6 \end{pmatrix}$$

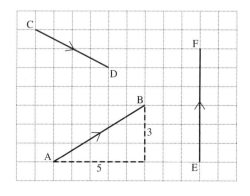

Addition of vectors

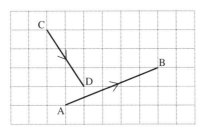

Figure 1

Suppose we wish to add vectors $\overrightarrow{AB}$ and $\overrightarrow{CD}$ in Figure 1.

First move $\overrightarrow{CD}$ so that $\overrightarrow{AB}$ and $\overrightarrow{CD}$ join 'nose to tail' as in Figure 2. Remember that changing the *position* of a vector does not change the vector. A vector is described by its length and direction.

The broken line shows the result of adding $\overrightarrow{AB}$ and $\overrightarrow{CD}$.

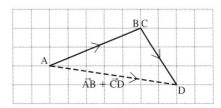

Figure 2

In column vectors,

$$\overrightarrow{AB} + \overrightarrow{CD} = \begin{pmatrix} 5 \\ 2 \end{pmatrix} + \begin{pmatrix} 2 \\ -3 \end{pmatrix}$$

We see that the column vector for the broken line is $\begin{pmatrix} 7 \\ -1 \end{pmatrix}$. So we perform addition with vectors by adding together the corresponding components of the vectors.

Subtraction of vectors

Figure 3 shows $\overrightarrow{AB} - \overrightarrow{CD}$.

To subtract vector $\overrightarrow{CD}$ from $\overrightarrow{AB}$ we *add* the *negative* of $\overrightarrow{CD}$ to $\overrightarrow{AB}$.
So $\overrightarrow{AB} - \overrightarrow{CD} = \overrightarrow{AB} + (-\overrightarrow{CD})$

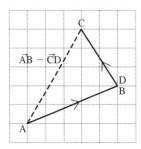

Figure 3

In column vectors,

$$\overrightarrow{AB} + (-\overrightarrow{CD}) = \begin{pmatrix} 5 \\ 2 \end{pmatrix} + \begin{pmatrix} -2 \\ 3 \end{pmatrix} = \begin{pmatrix} 3 \\ 5 \end{pmatrix}$$

Multiplication by a scalar

If $\mathbf{a} = \begin{pmatrix} 3 \\ -4 \end{pmatrix}$ then $2\mathbf{a} = 2\begin{pmatrix} 3 \\ -4 \end{pmatrix} = \begin{pmatrix} 6 \\ -8 \end{pmatrix}$.

Each component is multiplied by the number 2.

Parallel vectors

Vectors are parallel if they have the same direction. Both components of one vector must be in the same ratio to the corresponding components of the parallel vector.

e.g. $\begin{pmatrix} 3 \\ -5 \end{pmatrix}$ is parallel to $\begin{pmatrix} 6 \\ -10 \end{pmatrix}$,

because $\begin{pmatrix} 6 \\ -10 \end{pmatrix}$ may be written $2\begin{pmatrix} 3 \\ -5 \end{pmatrix}$.

In general the vector $k\begin{pmatrix} a \\ b \end{pmatrix}$ is parallel to $\begin{pmatrix} a \\ b \end{pmatrix}$.

Exercise 7

Questions **1** to **36** refer to the following vectors.

$$\mathbf{a} = \begin{pmatrix} 3 \\ 4 \end{pmatrix} \quad \mathbf{b} = \begin{pmatrix} 1 \\ 4 \end{pmatrix} \quad \mathbf{c} = \begin{pmatrix} 4 \\ -3 \end{pmatrix} \quad \mathbf{d} = \begin{pmatrix} -1 \\ 1 \end{pmatrix}$$

$$\mathbf{e} = \begin{pmatrix} 5 \\ 12 \end{pmatrix} \quad \mathbf{f} = \begin{pmatrix} 3 \\ -2 \end{pmatrix} \quad \mathbf{g} = \begin{pmatrix} -4 \\ -2 \end{pmatrix} \quad \mathbf{h} = \begin{pmatrix} -12 \\ 5 \end{pmatrix}$$

Draw and label the following vectors on graph paper (take 1 cm to 1 unit).

1. c	**2.** f	**3.** 2b	**4.** $-$a
5. $-$g	**6.** 3a	**7.** $\frac{1}{2}$e	**8.** 5d
9. $-\frac{1}{2}$h	**10.** $\frac{3}{2}$g	**11.** $\frac{1}{5}$h	**12.** -3b

Find the following vectors in component form.

13. b $+$ h	**14.** f $+$ g	**15.** e $-$ b
16. a $-$ d	**17.** g $-$ h	**18.** 2a $+$ 3c
19. 3f $+$ 2d	**20.** 4g $-$ 2b	**21.** 5a $+$ $\frac{1}{2}$g
22. a $+$ b $+$ c	**23.** 3f $-$ a $+$ c	**24.** c $+$ 2d $+$ 3e

In each of the following, find **x** in component form.

25. x $+$ b $=$ e	**26.** x $+$ d $=$ a	**27.** c $+$ x $=$ f
28. x $-$ g $=$ h	**29.** 2x $+$ b $=$ g	**30.** 2x $-$ 3d $=$ g
31. 2b $=$ d $-$ x	**32.** f $-$ g $=$ e $-$ x	**33.** 2x $+$ b $=$ x $+$ e
34. 3x $-$ b $=$ x $+$ h	**35.** a $+$ b $+$ x $=$ b $+$ a	**36.** 2x $+$ e $=$ 0 (zero vector)

37. (a) Draw and label each of the following vectors on graph paper.

$$\mathbf{l} = \begin{pmatrix} -3 \\ -3 \end{pmatrix}; \ \mathbf{m} = \begin{pmatrix} 2 \\ 0 \end{pmatrix}; \ \mathbf{n} = \begin{pmatrix} 3 \\ 2 \end{pmatrix}; \quad \mathbf{p} = \begin{pmatrix} 1 \\ -2 \end{pmatrix}; \ \mathbf{q} = \begin{pmatrix} 3 \\ 0 \end{pmatrix};$$

$$\mathbf{r} = \begin{pmatrix} 6 \\ 4 \end{pmatrix}; \quad \mathbf{s} = \begin{pmatrix} 2 \\ 2 \end{pmatrix}; \ \mathbf{t} = \begin{pmatrix} 2 \\ -4 \end{pmatrix}; \ \mathbf{u} = \begin{pmatrix} -1 \\ -3 \end{pmatrix}; \ \mathbf{v} = \begin{pmatrix} 0 \\ 3 \end{pmatrix}$$

(b) Find four pairs of parallel vectors amongst the ten vectors.

38. State whether 'true' or 'false'.

(a) $\begin{pmatrix} 3 \\ -1 \end{pmatrix}$ is parallel to $\begin{pmatrix} 9 \\ -3 \end{pmatrix}$ (b) $\begin{pmatrix} -2 \\ 0 \end{pmatrix}$ is parallel to $\begin{pmatrix} 4 \\ 0 \end{pmatrix}$

(c) $\begin{pmatrix} -1 \\ 1 \end{pmatrix}$ is parallel to $\begin{pmatrix} 1 \\ -1 \end{pmatrix}$ (d) $\begin{pmatrix} 5 \\ -15 \end{pmatrix} = 5\begin{pmatrix} 1 \\ -3 \end{pmatrix}$

(e) $\begin{pmatrix} 4 \\ 0 \end{pmatrix}$ is parallel to $\begin{pmatrix} 0 \\ 6 \end{pmatrix}$ (f) $\begin{pmatrix} 3 \\ -1 \end{pmatrix} + \begin{pmatrix} -4 \\ -2 \end{pmatrix} = \begin{pmatrix} -1 \\ 1 \end{pmatrix}$

39. (a) Draw a diagram to illustrate the vector addition $\overrightarrow{AB} + \overrightarrow{CD}$.

(b) Draw a diagram to illustrate $\overrightarrow{AB} - \overrightarrow{CD}$.

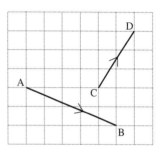

40. Draw separate diagrams to illustrate the following.

(a) $\overrightarrow{FE} + \overrightarrow{JI}$

(b) $\overrightarrow{HG} + \overrightarrow{FE}$

(c) $\overrightarrow{JI} - \overrightarrow{FE}$

(d) $\overrightarrow{HG} + \overrightarrow{JI}$

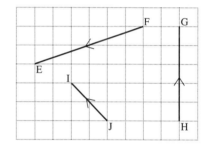

Exercise 8

1. If D has coordinates (7, 2) and E has coordinates (9, 0), find the column vector for $\overrightarrow{DE}$.

2. Find the column vector $\overrightarrow{XY}$ where X and Y have coordinates (−1, 4) and (5, 2) respectively.

3. In the diagram $\overrightarrow{AB}$ represents the vector $\begin{pmatrix} 5 \\ 2 \end{pmatrix}$ and $\overrightarrow{BC}$ represents the vector $\begin{pmatrix} 0 \\ 3 \end{pmatrix}$.

(a) Copy the diagram and mark point D such that ABCD is a parallelogram.

(b) Write $\overrightarrow{AD}$ and $\overrightarrow{CA}$ as column vectors.

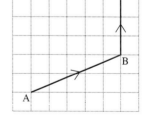

4. (a) On squared paper draw $\overrightarrow{AB} = \begin{pmatrix} 3 \\ -2 \end{pmatrix}$ and $\overrightarrow{BC} = \begin{pmatrix} 4 \\ 2 \end{pmatrix}$ and mark point D such that ABCD is a parallelogram.

(b) Write $\overrightarrow{AD}$ and $\overrightarrow{CA}$ as column vectors.

5. Copy the diagram in which $\overrightarrow{OA} = \begin{pmatrix} 5 \\ 2 \end{pmatrix}$ and $\overrightarrow{OB} = \begin{pmatrix} 2 \\ 5 \end{pmatrix}$.

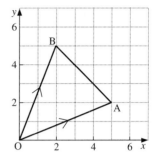

M is the mid-point of AB. Express the following as column vectors:

(a) $\overrightarrow{BA}$　　　　(b) $\overrightarrow{BM}$　　　　(c) $\overrightarrow{OM}$ (use $\overrightarrow{OM} = \overrightarrow{OB} + \overrightarrow{BM}$)

Hence write down the coordinates of M.

6. On a graph with origin at O, draw $\overrightarrow{OA} = \begin{pmatrix} 5 \\ -1 \end{pmatrix}$ and

$\overrightarrow{OB} = \begin{pmatrix} 6 \\ -7 \end{pmatrix}$. Given that M is the mid-point of AB express the

following as column vectors:

(a) $\overrightarrow{BA}$　　　　(b) $\overrightarrow{BM}$　　　　(c) $\overrightarrow{OM}$

Hence write down the coordinates of M.

7. On a graph with origin at O, draw $\overrightarrow{OA} = \begin{pmatrix} -2 \\ 5 \end{pmatrix}$, $\overrightarrow{OB} = \begin{pmatrix} 4 \\ 2 \end{pmatrix}$

and $\overrightarrow{OC} = \begin{pmatrix} -2 \\ -4 \end{pmatrix}$.

(a) Given that M divides AB such that $AM : MB = 2 : 1$, express the
following as column vectors:

(i) $\overrightarrow{BA}$　　　　(ii) $\overrightarrow{BM}$　　　　(iii) $\overrightarrow{OM}$

(b) Given that N divides AC such that $AN : NC = 1 : 2$, express the
following as column vectors:

(i) $\overrightarrow{AC}$　　　　(ii) $\overrightarrow{AN}$　　　　(iii) $\overrightarrow{ON}$

8. In square ABCD, side AB has column vector $\begin{pmatrix} 2 \\ 1 \end{pmatrix}$. Find two

possible column vectors for $\overrightarrow{BC}$.

9. Rectangle KLMN has an area of 10 square units and $\overrightarrow{KL}$ has

column vector $\begin{pmatrix} 5 \\ 0 \end{pmatrix}$. Find two possible column vectors for $\overrightarrow{LM}$.

10. In the diagram, ABCD is a trapezium in which $\overrightarrow{DC} = 2\overrightarrow{AB}$.
If $\overrightarrow{AB} = \mathbf{p}$ and $\overrightarrow{AD} = \mathbf{q}$ express in terms of $\mathbf{p}$ and $\mathbf{q}$:

(a) $\overrightarrow{BD}$　　　　(b) $\overrightarrow{AC}$　　　　(c) $\overrightarrow{BC}$

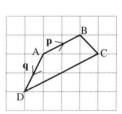

11. Find the image of the vector $\begin{pmatrix} 1 \\ 3 \end{pmatrix}$ after reflection in the following lines:

(a) $y = 0$　　　(b) $x = 0$　　　(c) $y = x$　　　(d) $y = -x$

Modulus of a vector

The modulus of a vector **a** is written $|\mathbf{a}|$ and represents the length (or magnitude) of the vector.

In the diagram above, $\mathbf{a} = \begin{pmatrix} 5 \\ 3 \end{pmatrix}$.

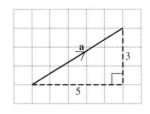

By Pythagoras' theorem, $|\mathbf{a}| = \sqrt{(5^2 + 3^2)}$
$$|\mathbf{a}| = \sqrt{34} \text{ units}$$

In general if $\mathbf{x} = \begin{pmatrix} m \\ n \end{pmatrix}$, $|\mathbf{x}| = \sqrt{(m^2 + n^2)}$

Exercise 9

Questions **1** to **12** refer to the following vectors:

$$\mathbf{a} = \begin{pmatrix} 3 \\ 4 \end{pmatrix} \quad \mathbf{b} = \begin{pmatrix} 4 \\ 1 \end{pmatrix} \quad \mathbf{c} = \begin{pmatrix} 5 \\ 12 \end{pmatrix} \quad \mathbf{d} = \begin{pmatrix} -3 \\ 0 \end{pmatrix}$$

$$\mathbf{e} = \begin{pmatrix} -4 \\ -3 \end{pmatrix} \quad \mathbf{f} = \begin{pmatrix} -3 \\ 6 \end{pmatrix}$$

Find the following, leaving the answer in square root form where necessary.

1. $|\mathbf{a}|$ **2.** $|\mathbf{b}|$ **3.** $|\mathbf{c}|$ **4.** $|\mathbf{d}|$

5. $|\mathbf{e}|$ **6.** $|\mathbf{f}|$ **7.** $|\mathbf{a} + \mathbf{b}|$ **8.** $|\mathbf{c} - \mathbf{d}|$

9. $|2\mathbf{e}|$ **10.** $|\mathbf{f} + 2\mathbf{b}|$

11. (a) Find $|\mathbf{a} + \mathbf{c}|$.   (b) Is $|\mathbf{a} + \mathbf{c}|$ equal to $|\mathbf{a}| + |\mathbf{c}|$?

12. (a) Find $|\mathbf{c} + \mathbf{d}|$.   (b) Is $|\mathbf{c} + \mathbf{d}|$ equal to $|\mathbf{c}| + |\mathbf{d}|$?

13. If $\overrightarrow{AB} = \begin{pmatrix} 3 \\ -1 \end{pmatrix}$ and $\overrightarrow{BC} = \begin{pmatrix} 2 \\ 3 \end{pmatrix}$, find $|\overrightarrow{AC}|$.

14. If $\overrightarrow{PQ} = \begin{pmatrix} 5 \\ -2 \end{pmatrix}$ and $\overrightarrow{QR} = \begin{pmatrix} 0 \\ 1 \end{pmatrix}$, find $|\overrightarrow{PR}|$.

15. If $\overrightarrow{WX} = \begin{pmatrix} 1 \\ 3 \end{pmatrix}$, $\overrightarrow{XY} = \begin{pmatrix} -2 \\ 1 \end{pmatrix}$ and $\overrightarrow{YZ} = \begin{pmatrix} 2 \\ -1 \end{pmatrix}$, find $|\overrightarrow{WZ}|$.

16. Given that $\overrightarrow{OP} = \begin{pmatrix} 0 \\ 5 \end{pmatrix}$ and $\overrightarrow{OQ} = \begin{pmatrix} n \\ 3 \end{pmatrix}$, find:

     (a) $|\overrightarrow{OP}|$ (b) a value for n if $|\overrightarrow{OP}| = |\overrightarrow{OQ}|$

17. Given that $\overrightarrow{OA} = \begin{pmatrix} 5 \\ 12 \end{pmatrix}$ and $\overrightarrow{OB} = \begin{pmatrix} 0 \\ m \end{pmatrix}$, find:

     (a) $|\overrightarrow{OA}|$ (b) a value for m if $|\overrightarrow{OA}| = |\overrightarrow{OB}|$

18. Given that $\overrightarrow{LM} = \begin{pmatrix} -3 \\ 4 \end{pmatrix}$ and $\overrightarrow{MN} = \begin{pmatrix} -15 \\ p \end{pmatrix}$, find:

   (a) $|\overrightarrow{LM}|$ (b) a value for p if $|\overrightarrow{MN}| = 3|\overrightarrow{LM}|$

19. **a** and **b** are two vectors and $|\mathbf{a}| = 3$.
   Find the value of $|\mathbf{a} + \mathbf{b}|$ when:
   (a) $\mathbf{b} = 2\mathbf{a}$
   (b) $\mathbf{b} = -3\mathbf{a}$
   (c) **b** is perpendicular to **a** and $|\mathbf{b}| = 4$

20. **r** and **s** are two vectors and $|\mathbf{r}| = 5$.
   Find the value of $|\mathbf{r} + \mathbf{s}|$ when:
   (a) $\mathbf{s} = 5\mathbf{r}$
   (b) $\mathbf{s} = -2\mathbf{r}$
   (c) **r** is perpendicular to **s** and $|\mathbf{s}| = 5$
   (d) **s** is perpendicular to $(\mathbf{r} + \mathbf{s})$ and $|\mathbf{s}| = 3$

8.5 Vector geometry

Example

In the diagram, $\overrightarrow{OD} = 2\overrightarrow{OA}$, $\overrightarrow{OE} = 4\overrightarrow{OB}$, $\overrightarrow{OA} = \mathbf{a}$ and $\overrightarrow{OB} = \mathbf{b}$.
(a) Express $\overrightarrow{OD}$ and $\overrightarrow{OE}$ in terms of **a** and **b** respectively.
(b) Express $\overrightarrow{BA}$ in terms of **a** and **b**.
(c) Express $\overrightarrow{ED}$ in terms of **a** and **b**.
(d) Given that $\overrightarrow{BC} = 3\overrightarrow{BA}$, express $\overrightarrow{OC}$ in terms of **a** and **b**.
(e) Express $\overrightarrow{EC}$ in terms of **a** and **b**.
(f) Hence show that the points E, D and C lie on a straight line.

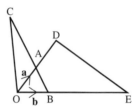

(a) $\overrightarrow{OD} = 2\mathbf{a}$
   $\overrightarrow{OE} = 4\mathbf{b}$
(b) $\overrightarrow{BA} = -\mathbf{b} + \mathbf{a}$
(c) $\overrightarrow{ED} = -4\mathbf{b} + 2\mathbf{a}$
(d) $\overrightarrow{OC} = \overrightarrow{OB} + \overrightarrow{BC}$
   $\overrightarrow{OC} = \mathbf{b} + 3(-\mathbf{b} + \mathbf{a})$
   $\overrightarrow{OC} = -2\mathbf{b} + 3\mathbf{a}$
(e) $\overrightarrow{EC} = \overrightarrow{EO} + \overrightarrow{OC}$
   $\overrightarrow{EC} = -4\mathbf{b} + (-2\mathbf{b} + 3\mathbf{a})$
   $\overrightarrow{EC} = -6\mathbf{b} + 3\mathbf{a}$
(f) Using the results for $\overrightarrow{ED}$ and $\overrightarrow{EC}$, we see that $\overrightarrow{EC} = \frac{3}{2}\overrightarrow{ED}$.

Since $\overrightarrow{EC}$ and $\overrightarrow{ED}$ are parallel vectors which both pass through the point E, the points E, D and C must lie on a straight line.

> $\overrightarrow{OD}$, $\overrightarrow{OC}$ and $\overrightarrow{OB}$ are called **position vectors**.
> The **position vector** $\overrightarrow{OB}$ gives the position of B relative to the origin, O.

Exercise 10

1. $\overrightarrow{OD} = 2\overrightarrow{OA}$,
$\overrightarrow{OE} = 3\overrightarrow{OB}$,
$\overrightarrow{OA} = \mathbf{a}$ and
$\overrightarrow{OB} = \mathbf{b}$.

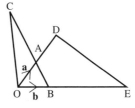

(a) Express $\overrightarrow{OD}$ and $\overrightarrow{OE}$ in terms of $\mathbf{a}$ and $\mathbf{b}$ respectively.

(b) Express $\overrightarrow{BA}$ in terms of $\mathbf{a}$ and $\mathbf{b}$.

(c) Express $\overrightarrow{ED}$ in terms of $\mathbf{a}$ and $\mathbf{b}$.

(d) Given that $\overrightarrow{BC} = 4\overrightarrow{BA}$, express $\overrightarrow{OC}$ in terms of $\mathbf{a}$ and $\mathbf{b}$.

(e) Express $\overrightarrow{EC}$ in terms of $\mathbf{a}$ and $\mathbf{b}$.

(f) Use the results for $\overrightarrow{ED}$ and $\overrightarrow{EC}$ to show that points E, D and C lie on a straight line.

2. $\overrightarrow{OY} = 2\overrightarrow{OB}$,
$\overrightarrow{OX} = \frac{5}{2}\overrightarrow{OA}$,
$\overrightarrow{OA} = \mathbf{a}$ and
$\overrightarrow{OB} = \mathbf{b}$.

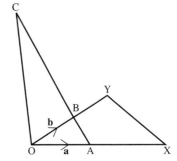

(a) Express $\overrightarrow{OY}$ and $\overrightarrow{OX}$ in terms of $\mathbf{b}$ and $\mathbf{a}$ respectively.

(b) Express $\overrightarrow{AB}$ in terms of $\mathbf{a}$ and $\mathbf{b}$.

(c) Express $\overrightarrow{XY}$ in terms of $\mathbf{a}$ and $\mathbf{b}$.

(d) Given that $\overrightarrow{AC} = 6\overrightarrow{AB}$, express $\overrightarrow{OC}$ in terms of $\mathbf{a}$ and $\mathbf{b}$.

(e) Express $\overrightarrow{XC}$ in terms of $\mathbf{a}$ and $\mathbf{b}$.

(f) Use the results for $\overrightarrow{XY}$ and $\overrightarrow{XC}$ to show that points X, Y and C lie on a straight line.

3. $\overrightarrow{OA} = \mathbf{a}$,
$\overrightarrow{OB} = \mathbf{b}$,
$\overrightarrow{AQ} = \frac{1}{2}\mathbf{a}$,
$\overrightarrow{BR} = \mathbf{b}$ and
$\overrightarrow{AP} = 2\overrightarrow{BA}$.

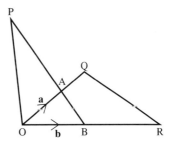

(a) Express $\overrightarrow{BA}$ and $\overrightarrow{BP}$ in terms of $\mathbf{a}$ and $\mathbf{b}$.

(b) Express $\overrightarrow{RQ}$ in terms of $\mathbf{a}$ and $\mathbf{b}$.

(c) Express $\overrightarrow{QA}$ and $\overrightarrow{QP}$ in terms of $\mathbf{a}$ and $\mathbf{b}$.

(d) Using the vectors for $\overrightarrow{RQ}$ and $\overrightarrow{QP}$, show that R, Q and P lie on a straight line.

4. In the diagram, $\overrightarrow{OA} = \mathbf{a}$ and $\overrightarrow{OB} = \mathbf{b}$, M is the mid-point of OA and P lies on AB such that $\overrightarrow{AP} = \frac{2}{3}\overrightarrow{AB}$.

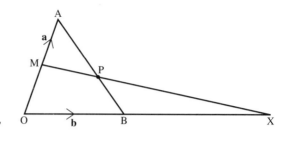

(a) Express $\overrightarrow{AB}$ and $\overrightarrow{AP}$ in terms of $\mathbf{a}$ and $\mathbf{b}$.

(b) Express $\overrightarrow{MA}$ and $\overrightarrow{MP}$ in terms of $\mathbf{a}$ and $\mathbf{b}$.

(c) If X lies on OB produced such that $OB = BX$, express $\overrightarrow{MX}$ in terms of $\mathbf{a}$ and $\mathbf{b}$.

(d) Show that MPX is a straight line.

5. $\overrightarrow{OP} = \mathbf{a}$,

$\overrightarrow{OA} = 3\mathbf{a}$,

$\overrightarrow{OB} = \mathbf{b}$ and

M is the mid-point of AB.

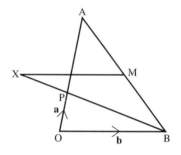

(a) Express $\overrightarrow{BP}$ and $\overrightarrow{AB}$ in terms of $\mathbf{a}$ and $\mathbf{b}$.

(b) Express $\overrightarrow{MB}$ in terms of $\mathbf{a}$ and $\mathbf{b}$.

(c) If X lies on BP produced so that $\overrightarrow{BX} = k\overrightarrow{BP}$, express $\overrightarrow{MX}$ in terms of $\mathbf{a}$, $\mathbf{b}$ and k.

(d) Find the value of k if MX is parallel to BO.

6. AC is parallel to OB,

$\overrightarrow{AX} = \frac{1}{4}\overrightarrow{AB}$,

$\overrightarrow{OA} = \mathbf{a}$,

$\overrightarrow{OB} = \mathbf{b}$ and

$\overrightarrow{AC} = m\mathbf{b}$.

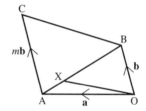

(a) Express $\overrightarrow{AB}$ in terms of $\mathbf{a}$ and $\mathbf{b}$.

(b) Express $\overrightarrow{AX}$ in terms of $\mathbf{a}$ and $\mathbf{b}$.

(c) Express $\overrightarrow{BC}$ in terms of $\mathbf{a}$, $\mathbf{b}$ and m.

(d) Given that OX is parallel to BC, find the value of m.

7. CY is parallel to OD,

$\overrightarrow{CX} = \frac{1}{5}\overrightarrow{CD}$,

$\overrightarrow{OC} = \mathbf{c}$,

$\overrightarrow{OD} = \mathbf{d}$ and

$\overrightarrow{CY} = n\mathbf{d}$.

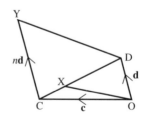

(a) Express $\overrightarrow{CD}$ in terms of $\mathbf{c}$ and $\mathbf{d}$.

(b) Express $\overrightarrow{CX}$ in terms of $\mathbf{c}$ and $\mathbf{d}$.

(c) Express $\overrightarrow{OX}$ in terms of $\mathbf{c}$ and $\mathbf{d}$.

(d) Express $\overrightarrow{DY}$ in terms of $\mathbf{c}$, $\mathbf{d}$ and n.

(e) Given that OX is parallel to DY, find the value of n.

8. M is the mid-point of AB,
N is the mid-point of OB,
$\overrightarrow{OA} = \mathbf{a}$ and
$\overrightarrow{OB} = \mathbf{b}$.

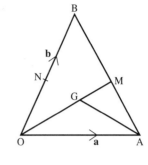

(a) Express $\overrightarrow{AB}$, $\overrightarrow{AM}$ and $\overrightarrow{OM}$ in terms of $\mathbf{a}$ and $\mathbf{b}$.

(b) Given that G lies on OM such that
OG : GM = 2 : 1, express $\overrightarrow{OG}$ in terms of $\mathbf{a}$ and $\mathbf{b}$.

(c) Express $\overrightarrow{AG}$ in terms of $\mathbf{a}$ and $\mathbf{b}$.

(d) Express $\overrightarrow{AN}$ in terms of $\mathbf{a}$ and $\mathbf{b}$.

(e) Show that $\overrightarrow{AG} = m\overrightarrow{AN}$ and find the value of m.

9. M is the mid-point of AC and N is the mid-point of OB,
$\overrightarrow{OA} = \mathbf{a}$,
$\overrightarrow{OB} = \mathbf{b}$ and
$\overrightarrow{OC} = \mathbf{c}$.

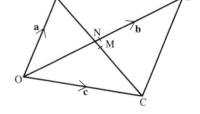

(a) Express $\overrightarrow{AB}$ in terms of $\mathbf{a}$ and $\mathbf{b}$.

(b) Express $\overrightarrow{ON}$ in terms of $\mathbf{b}$.

(c) Express $\overrightarrow{AC}$ in terms of $\mathbf{a}$ and $\mathbf{c}$.

(d) Express $\overrightarrow{AM}$ in terms of $\mathbf{a}$ and $\mathbf{c}$.

(e) Express $\overrightarrow{OM}$ in terms of $\mathbf{a}$ and $\mathbf{c}$.

(f) Express $\overrightarrow{NM}$ in terms of $\mathbf{a}$, $\mathbf{b}$ and $\mathbf{c}$.

(g) If N and M coincide, write down an equation connecting $\mathbf{a}$, $\mathbf{b}$
and $\mathbf{c}$.

10. $\overrightarrow{OA} = \mathbf{a}$ and
$\overrightarrow{OB} = \mathbf{b}$.

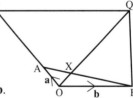

(a) Express $\overrightarrow{BA}$ in terms of $\mathbf{a}$ and $\mathbf{b}$.

(b) Given that $\overrightarrow{BX} = m\overrightarrow{BA}$, show that $\overrightarrow{OX} = m\mathbf{a} + (1 - m)\mathbf{b}$.

(c) Given that $OP = 4\mathbf{a}$ and $\overrightarrow{PQ} = 2\mathbf{b}$, express $\overrightarrow{OQ}$ in terms of $\mathbf{a}$ and $\mathbf{b}$.

(d) Given that $\overrightarrow{OX} = n\overrightarrow{OQ}$ use the results for $\overrightarrow{OX}$ and $\overrightarrow{OQ}$ to find
the values of m and n.

11. X is the mid-point of OD, Y lies on CD such that
$\overrightarrow{CY} = \frac{1}{4}\overrightarrow{CD}$,
$\overrightarrow{OC} = \mathbf{c}$ and
$\overrightarrow{OD} = \mathbf{d}$.

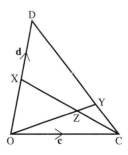

(a) Express $\overrightarrow{CD}$, $\overrightarrow{CY}$ and $\overrightarrow{OY}$ in terms of $\mathbf{c}$ and $\mathbf{d}$.

(b) Express $\overrightarrow{CX}$ in terms of $\mathbf{c}$ and $\mathbf{d}$.

(c) Given that $\overrightarrow{CZ} = h\overrightarrow{CX}$, express $\overrightarrow{OZ}$ in terms of $\mathbf{c}$, $\mathbf{d}$ and h.

(d) If $\overrightarrow{OZ} = k\overrightarrow{OY}$, form an equation and hence find the values of h and k.

8.6 Functions

The idea of a function is used in almost every branch of mathematics.
The two common notations used are:
(a) $f(x) = x^2 + 4$ (b) $f : x \mapsto x^2 + 4$

We may interpret (b) as follows: 'function f such that x is mapped onto $x^2 + 4$'.

Example

If $f(x) = 3x - 1$ and $g(x) = 1 - x^2$ find:

(a) $f(2)$ (b) $f(-2)$ (c) $g(0)$ (d) $g(3)$ (e) x if $f(x) = 1$

(a) $f(2) = 5$ (b) $f(-2) = -7$ (c) $g(0) = 1$ (d) $g(3) = -8$

(e) If $\qquad f(x) = 1$
Then $\quad 3x - 1 = 1$
$$3x = 2$$
$$x = \tfrac{2}{3}$$

Flow diagrams

The function f in the example consisted of two simpler functions as
illustrated by a flow diagram.

$$x \longrightarrow \boxed{\text{multiply by 3}} \xrightarrow{\ 3x\ } \boxed{\text{subtract 1}} \xrightarrow{\ 3x - 1\ }$$

It is obviously important to 'multiply by 3' and 'subtract 1' in the
correct order.

Example

Draw flow diagrams for the functions:

(a) $f : x \mapsto (2x + 5)^2$,   (b) $g(x) = \dfrac{5 - 7x}{3}$

(a) $x \rightarrow \boxed{\text{multiply by 2}} \xrightarrow{\ 2x\ } \boxed{\text{add 5}} \xrightarrow{\ 2x + 5\ } \boxed{\text{square}} \xrightarrow{\ (2x + 5)^2\ }$

(b)
$$x \rightarrow \boxed{\text{multiply by }(-7)} \xrightarrow{\ -7x\ } \boxed{\text{add 5}} \xrightarrow{\ 5 - 7x\ } \boxed{\text{divide by 3}} \xrightarrow{\ \frac{5 - 7x}{3}\ }$$

Exercise 11

1. Given the functions $h : x \mapsto x^2 + 1$ and $g : x \mapsto 10x + 1$. Find:
   (a) $h(2), h(-3), h(0)$ (b) $g(2), g(10), g(-3)$

In questions **2** to **15**, draw a flow diagram for each function.

2. $f : x \mapsto 5x + 4$ 3. $f : x \mapsto 3(x - 4)$ 4. $f : x \mapsto (2x + 7)^2$

5. $f : x \mapsto \left(\dfrac{9 + 5x}{4} \right)$ 6. $f : x \mapsto \dfrac{4 - 3x}{5}$ 7. $f : x \mapsto 2x^2 + 1$

8. $f : x \mapsto \dfrac{3x^2}{2} + 5$ 9. $f : x \mapsto \sqrt{(4x - 5)}$ 10. $f : x \mapsto 4\sqrt{(x^2 + 10)}$

11. $f : x \mapsto (7 - 3x)^2$

12. $f : x \mapsto 4(3x + 1)^2 + 5$

13. $f : x \mapsto 5 - x^2$

14. $f : x \mapsto \dfrac{10\sqrt{(x^2 + 1)} + 6}{4}$

15. $f : x \mapsto \left(\dfrac{x^3}{4} + 1\right)^2 - 6$

For questions **16**, **17** and **18**, the functions f, g and h are defined as follows:

$f : x \mapsto 1 - 2x$

$g : x \mapsto \dfrac{x^3}{10}$

$h : x \mapsto \dfrac{12}{x}$

16. Find:
  (a) $f(5)$, $f(-5)$, $f(\tfrac{1}{4})$
  (b) $g(2)$, $g(-3)$, $g(\tfrac{1}{2})$
  (c) $h(3)$, $h(10)$, $h(\tfrac{1}{3})$

17. Find:
  (a) x if $f(x) = 1$ (b) x if $f(x) = -11$ (c) x if $h(x) = 1$

18. Find:
  (a) y if $g(y) = 100$ (b) z if $h(z) = 24$ (c) w if $g(w) = 0 \cdot 8$

For questions **19** and **20**, the functions k, l and m are defined as follows:

$k : x \mapsto \dfrac{2x^2}{3}$

$l : x \mapsto \sqrt{[(y - 1)(y - 2)]}$
$m : x \mapsto 10 - x^2$

19. Find:
  (a) $k(3)$, $k(6)$, $k(-3)$
  (b) $l(2)$, $l(0)$, $l(4)$
  (c) $m(4)$, $m(-2)$, $m(\tfrac{1}{2})$

20. Find:
  (a) x if $k(x) = 6$ (b) x if $m(x) = 1$
  (c) y if $k(y) = 2\tfrac{2}{3}$ (d) p if $m(p) = -26$

21. $f(x)$ is defined as the product of the digits of x,
  e.g. $f(12) = 1 \times 2 = 2$
  (a) Find: (i) $f(25)$ (ii) $f(713)$
  (b) If x is an integer with three digits, find:
      (i) x such that $f(x) = 1$
      (ii) the largest x such that $f(x) = 4$
      (iii) the largest x such that $f(x) = 0$
      (iv) the smallest x such that $f(x) = 2$

22. g(x) is defined as the sum of the prime factors of x,
e.g. g(12) = 2 + 3 = 5. Find:
(a) g(10)   (b) g(21)   (c) g(36)
(d) g(99)   (e) g(100)   (f) g(1000)

23. h(x) is defined as the number of letters in the English word
describing the number x, e.g. h(1) = 3. Find:
(a) h(2)   (b) h(11)   (c) h(18)
(d) the largest value of x for which h(x) = 3

24. If f : x ↦ next prime number greater than x, find:
(a) f(7)   (b) f(14)   (c) f[f(3)]

25. If g : x → 2^x + 1, find:
(a) g(2)   (b) g(4)   (c) g(−1)
(d) the value of x if g(x) = 9

26. The function f is defined as f : x → ax + b where a and b are
constants.
If f(1) = 8 and f(4) = 17, find the values of a and b.

27. The function g is defined as g(x) = ax^2 + b where a and b are constants.
If g(2) = 3 and g(−3) = 13, find the values of a and b.

28. Functions h and k are defined as follows:
h : x ↦ x^2 + 1, k : x ↦ ax + b, where a and b are constants.
If h(0) = k(0) and k(2) = 15, find the values of a and b.

Composite functions

The function f(x) = 3x + 2 is itself a composite function, consisting of
two simpler functions: 'multiply by 3' and 'add 2'.

If f(x) = 3x + 2 and g(x) = x^2 then f [g(x)] is a composite function
where g is performed first and then f is performed on the result of g.
f [g(x)] is usually abbreviated to fg(x).

The function fg may be found using a flow diagram.

Thus

$$x \rightarrow \boxed{\text{square}} \xrightarrow{x^2} \boxed{\text{multiply by 3}} \xrightarrow{3x^2} \boxed{\text{add 2}} \xrightarrow{3x^2 + 2}$$
$$\quad\quad\quad g \quad\quad\quad\quad\quad\quad f$$

fg(x) = $3x^2$ + 2

Inverse functions

If a function f maps a number n onto m, then the inverse function f^{-1}
maps m onto n. The inverse of a given function is found using a flow
diagram.

Example

Method 1

Find the inverse of f where $f : x \rightarrow \dfrac{5x - 2}{3}$.

(a) Draw a flow diagram for f.

$$\xrightarrow{\;x\;} \boxed{\text{multiply by 5}} \xrightarrow{5x} \boxed{\text{subtract 2}} \xrightarrow{5x-2} \boxed{\text{divide by 3}} \xrightarrow{\;\frac{5x-2}{3}\;}$$

(b) Draw a new flow diagram with each operation replaced by its inverse. Start with x on the right.

$$\xleftarrow{\;\frac{3x+2}{5}\;} \boxed{\text{divide by 5}} \xleftarrow{3x+2} \boxed{\text{add 2}} \xleftarrow{3x} \boxed{\text{multiply by 3}} \xleftarrow{\;x\;}$$

Thus the inverse of f is given by

$$f^{-1} : x \mapsto \frac{3x + 2}{5} \quad \text{or} \quad f^{-1}(x) = \frac{3x + 2}{5}$$

Method 2

Again find the inverse of f where $f : x \rightarrow \dfrac{5x - 2}{3}$

Let $y = \dfrac{5x - 2}{3}$

Rearrange this equation to make x the subject:

$$3y = 5x - 2$$

$$3y + 2 = 5x$$

$$x = \frac{3y + 2}{5}$$

For an inverse function we interchange x and y.

So the inverse function is $\dfrac{3x + 2}{5}$.

Many people prefer this algebraic method. You should use the method which you find easier.

Exercise 12

For questions **1** and **2**, the functions f, g and h are as follows:

$f : x \mapsto 4x$
$g : x \mapsto x + 5$
$h : x \mapsto x^2$

1. Find the following in the form '$x \mapsto \ldots$'

   (a) fg   (b) gf   (c) hf   (d) fh
   (e) gh   (f) fgh   (g) hfg

2. Find:

(a) x if $hg(x) = h(x)$ (b) x if $fh(x) = gh(x)$

For questions **3**, **4** and **5**, the functions f, g and h are as follows:

$f : x \mapsto 2x$
$g : x \mapsto x - 3$
$h : x \mapsto x^2$

3. Find the following in the form '$x \mapsto \ldots$'

(a) fg   (b) gf   (c) gh
(d) hf   (e) ghf   (f) hgf

4. Evaluate:

(a) fg(4)   (b) gf(7)   (c) gh(−3)
(d) fgf(2)   (e) ggg(10)   (f) hfh(−2)

5. Find:

(a) x if $f(x) = g(x)$ (b) x if $hg(x) = gh(x)$
(c) x if $gf(x) = 0$ (d) x if $fg(x) = 4$

For questions **6**, **7** and **8**, the functions l, m and n are as follows:

$l : x \mapsto 2x + 1$
$m : x \mapsto 3x - 1$
$n : x \mapsto x^2$

6. Find the following in the form '$x \mapsto \ldots$'

(a) lm   (b) ml   (c) ln
(d) nm   (e) lnm   (f) mln

7. Find:

(a) lm(2)   (b) nl(1)   (c) mn(−2)
(d) mm(2)   (e) nln(2)   (f) llm(0)

8. Find:

(a) x if $l(x) = m(x)$
(b) two values of x if $nl(x) = nm(x)$
(c) x if $ln(x) = mn(x)$

In questions **9** to **22**, find the inverse of each function in the form
'$x \mapsto \ldots$'

9. $f : x \mapsto 5x - 2$ **10.** $f : x \mapsto 5(x - 2)$ **11.** $f : x \mapsto 3(2x + 4)$

12. $g : x \mapsto \dfrac{2x + 1}{3}$ **13.** $f : x \mapsto \dfrac{3(x - 1)}{4}$ **14.** $g : x \mapsto 2(3x + 4) - 6$

15. $h : x \mapsto \frac{1}{2}(4 + 5x) + 10$ **16.** $k : x \mapsto -7x + 3$ **17.** $j : x \mapsto \dfrac{12 - 5x}{3}$

18. $l : x \mapsto \dfrac{4-x}{3} + 2$

19. $m : x \mapsto \dfrac{\left[\dfrac{(2x-1)}{4}\right] - 3}{5}$

20. $f : x \mapsto \dfrac{3(10-2x)}{7}$

21. $g : x \mapsto \dfrac{\left[\dfrac{x}{4}+6\right]}{5} + 7$

22. A calculator has the following function buttons:

$x \mapsto x^2; \quad x \mapsto \sqrt{x}; \quad x \mapsto \frac{1}{x}; \quad x \mapsto \log x;$

$x \mapsto \ln x; \quad x \mapsto \sin x; \quad x \mapsto \cos x; \quad x \mapsto \tan x; \quad x \mapsto x!$

Find which button was used for the following input/outputs:

	x! is 'x factorial'
	$4! = 4 \times 3 \times 2 \times 1$
	$3! = 3 \times 2 \times 1$ etc

(a) $1\,000\,000 \rightarrow 1000$

(b) $1000 \rightarrow 3$

(c) $3 \rightarrow 6$

(d) $0{\cdot}2 \rightarrow 0{\cdot}04$

(e) $10 \rightarrow 0{\cdot}1$

(f) $45 \rightarrow 1$

(g) $0{\cdot}5 \rightarrow 2$

(h) $64 \rightarrow 8$

(i) $60 \rightarrow 0{\cdot}5$

(j) $1 \rightarrow 0$

(k) $135 \rightarrow -1$

(l) $10 \rightarrow 3\,628\,800$

(m) $0 \rightarrow 1$

(n) $30 \rightarrow 0{\cdot}5$

(o) $90 \rightarrow 0$

(p) $0{\cdot}4 \rightarrow 2{\cdot}5$

(q) $4 \rightarrow 24$

(r) $1\,000\,000 \rightarrow 6$

Revision exercise 8A

1. Given that $\mathscr{E} = \{1, 2, 3, 4, 5, 6, 7, 8\}$,
A = $\{1, 3, 5\}$, B = $\{5, 6, 7\}$, list the members of the sets:

(a) $A \cap B$ (b) $A \cup B$ (c) A'

(d) $A' \cap B'$ (e) $A \cup B'$

2. The sets P and Q are such that $n(P \cup Q) = 50$, $n(P \cap Q) = 9$ and $n(P) = 27$. Find the value of $n(Q)$.

3. Draw three diagrams similar to Figure 1, and shade the following

(a) $Q \cap R'$ (b) $(P \cup Q) \cap R$ (c) $(P \cap Q) \cap R'$

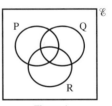

Figure 1

4. Describe the shaded regions in Figures 2 and 3.

(a)

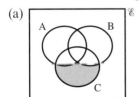

Figure 2

(b)

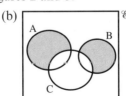

Figure 3

5. Given that $\mathscr{E} = \{\text{people on a train}\}$, $M = \{\text{males}\}$,
$T = \{\text{people over 25 years old}\}$ and $S = \{\text{snooker players}\}$,
  (a) express in set notation:
      (i) all the snooker players are over 25
      (ii) some snooker players are women
  (b) express in words: $T \cap M' = \varnothing$

6. The figures in the diagram indicate the number of elements in each
subset of $\mathscr{E}$.
  (a) Find $n(P \cap R)$.
  (b) Find $n(Q \cup R)'$.
  (c) Find $n(P' \cap Q')$.

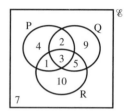

7. In $\triangle OPR$, the mid-point of PR is M.
If $\overrightarrow{OP} = \mathbf{p}$ and $\overrightarrow{OR} = \mathbf{r}$, find in terms of $\mathbf{p}$ and $\mathbf{r}$:
  (a) $\overrightarrow{PR}$ (b) $\overrightarrow{PM}$ (c) $\overrightarrow{OM}$

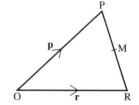

8. If $\mathbf{a} = \begin{pmatrix} 1 \\ 4 \end{pmatrix}$ and $\mathbf{b} = \begin{pmatrix} -3 \\ 4 \end{pmatrix}$, find:
  (a) $|\mathbf{b}|$ (b) $|\mathbf{a} + \mathbf{b}|$ (c) $|2\mathbf{a} - \mathbf{b}|$

9. If $4\begin{pmatrix} 1 \\ 3 \end{pmatrix} + 2\begin{pmatrix} 1 \\ m \end{pmatrix} = 3\begin{pmatrix} n \\ -6 \end{pmatrix}$, find the values of m and n.

10. The points O, A and B have coordinates $(0, 0)$, $(5, 0)$ and $(-1, 4)$
respectively. Write as column vectors.
  (a) $\overrightarrow{OB}$ (b) $\overrightarrow{OA} + \overrightarrow{OB}$ (c) $\overrightarrow{OA} - \overrightarrow{OB}$
  (d) $\overrightarrow{OM}$ where M is the mid-point of AB.

11. In the parallelogram OABC, M is the mid-point of AB and N is the
mid-point of BC.
If $\overrightarrow{OA} = \mathbf{a}$ and $\overrightarrow{OC} = \mathbf{c}$, express in terms of $\mathbf{a}$ and $\mathbf{c}$:
  (a) $\overrightarrow{CA}$ (b) $\overrightarrow{ON}$ (c) $\overrightarrow{NM}$
  Describe the relationship between CA and NM.

12. The vectors $\mathbf{a}$, $\mathbf{b}$, $\mathbf{c}$ are given by:
$$\mathbf{a} = \begin{pmatrix} 1 \\ 5 \end{pmatrix}, \mathbf{b} = \begin{pmatrix} -2 \\ 1 \end{pmatrix}, \mathbf{c} = \begin{pmatrix} -1 \\ 17 \end{pmatrix}$$
Find numbers m and n so that $m\mathbf{a} + n\mathbf{b} = \mathbf{c}$.

13. Given that $\overrightarrow{OP} = \begin{pmatrix} 3 \\ 2 \end{pmatrix}$, $\overrightarrow{OQ} = \begin{pmatrix} 0 \\ 4 \end{pmatrix}$

and that M is the mid-point

of PQ, express as column vectors:

(a) $\overrightarrow{PQ}$ (b) $\overrightarrow{PM}$ (c) $\overrightarrow{OM}$

14. Given $f : x \mapsto 2x - 3$ and $g : x \mapsto x^2 - 1$, find:

(a) $f(-1)$ (b) $g(-1)$

(c) $fg(-1)$ (d) $gf(3)$

Write the function ff in the form

'$ff : x \mapsto \ldots$'

15. If $f : x \mapsto 3x + 4$ and $h : x \mapsto \dfrac{x - 2}{5}$

express f^{-1} and h^{-1} in the form '$x \mapsto \ldots$'.

Find: (a) $f^{-1}(13)$ (b) the value of z if $f(z) = 20$

16. Given that $f(x) = x - 5$, find:

(a) the value of s such that $f(s) = -2$

(b) the values of t such that $t \times f(t) = 0$

Examination exercise 8B

1. (a) Shade the region $A \cap B$.

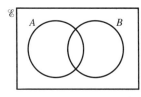

[1]

(b) Shade the region $(A \cup B)'$.

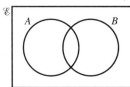

[1]

(c) Shade the complement of set B.

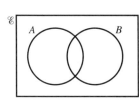

[1]

Cambridge IGCSE Mathematics 0580
Paper 2 Q11 November 2006

2. $\mathscr{E} = \{1, 2, 3, 4, 5, 6, 7, 9, 11, 16\}$ $P = \{2, 3, 5, 7, 11\}$ $S = \{1, 4, 9, 16\}$ $M = \{3, 6, 9\}$

(a) Draw a Venn diagram to show this information.   [2]

(b) Write down the value of $n(M' \cap P)$.   [1]

Cambridge IGCSE Mathematics 0580
Paper 22 Q12 June 2008

3. A and B are sets.

Write the following sets in their simplest form.

(a) $A \cap A'$.   [1]

(b) $A \cup A'$.   [1]

(c) $(A \cap B) \cup (A \cap B)'$.   [1]

Cambridge IGCSE Mathematics 0580
Paper 2 Q12 November 2007

4.

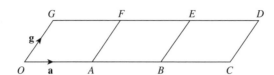

The diagram is made from three identical parallelograms.

O is the origin. $\overrightarrow{OA} = \mathbf{a}$ and $\overrightarrow{OG} = \mathbf{g}$.
Write down in terms of $\mathbf{a}$ and $\mathbf{g}$

(a) $\overrightarrow{GB}$,   [1]

(b) the position vector of the centre of the parallelogram
    $BCDE$.   [1]

Cambridge IGCSE Mathematics 0580
Paper 21 Q8 June 2009

5.

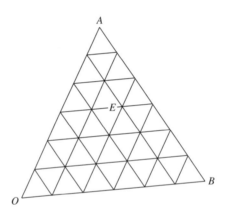

O is the origin, $\overrightarrow{OA} = \mathbf{a}$, and $\overrightarrow{OB} = \mathbf{b}$.

(a) C has position vector $\dfrac{1}{3}\mathbf{a} + \dfrac{2}{3}\mathbf{b}$.

    Mark the point C on the diagram.   [1]

(b) Write down, in terms of $\mathbf{a}$ and $\mathbf{b}$, the position vector of the
    point E.   [1]

(c) Find, in terms of $\mathbf{a}$ and $\mathbf{b}$, the vector $\overrightarrow{EB}$.   [2]

Cambridge IGCSE Mathematics 0580
Paper 2 Q15 November 2007

6.

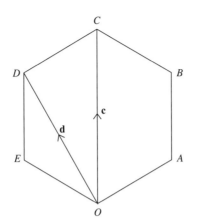

Not to scale

OABCDE is a regular hexagon.
With *O* as origin the position vector of *C* is **c** and the position
vector of *D* is **d**.

(a) Find, in terms of **c** and **d**,

    (i) $\overrightarrow{DC}$,   [1]

    (ii) $\overrightarrow{OE}$,   [2]

    (iii) the position vector of *B*.   [2]

(b) The sides of the hexagon are each of length 8 cm.
    Calculate

    (i) the size of angle *ABC*,   [1]

    (ii) the area of triangle *ABC*,   [2]

    (iii) the length of the straight line *AC*,   [3]

    (iv) the area of the hexagon.   [3]

Cambridge IGCSE Mathematics 0580
Paper 4 Q5 June 2005

7. $f(x) = x^3 - 3x^2 + 6x - 4$ and $g(x) = 2x - 1$.
Find:

(a) $f(-1)$,   [1]

(b) $gf(x)$,   [2]

(c) $g^{-1}(x)$.   [2]

Cambridge IGCSE Mathematics 0580
Paper 21 Q18 June 2008

8. f: $x \mapsto 5 - 3x$.
   (a) Find $f(-1)$.   [1]
   (b) Find $f^{-1}(x)$.   [2]
   (c) Find $ff^{-1}(8)$.   [1]

Cambridge IGCSE Mathematics 0580
Paper 2 Q15 November 2006

9. $f(x) = \tan x°$, $g(x) = 2x + 6$.
   Find
   (a) $f(45)$,   [1]
   (b) $fg(87)$,   [2]
   (c) $g^{-1}(f(x))$.   [2]

Cambridge IGCSE Mathematics 0580
Paper 22 Q15 November 2008

10. $f(x) = 2x - 1$, $g(x) = \dfrac{3}{x} + 1$, $h(x) = 2^x$.
   (a) Find the value of $fg(6)$.   [1]
   (b) Write, as **single fraction**, $gf(x)$ in terms of x.   [3]
   (c) Find $g^{-1}(x)$.   [3]
   (d) Find $hh(3)$.   [2]
   (e) Find x when $h(x) = g\left(-\dfrac{24}{7}\right)$ [2]

Cambridge IGCSE Mathematics 0580
Paper 4 Q8 November 2007

9 MATRICES AND TRANSFORMATIONS

Albert Einstein (1879–1955) working as a patent office clerk in Berne, was responsible for the greatest advance in mathematical physics of this century. His theories of relativity, put forward in 1905 and 1915 were based on the postulate that the velocity of light is absolute: mass, length and even time can only be measured relative to the observer and undergo transformation when studied by another observer. His formula $E = mc^2$ laid the foundations of nuclear physics, a fact that he came to deplore in its application to warfare. In 1933 he moved from Nazi Germany and settled in America.

36 Display information in the form of a matrix; calculate the sum and product of two matrices; calculate the product of a matrix and a scalar quantity; calculate the determinant and inverse

37 Use the following transformations: reflection, rotation, translation, enlargement, shear, stretching and their combinations; identify and give descriptions of transformations connecting given figures; describe transformations using coordinates and matrices

9.1 Matrix operations

Addition and subtraction

Matrices of the same order are added (or subtracted) by adding (or subtracting) the corresponding elements in each matrix.

Example

$$\begin{pmatrix} 2 & -4 \\ 3 & 0 \end{pmatrix} + \begin{pmatrix} 3 & 5 \\ -1 & 7 \end{pmatrix} = \begin{pmatrix} 5 & 1 \\ 2 & 7 \end{pmatrix}$$

Multiplication by a number

Each element of the matrix is multiplied by the multiplying number.

Example

$$3 \times \begin{pmatrix} 2 & -1 \\ 1 & 4 \end{pmatrix} = \begin{pmatrix} 6 & -3 \\ 3 & 12 \end{pmatrix}$$

Multiplication by another matrix

For 2×2 matrices,

$$\begin{pmatrix} a & b \\ c & d \end{pmatrix} \begin{pmatrix} w & x \\ y & z \end{pmatrix} = \begin{pmatrix} aw + by & ax + bz \\ cw + dy & cx + dz \end{pmatrix}$$

The same process is used for matrices of other orders.

Example

Perform the following multiplications.

(a) $\begin{pmatrix} 3 & 2 \\ 4 & 1 \end{pmatrix} \begin{pmatrix} 2 & 1 \\ 1 & 5 \end{pmatrix} = \begin{pmatrix} 6+2 & 3+10 \\ 8+1 & 4+5 \end{pmatrix} = \begin{pmatrix} 8 & 13 \\ 9 & 9 \end{pmatrix}$

(b) $\begin{pmatrix} 2 & 1 & -2 \\ 0 & 1 & 3 \end{pmatrix} \begin{pmatrix} 1 & 0 \\ 1 & -2 \\ 4 & 3 \end{pmatrix} = \begin{pmatrix} 2+1-8 & 0-2-6 \\ 0+1+12 & 0-2+9 \end{pmatrix}$

$$= \begin{pmatrix} -5 & -8 \\ 13 & 7 \end{pmatrix}$$

Matrices may be multiplied only if they are *compatible*. The number of *columns* in the left-hand matrix must equal the number of *rows* in the right-hand matrix.

Matrix multiplication is not commutative, i.e. for square matrices **A** and **B**, the product **AB** does not necessarily equal the product **BA**.

Note
In matrices, A^2 means $A \times A$. You must multiply the matrices together.

Exercise 1

In questions **1** to **36**, the matrices have the following values:

$$A = \begin{pmatrix} 2 & -1 \\ 3 & 4 \end{pmatrix}; \quad B = \begin{pmatrix} 0 & 5 \\ 1 & -2 \end{pmatrix}; \quad C = \begin{pmatrix} 4 & 3 \\ 1 & -2 \end{pmatrix}; \quad D = \begin{pmatrix} 1 & 5 & 1 \\ 4 & -6 & 1 \end{pmatrix}; \quad E = \begin{pmatrix} 1 & 0 \\ -1 & 1 \\ 2 & 5 \end{pmatrix};$$

$$F = (4 \quad 5); \quad G = \begin{pmatrix} 4 \\ 1 \\ 3 \end{pmatrix}; \quad H = \begin{pmatrix} 0 & 1 & -2 \\ 3 & -4 & 5 \end{pmatrix}; \quad J = \begin{pmatrix} 3 \\ 1 \end{pmatrix}; \quad K = \begin{pmatrix} 1 & -3 \\ 0 & 1 \\ -7 & 0 \end{pmatrix}$$

Calculate the resultant value for each question where possible.

1. $A + B$	**2.** $D + H$	**3.** $J + F$
4. $B - C$	**5.** $2F$	**6.** $3B$

7. $\mathbf{K} - \mathbf{E}$ **8.** $2\mathbf{A} + \mathbf{B}$ **9.** $\mathbf{G} - \mathbf{J}$

10. $\mathbf{C} + \mathbf{B} + \mathbf{A}$ **11.** $2\mathbf{E} - 3\mathbf{K}$ **12.** $\frac{1}{2}\mathbf{A} - \mathbf{B}$

13. $\mathbf{AB}$ **14.** $\mathbf{BA}$ **15.** $\mathbf{BC}$

16. $\mathbf{CB}$ **17.** $\mathbf{DG}$ **18.** $\mathbf{AJ}$

19. $\mathbf{HK}$ **20.** $(\mathbf{AB})\mathbf{C}$ **21.** $\mathbf{A}(\mathbf{BC})$

22. $\mathbf{AF}$ **23.** $\mathbf{CK}$ **24.** $\mathbf{GF}$

25. $\mathbf{B}(2\mathbf{A})$ **26.** $(\mathbf{D} + \mathbf{H})\mathbf{G}$ **27.** $\mathbf{JF}$

28. $\mathbf{FJ}$ **29.** $(\mathbf{A} - \mathbf{C})\mathbf{D}$ **30.** $\mathbf{A}^2$

31. $\mathbf{A}^4$ **32.** $\mathbf{E}^2$ **33.** $\mathbf{KH}$

34. $(\mathbf{CA})\mathbf{J}$ **35.** $\mathbf{ED}$ **36.** $\mathbf{B}^4$

In questions **37** to **46**, find the value of the letters.

37. $\begin{pmatrix} 2 & x \\ y & 7 \end{pmatrix} + \begin{pmatrix} 4 & y \\ -3 & 2 \end{pmatrix} = \begin{pmatrix} x & 9 \\ z & 9 \end{pmatrix}$

38. $\begin{pmatrix} x & 2 \\ -1 & -2 \\ w & 3 \end{pmatrix} + \begin{pmatrix} x & y \\ y & -3 \\ v & 5 \end{pmatrix} = \begin{pmatrix} 8 & z \\ x & w \\ w & 8 \end{pmatrix}$

39. $\begin{pmatrix} a & b \\ c & 0 \end{pmatrix} - \begin{pmatrix} 2 & 5 \\ -3 & d \end{pmatrix} = 2\begin{pmatrix} 1 & a \\ b & -1 \end{pmatrix}$

40. $\begin{pmatrix} x & 3 \\ -2 & y \end{pmatrix}\begin{pmatrix} 2 \\ 1 \end{pmatrix} = \begin{pmatrix} 5 \\ 0 \end{pmatrix}$

41. $\begin{pmatrix} 2 & 0 \\ 0 & -3 \end{pmatrix}\begin{pmatrix} m \\ n \end{pmatrix} = \begin{pmatrix} 10 \\ 1 \end{pmatrix}$

42. $\begin{pmatrix} p & 2 & -1 \\ q & -2 & 2q \end{pmatrix}\begin{pmatrix} 2 \\ 1 \\ 3 \end{pmatrix} = \begin{pmatrix} 5 \\ -10 \end{pmatrix}$

43. $\begin{pmatrix} 3 & 0 \\ 2 & x \end{pmatrix}\begin{pmatrix} y & z \\ 4 & 0 \end{pmatrix} = \begin{pmatrix} 6 & -3 \\ 8 & w \end{pmatrix}$

44. $\begin{pmatrix} 3y & 3z \\ 2y + 4x & 2z \end{pmatrix} = \begin{pmatrix} 6 & -3 \\ 8 & w \end{pmatrix}$

45. $\begin{pmatrix} 2 & e \\ a & 3 \end{pmatrix} + k\begin{pmatrix} 3 & 1 \\ 0 & -2 \end{pmatrix} = \begin{pmatrix} 8 & 6 \\ -3 & -1 \end{pmatrix}$

46. $\begin{pmatrix} 4 & 0 \\ 1 & m \end{pmatrix}\begin{pmatrix} n & p \\ -2 & 0 \end{pmatrix} = \begin{pmatrix} 20 & 12 \\ -1 & q \end{pmatrix}$

47. If $\mathbf{A} = \begin{pmatrix} 1 & 0 \\ 3 & 2 \end{pmatrix}$, $\mathbf{B} = \begin{pmatrix} x & 0 \\ 1 & 3 \end{pmatrix}$, and $\mathbf{AB} = \mathbf{BA}$, find x.

48. If $\mathbf{X} = \begin{pmatrix} k & 2 \\ 2 & -k \end{pmatrix}$ and $\mathbf{X}^2 = 5\begin{pmatrix} 1 & 0 \\ 0 & 1 \end{pmatrix}$, find k.

49. $\mathbf{B} = \begin{pmatrix} 3 & 3 \\ -1 & -1 \end{pmatrix}$

    (a) Find k if $\mathbf{B}^2 = k\mathbf{B}$
    (b) Find m if $\mathbf{B}^4 = m\mathbf{B}$

50. $\mathbf{A} = \begin{pmatrix} 5 & 5 \\ -2 & -2 \end{pmatrix}$

    (a) Find n if $\mathbf{A}^2 = n\mathbf{A}$
    (b) Find q if $\mathbf{A}^3 = q\mathbf{A}$

9.2 The inverse of a matrix

The inverse of a matrix $\mathbf{A}$ is written $\mathbf{A}^{-1}$, and the inverse exists if

$$\mathbf{AA}^{-1} = \mathbf{A}^{-1}\mathbf{A} = \mathbf{I}$$

where $\mathbf{I}$ is called the identity matrix.

Only square matrices possess an inverse.

For 2×2 matrices, $\mathbf{I} = \begin{pmatrix} 1 & 0 \\ 0 & 1 \end{pmatrix}$

For 3×3 matrices, $\mathbf{I} = \begin{pmatrix} 1 & 0 & 0 \\ 0 & 1 & 0 \\ 0 & 0 & 1 \end{pmatrix}$, etc.

If $\mathbf{A} = \begin{pmatrix} a & b \\ c & d \end{pmatrix}$, the inverse $\mathbf{A}^{-1}$ is given by $\mathbf{A}^{-1} = \dfrac{1}{(ad - cb)} \begin{pmatrix} d & -b \\ -c & a \end{pmatrix}$

Here, the number $(ad - cb)$ is called the *determinant* of the matrix and is written $|\mathbf{A}|$.

If $|\mathbf{A}| = 0$, then the matrix has no inverse.

Example
Find the inverse of $\mathbf{A} = \begin{pmatrix} 3 & -4 \\ 1 & -2 \end{pmatrix}$.

$$\mathbf{A}^{-1} = \frac{1}{[3(-2) - 1(-4)]} \begin{pmatrix} -2 & 4 \\ -1 & 3 \end{pmatrix} = \frac{1}{-2} \begin{pmatrix} -2 & 4 \\ -1 & 3 \end{pmatrix}$$

Check: $\mathbf{A}^{-1}\mathbf{A} = \dfrac{1}{-2} \begin{pmatrix} -2 & 4 \\ -1 & 3 \end{pmatrix} \begin{pmatrix} 3 & -4 \\ 1 & -2 \end{pmatrix} = \dfrac{-1}{2} \begin{pmatrix} -2 & 0 \\ 0 & -2 \end{pmatrix}$

$$= \begin{pmatrix} 1 & 0 \\ 0 & 1 \end{pmatrix}$$

Multiplying by the inverse of a matrix gives the same result as dividing by the matrix: the effect is similar to ordinary algebraic operations.

e.g.   if $\mathbf{AB} = \mathbf{C}$

$\mathbf{A}^{-1}\mathbf{AB} = \mathbf{A}^{-1}\mathbf{C}$

$\mathbf{B} = \mathbf{A}^{-1}\mathbf{C}$

Exercise 2

In questions **1** to **15**, find the inverse of the matrix.

1. $\begin{pmatrix} 4 & 1 \\ 3 & 1 \end{pmatrix}$ **2.** $\begin{pmatrix} 1 & 2 \\ 2 & 5 \end{pmatrix}$ **3.** $\begin{pmatrix} 3 & 4 \\ 1 & 2 \end{pmatrix}$

4. $\begin{pmatrix} 5 & 2 \\ 1 & 1 \end{pmatrix}$ **5.** $\begin{pmatrix} 2 & -2 \\ -1 & 2 \end{pmatrix}$ **6.** $\begin{pmatrix} 4 & -3 \\ -1 & 2 \end{pmatrix}$

7. $\begin{pmatrix} 2 & 1 \\ -2 & 3 \end{pmatrix}$ **8.** $\begin{pmatrix} 0 & -3 \\ 2 & 4 \end{pmatrix}$ **9.** $\begin{pmatrix} -1 & -2 \\ 1 & -3 \end{pmatrix}$

10. $\begin{pmatrix} 2 & 4 \\ 1 & 2 \end{pmatrix}$
11. $\begin{pmatrix} 3 & -2 \\ 1 & 4 \end{pmatrix}$
12. $\begin{pmatrix} -3 & 1 \\ 2 & 1 \end{pmatrix}$

13. $\begin{pmatrix} 2 & -3 \\ 1 & -4 \end{pmatrix}$
14. $\begin{pmatrix} 7 & 0 \\ -5 & 1 \end{pmatrix}$
15. $\begin{pmatrix} 2 & 1 \\ -2 & -4 \end{pmatrix}$

16. If $\mathbf{B} = \begin{pmatrix} 2 & 4 \\ 1 & 3 \end{pmatrix}$ and $\mathbf{AB} = \mathbf{I}$, find $\mathbf{A}$.

17. Find $\mathbf{Y}$ if $\mathbf{Y}\begin{pmatrix} -2 & 0 \\ 3 & 1 \end{pmatrix} = \begin{pmatrix} 1 & 0 \\ 0 & 1 \end{pmatrix}$.

18. If $\begin{pmatrix} 2 & -3 \\ 0 & 4 \end{pmatrix} + \mathbf{X} = \begin{pmatrix} 1 & 0 \\ 0 & 1 \end{pmatrix}$, find $\mathbf{X}$.

19. Find $\mathbf{B}$ if $\mathbf{A} = \begin{pmatrix} 2 & -2 \\ -1 & 3 \end{pmatrix}$ and $\mathbf{AB} = \begin{pmatrix} 4 & -2 \\ 0 & 7 \end{pmatrix}$.

20. If $\begin{pmatrix} 3 & -3 \\ 2 & 5 \end{pmatrix} - \mathbf{X} = \begin{pmatrix} 1 & 0 \\ 0 & 1 \end{pmatrix}$, find $\mathbf{X}$.

21. Find $\mathbf{M}$ if $\begin{pmatrix} 1 & 1 \\ -2 & 1 \end{pmatrix}\mathbf{M} = 2\begin{pmatrix} 1 & 0 \\ 0 & 1 \end{pmatrix}$.

22. $\mathbf{A} = \begin{pmatrix} 2 & -3 \\ 0 & 1 \end{pmatrix}$ and $\mathbf{B} = \begin{pmatrix} 1 & -1 \\ -1 & 3 \end{pmatrix}$.

Find: (a) $\mathbf{AB}$ (b) $\mathbf{A}^{-1}$ (c) $\mathbf{B}^{-1}$

Show that $(\mathbf{AB})^{-1} = \mathbf{B}^{-1}\mathbf{A}^{-1}$.

23. If $\mathbf{M} = \begin{pmatrix} 3 & 1 \\ 2 & -1 \end{pmatrix}$ and $\mathbf{MN} = \begin{pmatrix} 7 & -9 \\ -2 & -6 \end{pmatrix}$, find $\mathbf{N}$.

24. $\mathbf{A} = \begin{pmatrix} 2 & 1 \\ 1 & 1 \end{pmatrix}$; $\mathbf{C} = \begin{pmatrix} 11 \\ 7 \end{pmatrix}$. If $\mathbf{B}$ is a (2×1)

matrix such that $\mathbf{AB} = \mathbf{C}$, find $\mathbf{B}$.

25. Find x if the determinant of $\begin{pmatrix} x & 3 \\ 1 & 2 \end{pmatrix}$ is

(a) 5   (b) -1 (c) 0

26. If the matrix $\begin{pmatrix} 1 & -2 \\ x & 4 \end{pmatrix}$ has no inverse, what is the value of x?

27. The elements of a (2×2) matrix consist of four different numbers. Find the largest possible value of the determinant of this matrix if the numbers are:

(a) 1, 3, 5, 9   (b) $-1, 2, 3, 4$

28. A business makes floor tiles and wall tiles.
The table below is used in calculating the cost of making each sort of tile.

	Labour (hours)	Material/tile	Paint (tins)
Floor tile	3	4	2
Wall tile	1	m	1

Labour costs $15 per hour, material costs $2 and painting costs $n per tin.
The information above can be summarised in the matrices **A** and **B**.

where $\mathbf{A} = \begin{pmatrix} 3 & 4 & 2 \\ 1 & m & 1 \end{pmatrix}$ and $\mathbf{B} = \begin{pmatrix} 15 \\ 2 \\ n \end{pmatrix}$

(a) Given that $AB = \begin{pmatrix} 59 \\ 22 \end{pmatrix}$, find

    (i) n
    (ii) m

(b) Evaluate $(50 \quad 300) \begin{pmatrix} 59 \\ 22 \end{pmatrix}$

(c) Explain what your answer to (b) represents.

9.3 Simple transformations

Reflection

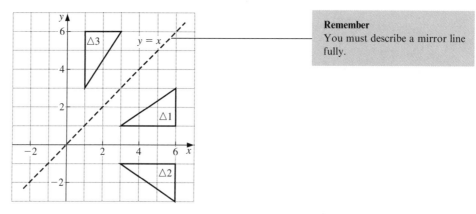

$\triangle2$ is the image of $\triangle1$ after reflection in the x-axis.

$\triangle3$ is the image of $\triangle1$ after reflection in the line $y = x$.

Exercise 3

In questions **1** to **6** draw the object and its image after reflection in the broken line.

1.
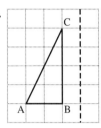

2.
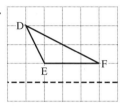

3.
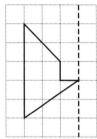

4.
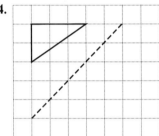

5.
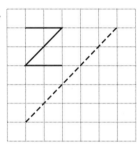

6.
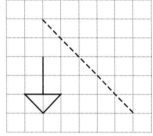

In questions **7**, **8**, **9** draw the image of the given shape after reflection in line M_1 and then reflect this new shape in line M_2.

7.
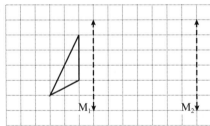

8.
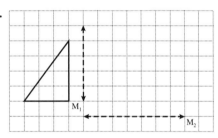

9.
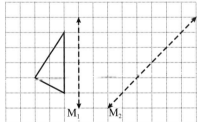

Exercise 4

For each question draw x- and y-axes with values from -8 to 8.

1. (a) Draw the triangle ABC at A(6, 8), B(2, 8), C(2, 6). Draw the lines $y = 2$ and $y = x$.

   (b) Draw the image of $\triangle$ABC after reflection in:
      (i) the y-axis. Label it $\triangle 1$.
      (ii) the line $y = 2$. Label it $\triangle 2$.
      (iii) the line $y = x$. Label it $\triangle 3$.

   (c) Write down the coordinates of the image of point A in each case.

2. (a) Draw the triangle DEF at D(-6, 8), E(-2, 8), F(-2, 6).
      Draw the lines $x = 1$, $y = x$, $y = -x$.

   (b) Draw the image of $\triangle$DEF after reflection in:
      (i) the line $x = 1$. Label it $\triangle 1$.
      (ii) the line $y = x$. Label it $\triangle 2$.
      (iii) the line $y = -x$. Label it $\triangle 3$.

   (c) Write down the coordinates of the image of point D in each case.

3. (a) Draw the triangle ABC at A(5, 1), B(8, 1), C(8, 3). Draw the lines $x + y = 4$, $y = x - 3$, $x = 2$.

   (b) Draw the image of $\triangle$ABC after reflection in:
      (i) the line $x + y = 4$. Label it $\triangle 1$.
      (ii) the line $y = x - 3$. Label it $\triangle 2$.
      (iii) the line $x = 2$. Label it $\triangle 3$.

   (c) Write down the coordinates of the image of point A in each case.

4. (a) Draw and label the following triangles:
      (i) $\triangle 1$: (3, 3), (3, 6), (1, 6)
         $\triangle 2$: (3, -1), (3, -4), (1, -4)
         $\triangle 3$: (3, 3), (6, 3), (6, 1)
         $\triangle 4$: (-6, -1), (-6, -3), (-3, -3)
         $\triangle 5$: (-6, 5), (-6, 7), (-3, 7)

   (b) Find the equation of the mirror line for the reflection:
      (i) $\triangle 1$ onto $\triangle 2$ (ii) $\triangle 1$ onto $\triangle 3$
      (iii) $\triangle 1$ onto $\triangle 4$ (iv) $\triangle 4$ onto $\triangle 5$

5. (a) Draw $\triangle 1$ at (3, 1), (7, 1), (7, 3).

   (b) Reflect $\triangle 1$ in the line $y = x$ onto $\triangle 2$.

   (c) Reflect $\triangle 2$ in the x-axis onto $\triangle 3$.

   (d) Reflect $\triangle 3$ in the line $y = -x$ onto $\triangle 4$.

   (e) Reflect $\triangle 4$ in the line $x = 2$ onto $\triangle 5$.

   (f) Write down the coordinates of $\triangle 5$.

6. (a) Draw △1 at (2, 6), (2, 8), (6, 6).

(b) Reflect △1 in the line $x + y = 6$ onto △2.

(c) Reflect △2 in the line $x = 3$ onto △3.

(d) Reflect △3 in the line $x + y = 6$ onto △4.

(e) Reflect △4 in the line $y = x - 8$ onto △5.

(f) Write down the coordinates of △5.

Rotation

Example

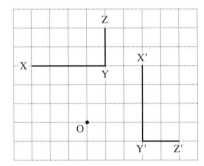

The letter L has been rotated through 90° clockwise about the centre
O. The angle, direction, and centre are needed to fully describe a
rotation.

We say that the object *maps* onto the image. Here,

  X maps onto X′
  Y maps onto Y′
  Z maps onto Z′

In this work, a clockwise rotation is *negative* and an anticlockwise
rotation is *positive*: in this example, the letter L has been rotated
through −90°. The angle, the direction, and the centre of rotation can
be found using tracing paper and a sharp pencil placed where you think
the centre of rotation is.

For more accurate work, draw the perpendicular bisector of the line
joining two corresponding points, e.g. Y and Y′. Repeat for another
pair of corresponding points. The centre of rotation is at the
intersection of the two perpendicular bisectors.

Exercise 5

In questions **1** to **4** draw the object and its image under the rotation
given. Take O as the centre of rotation in each case.

1.

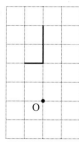

90° clockwise

2.

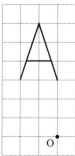

90° anticlockwise

3.

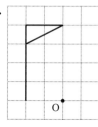

180°

4.

90° clockwise

In questions **5** to **8**, copy the diagram on squared paper and find the angle, the direction, and the centre of the rotation.

5.

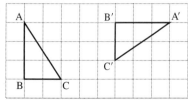

6.

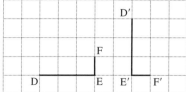

7.

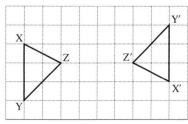

8.

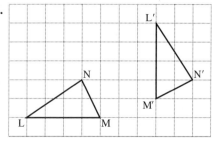

Exercise 6

For all questions draw *x*- and *y*-axes for values from −8 to +8.

1. (a) Draw the object triangle ABC at A(1, 3), B(1, 6), C(3, 6), rotate
       ABC through 90° clockwise about (0, 0), mark A′B′C′.
   (b) Draw the object triangle DEF at D(3, 3), E(6, 3), F(6, 1), rotate
       DEF through 90° clockwise about (0, 0), mark D′E′F′.

(c) Draw the object triangle PQR at P(−4, 7), Q(−4, 5), R(−1, 5), rotate PQR through 90° anticlockwise about (0, 0), mark P′Q′R′.

2. (a) Draw △1 at (1, 4), (1, 7), (3, 7).
   (b) Draw the images of △1 under the following rotations:
       (i) 90° clockwise, centre (0, 0). Label it △2.
       (ii) 180°, centre (0, 0). Label it △3.
       (iii) 90° anticlockwise, centre (0, 0). Label it △4.

3. (a) Draw triangle PQR at P(1, 2), Q(3, 5), R(6, 2).
   (b) Find the image of PQR under the following rotations:
       (i) 90° anticlockwise, centre (0, 0); label the image P′Q′R′
       (ii) 90° clockwise, centre (−2, 2); label the image P″Q″R″
       (iii) 180°, centre (1, 0); label the image P*Q*R*.
   (c) Write down the coordinates of P′, P″, P*.

4. (a) Draw △1 at (1, 2), (1, 6), (3, 5).
   (b) Rotate △1 90° clockwise, centre (1, 2) onto △2.
   (c) Rotate △2 180°, centre (2, −1) onto △3.
   (d) Rotate △3 90° clockwise, centre (2, 3) onto △4.
   (e) Write down the coordinates of △4.

5. (a) Draw and label the following triangles:
       △1 : (3, 1), (6, 1), (6, 3)
       △2 : (−1, 3), (−1, 6), (−3, 6)
       △3 : (1, 1), (−2, 1), (−2, −1)
       △4 : (3, −1), (3, −4), (5, −4)
       △5 : (4, 4), (1, 4), (1, 2)
   (b) Describe fully the following rotations:
       (i) △1 onto △2   (ii) △1 onto △3
       (iii) △1 onto △4   (iv) △1 onto △5
       (v) △5 onto △4   (vi) △3 onto △2

6. (a) Draw △1 at (4, 7), (8, 5), (8, 7).
   (b) Rotate △1 90° clockwise, centre (4, 3) onto △2.
   (c) Rotate △2 180°, centre (5, −1) onto △3.
   (d) Rotate △3 90° anticlockwise, centre (0, −8) onto △4.
   (e) Describe fully the following rotations:
       (i) △4 onto △1
       (ii) △4 onto △2

Translation

The triangle ABC below has been transformed onto the triangle A′B′C′ by a *translation*.

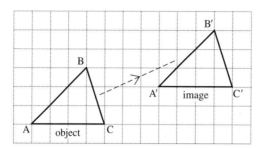

Here the translation is 7 squares to the right and 2 squares up the page.
The translation can be described by a column vector.

In this case the translation is $\begin{pmatrix} 7 \\ 2 \end{pmatrix}$.

Exercise 7

1. Make a copy of the diagram below and write down the column
   vector for each of the following translations:

   (a) D onto A   (b) B onto F

   (c) E onto A   (d) A onto C

   (e) E onto C   (f) C onto B

   (g) F onto E   (h) B onto C.

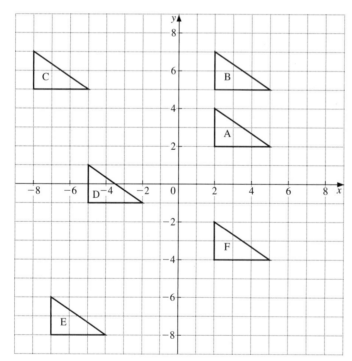

For questions **2** to **11** draw x and y axes with values from -8 to 8.
Draw object triangle ABC at A(-4, -1), B(-4, 1), C(-1, -1) and
shade it.

Draw the image of ABC under the translations described by the vectors below. For each question, write down the new coordinates of point C.

2. $\begin{pmatrix} 6 \\ 3 \end{pmatrix}$ **3.** $\begin{pmatrix} 6 \\ 7 \end{pmatrix}$ **4.** $\begin{pmatrix} 9 \\ -4 \end{pmatrix}$ **5.** $\begin{pmatrix} 1 \\ 7 \end{pmatrix}$

6. $\begin{pmatrix} 5 \\ -6 \end{pmatrix}$ **7.** $\begin{pmatrix} -2 \\ 5 \end{pmatrix}$ **8.** $\begin{pmatrix} -2 \\ -4 \end{pmatrix}$ **9.** $\begin{pmatrix} 0 \\ -7 \end{pmatrix}$

10. $\begin{pmatrix} 3 \\ 1 \end{pmatrix}$ followed by $\begin{pmatrix} 3 \\ 2 \end{pmatrix}$

11. $\begin{pmatrix} -2 \\ 0 \end{pmatrix}$ followed by $\begin{pmatrix} 0 \\ 3 \end{pmatrix}$ followed by $\begin{pmatrix} 1 \\ -1 \end{pmatrix}$

Enlargement

In the diagram below, the letter T has been enlarged by a scale factor of 2 using the point O as the centre of the enlargement.

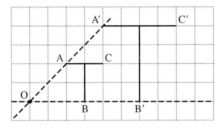

Notice that $OA' = 2 \times OA$
$OB' = 2 \times OB$

The scale factor and the centre of enlargement are both required to describe an enlargement.

Example 1

Draw the image of triangle ABC under an enlargement scale factor of $\frac{1}{2}$ using O as centre of enlargement.

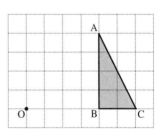

(a) Draw lines through OA, OB and OC.
(b) Mark A' so that $OA' = \frac{1}{2}OA$
     Mark B' so that $OB' = \frac{1}{2}OB$
     Mark C' so that $OC' = \frac{1}{2}OC$.
(c) Join A'B'C as shown.

Remember always to measure the lengths from O, not from A, B or C.

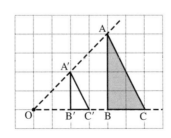

Example 2

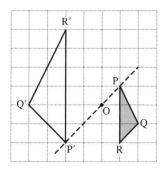

P'Q'R' is the image of PQR after enlargement with scale factor −2 and centre O. Notice that P' and P are on opposite sides of point O. Similarly Q' and Q, R' and R.

Exercise 8

In questions **1** to **6** copy the diagram and draw an enlargement using the centre O and the scale factor given.

1. Scale factor 2

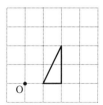

2. Scale factor 3

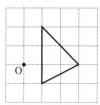

3. Scale factor 3

4. Scale factor −2

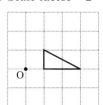

5. Scale factor −3

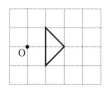

6. Scale factor $1\frac{1}{2}$

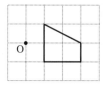

Answer questions **7** to **19** on graph paper taking x and y from 0 to 15. The vertices of the object are given in coordinate form.

In questions **7** to **10**, enlarge the object with the centre of enlargement and scale factor indicated.

	object	centre	scale factor
7.	(2, 4) (4, 2) (5, 5)	(0, 0)	+2
8.	(2, 4) (4, 2) (5, 5)	(1, 2)	+2
9.	(1, 1) (4, 2) (2, 3)	(1, 1)	+3
10.	(4, 4) (7, 6) (9, 3)	(7, 4)	+2

In questions **11** to **14** plot the object and image and find the centre of enlargement and the scale factor.

11. object   A(2, 1), B(5, 1), C(3, 3)
    image   A′(2, 1), B′(11, 1), C′(5, 7)

12. object   A(2, 5), B(9, 3), C(5, 9)
    image   A′(6½, 7), B′(10, 6), C′(8, 9)

13. object   A(2, 2), B(4, 4), C(2, 6)
    image   A′(11, 8), B′(7, 4), C′(11, 0)

14. object   A(0, 6), B(4, 6), C(3, 0)
    image   A′(12, 6), B′(8, 6), C′(9, 12)

In questions **15** to **19** enlarge the object using the centre of enlargement and scale factor indicated.

object	*centre*	*S.F.*
15. (1, 2), (13, 2), (1, 10)	(0, 0)	$+\frac{1}{2}$
16. (5, 10), (5, 7), (11, 7)	(2, 1)	$+\frac{1}{3}$
17. (7, 3), (9, 3), (7, 8)	(5, 5)	-1
18. (1, 1), (3, 1), (3, 2)	(4, 3)	-2
19. (9, 2), (14, 2), (14, 6)	(7, 4)	$-\frac{1}{2}$

The next exercise contains questions involving the four basic transformations: reflection, rotation, translation, enlargement.

Exercise 9

1. (a) Copy the diagram below.

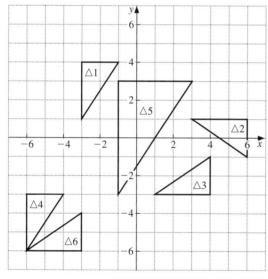

(b) Describe fully the following transformations:
    (i) △1 → △2   (ii) △1 → △3
    (iii) △4 → △1   (iv) △1 → △5
    (v) △3 → △6   (vi) △6 → △4

2. Plot and label the following triangles:
    △1:(−5, −5), (−1, −5), (−1, −3)   △2:(1, 7), (1, 3), (3, 3)
    △3:(3, −3), (7, −3), (7, −1)   △4:(−5, −5), (−5, −1), (−3, −1)
    △5:(1, −6), (3, −6), (3, −5)   △6:(−3, 3), (−3, 7), (−5, 7)

Describe fully the following transformations:
(a) $\triangle 1 \rightarrow \triangle 2$ (b) $\triangle 1 \rightarrow \triangle 3$
(c) $\triangle 1 \rightarrow \triangle 4$ (d) $\triangle 1 \rightarrow \triangle 5$
(e) $\triangle 1 \rightarrow \triangle 6$ (f) $\triangle 5 \rightarrow \triangle 3$
(g) $\triangle 2 \rightarrow \triangle 3$

3. Plot and label the following triangles:
$\triangle 1 : (-3, -6), (-3, -2), (-5, -2)$ $\triangle 2 : (-5, -1), (-5, -7), (-8, -1)$
$\triangle 3 : (-2, -1), (2, -1), (2, 1)$ $\triangle 4 : (6, 3), (2, 3), (2, 5)$
$\triangle 5 : (8, 4), (8, 8), (6, 8)$ $\triangle 6 : (-3, 1), (-3, 3), (-4, 3)$

Describe fully the following transformations:
(a) $\triangle 1 \rightarrow \triangle 2$ (b) $\triangle 1 \rightarrow \triangle 3$
(c) $\triangle 1 \rightarrow \triangle 4$ (d) $\triangle 1 \rightarrow \triangle 5$
(e) $\triangle 1 \rightarrow \triangle 6$ (f) $\triangle 3 \rightarrow \triangle 5$
(g) $\triangle 6 \rightarrow \triangle 2$

9.4 Combined transformations

It is convenient to denote transformations by a symbol.

Let **A** denote 'reflection in line $x = 3$' and

B denote 'translation $\begin{pmatrix} 2 \\ 1 \end{pmatrix}$'.

Perform **A** on $\triangle 1$.

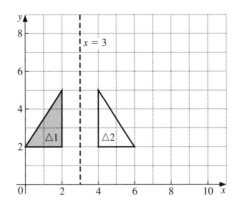

Figure 1

$\triangle 2$ is the image of $\triangle 1$ under the reflection in $x = 3$

i.e. $\mathbf{A}(\triangle 1) = \triangle 2$

$\mathbf{A}(\triangle 1)$ means 'perform the transformation **A** on triangle $\triangle 1$'

Perform **B** on $\triangle 2$.

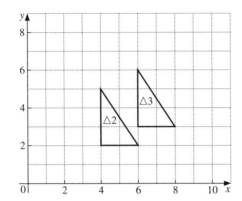

Figure 2

From Figure 2 we can see that

   B($\triangle$2) = $\triangle$3

The effect of going from $\triangle$1 to $\triangle$3 may be written

   BA($\triangle$1) = $\triangle$3

It is very important to notice that **BA**($\triangle$1) means do **A** first and then **B**.

Repeated transformations

XX(P) means 'perform transformation **X** on P and then perform **X** on the image'.

   It may be written $\mathbf{X}^2$(P)

Similarly **TTT**(P) = $\mathbf{T}^3$(P).

Inverse transformations

If translation **T** has vector $\begin{pmatrix} 3 \\ -2 \end{pmatrix}$, the translation which has the

opposite effect has vector $\begin{pmatrix} -3 \\ 2 \end{pmatrix}$. This is written $\mathbf{T}^{-1}$.

If rotation **R** denotes 90° clockwise rotation about (0, 0), then $\mathbf{R}^{-1}$ denotes 90° *anti*clockwise rotation about (0, 0).

The *inverse* of a transformation is the transformation which takes the *image* back to the object.

Note:
For all reflections, the inverse is the same reflection.

e.g. if **X** is reflection in $x = 0$, then $\mathbf{X}^{-1}$ is also reflection in $x = 0$.

The symbol $\mathbf{T}^{-3}$ means $(\mathbf{T}^{-1})^3$ i.e. perform $\mathbf{T}^{-1}$ three times.

Exercise 10

Draw x- and y-axes with values from -8 to $+8$ and plot the point P(3, 2).

R denotes 90° clockwise rotation about (0, 0);

X denotes reflection in $x = 0$.

H denotes 180° rotation about (0, 0);

T denotes translation $\begin{pmatrix} 3 \\ 2 \end{pmatrix}$.

For each question, write down the coordinates of the final image of P.

1. R(P)	**2. TR**(P)	**3. T**(P)	**4. RT**(P)
5. TH(P)	**6. XT**(P)	**7. HX**(P)	**8. XX**(P)
9. R^{-1}(P)	**10. T^{-1}**(P)	**11. X^3**(P)	**12. T^{-2}**(P)
13. R^2(P)	**14. T^{-1}R^2**(P)	**15. THX**(P)	**16. R^3**(P)
17. TX^{-1}(P)	**18. T^{3}X**(P)	**19. T^{2}H^{-1}**(P)	**20. XTH**(P)

Exercise 11

In this exercise, transformations **A**, **B**, ..., **H**, are as follows:

A denotes reflection in $x = 2$

B denotes 180° rotation, centre (1, 1)

C denotes translation $\begin{pmatrix} -6 \\ 2 \end{pmatrix}$

D denotes reflection in $y = x$

E denotes reflection in $y = 0$

F denotes translation $\begin{pmatrix} 4 \\ 3 \end{pmatrix}$

G denotes 90° rotation clockwise, centre (0, 0)

H denotes enlargement, scale factor $+\frac{1}{2}$, centre (0, 0)

Draw x- and y-axes with values from -8 to $+8$.

1. Draw triangle LMN at L(2, 2), M(6, 2), N(6, 4).
   Find the image of LMN under the following combinations of
   transformations. Write down the coordinates of the image of point
   L in each case:
   (a) **CA**(LMN)   (b) **ED**(LMN)   (c) **DB**(LMN)
   (d) **BE**(LMN)   (e) **EB**(LMN)

2. Draw triangle PQR at P(2, 2), Q(6, 2), R(6, 4).
   Find the image of PQR under the following combinations of
   transformations. Write down the coordinates of the image of point
   P in each case:
   (a) **AF**(PQR)   (b) **CG**(PQR)
   (c) **AG**(PQR)   (d) **HE**(PQR)

3. Draw triangle XYZ at X(−2, 4), Y(−2, 1), Z(−4, 1). Find the image of XYZ under the following combinations of transformations and state the equivalent single transformation in each case:
   (a) **G²E**(XYZ)   (b) **CB**(XYZ)   (c) **DA**(XYZ)

4. Draw triangle OPQ at O(0, 0), P(0, 2), Q(3, 2).
   Find the image of OPQ under the following combinations of transformations and state the equivalent single transformation in each case:
   (a) **DE**(OPQ)   (b) **FC**(OPQ)
   (c) **DEC**(OPQ)   (d) **DFE**(OPQ)

5. Draw triangle RST at R(−4, −1), S(−2½, −2), T(−4, −4). Find the image of RST under the following combinations of transformations and state the equivalent single transformation in each case:
   (a) **EAG**(RST)   (b) **FH**(RST)   (c) **GF**(RST)

6. Write down the inverses of the transformations **A**, **B**, . . ., **H**.

7. Draw triangle JKL at J(−2, 2), K(−2, 5), L(−4, 5). Find the image of JKL under the following transformations. Write down the coordinates of the image of point J in each case:
   (a) **C**⁻¹   (b) **F**⁻¹   (c) **G**⁻¹   (d) **D**⁻¹   (e) **A**⁻¹

8. Draw triangle PQR at P(−2, 4), Q(−2, 1), R(−4, 1). Find the image of PQR under the following combinations of transformations. Write down the coordinates of the image of point P in each case:
   (a) **DF**⁻¹(PQR)   (b) **EC**⁻¹(PQR)   (c) **D²F**(PQR)
   (d) **GA**(PQR)   (e) **C**⁻¹**G**⁻¹(PQR)

9. Draw triangle LMN at L(−2, 4), M(−4, 1), N(−2, 1). Find the image of LMN under the following combinations of transformations. Write down the coordinates of the image of point L in each case:
   (a) **HE**(LMN)   (b) **EAG**⁻¹(LMN)
   (c) **EDA**(LMN)   (d) **BG²E**(LMN)

10. Draw triangle XYZ at X(1, 2), Y(1, 6), Z(3, 6).
    (a) Find the image of XYZ under each of the transformations **BC** and **CB**.
    (b) Describe fully the single transformation equivalent to **BC**.
    (c) Describe fully the transformation **M** such that **MCB** = **BC**.

9.5 Transformations using matrices

Example 1

Find the image of triangle ABC, with A(1, 1), B(3, 1), C(3, 2), under the transformation represented by the matrix $\mathbf{M} = \begin{pmatrix} 1 & 0 \\ 0 & -1 \end{pmatrix}$.

(a) Write the coordinates of A as a column vector and multiply this vector by **M**.

$$\overset{\mathbf{M}}{\begin{pmatrix} 1 & 0 \\ 0 & -1 \end{pmatrix}} \overset{\mathbf{A}}{\begin{pmatrix} 1 \\ 1 \end{pmatrix}} = \overset{\mathbf{A'}}{\begin{pmatrix} 1 \\ -1 \end{pmatrix}}$$

A', the image of A, has coordinates $(1, -1)$.

(b) Repeat for B and C.

$$\overset{\mathbf{M}}{\begin{pmatrix} 1 & 0 \\ 0 & -1 \end{pmatrix}} \overset{\mathbf{B}}{\begin{pmatrix} 3 \\ 1 \end{pmatrix}} = \overset{\mathbf{B'}}{\begin{pmatrix} 3 \\ -1 \end{pmatrix}}$$

$$\overset{\mathbf{M}}{\begin{pmatrix} 1 & 0 \\ 0 & -1 \end{pmatrix}} \overset{\mathbf{C}}{\begin{pmatrix} 3 \\ 2 \end{pmatrix}} = \overset{\mathbf{C'}}{\begin{pmatrix} 3 \\ -2 \end{pmatrix}}$$

(c) Plot A'$(1, -1)$, B'$(3, -1)$ and C'$(3, -2)$.

The transformation is a reflection in the x-axis.

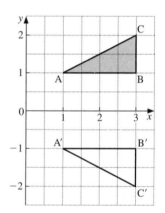

Example 2

Find the image of L(1, 1), M(1, 3), N(2, 3) under the transformation represented by the matrix $\begin{pmatrix} 0 & -1 \\ 1 & 0 \end{pmatrix}$.

A quicker method is to write the three vectors for L, M and N in a single 2×3 matrix, and then perform the multiplication.

$$\begin{pmatrix} 0 & -1 \\ 1 & 0 \end{pmatrix} \overset{\text{L M N}}{\begin{pmatrix} 1 & 1 & 2 \\ 1 & 3 & 3 \end{pmatrix}} = \overset{\text{L' M' N'}}{\begin{pmatrix} -1 & -3 & -3 \\ 1 & 1 & 2 \end{pmatrix}}$$

The transformation is a rotation, $+90°$, centre $(0, 0)$.

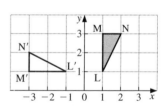

Exercise 12

For questions **1** to **5** draw x- and y-axes with values from -8 to 8. Do all parts of each question on one graph.

1. Draw the triangle A(2, 2), B(6, 2), C(6, 4). Find its image under the transformations represented by the following matrices:

(a) $\begin{pmatrix} 0 & -1 \\ 1 & 0 \end{pmatrix}$ (b) $\begin{pmatrix} -1 & 0 \\ 0 & 1 \end{pmatrix}$ (c) $\begin{pmatrix} 1 & 0 \\ 0 & -1 \end{pmatrix}$

(d) $\begin{pmatrix} 0 & 1 \\ 1 & 0 \end{pmatrix}$ (e) $\begin{pmatrix} \frac{1}{2} & 0 \\ 0 & \frac{1}{2} \end{pmatrix}$

2. Plot the object and image for the following:

	Object	Matrix
(a)	P(4, 2), Q(4, 4), R(0, 4)	$\begin{pmatrix} 2 & 0 \\ 0 & 2 \end{pmatrix}$
(b)	P(4, 2), Q(4, 4), R(0, 4)	$\begin{pmatrix} -\frac{1}{2} & 0 \\ 0 & -\frac{1}{2} \end{pmatrix}$
(c)	A(−6, 8), B(−2, 8), C(−2, 6)	$\begin{pmatrix} 0 & -1 \\ -1 & 0 \end{pmatrix}$
(d)	P(4, 2), Q(4, 4), R(0, 4)	$\begin{pmatrix} -2 & 0 \\ 0 & -2 \end{pmatrix}$

Describe each as a *single* transformation.

3. Draw a trapezium at K(2, 2), L(2, 5), M(5, 8), N(8, 8). Find the image of KLMN under the transformations described by the following matrices:

$A = \begin{pmatrix} 1 & 0 \\ 0 & -1 \end{pmatrix}$ $E = \begin{pmatrix} 0 & -1 \\ -1 & 0 \end{pmatrix}$

$B = \begin{pmatrix} -1 & 0 \\ 0 & 1 \end{pmatrix}$ $F = \begin{pmatrix} -1 & 0 \\ 0 & -1 \end{pmatrix}$

$C = \begin{pmatrix} 0 & 1 \\ 1 & 0 \end{pmatrix}$ $G = \begin{pmatrix} 0 & -1 \\ 1 & 0 \end{pmatrix}$

$D = \begin{pmatrix} 0 & 1 \\ -1 & 0 \end{pmatrix}$ $H = \begin{pmatrix} 1 & 0 \\ 0 & 1 \end{pmatrix}$

Describe fully each of the eight transformations.

4. (a) Draw a quadrilateral at A(3, 4), B(4, 0), C(3, 1), D(0, 0). Find the image of ABCD under the transformation represented by the matrix $\begin{pmatrix} -2 & 0 \\ 0 & -2 \end{pmatrix}$.

   (b) Find the ratio $\left(\dfrac{\text{area of image}}{\text{area of object}} \right)$.

5. (a) Draw axes so that both x and y can take values from −2 to +8.
   (b) Draw triangle ABC at A(2, 1), B(7, 1), C(2, 4).
   (c) Find the image of ABC under the transformation represented by the matrix $\begin{pmatrix} 1 & -1 \\ 1 & 1 \end{pmatrix}$ and plot the image on the graph.

(d) The transformation is a rotation followed by an enlargement. Calculate the angle of the rotation and the scale factor of the enlargement.

6. (a) Draw axes to that x can take values from 0 to 15 and y can take values from -6 to $+6$.
   (b) Draw triangle PQR at P(2, 1), Q(7, 1), R(2, 4).
   (c) Find the image of PQR under the transformation represented by the matrix $\begin{pmatrix} 2 & 1 \\ -1 & 2 \end{pmatrix}$ and plot the image on the graph.
   (d) The transformation is a rotation followed by an enlargement. Calculate the angle of the rotation and the scale factor of the enlargement.

7. (a) On graph paper, draw the triangle T whose vertices are (2, 2), (6, 2) and (6, 4).
   (b) Draw the image U of T under the transformation whose matrix is $\begin{pmatrix} 0 & 1 \\ 1 & 0 \end{pmatrix}$.
   (c) Draw the image V of T under the transformation whose matrix is $\begin{pmatrix} 1 & 0 \\ 0 & -1 \end{pmatrix}$.
   (d) Describe the single transformation which would map U onto V.

8. (a) Find the images of the points (1, 0), (2, 1), (3, -1), (-2, 3) under the transformation with matrix $\begin{pmatrix} 1 & 3 \\ 2 & 6 \end{pmatrix}$.
   (b) Show that the images lie on a straight line, and find its equation.

9. The transformation with matrix $\begin{pmatrix} 2 & 3 \\ 6 & 9 \end{pmatrix}$ maps every point in the plane onto a line. Find the equation of the line.

10. Using a scale of 1 cm to one unit in each case, draw x- and y-axes, taking values of x from -4 to $+6$ and values of y from 0 to 12.
    (a) Draw and label the quadrilateral OABC with O(0, 0), A(2, 0), B(4, 2), C(0, 2).
    (b) Find and draw the image of OABC under the transformation whose matrix is $\mathbf{R}$, where $\mathbf{R} = \begin{pmatrix} 2\cdot4 & -1\cdot8 \\ 1\cdot8 & 2\cdot4 \end{pmatrix}$.
    (c) Calculate, in surd form, the lengths OB and O'B'.
    (d) Calculate the angle AOA'.
    (e) Given that the transformation $\mathbf{R}$ consists of a rotation about O followed by an enlargement, state the angle of the rotation and the scale factor of the enlargement.

11. The matrix $\mathbf{R} = \begin{pmatrix} \cos\theta & -\sin\theta \\ \sin\theta & \cos\theta \end{pmatrix}$ represents a positive rotation of $\theta°$ about the origin. Find the matrix which represents a rotation of:

(a) 90°   (b) 180°   (c) 30°   (d) −90°
(e) 60°   (f) 150°   (g) 45°   (h) 53·1°
Confirm your results for parts (a), (e), (h) by applying the matrix to
the quadrilateral O(0, 0), A(0, 2), B(4, 2), C(4, 0).

12. Using the matrix **R** given in question **11**, find the angle of rotation
for the following:

(a) $\begin{pmatrix} 0 & -1 \\ 1 & 0 \end{pmatrix}$
(b) $\begin{pmatrix} 0·8 & -0·6 \\ 0·6 & 0·8 \end{pmatrix}$

(c) $\begin{pmatrix} 0·5 & 0·866 \\ -0·866 & 0·5 \end{pmatrix}$
(d) $\begin{pmatrix} 0·6 & 0·8 \\ -0·8 & 0·6 \end{pmatrix}$

Confirm your results by applying each matrix to the quadrilateral
O(0, 0), A(0, 2), B(4, 2), C(4, 0).

Exercise 13

1. Draw the rectangle (0, 0), (0, 1), (2, 1), (2, 0) and its image under
the following transformations and describe the *single*
transformation which each represents:

(a) $\begin{pmatrix} 0 & 1 \\ 1 & 0 \end{pmatrix}\begin{pmatrix} x \\ y \end{pmatrix} + \begin{pmatrix} 1 \\ -1 \end{pmatrix}$
(b) $\begin{pmatrix} 1 & 0 \\ 0 & -1 \end{pmatrix}\begin{pmatrix} x \\ y \end{pmatrix} + \begin{pmatrix} 0 \\ 2 \end{pmatrix}$

(c) $\begin{pmatrix} 0 & 1 \\ -1 & 0 \end{pmatrix}\begin{pmatrix} x \\ y \end{pmatrix} + \begin{pmatrix} 4 \\ 0 \end{pmatrix}$
(d) $\begin{pmatrix} 3 & 0 \\ 0 & 3 \end{pmatrix}\begin{pmatrix} x \\ y \end{pmatrix} + \begin{pmatrix} -4 \\ 2 \end{pmatrix}$

2. (a) Draw L(1, 1), M(3, 3), N(4, 1) and its image L′M′N′ under the
matrix $\mathbf{A} = \begin{pmatrix} 1 & 0 \\ 0 & -1 \end{pmatrix}$.

(b) Find and draw the image of L′M′N′ under matrix
$\mathbf{B} = \begin{pmatrix} 0 & 1 \\ -1 & 0 \end{pmatrix}$ and label it L″M″N″.

(c) Calculate the matrix product **BA**.

(d) Find the image of LMN under the matrix **BA**, and compare
with the result of performing **A** and then **B**.

3. (a) Draw P(0, 0), Q(2, 2), R(4, 0) and its image P′Q′R′
under matrix $\mathbf{A} = \begin{pmatrix} 2 & 0 \\ 0 & 2 \end{pmatrix}$.

(b) Find and draw the image of P′Q′R′ under
matrix $\mathbf{B} = \begin{pmatrix} 1 & 1 \\ 0 & 1 \end{pmatrix}$ and label it P″Q″R″.

(c) Calculate the matrix product **BA**.

(d) Find the image of PQR under the matrix **BA**, and compare with
the result of performing **A** and then **B**.

4. (a) Draw L(1, 1), M(3, 3), N(4, 1) and its image L′M′N′ under
matrix $\mathbf{K} = \begin{pmatrix} 2 & 0 \\ 0 & 2 \end{pmatrix}$.

Find $\mathbf{K}^{-1}$, the inverse of $\mathbf{K}$, and now find the image of L'M'N' under $\mathbf{K}^{-1}$.

(b) Repeat part (a) with $\mathbf{K} = \begin{pmatrix} 1 & 2 \\ 0 & 1 \end{pmatrix}$.

(c) Repeat part (a) with $\mathbf{K} = \begin{pmatrix} 3 & 0 \\ 0 & 1 \end{pmatrix}$.

5. The image (x', y') of a point (x, y) under a transformation is given

by $\begin{pmatrix} x' \\ y' \end{pmatrix} = \begin{pmatrix} 3 & 0 \\ 1 & -2 \end{pmatrix} \begin{pmatrix} x \\ y \end{pmatrix} + \begin{pmatrix} 2 \\ 5 \end{pmatrix}$

(a) Find the coordinates of the image of the point (4, 3).
(b) The image of the point (m, n) is the point (11, 7). Write down two equations involving m and n and hence find the values of m and n.
(c) The image of the point (h, k) is the point (5, 10). Find the values of h and k.

6. Draw A(0, 2), B(2, 2), C(0, 4) and its image under an enlargement, A'(2, 2), B'(6, 2), C'(2, 6).
(a) What is the centre of enlargement?
(b) Find the image of ABC under an enlargement, scale factor 2, centre (0, 0).
(c) Find the translation which maps this image onto A'B'C'.
(d) What is the matrix $\mathbf{X}$ and vector $\mathbf{v}$ which represents an enlargement scale factor 2, centre $(-2, 2)$?

7. Draw A(0, 1), B(1, 1), C(1, 3) and its image under a reflection A'(4, 1), B'(3, 1), C'(3, 3).
(a) What is the equation of the mirror line?
(b) Find the image of ABC under a reflection in the line $x = 0$.
(c) Find the translation which maps this image onto A'B'C'.
(d) What is the matrix $\mathbf{X}$ and vector $\mathbf{v}$ which represents a reflection in the line $x = 2$?

8. Use the same approach as in questions **6** and **7** to find the matrix $\mathbf{X}$ and vector $\mathbf{v}$ which represents each of the following transformations. (Start by drawing an object and its image under the transformation.)
(a) Enlargement scale factor 2, centre (1, 3)
(b) Enlargement scale factor 2, centre $(\frac{1}{2}, 1)$
(c) Reflection in $y = x + 3$
(d) Rotation 180°, centre $(1\frac{1}{2}, 2\frac{1}{2})$
(e) Reflection in $y = 1$
(f) Rotation $-90°$, centre $(2, -2)$

Describing a transformation using base vectors

It is possible to describe a transformation in matrix form by considering the effect on the *base vectors* $\begin{pmatrix} 1 \\ 0 \end{pmatrix}$ and $\begin{pmatrix} 0 \\ 1 \end{pmatrix}$.

We will let $\begin{pmatrix} 1 \\ 0 \end{pmatrix}$ be I and $\begin{pmatrix} 0 \\ 1 \end{pmatrix}$ be J.

The *columns* of a matrix give us the images of I and J after the transformation.

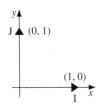

Example

Describe the transformation with matrix $\begin{pmatrix} 0 & 1 \\ -1 & 0 \end{pmatrix}$.

Column $\begin{pmatrix} 0 \\ -1 \end{pmatrix}$ represents I′ (the image of I).

Column $\begin{pmatrix} 1 \\ 0 \end{pmatrix}$ represents J′ (the image of J)

$$\begin{array}{cc} \text{I}' & \text{J}' \\ \begin{pmatrix} 0 & 1 \\ -1 & 0 \end{pmatrix} \end{array}$$

Draw I, J, I′ and J′ on a diagram.

Clearly both I and J have been rotated 90°

clockwise about the origin. $\begin{pmatrix} 0 & 1 \\ -1 & 0 \end{pmatrix}$ represents

a rotation of −90°.

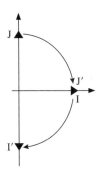

This method can be used to describe a reflection, rotation, enlargement, shear or stretch in which the origin remains fixed.

Exercise 14

In questions **1** to **12**, use base vectors to describe the transformation represented by each matrix.

1. $\begin{pmatrix} 0 & -1 \\ 1 & 0 \end{pmatrix}$ **2.** $\begin{pmatrix} -1 & 0 \\ 0 & 1 \end{pmatrix}$ **3.** $\begin{pmatrix} 0 & -1 \\ -1 & 0 \end{pmatrix}$ **4.** $\begin{pmatrix} 0 & 1 \\ 1 & 0 \end{pmatrix}$

5. $\begin{pmatrix} 2 & 0 \\ 0 & 2 \end{pmatrix}$ **6.** $\begin{pmatrix} \frac{1}{2} & 0 \\ 0 & \frac{1}{2} \end{pmatrix}$ **7.** $\begin{pmatrix} -2 & 0 \\ 0 & -2 \end{pmatrix}$ **8.** $\begin{pmatrix} \frac{1}{2} & 0 \\ 0 & -\frac{1}{2} \end{pmatrix}$

In questions **9** to **18**, use base vectors to write down the matrix which represents each of the transformations:

9. Rotation +90° about (0, 0)

10. Reflection in $y = x$

11. Reflection in y-axis

12. Rotation 180° about (0, 0)

13. Enlargement, centre (0, 0), scale factor 3

14. Reflection in $y = -x$

15. Enlargement, centre (0, 0), scale factor −2

16. Reflection in x-axis

17. Rotation $-90°$ about $(0, 0)$

18. Enlargement, centre $(0, 0)$, scale factor $\frac{1}{2}$

Shear

Figure 1 shows a pack of cards stacked neatly into a pile.
Figure 2 shows the same pack after a *shear* has been
performed.

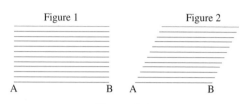

The area is the
same after a shear.

Note:
(a) the card AB, at the bottom has not moved (we say
the line AB is invariant).
(b) the distance moved by any card depends on its distance from
the base card.
The card at the top moves twice as far as the card in the middle.

In this example ABCD → A′B′C′D′ by a shear.
PQ lies on the invariant line. Notice that in
this case the invariant line is not on the object.
Notice also that area ABCD = area A′B′C′D′.

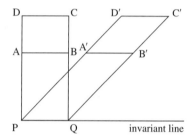

Here the squares A and B are
transformed by the given matrices.

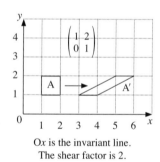

Ox is the invariant line.
The shear factor is 2.

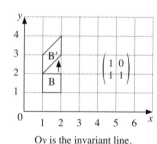

Oy is the invariant line.
The shear factor is 1.

Remember: $\begin{pmatrix} 1 & k \\ 0 & 1 \end{pmatrix}$ is a shear with Ox the invariant line

$\begin{pmatrix} 1 & 0 \\ k & 1 \end{pmatrix}$ is a shear with Oy the invariant line

k is the shear factor.

Stretch

The rectangle ABCD has been *stretched* in the direction of the y-axis so that A′B′ is twice AB.

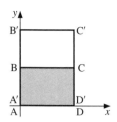

Stretch factor 2 parallel to the y-axis, invariant line y = 0.

A stretch is fully described if we know:
(a) the direction of the stretch and the invariant line
(b) the ratio of corresponding lengths, the stretch factor.

The matrix $\begin{pmatrix} k & 0 \\ 0 & 1 \end{pmatrix}$ represents a stretch parallel to the x-axis, invariant line $x = 0$, where the ratio of corresponding lengths (stretch factor) is k.

The matrix $\begin{pmatrix} 1 & 0 \\ 0 & k \end{pmatrix}$ represents a stretch parallel to the y-axis, invariant line $y = 0$, with a stretch factor k.

Exercise 15

1. In the diagram, OABC has been mapped onto OA′B′C by a shear. What is the invariant line of the shear?

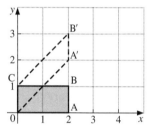

2. What is the invariant line of this shear?

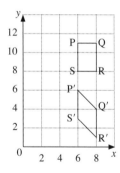

3. (a) Draw axes for values of x from 0 to 6 and for values of y from 0 to 16.
   (b) Draw triangle P with vertices (2, 2), (2, 4) and (6, 4).
   (c) Draw triangle Q which is the image of triangle P under the transformation represented by the matrix $\mathbf{M} = \begin{pmatrix} 1 & 0 \\ 2 & 1 \end{pmatrix}$.
   (d) Describe fully the single transformation represented by the matrix $\mathbf{M}$.

4. Draw axes for values of x from -6 to $+9$ and for values of y from -2 to $+5$.

   Find the coordinates of the image of each of the following shapes under the shear represented by the matrix $\begin{pmatrix} 1 & 2 \\ 0 & 1 \end{pmatrix}$.

Draw each object and image together on a diagram.

(a) (0, 0)(0, 3)(2, 3)(2, 0)   (b) (0, 0)(−2, 0)(−2, −2)(0, −2)

(c) (0, 0)(2, 2)(3, 0)   (d) (1, 1)(1, 3)(3, 3)(3, 1)

What is the invariant line for this shear?

5. Use base vectors to describe the transformation represented by each matrix:

(a) $\begin{pmatrix} 3 & 0 \\ 0 & 1 \end{pmatrix}$ (b) $\begin{pmatrix} 1 & 1 \\ 0 & 1 \end{pmatrix}$ (c) $\begin{pmatrix} 1 & 0 \\ 2 & 1 \end{pmatrix}$ (d) $\begin{pmatrix} 1 & 0 \\ 0 & 2 \end{pmatrix}$

In questions **6** to **13**, plot the rectangle ABCD at A(0, 0), B(0, 2), C(3, 2), D(3, 0). Find and draw the image of ABCD under the transformation given and describe the transformation fully.

6. $\begin{pmatrix} 2 & 0 \\ 0 & 1 \end{pmatrix}$ **7.** $\begin{pmatrix} 3 & 0 \\ 0 & 1 \end{pmatrix}$ **8.** $\begin{pmatrix} 1 & 0 \\ 0 & 2 \end{pmatrix}$ **9.** $\begin{pmatrix} 1\frac{1}{2} & 0 \\ 0 & 1 \end{pmatrix}$

10. $\begin{pmatrix} 1 & 1 \\ 0 & 1 \end{pmatrix}$ **11.** $\begin{pmatrix} -2 & 0 \\ 0 & 1 \end{pmatrix}$ **12.** $\begin{pmatrix} 1 & 0 \\ 0 & 3 \end{pmatrix}$ **13.** $\begin{pmatrix} \frac{1}{2} & 0 \\ 0 & 1 \end{pmatrix}$

14. (a) Find and draw the image of the square (0, 0), (1, 1), (0, 2), (−1, 1) under the transformation represented by the matrix $\begin{pmatrix} 4 & 3 \\ -3 & -2 \end{pmatrix}$.

(b) Show that the transformation is a shear and find the equation of the invariant line.

15. (a) Find and draw the image of the square (0, 0), (1, 1), (0, 2), (−1, 1) under the shear represented by the matrix $\begin{pmatrix} 0.5 & -0.5 \\ 0.5 & 1.5 \end{pmatrix}$.

(b) Find the equation of the invariant line.

16. Find and draw the image of the square (0, 0), (1, 0), (1, 1), (0, 1) under the transformation represented by the matrix $\begin{pmatrix} 3 & 0 \\ 0 & 2 \end{pmatrix}$.

This transformation is called a two-way stretch.

Revision exercise 9A

1. $A = \begin{pmatrix} 3 & 2 \\ 1 & 4 \end{pmatrix}$, $B = \begin{pmatrix} -1 & 3 \\ 0 & 2 \end{pmatrix}$.

Express as a single matrix:

(a) 2A   (b) A − B   (c) $\frac{1}{2}$A   (d) AB   (e) B^2

2. Evaluate:

(a) $\begin{pmatrix} -3 & 0 \\ 1 & 2 \end{pmatrix}\begin{pmatrix} 3 & \frac{1}{3} \\ 1 & \frac{1}{2} \end{pmatrix}$ (b) $\begin{pmatrix} 3 \\ 1 \end{pmatrix}(4 \quad 2)$

(c) $\begin{pmatrix} 3 & -2 \\ 4 & 1 \end{pmatrix} + 2 \begin{pmatrix} 3 & 0 \\ -1 & -4 \end{pmatrix}$

3. $\mathbf{A} = \begin{pmatrix} 4 & 2 \\ 1 & 1 \end{pmatrix}$, $\mathbf{B} = (1 \quad 5)$, $\mathbf{C} = \begin{pmatrix} -1 \\ 3 \end{pmatrix}$.

(a) Determine $\mathbf{BC}$ and $\mathbf{CB}$.

(b) If $\mathbf{AX} = \begin{pmatrix} 8 & 20 \\ 3 & 7 \end{pmatrix}$, where $\mathbf{X}$ is a (2×2) matrix,

determine $\mathbf{X}$.

4. Find the inverse of the matrix $\begin{pmatrix} 2 & -1 \\ 3 & 5 \end{pmatrix}$.

5. The determinant of the matrix $\begin{pmatrix} 3 & 2 \\ x & -1 \end{pmatrix}$ is -9.

Find the value of x and write down the inverse of the matrix.

6. $\mathbf{A} = \begin{pmatrix} 2 & 0 \\ 1 & 2 \end{pmatrix}$; h and k are numbers so that

$\mathbf{A}^2 = h\mathbf{A} + k\mathbf{I}$, where $\mathbf{I} = \begin{pmatrix} 1 & 0 \\ 0 & 1 \end{pmatrix}$.

Find the values of h and k.

7. $\mathbf{M} = \begin{pmatrix} a & 1 \\ 1 & -a \end{pmatrix}$.

(a) Find the values of a if $\mathbf{M}^2 = 17 \begin{pmatrix} 1 & 0 \\ 0 & 1 \end{pmatrix}$.

(b) Find the values of a if $|\mathbf{M}| = -10$.

8. Find the coordinates of the image of $(1, 4)$ under:

(a) a clockwise rotation of $90°$ about $(0, 0)$

(b) a reflection in the line $y = x$

(c) a translation which maps $(5, 3)$ onto $(1, 1)$

9. Draw x- and y-axes with values from -8 to $+8$. Draw triangle
A$(1, -1)$, B$(3, -1)$, C$(1, -4)$. Find the image of ABC under the
following enlargements:

(a) scale factor 2, centre $(5, -1)$

(b) scale factor 2, centre $(0, 0)$

(c) scale factor $\frac{1}{2}$, centre $(1, 3)$

(d) scale factor $-\frac{1}{2}$, centre $(3, 1)$

(e) scale factor -2, centre $(0, 0)$

10. Using the diagram on the right, describe the
    transformations for the following:
    (a) $T_1 \rightarrow T_6$ (b) $T_4 \rightarrow T_5$
    (c) $T_8 \rightarrow T_2$ (d) $T_4 \rightarrow T_1$
    (e) $T_8 \rightarrow T_4$ (f) $T_6 \rightarrow T_8$

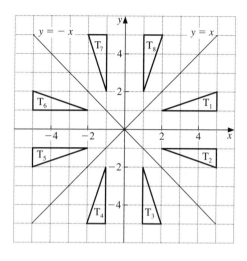

11. Describe the single transformation which maps:
    (a) $\triangle ABC$ onto $\triangle DEF$
    (b) $\triangle ABC$ onto $\triangle PQR$
    (c) $\triangle ABC$ onto $\triangle XYZ$

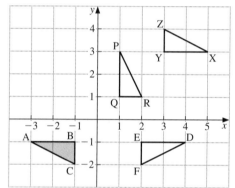

12. **M** is a reflection in the line $x + y = 0$.
    R is an anticlockwise rotation of $90°$ about $(0, 0)$.
    T is a translation which maps $(-1, -1)$ onto $(2, 0)$.
    Find the image of the point $(3, 1)$ under:
    (a) **M**   (b) **R**   (c) **T**
    (d) **MR**   (e) **RT**   (f) **TMR**

13. **A** is a rotation of $180°$ about $(0, 0)$.
    B is a reflection in the line $x = 3$.
    C is a translation which maps $(3, -1)$ onto $(-2, -1)$.
    Find the image of the point $(1, -2)$ under:
    (a) **A**   (b) $\mathbf{A}^2$ (c) **BC**
    (d) $\mathbf{C}^{-1}$ (e) **ABC**   (f) $\mathbf{C}^{-1}\mathbf{B}^{-1}\mathbf{A}^{-1}$

14. The matrix $\begin{pmatrix} 0 & -1 \\ 1 & 0 \end{pmatrix}$ represents the
    transformation **X**.
    (a) Find the image of $(5, 2)$ under **X**.
    (b) Find the image of $(-3, 4)$ under **X**.
    (c) Describe the transformation **X**.

15. Draw x- and y-axes with values from -8 to $+8$.
Draw triangle A(2, 2), B(6, 2), C(6, 4).
Find the image of ABC under the transformations represented by
the matrices:

(a) $\begin{pmatrix} 0 & -1 \\ 1 & 0 \end{pmatrix}$

(b) $\begin{pmatrix} 1 & 0 \\ 0 & -1 \end{pmatrix}$

(c) $\begin{pmatrix} -1 & 0 \\ 0 & -1 \end{pmatrix}$

(d) $\begin{pmatrix} 0 & 1 \\ -1 & 0 \end{pmatrix}$

(e) $\begin{pmatrix} 0 & -1 \\ -1 & 0 \end{pmatrix}$

Describe each transformation.

16. Using base vectors, describe the transformations represented by the
following matrices:

(a) $\begin{pmatrix} 0 & 1 \\ 1 & 0 \end{pmatrix}$

(b) $\begin{pmatrix} -1 & 0 \\ 0 & 1 \end{pmatrix}$

(c) $\begin{pmatrix} 3 & 0 \\ 0 & 3 \end{pmatrix}$

17. Using base vectors, write down the matrices which describe the
following transformations:
(a) Rotation 180°, centre (0, 0)
(b) Reflection in the line $y = 0$
(c) Enlargement scale factor 4, centre (0, 0)
(d) Reflection in the line $x = -y$
(e) Clockwise rotation 90°, centre (0, 0)

18. Transformation **N**, which is given by

$$\begin{pmatrix} x' \\ y' \end{pmatrix} = \begin{pmatrix} 2 & 0 \\ 0 & 2 \end{pmatrix}\begin{pmatrix} x \\ y \end{pmatrix} + \begin{pmatrix} 5 \\ -2 \end{pmatrix},$$

is composed of two single transformations.
(a) Describe each of the transformations.
(b) Find the image of the point (3, -1) under **N**.
(c) Find the image of the point (-1, $\frac{1}{2}$) under **N**.
(d) Find the point which is mapped by **N** onto the point (7, 4).

19. A is the reflection in the line $y = x$.
B is the reflection in the y-axis.
Find the matrix which represents:

(a) **A**   (b) **B**   (c) **AB**   (d) **BA**

Describe the single transformations **AB** and **BA**.

Examination exercise 9B

1.

$$\mathbf{A} = \begin{pmatrix} x & 6 \\ 4 & 3 \end{pmatrix} \qquad\qquad \mathbf{B} = \begin{pmatrix} 2 & 3 \\ 2 & 1 \end{pmatrix}$$

(a) Find **AB**.   [2]
(b) When **AB** = **BA**, find the value of x.   [3]

Cambridge IGCSE Mathematics 0580
Paper 22 Q21 June 2009

2. $\begin{pmatrix} 1 & -2 \\ 0 & 1 \\ 5 & 6 \end{pmatrix} \begin{pmatrix} 3 & 4 & 8 & 7 \\ 1 & 1 & 3 & 3 \end{pmatrix}$

The answer to this matrix multiplication is of order $a \times b$.
Find the values of a and b.   [2]

Cambridge IGCSE Mathematics 0580
Paper 21 Q2 November 2008

3. A $= \begin{pmatrix} -2 & 3 \\ -4 & 5 \end{pmatrix}$

Find $\mathbf{A}^{-1}$, the inverse of the matrix **A**.   [2]

Cambridge IGCSE Mathematics 0580
Paper 21 Q5 June 2009

4. M $= \begin{pmatrix} 1 & 1 \\ 1 & 2 \end{pmatrix} \qquad \mathbf{M}^2 = \begin{pmatrix} 2 & 3 \\ 3 & 5 \end{pmatrix} \qquad \mathbf{M}^3 = \begin{pmatrix} 5 & 8 \\ 8 & 13 \end{pmatrix}$

Find $\mathbf{M}^4$.   [2]

Cambridge IGCSE Mathematics 0580
Paper 2 Q7 June 2007

5. Work out

$$\begin{pmatrix} 2 & 1 & 2 \\ 1 & 5 & 0 \\ 3 & -2 & 4 \end{pmatrix} \begin{pmatrix} 4 \\ -3 \\ -8 \end{pmatrix}.$$ [3]

Cambridge IGCSE Mathematics 0580
Paper 22 Q15 June 2008

6.

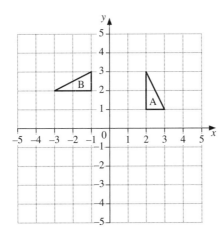

(a) A transformation is represented by the matrix $\begin{pmatrix} 0 & -1 \\ -1 & 0 \end{pmatrix}$.

    (i) **On the grid above**, draw the image of triangle A after
       this transformation.   [2]
    (ii) Describe fully this transformation.   [2]
(b) Find the 2 by 2 matrix representing the transformation which
    maps triangle A onto triangle B.   [2]

Cambridge IGCSE Mathematics 0580
Paper 21 Q19 June 2008

7. Transformation T is translation by the vector $\begin{pmatrix} 3 \\ 2 \end{pmatrix}$.

Transformation M is reflection in the line $y = x$.

(a) The point A has coordinates (2, 1).
    Find the coordinates of
    (i) T(A),   [1]
    (ii) MT(A).   [2]
(b) Find the 2 by 2 matrix **M**, which represents the transformation M.   [2]
(c) Show that, for any value of k, the point $Q(k - 2, k - 3)$ maps onto a point on the
    line $y = x$ following the transformation TM(Q).   [3]
(d) Find $\mathbf{M}^{-1}$, the inverse of the matrix **M**.   [2]
(e) **N** is the matrix such that $\mathbf{N} + \begin{pmatrix} 0 & 3 \\ 1 & 0 \end{pmatrix} = \begin{pmatrix} 0 & 4 \\ 0 & 0 \end{pmatrix}$.

    (i) Write down the matrix **N**.   [2]
    (ii) Describe completely the **single** transformation represented by **N**.   [3]

Cambridge IGCSE Mathematics 0580
Paper 4 Q7 June 2006

8. **Answer the whole of this question on one sheet of graph paper.**

(a) Draw and label x and y axes from -8 to $+8$, using a scale of
    1 cm to 1 unit on each axis.   [1]
(b) Draw and label triangle ABC with $A(2,2)$, $B(5,2)$ and $C(5,4)$.   [1]
(c) On your grid:

   (i) translate **triangle ABC** by the vector $\begin{pmatrix} 3 \\ -9 \end{pmatrix}$ and label this
    image $A_1B_1C_1$;   [2]
  (ii) reflect **triangle ABC** in the line $x = -1$ and label this image
    $A_2B_2C_2$;   [2]
 (iii) rotate **triangle ABC** by $180°$ about $(0, 0)$ and label this image $A_3B_3C_3$.   [2]

(d) A stretch is represented by the matrix $\begin{pmatrix} 1.5 & 0 \\ 0 & 1 \end{pmatrix}$.

   (i) Draw the image of **triangle ABC** under this transformation.
    Label this image $A_4B_4C_4$.   [3]
  (ii) Work out the inverse of the matrix $\begin{pmatrix} 1.5 & 0 \\ 0 & 1 \end{pmatrix}$.   [2]
 (iii) Describe **fully** the single transformation represented by this inverse.   [3]

Cambridge IGCSE Mathematics 0580
Paper 4 Q2 November 2005

9. **Answer the whole of this question on a sheet of graph paper.**

(a) Draw a label x and y axes from -6 to 6, using a scale of 1 cm
    to 1 unit.   [1]
(b) Draw triangle ABC with $A(2, 1)$, $B(3, 3)$ and $C(5, 1)$.   [1]
(c) Draw the reflection of triangle ABC in the line $y = x$. Label this $A_1B_1C_1$.   [2]
(d) Rotate **triangle $A_1B_1C_1$** about $(0, 0)$ through $90°$ anticlockwise. Label this $A_2B_2C_2$.   [2]
(e) Describe fully the single transformation which maps triangle ABC onto triangle $A_2B_2C_2$.   [2]
(f) A transformation is represented by the matrix $\begin{pmatrix} 1 & 0 \\ -1 & 1 \end{pmatrix}$.

   (i) Draw the image of triangle ABC under this transformation. Label this $A_3B_3C_3$.   [3]
  (ii) Describe fully the single transformation represented by the matrix $\begin{pmatrix} 1 & 0 \\ -1 & 1 \end{pmatrix}$.   [2]

 (iii) Find the matrix which represents the transformation that maps triangle $A_3B_3C_3$
    onto triangle ABC.   [2]

Cambridge IGCSE Mathematics 0580
Paper 4 Q2 June 2007

10 STATISTICS AND PROBABILITY

Blaise Pascal (1623–1662) suffered the most appalling ill-health throughout his short life. He is best known for his work with Fermat on probability. This followed correspondence with a gentleman gambler who was puzzled as to why he lost so much in betting on the basis of the appearance of a throw of dice. Pascal's work on probability became of enormous importance and showed for the first time that absolute certainty is not a necessity in mathematics and science. He also studied physics, but his last years were spent in religious meditation and illness.

33 Construct and read bar charts, histograms, scatter diagrams and cumulative frequency diagrams; calculate the mean, median and mode

34 Calculate the probability of a single event, and simple combined events

10.1 Data display

Bar chart

The length of each bar represents the quantity in question. The width of each bar has no significance. In this bar chart, the number of the cars of each colour in a car park is shown. The bars can be joined together or separated.

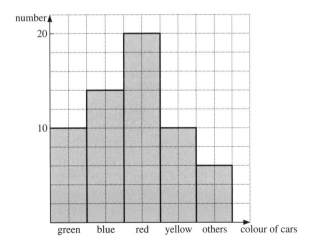

Example

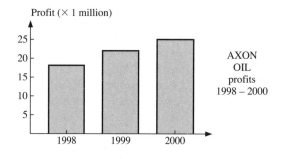

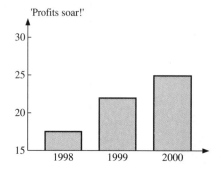

The bar charts show the profits of a company over a 3-year period. The second graph has been drawn to give the impression that the profits have increased dramatically over the last three years.
How has this been done?

Pie chart

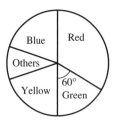

The information is displayed using sectors of a circle.
This pie chart shows the same information as the bar chart on the previous page.
The angles of the sectors are calculated as follows:

Total number of cars $= 10 + 14 + 20 + 10 + 6 = 60$

Angle representing green cars $= \dfrac{10}{60} \times 360° = 60°$

Angle representing blue cars $= \dfrac{14}{60} \times 360°$, etc.

Exercise 1

1. The bar chart shows the number of children playing various games on a given day.

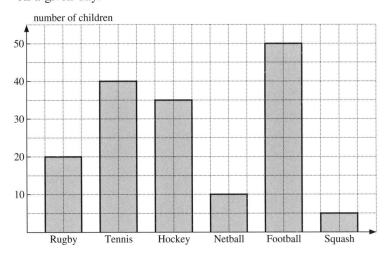

number of children

    Rugby   Tennis   Hockey   Netball   Football   Squash

(a) Which game had the least number of players?
(b) What was the total number of children playing all the games?
(c) How many more footballers were there than tennis players?

2. The table shows the number of cars of different makes in a car park. Illustrate this data on a bar chart.

make	Skoda	Renault	Saab	Kia	Subaru	Lexus
number	14	23	37	5	42	18

3. The pie chart illustrates the values of various goods sold by a certain shop. If the total value of the sales was $24 000, find the sales value of:
(a) toys
(b) grass seed
(c) records
(d) food.

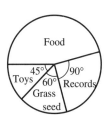

4. The table shows the colours of a random selection of sweets. Calculate the angles on a pie chart corresponding to each colour.

colour	red	green	blue	yellow	pink
number	5	7	11	4	9

5. A quantity of scrambled eggs is made using the following recipe:

ingredient	eggs	milk	butter	cheese	salt/pepper
mass	450 g	20 g	39 g	90 g	1 g

Calculate the angles on a pie chart corresponding to each ingredient.

6. Calculate the angles on a pie chart corresponding to quantities A,
B, C, D and E given in the tables.

(a)

quantity	A	B	C	D	E
number	3	5	3	7	0

(b)

quantity	A	B	C	D	E
mass	10 g	15 g	34 g	8 g	5 g

(c)

quantity	A	B	C	D	E
length	7	11	9	14	11

7. A firm making artificial sand sold its products in four countries:
    5% were sold in Spain
   15% were sold in France
   15% were sold in Germany
   65% were sold in the U.K.
What would be the angles on a pie chart drawn to represent this
information?

8. The weights of A, B, C are in the ratio $2:3:4$. Calculate the angles
representing A, B and C on a pie-chart.

9. The cooking times for meals L, M and N are in the ratio $3:7:x$. On
a pie-chart, the angle corresponding to L is $60°$. Find x.

10. The results of an opinion poll of 2000 people are represented on a
pie chart. The angle corresponding to 'don't know' is $18°$. How
many people in the sample did not know?

11. The pie chart illustrates the sales of various makes of petrol.
(a) What percentage of sales does 'Esso' have?
(b) If 'Jet' accounts for $12\frac{1}{2}\%$ of total sales, calculate the angles x
and y.

12.

	% of total spent
Press	51
Television	40
Posters	
Cinema	
Radio	3
Total	100

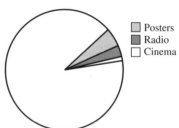

In Spain money was spent on advertisements in 1999 in the press,
TV, posters, etc. The incomplete table and pie-chart show the way
this was divided between the media.
(a) Calculate the angle of the sector representing television, and
complete the pie-chart.
(b) The angle of the sector representing posters is $18°$. Calculate the
percentage spent on posters, and hence complete the table.

13. The diagram illustrates the production of
apples in two countries.

U.K.
470
thousand
tonnes

FRANCE
950
thousand
tonnes

In what way could the pictorial display
be regarded as misleading?

14. The graph shows the performance of a company in
the year in which a new manager was appointed.
In what way is the graph misleading?

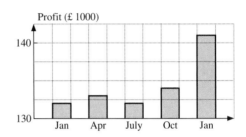

Frequency polygons

A frequency polygon can be drawn by joining the mid-points of the
tops of the bars on a frequency chart.
Frequency polygons are used mainly to compare data.

- Here is a frequency chart
  showing the heights (or lengths)
  of the babies treated at a
  hospital one day.

- Here is the corresponding frequency
  polygon, drawn by joining the
  mid-points of the tops of the bars.

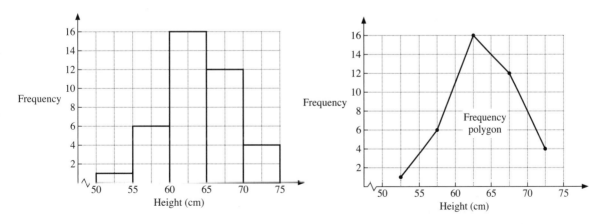

It is not necessary to draw the bars if you require only the frequency
polygon.

The diagram on the right shows the
frequency polygons for the exam results
of 34 pupils in two subjects, Maths and French.
Two main differences are apparent:
(a) The marks obtained in the Maths exam
    were significantly lower for most pupils.
(b) The marks obtained in the French
    exam were more spread out than
    the Maths marks. The French marks
    were distributed fairly evenly over the
    range from 0 to 100% whereas the Maths
    marks were mostly between 0 and 40%.

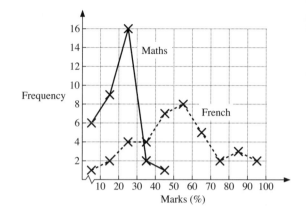

Exercise 2

1. Draw a frequency polygon
   for the distribution of masses
   of children drawn in the
   diagram.

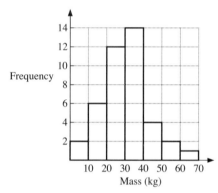

2. In a supermarket survey, shoppers were
   asked two questions as they left:
   (a) How much have you just spent?
   (b) How far away do you live?
   The results were separated into two
   groups: shoppers who lived less than
   2 miles from the supermarket and shoppers
   who lived further away. The frequency
   polygons show how much shoppers in each
   group had spent.
   Decide which polygon, P or Q, is most likely to represent
   shoppers who lived less than 2 miles from the supermarket.
   Give your reasons.

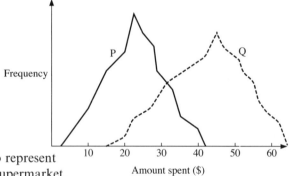

3. Scientists doing research in genetic engineering altered the genes of a
   certain kind of rabbit. Over a period of several years, measurements
   were made of the adult weight of the rabbits and their lifespans. The
   frequency polygons on the next page show the results.

   What can you deduce from the two frequency polygons? Write one
   sentence about weight and one sentence about lifespan.

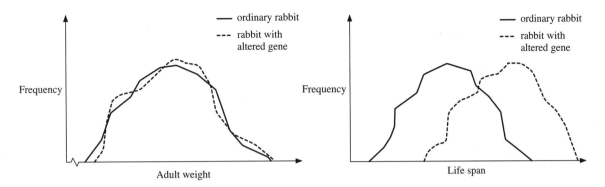

Histograms

In a histogram, the frequency of the data is shown by the *area* of each bar. Histograms resemble bar charts but are not to be confused with them: in bar charts the frequency is shown by the height of each bar. Histograms often have bars of varying widths. Because the area of the bar represents frequency, the height must be adjusted to correspond with the width of the bar. The vertical axis is not labelled frequency but frequency density.

$$\text{frequency density} = \frac{\text{frequency}}{\text{class width}}$$

Histograms can be used to represent both discrete data and continuous data, but their main purpose is for use with continuous data.

Example

Draw a histogram from the table shown for the distribution of ages of passengers travelling on a flight to New York.

Note that the data has been collected into class intervals of different widths.

ages	frequency
$0 \leqslant x < 20$	28
$20 \leqslant x < 40$	36
$40 \leqslant x < 50$	20
$50 \leqslant x < 70$	30
$70 \leqslant x < 100$	18

To draw the histogram, the heights of the bars must be adjusted by calculating frequency density.

ages	frequency	frequency density (f.d.)
$0 \leqslant x < 20$	28	$28 \div 20 = 1{\cdot}4$
$20 \leqslant x < 40$	36	$36 \div 20 = 1{\cdot}8$
$40 \leqslant x < 50$	20	$20 \div 10 = 2$
$50 \leqslant x < 70$	30	$30 \div 20 = 1{\cdot}5$
$70 \leqslant x < 100$	18	$18 \div 30 = 0{\cdot}6$

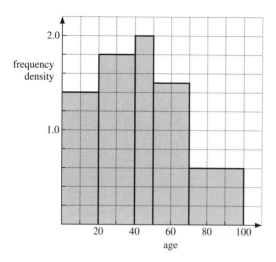

Exercise 3

1. The lengths of 20 copper nails were measured. The results are shown in the frequency table.

length l (in mm)	frequency	frequency density (f.d.)
$0 \leqslant L < 20$	5	$5 \div 20 = 0.25$
$20 \leqslant L < 25$	5	
$25 \leqslant L < 30$	7	
$30 \leqslant L < 40$	3	

Calculate the frequency densities and draw the histogram as started below.

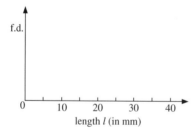

2. The volumes of 55 containers were measured and the results presented in a frequency table as shown in the table.

volume (mm^3)	frequency
$0 \leqslant V < 5$	5
$5 \leqslant V < 10$	3
$10 \leqslant V < 20$	12
$20 \leqslant V < 30$	17
$30 \leqslant V < 40$	13
$40 \leqslant V < 60$	5

Calculate the frequency densities and draw the histogram.

3. The masses of thirty students in a class are measured. Draw a histogram to represent this data.

mass (kg)	frequency
30–40	5
40–45	7
45–50	10
50–55	5
55–70	3

Note that the masses do not start a zero. This can be shown on the graph as follows:

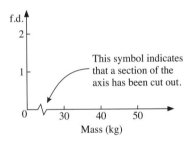

4. The ages of 120 people passing through a turnstyle were recorded and are shown in the frequency table.

age (yrs)	frequency
–10	18
–15	46
–20	35
–30	13
–40	8

The notation –10 means '0 < age ⩽ 10' and similarly –15 means '10 < age ⩽ 15'. The class boundaries are 0, 10, 15, 20, 30, 40. Draw the histogram for the data.

5. Another common notation is used here for the masses of plums picked in an orchard, shown in the table below.

mass (g)	20–	30–	40–	60–	80–
frequency	11	18	7	5	0

The notation 20– means $20\,\text{g} \leqslant \text{mass} < 30\,\text{g}$.
Draw a histogram with class boundaries at 20, 30, 40, 60, 80.

6. The heights of 50 Olympic athletes were measured as shown in the table.

height (cm)	170–174	175–179	180–184	185–194
frequency	8	17	14	11

These values were rounded off to the nearest cm. For example, an athlete whose height h is 181 cm could be entered anywhere in the class $180 \cdot 5\,\text{cm} \leqslant h < 181 \cdot 5\,\text{cm}$. So the table is as follows:

height	169·5–174·5	174·5–179·5	179·5–184·5	184·5–194·5
frequency	8	17	14	11

Draw a histogram with class boundaries at 169·5, 174·5, 179·5, ...

7. The number of people travelling in 33 vehicles one day was as shown in the table below.

number of people	1	2	3	4	5–6	7–10
frequency	8	11	6	4	2	2

In this case, the data is discrete. To represent this information on a histogram, draw the column for the value 2, for example, from 1·5 to 2·5, and that for the values 5–6 from 4·5 to 6·5 as shown below.

number of people	frequency	interval on histogram	width of interval	frequency density
1	8	0·5–1·5	1	8
2	11	1·5–2·5		
3	6			
4	4			
5–6	2	4·5–6·5	2	1
7–10	2			

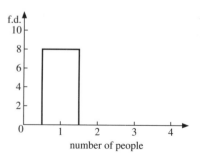

Copy and complete the above table and the histogram which has been started on the right.

10.2 Mean, median and mode

(a) The *mean* of a series of numbers is obtained by adding the numbers and dividing the result by the number of numbers.

(b) The *median* of a series of numbers is obtained by arranging the numbers in ascending order and then choosing the number in the 'middle'. If there are *two* 'middle' numbers the median is the average (mean) of these two numbers.

(c) The *mode* of a series of numbers is simply the number which occurs most often.

Example

Find the mean, median and mode of the following numbers:
5, 4, 10, 3, 3, 4, 7, 4, 6, 5.

(a) Mean $= \dfrac{(5 + 4 + 10 + 3 + 3 + 4 + 7 + 4 + 6 + 5)}{10} = \dfrac{51}{10} = 5 \cdot 1$

(b) Median: arranging numbers in order of size

   3, 3, 4, 4, 4, 5, 5, 6, 7, 10
   $\uparrow$

   The median is the 'average' of 4 and 5
   $\therefore$ median $= 4 \cdot 5$

(c) Mode $= 4$ (there are more 4's than any other number).

Frequency tables

A frequency table shows a number x such as a mark or a score, against the frequency f or number of times that x occurs.

The symbol Σ (or sigma) means 'the sum of'.

The next example shows how these symbols are used in calculating the mean, the median and the mode.

Example

The marks obtained by 100 students in a test were as follows:

mark (x)	0	1	2	3	4
frequency (f)	4	19	25	29	23

Find:
(a) the mean mark   (b) the median mark   (c) the modal mark

(a) Mean $= \dfrac{\Sigma xf}{\Sigma f}$

     where Σxf means 'the sum of the products'
        i.e. $\Sigma(\text{number} \times \text{frequency})$

     and Σf means 'the sum of the frequencies'.

$$\text{Mean} = \frac{(0 \times 4) + (1 \times 19) + (2 \times 25) + (3 \times 29) + (4 \times 23)}{100}$$

$$= \frac{248}{100} = 2 \cdot 48$$

(b) The median mark is the number between the 50th and 51st numbers. By inspection, both the 50th and 51st numbers are 3.

$\therefore$ Median = 3 marks

(c) The modal mark = 3

Exercise 4

1. Find the mean, median and mode of the following sets of numbers:
   (a) 3, 12, 4, 6, 8, 5, 4
   (b) 7, 21, 2, 17, 3, 13, 7, 4, 9, 7, 9
   (c) 12, 1, 10, 1, 9, 3, 4, 9, 7, 9
   (d) 8, 0, 3, 3, 1, 7, 4, 1, 4, 4

2. Find the mean, median and mode of the following sets of numbers:
   (a) 3, 3, 5, 7, 8, 8, 8, 9, 11, 12, 12
   (b) 7, 3, 4, 10, 1, 2, 1, 3, 4, 11, 10, 4
   (c) $-3, 4, 0, 4, -2, -5, 1, 7, 10, 5$
   (d) $1, \frac{1}{2}, \frac{1}{2}, \frac{3}{4}, \frac{1}{4}, 2, \frac{1}{2}, \frac{1}{4}, \frac{3}{4}$

3. The mean mass of five men is 76 kg. The masses of four of the men are 72 kg, 74 kg, 75 kg and 81 kg. What is the mass of the fifth man?

4. The mean length of 6 rods is 44·2 cm. The mean length of 5 of them is 46 cm. How long is the sixth rod?

5. (a) The mean of 3, 7, 8, 10 and x is 6. Find x.
   (b) The mean of 3, 3, 7, 8, 10, x and x is 7. Find x.

6. The mean height of 12 men is 1·70 m, and the mean height of 8 women is 1·60 m. Find:
   (a) the total height of the 12 men,
   (b) the total height of the 8 women,
   (c) the mean height of the 20 men and women.

7. The total mass of 6 rugby players is 540 kg and the mean mass of 14 ballet dancers is 40 kg. Find the mean mass of the group of 20 rugby players and ballet dancers.

8. The mean mass of 8 boys is 55 kg and the mean mass of a group of girls is 52 kg. The mean mass of all the children is 53·2 kg. How many girls are there?

9. For the set of numbers below, find the mean and the median.

1, 3, 3, 3, 4, 6, 99

   Which average best describes the set of numbers?

10. In a history test, Andrew got 62%. For the whole class, the mean mark was 64% and the median mark was 59%. Which 'average' tells Andrew whether he is in the 'top' half or the 'bottom' half of the class?

11. The mean age of three people is 22 and their median age is 20. The range of their ages is 16. How old is each person?

12. A group of 50 people were asked how many books they had read in the previous year; the results are shown in the frequency table below. Calculate the mean number of books read per person.

number of books	0	1	2	3	4	5	6	7	8
frequency	5	5	6	9	11	7	4	2	1

13. A number of people were asked how many coins they had in their pockets; the results are shown below. Calculate the mean number of coins per person.

number of coins	0	1	2	3	4	5	6	7
frequency	3	6	4	7	5	8	5	2

14. The following tables give the distribution of marks obtained by different classes in various tests. For each table, find the mean, median and mode.

(a)

mark	0	1	2	3	4	5	6
frequency	3	5	8	9	5	7	3

(b)

mark	15	16	17	18	19	20
frequency	1	3	7	1	5	3

(c)

mark	0	1	2	3	4	5	6
frequency	10	11	8	15	25	20	11

15. One hundred golfers play a certain hole and their scores are summarised below.

score	2	3	4	5	6	7	8
number of players	2	7	24	31	18	11	7

Find:
(a) the mean score   (b) the median score.

16. The number of goals scored in a series of football matches was as follows:

number of goals	1	2	3
number of matches	8	8	x

(a) If the mean number of goals is 2·04, find x.
(b) If the modal number of goals is 3, find the smallest possible value of x.
(c) If the median number of goals is 2, find the largest possible value of x.

17. In a survey of the number of occupants in a number of cars, the following data resulted.

number of occupants	1	2	3	4
number of cars	7	11	7	x

(a) If the mean number of occupants is $2\frac{1}{3}$, find x.
(b) If the mode is 2, find the largest possible value of x.
(c) If the median is 2, find the largest possible value of x.

18. The numbers 3, 5, 7, 8 and N are arranged in ascending order. If the mean of the numbers is equal to the median, find N.

19. The mean of 5 numbers is 11. The numbers are in the ratio $1:2:3:4:5$. Find the smallest number.

20. The mean of a set of 7 numbers is 3·6 and the mean of a different set of 18 numbers is 5·1. Calculate the mean of the 25 numbers.

21. The marks obtained by the members of a class are summarised in the table.

mark	x	y	z
frequency	a	b	c

Calculate the mean mark in terms of a, b, c, x, y and z.

Data in groups

Example

The results of 51 students in a test are given in the frequency table.
Find the (a) mean (b) median (c) modal class.

mark	30–39	40–49	50–59	60–69
frequency	7	14	21	9

Don't forget the mean is only an **estimate** because you do not have the raw data and you have made an assumption with the mid-point of each interval.

In order to find the mean you approximate by saying each interval is represented by its mid-point. For the 30–39 interval you say there are 7 marks of 34·5 [that is $(30 + 39) \div 2 = 34 \cdot 5$].

(a) $\text{Mean} = \dfrac{(34 \cdot 5 \times 7) + (44 \cdot 5 \times 14) + (54 \cdot 5 \times 21) + (64 \cdot 5 \times 9)}{(7 + 14 + 21 + 9)}$

$= 50 \cdot 7745098$

$= 51 \ (2 \ \text{s.f.})$

(b) The median is the 26th mark, which is in the interval 50–59. You cannot find the exact median.

(c) The **modal class** is 50–59. You cannot find an exact mode.

Later you will find out how to get an estimate of the median by drawing a cumulative frequency curve.

Exercise 5

1. The table gives the number of words in each sentence of a page in a book.
   (a) Copy and complete the table.
   (b) Work out an estimate for the mean number of words in a sentence.

number of words	frequency f	mid-point x	fx
1–5	6	3	18
6–10	5	8	40
11–15	4		
16–20	2		
21–25	3		
totals	20		–

2. The results of 24 students in a test are given in the table.

mark	frequency
85–99	4
70–84	7
55–69	8
40–54	5

(a) Find the mid-point of each group of marks and calculate an estimate of the mean mark.
(b) Explain why your answer is an estimate.

3. The results of 24 students in a test are given in the table.

mark	40–54	55–69	70–84	85–99
frequency	5	8	7	4

Find the mid-point of each group of marks and calculate an estimate of the mean mark.

4. The table shows the number of letters delivered to the 26 houses in a street.

Calculate an estimate of the mean number of letters delivered per house.

number of letters delivered	number of houses (frequency)
0–2	10
3–4	8
5–7	5
8–12	3

5. The histogram shows the heights of the 60 athletes in the Indian athletics team.
(a) Calculate an estimate for the mean height of the 60 athletes.
(b) Explain why your answer is an **estimate** for the mean height.
(c) What is the modal class for the heights of these athletes?

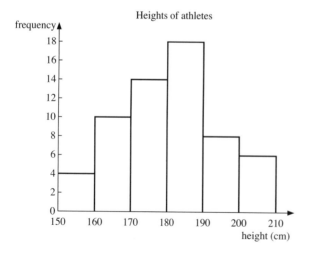

Heights of athletes

Scatter graphs

Sometimes it is important to discover if there is a connection or relationship between two sets of data.

Examples:

- Are more ice creams sold when the weather is hot?
- Do tall people have higher pulse rates?
- Are people who are good at maths also good at science?
- Does watching TV improve examination results?

If there is a relationship, it will be easy to spot if your data is plotted on a scatter diagram – that is a graph in which one set of data is plotted on the horizontal axis and the other on the vertical axis.

Here is a scatter graph showing the price of pears and the quantity sold.

We can see a *connection* – when the price was high the sales were low and when the price went down the sales increased.

This scatter graph shows the sales of a newspaper and the temperature. We can see there is *no connection* between the two variables.

Correlation

The word correlation describes how things *co-relate*. There is correlation between two sets of data if there is a connection or relationship.

The correlation between two sets of data can be positive or negative and it can be strong or weak as indicated by the scatter graphs below.

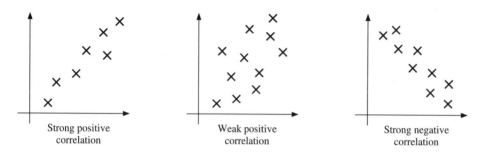

| Strong positive correlation | Weak positive correlation | Strong negative correlation |

When the correlation is positive the points are around a line which slopes upwards to the right. When the correlation is negative the 'line' slopes downwards to the right.

When the correlation is strong the points are bunched close to a line through their midst. When the correlation is weak the points are more scattered.

It is important to realise that often there is *no* correlation between two sets of data.

If, for example, we take a group of students and plot their maths test results against their time to run 800 m, the graph might look like the one on the right. A common mistake in this topic is to 'see' a correlation on a scatter graph where none exists.

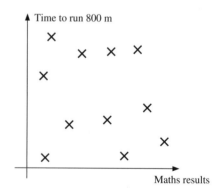

There is also *no* correlation in these two scatter graphs.

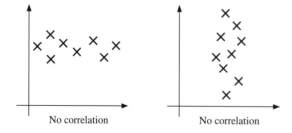

| No correlation | No correlation |

Line of best fit

When a scatter graph shows either positive or negative correlation, a *line of best fit* can be drawn. The sums of the distances to points on either side of the line are equal and there should be an equal number of points on each side of the line. The line is easier to draw when a transparent ruler is used.

Here are the marks obtained in two tests by 9 students.

Student	A	B	C	D	E	F	G	H	I
Maths mark	28	22	9	40	37	35	30	23	?
Physics mark	48	45	34	57	50	55	53	45	52

A line of best fit can be drawn as there is strong positive correlation between the two sets of marks.

The line of best fit can be used to estimate the maths result of student J, who missed the maths test but scored 52 in the physics test.

We can *estimate that student J would have scored about 33* in the maths test. It is not possible to be *very* accurate using scatter graphs. It is reasonable to state that student J 'might have scored between 30 and 36' in the maths test.

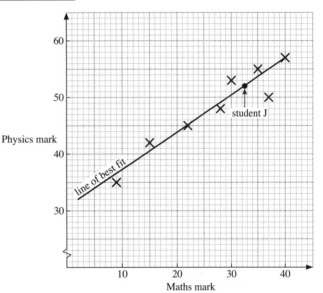

Here is a scatter graph in which the heights of boys of different ages is recorded. A line of best fit is drawn.

(a) We can estimate that the height of an 8-year-old boy might be about 123 cm [say between 120 and 126 cm].
(b) We can only predict a height within the range of values plotted. We could not extend the line of best and use it to predict the height of a 30 year old! Why not?

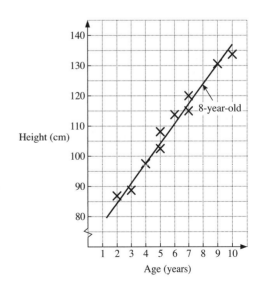

Exercise 6

1. Make the following measurements
   for everyone in your class:
   height   (nearest cm)
   arm span   (nearest cm)
   head circumference   (nearest cm)
   hand span   (nearest cm)
   pulse rate   (beats/minute)

For greater consistency of measuring,
one person (or perhaps two people)
should do all the measurements of
one kind (except on themselves!).

Enter all the measurements in a table,
either on the board or on a sheet of paper.

Name	Height	Armspan	Head
Roger	161	165	56
Liz	150	148	49
Gill			

(a) Draw the scatter graphs shown below:

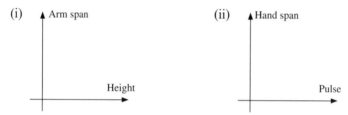

(b) Describe the correlation, if any, in the scatter graphs you drew
    in part (a).

(c) (i) Draw a scatter graph of two measurements where you
        think there might be positive correlation.
    (ii) Was there indeed a positive correlation?

2. Plot the points given on a scatter graph, with s across the page and
   p up the page. Draw axes with values from 0 to 20.
   Describe the correlation, if any, between the values of s and p.
   [i.e. 'strong negative', 'weak positive' etc.]

(a)

s	7	16	4	12	18	6	20	4	10	13
p	8	15	6	12	17	9	18	7	10	14

(b)

s	3	8	12	15	16	5	6	17	9
p	4	2	10	17	5	10	17	11	15

(c)

s	11	1	16	7	2	19	8	4	13	18
p	5	12	7	14	17	1	11	8	11	5

In questions **3**, **4** and **5** plot the points given on a scatter graph, with s across the page and p up the page.
Draw axes with the values from 0 to 20.
If possible draw a line of best fit on the graph.
Where possible estimate the value of p on the line of best fit where $s = 10$.

3.

s	2	14	14	4	12	18	12	6
p	5	15	16	6	12	18	13	7

4.

s	2	15	17	3	20	3	6
p	13	7	5	12	4	13	11

5.

s	4	10	15	18	19	4	19	5
p	19	16	11	19	15	3	1	9

6. The following data gives the marks of 11 students in a French test and in a German test.

French	15	36	36	22	23	27	43	22	43	40	26
German	6	28	35	18	28	28	37	9	41	45	17

(a) Plot this data on a scatter graph, with French marks on the horizontal axis.

(b) Draw the line of best fit.

(c) Estimate the German mark of a student who got 30 in French.

(d) Estimate the French mark of a student who got 45 in German.

7. The data below gives the petrol consumption figures of cars, with the same size engine, when driven at different speeds.

Speed (m.p.h.)	30	62	40	80	70	55	75
Petrol consumption (m.p.g.)	38	25	35	20	26	34	22

(a) Plot a scatter graph and draw a line of best fit.

(b) Estimate the petrol consumption of a car travelling at 45 m.p.h.

(c) Estimate the speed of a car whose petrol consumption is 27 m.p.g.

10.3 Cumulative frequency

Cumulative frequency is the total frequency up to a given point.
A cumulative frequency curve (or ogive) shows the *median* at the 50th
percentile of the cumulative frequency.
The value at the 25th percentile is known as the *lower quartile*, and that
at the 75th percentile as the *upper quartile*.
A measure of the spread or dispersion of the data is given by the
inter-quartile range where

inter-quartile range = upper quartile − lower quartile

Example
The marks obtained by 80 students in an examination are shown below.

mark	frequency	cumulative frequency	marks represented by cumulative frequency
1–10	3	3	$\leqslant 10$
11–20	5	8	$\leqslant 20$
21–30	5	13	$\leqslant 30$
31–40	9	22	$\leqslant 40$
41–50	11	33	$\leqslant 50$
51–60	15	48	$\leqslant 60$
61–70	14	62	$\leqslant 70$
71–80	8	70	$\leqslant 80$
81–90	6	76	$\leqslant 90$
91–100	4	80	$\leqslant 100$

The table also shows the cumulative frequency.
Plot a cumulative frequency curve and hence estimate:
(a) the median
(b) the inter-quartile range.

The points on the graph are plotted at the upper
limit of each group of marks.

From the cumulative frequency curve

median = 55 marks
lower quartile = 37·5 marks
upper quartile = 68 marks
∴   inter-quartile range = 68 − 37·5
= 30·5 marks.

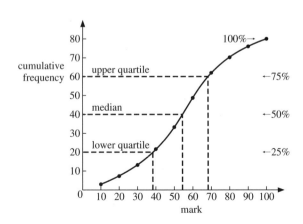

Exercise 7

1. Figure 1 shows the cumulative frequency curve for the marks of 60 students in an examination.

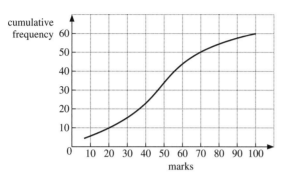

Figure 1

From the graph, estimate:
(a) the median mark,
(b) the mark at the lower quartile and at the upper quartile,
(c) the inter-quartile range,
(d) the pass mark if two-thirds of the students passed,
(e) the number of students achieving less than 40 marks.

2. Figure 2 shows the cumulative frequency curve for the marks of 140 students in an examination.

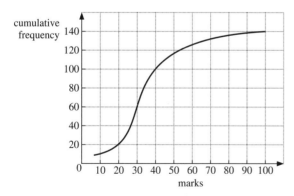

Figure 2

From the graph, estimate:
(a) the median mark,
(b) the mark at the lower quartile and at the upper quartile,
(c) the inter-quartile range,
(d) the pass mark if three-fifths of the students passed,
(e) the number of students achieving more than 0 marks.

In questions **3** to **6**, draw a cumulative frequency curve, and find:

(a) the median,   (b) the interquartile range.

3.

mass (kg)	frequency
1–5	4
6–10	7
11–15	11
16–20	18
21–25	22
26–30	10
31–35	5
36–40	3

4.

length (cm)	frequency
41–50	6
51–60	8
61–70	14
71–80	21
81–90	26
91–100	14
101–110	7
111–120	4

5.

time (seconds)	frequency
36–45	3
46–55	7
56–65	10
66–75	18
76–85	12
86–95	6
96–105	4

6.

number of marks	frequency
1–10	0
11–20	2
21–30	4
31–40	10
41–50	17
51–60	11
61–70	3
71–80	3

7. In an experiment, 50 people were asked to guess the mass of a bunch of flowers in grams. The guesses were as follows:

```
47  39  21  30  42  35  44  36  19  52
23  32  66  29   5  40  33  11  44  22
27  58  38  37  48  63  23  40  53  24
47  22  44  33  13  59  33  49  57  30
17  45  38  33  25  40  51  56  28  64
```

Construct a frequency table using intervals 0–9, 10–19, 20–29, etc. Hence draw a cumulative frequency curve and estimate:

(a) the median mass,

(b) the inter-quartile range,

(c) the number of people who guessed a mass within 10 grams of the median.

8. In a competition, 30 children had to pick up as many paper clips as possible in one minute using a pair of tweezers. The results were as follows:

```
 3  17   8  11  26  23  18  28  33  38
12  38  22  50   5  35  39  30  31  43
27  34   9  25  39  14  27  16  33  49
```

Construct a frequency table using intervals 1–10, 11–20, etc. and hence draw a cumulative frequency curve.
(a) From the curve, estimate the median number of clips picked up.
(b) From the frequency table, estimate the mean of the distribution using the mid-interval values 5·5, 15·5, etc.
(c) Calculate the exact value of the mean using the original data.
(d) Why is it possible only to estimate the mean in part (b)?

9. The children in two schools took the same test in mathematics and their results are shown.

School A	School B
median mark = 52%	median mark = 51.8%
IQR = 7·2	IQR = 11·2

Note: IQR is shorthand for inter-quartile range.

What can you say about these two sets of results?

10. As part of a health improvement programme, people from one town and from one village in Gambia were measured. Here are the results.

People in town	People in village
median height = 171 cm	median height = 163 cm
IQR = 8·4	IQR = 3·7

What can you say about these two sets of results?

10.4 Simple probability

Probability theory is not the sole concern of people interested in betting, although it is true to say that a 'lucky' poker player is likely to be a player with a sound understanding of probability. All major airlines regularly overbook aircraft because they can usually predict with accuracy the probability that a certain number of passengers will fail to arrive for the flight.

Suppose a 'trial' can have n equally likely results and suppose that a 'success' can occur in s ways (from the n). Then the probability of a 'success' $= \dfrac{s}{n}$.

- If an event **cannot** happen the probability of it occurring is 0.
- If an event is **certain** to happen the probability of it occurring is 1.
- All probabilities lie between 0 and 1.
  You write probabilities using fractions or decimals.

Example 1

The numbers 1 to 20 are each written on a card.
The 20 cards are mixed together.
One card is chosen at random from the pack.
Find the probability that the number on the card is:

(a) even   (b) a factor of 24   (c) prime.

We will use '$p(x)$' to mean 'the probability of x'.

(a) $p(\text{even}) = \dfrac{10}{20}$ (b) $p(\text{factor of 24})$ (c) $p(\text{prime})$

$\qquad\qquad = \dfrac{1}{2}$ $\qquad = p(1, 2, 3, 4, 6, 8, 12)$ $\qquad = p(2, 3, 5, 7, 11, 13, 17, 19)$

$\qquad\qquad\qquad\qquad\qquad\qquad = \dfrac{7}{20}$ $\qquad\qquad = \dfrac{8}{20} = \dfrac{2}{5}$

In each case, we have counted the number of ways in which a 'success'
can occur and divided by the number of possible results of a 'trial'.

Example 2

A black die and a white die are thrown at the same time. Display all the
possible outcomes. Find the probability of obtaining:

(a) a total of 5,
(b) a total of 11,
(c) a 'two' on the black die and a 'six' on the white die.

It is convenient to display all the possible outcomes on a grid.

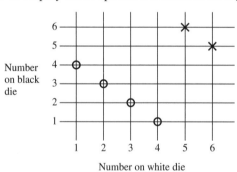

There are 36 possible outcomes, shown where the lines cross.

(a) There are four ways of obtaining a total of 5 on the
    two dice. They are shown circled on the diagram.

$\qquad \therefore \quad$ Probability of obtaining a total of $5 = \dfrac{4}{36}$

(b) There are two ways of obtaining a total of 11. They are shown
    with a cross on the diagram.

$\qquad \therefore \quad p\,(\text{total of } 11) = \dfrac{2}{36} = \dfrac{1}{18}$

(c) There is only one way of obtaining a 'two' on the black die and a 'six' on the white die.

$$\therefore \quad p \text{ (2 on black and 6 on white)} = \frac{1}{36}$$

Exercise 8

In this exercise, all dice are normal cubic dice with faces numbered 1 to 6.

1. A fair die is thrown once. Find the probability of obtaining:
   (a) a six,
   (b) an even number,
   (c) a number greater than 3,
   (d) a three or a five.

2. The two sides of a coin are known as 'head' and 'tail'. A 10c and a 5c coin are tossed at the same time. List all the possible outcomes. Find the probability of obtaining:
   (a) two heads,   (b) a head and a tail.

3. A bag contains 6 red balls and 4 green balls.
   (a) Find the probability of selecting at random:
      (i) a red ball   (ii) a green ball.
   (b) One red ball is removed from the bag. Find the new probability of selecting at random
      (i) a red ball   (ii) a green ball.

4. One letter is selected at random from the word 'UNNECESSARY'. Find the probability of selecting:
   (a) an R   (b) an E
   (c) an O   (d) a C

5. Three coins are tossed at the same time. List all the possible outcomes. Find the probability of obtaining:
   (a) three heads,
   (b) two heads and one tail,
   (c) no heads,
   (d) at least one head.

6. A bag contains 10 red balls, 5 blue balls and 7 green balls. Find the probability of selecting at random:
   (a) a red ball,
   (b) a green ball,
   (c) a blue *or* a red ball,
   (d) a red *or* a green ball.

7. Cards with the numbers 2 to 101 are placed in a hat. Find the probability of selecting:
   (a) an even number,
   (b) a number less than 14,
   (c) a square number,
   (d) a prime number less than 20.

8. A red die and a blue die are thrown at the same time. List all the possible outcomes in a systematic way. Find the probability of obtaining:
(a) a total of 10,
(b) a total of 12,
(c) a total less than 6,
(d) the same number on both dice,
(e) a total more than 9.
What is the most likely total?

9. A die is thrown; when the result has been recorded, the die is thrown a second time. Display all the possible outcomes of the two throws. Find the probability of obtaining:
(a) a total of 4 from the two throws,
(b) a total of 8 from the two throws,
(c) a total between 5 and 9 inclusive from the two throws,
(d) a number on the second throw which is double the number on the first throw,
(e) a number on the second throw which is four times the number on the first throw.

10. Find the probability of the following:
(a) throwing a number less than 8 on a single die,
(b) obtaining the same number of heads and tails when five coins are tossed,
(c) selecting a square number from the set
A = {4, 9, 16, 25, 36, 49},
(d) selecting a prime number from the set A.

11. Four coins are tossed at the same time. List all the possible outcomes in a systematic way. Find the probability of obtaining:
(a) two heads and two tails,
(b) four tails,
(c) at least one tail,
(d) three heads and one tail.

> **Remember**
> 'head' and 'tail' are the two sides of a coin.

12. Cards numbered 1 to 1000 were put in a box. Ali selects a card at random. What is the probability that Ali selects a card containing at least one '3'?

13. One ball is selected at random from a bag containing 12 balls of which x are white.
(a) What is the probability of selecting a white ball?
When a further 6 white balls are added the probability of selecting a white ball is doubled.
(b) Find x.

14. Two dice and two coins are thrown at the same time. Find the probability of obtaining:
(a) two heads and a total of 12 on the dice,
(b) a head, a tail and a total of 9 on the dice,
(c) two tails and a total of 3 on the dice.
What is the most likely outcome?

15. A red, a blue and a green die are all thrown at the same time. Display all the possible outcomes in a suitable way. Find the probability of obtaining:
(a) a total of 18 on the three dice,
(b) a total of 4 on the three dice,
(c) a total of 10 on the three dice,
(d) a total of 15 on the three dice,
(e) a total of 7 on the three dice,
(f) the same number on each die.

10.5 Exclusive and independent events

Two events are *exclusive* if they cannot occur at the same time:
e.g. Selecting an 'even number' or selecting a 'one' from a set of numbers.

The 'OR' rule:

For exclusive events A and B

$$p(A \text{ or } B) = p(A) + p(B)$$

Two events are *independent* if the occurrence of one event is unaffected by the occurrence of the other.
e.g. Obtaining a 'head' on one coin, and a 'tail' on another coin when the coins are tossed at the same time.

The 'AND' rule:

$$p(A \text{ and } B) = p(A) \times p(B)$$

where $p(A)$ = probability of A occurring etc. This is the multiplication law.

Example 1

One ball is selected at random from a bag containing 5 red balls, 2 yellow balls and 4 white balls. Find the probability of selecting a red ball or a white ball.

The two events are exclusive.

$$p(\text{red ball } or \text{ white ball}) = p(\text{red}) + p(\text{white})$$
$$= \frac{5}{11} + \frac{4}{11}$$
$$= \frac{9}{11}$$

Example 2

A fair coin is tossed and a fair die is rolled. Find the probability of obtaining a 'head' and a 'six'.

The two events are independent.

p (head *and* six) $= p$ (head) $\times p$ (six)

$$= \tfrac{1}{2} \times \tfrac{1}{6}$$

$$= \tfrac{1}{12}$$

Exercise 9

1. A coin is tossed and a die is thrown. Write down the probability of obtaining:
   (a) a 'head' on the coin,
   (b) an odd number on the die,
   (c) a 'head' on the coin and an odd number on the die.

2. A ball is selected at random from a bag containing 3 red balls, 4 black balls and 5 green balls. The first ball is replaced and a second is selected. Find the probability of obtaining:
   (a) two red balls,   (b) two green balls.

3. The letters of the word 'INDEPENDENT' are written on individual cards and the cards are put into a box. A card is selected and then replaced and then a second card is selected. Find the probability of obtaining:
   (a) the letter 'P' twice,   (b) the letter 'E' twice.

4. Three coins are tossed and two dice are thrown at the same time. Find the probability of obtaining:
   (a) three heads and a total of 12 on the dice,
   (b) three tails and a total of 9 on the dice.

5. When a golfer plays any hole, he will take 3, 4, 5, 6, or 7 strokes with probabilities of $\tfrac{1}{10}, \tfrac{1}{5}, \tfrac{2}{5}, \tfrac{1}{5}$ and $\tfrac{1}{10}$ respectively. He never takes more than 7 strokes. Find the probability of the following events:
   (a) scoring 4 on each of the first three holes,
   (b) scoring 3, 4 and 5 (in that order) on the first three holes,
   (c) scoring a total of 28 for the first four holes,
   (d) scoring a total of 10 for the first three holes,
   (e) scoring a total of 20 for the first three holes.

6. A coin is biased so that it shows 'heads' with a probability of $\tfrac{2}{3}$. The same coin is tossed three times. Find the probability of obtaining:
   (a) two tails on the first two tosses,
   (b) a head, a tail and a head (in that order),
   (c) two heads and one tail (in any order).

10.6 Tree diagrams

Example 1

A bag contains 5 red balls and 3 green balls. A ball is drawn at random and then replaced. Another ball is drawn.
What is the probability that both balls are green?

The branch marked * involves the selection of a green ball twice.
The probability of this event is obtained by simply multiplying the fractions on the two branches.

$$\therefore \quad p(\text{two green balls}) = \tfrac{3}{8} \times \tfrac{3}{8} = \tfrac{9}{64}$$

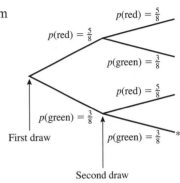

Example 2

A bag contains 5 red balls and 3 green balls. A ball is selected at random and not replaced. A second ball is then selected. Find the probability of selecting:
(a) two green balls
(b) one red ball and one green ball.

(a) $p(\text{two green balls}) = \tfrac{3}{8} \times \tfrac{2}{7}$
$$= \tfrac{3}{28}$$
(b) $p(\text{one red, one green}) = (\tfrac{5}{8} \times \tfrac{3}{7}) + (\tfrac{3}{8} \times \tfrac{5}{7})$
$$= \tfrac{15}{28}$$

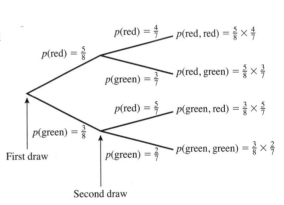

Exercise 10

1. A bag contains 10 discs; 7 are black and 3 white. A disc is selected, and then replaced. A second disc is selected. Copy and complete the tree diagram showing all the probabilities and outcomes.
   Find the probability of the following:
   (a) both discs are black,   (b) both discs are white.

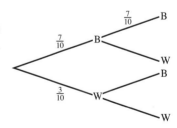

2. A bag contains 5 red balls and 3 green balls. A ball is drawn and then replaced before a ball is drawn again. Draw a tree diagram to show all the possible outcomes. Find the probability that:
   (a) two green balls are drawn,
   (b) the first ball is red and the second is green.

3. A bag contains 7 green discs and 3 blue discs. A disc is drawn and
   not replaced.
   A second disc is drawn. Copy and complete the tree diagram.
   Find the probability that:
   (a) both discs are green,
   (b) both discs are blue.

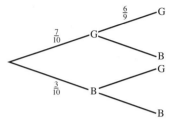

4. A bag contains 5 red balls, 3 blue balls and 2 yellow balls. A ball is
   drawn and not replaced. A second ball is drawn. Find the
   probability of drawing:
   (a) two red balls,
   (b) one blue ball and one yellow ball,
   (c) two yellow balls,
   (d) two balls of the same colour.

5. A bag contains 4 red balls, 2 green balls and 3 blue balls. A ball is
   drawn and not replaced. A second ball is drawn. Find the
   probability of drawing:
   (a) two blue balls,
   (b) two red balls,
   (c) one red ball and one blue ball,
   (d) one green ball and one red ball.

6. A six-sided die is thrown three times. Draw a tree diagram, showing
   at each branch the two events: 'six' and 'not six'. What is the
   probability of throwing a total of:
   (a) three sixes,
   (b) no sixes,
   (c) one six,
   (d) at least one six (use part (b)).

7. A bag contains 6 red marbles and 4 blue marbles. A marble is
   drawn at random and not replaced. Two further draws are made,
   again without replacement. Find the probability of drawing:
   (a) three red marbles,
   (b) three blue marbles,
   (c) no red marbles,
   (d) at least one red marble.

8. When a cutting is taken from a geranium the probability that it
   grows is $\frac{3}{4}$. Three cuttings are taken. What is the probability that:
   (a) all three grow,
   (b) none of them grow?

9. A die has its six faces marked 0, 1, 1, 1, 6, 6. Two of these dice are thrown together and the total score is recorded. Draw a tree diagram.
(a) How many different totals are possible?
(b) What is the probability of obtaining a total of 7?

10. A coin is biased so that the probability of a 'head' is $\frac{3}{4}$. Find the probability that, when tossed three times, it shows:
(a) three tails,
(b) two heads and one tail,
(c) one head and two tails,
(d) no tails.
Write down the sum of the probabilities in (a), (b), (c) and (d).

11. A teacher decides to award exam grades A, B or C by a new method. Out of 20 children, three are to receive A's, five B's and the rest C's. She writes the letters A, B and C on 20 pieces of paper and invites the pupils to draw their exam result, going through the class in alphabetical order. Find the probability that:
(a) the first three pupils all get grade 'A'.
(b) the first three pupils all get grade 'B',
(c) the first three pupils all get different grades,
(d) the first four pupils all get grade B.
(Do not cancel down the fractions.)

12. The probability that an amateur golfer actually hits the ball is $\frac{1}{10}$. If four separate attempts are made, find the probability that the ball will be hit:
(a) four times,
(b) at least twice,
(c) not at all.

13. A box contains x milk chocolates and y plain chocolates. Two chocolates are selected at random. Find, in terms of x and y, the probability of choosing:
(a) a milk chocolate on the first choice,
(b) two milk chocolates,
(c) one of each sort,
(d) two plain chocolates.

14. If a hedgehog crosses a certain road before 7.00 a.m., the probability of being run over is $\frac{1}{10}$. After 7.00 a.m., the corresponding probability is $\frac{3}{4}$. The probability of the hedgehog waking up early enough to cross before 7.00 a.m., is $\frac{4}{5}$.
What is the probability of the following events:
(a) the hedgehog waking up too late to reach the road before 7.00 a.m.,
(b) the hedgehog waking up early and crossing the road in safety,
(c) the hedgehog waking up late and crossing the road in safety,
(d) the hedgehog waking up early and being run over,
(e) the hedgehog crossing the road in safety.

15. Bag A contains 3 red balls and 3 blue balls.
    Bag B contains 1 red ball and 3 blue balls.
    A ball is taken at random from bag A and placed in
    bag B. A ball is then chosen from bag B. What is the
    probability that the ball taken from B is red?

16. On a Monday or a Thursday, Ceren paints a 'masterpiece' with a
    probability of $\frac{1}{5}$. On any other day, the probability of
    producing a 'masterpiece' is $\frac{1}{100}$. Find the probability that on one
    day chosen at random, she will paint a masterpiece.

17. Two dice, each with four faces marked 1, 2, 3 and 4, are thrown
    together.
    (a) What is the most likely total score on the faces pointing
    downwards?
    (b) What is the probability of obtaining this score on three
    successive throws of the two dice?

18. In the Venn diagram, $\mathcal{E} = \{$pupils in a class of 15$\}$, $G = \{$girls$\}$,
    $S = \{$swimmers$\}$, $F = \{$pupils who were born on a Friday$\}$.
    A pupil is chosen at random. Find the probability that the pupil:
    (a) can swim,
    (b) is a girl swimmer,
    (c) is a boy swimmer who was born on a Friday.
    Two pupils are chosen at random. Find the probability that:
    (d) both are boys,
    (e) neither can swim,
    (f) both are girl swimmers who were born on a Friday.

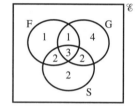

19. A bag contains 3 red, 4 white and 5 green balls. Three balls are
    selected without replacement. Find the probability that the three
    balls chosen are:
    (a) all red,
    (b) all green,
    (c) one of each colour.
    If the selection of the three balls was carried out 1100 times, how
    often would you expect to choose:
    (d) three red balls?
    (e) one of each colour?

20. There are 1000 components in a box of which 10 are known to be
    defective. Two components are selected at random. What is the
    probability that:
    (a) both are defective,
    (b) neither are defective,
    (c) just one is defective?
    (Do *not* simplify your answers.)

21. There are 10 boys and 15 girls in a class. Two children are chosen at random. What is the probability that:
    (a) both are boys,
    (b) both are girls,
    (c) one is a boy and one is a girl?

22. There are 500 ball bearings in a box of which 100 are known to be undersize. Three ball bearings are selected at random. What is the probability that:
    (a) all three are undersize,   (b) none are undersize?
    Give your answers as decimals correct to three significant figures.

23. There are 9 boys and 15 girls in a class. Three children are chosen at random. What is the probability that:
    (a) all three are boys,
    (b) all three are girls,
    (c) one is a boy and two are girls?
    Give your answers as fractions.

Revision exercise 10A

1. A pie chart is drawn with sectors to represent the following percentages:

   20%, 45%, 30%, 5%.

   What is the angle of the sector which represents 45%?

2. The pie chart shows the numbers of votes for candidates A, B and C in an election.
   What percentage of the votes were cast in favour of candidate C?

   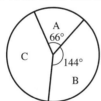

3. A pie chart is drawn showing the expenditure of a football club as follows:

Wages	$41 000
Travel	$9000
Rates	$6000
Miscellaneous	$4000

   What is the angle of the sector showing the expenditure on travel?

4. The mean of four numbers is 21.
   (a) Calculate the sum of the four numbers.
   Six other numbers have a mean of 18.
   (b) Calculate the mean of the ten numbers.

5. Find:
  (a) the mean,   (b) the median,   (c) the mode,
  of the numbers 3, 1, 5, 4, 3, 8, 2, 3, 4, 1.

6.

marks	3	4	5	6	7	8
number of pupils	2	3	6	4	3	2

The table shows the number of pupils in a class who scored marks 3
to 8 in a test. Find:
  (a) the mean mark,
  (b) the modal mark,
  (c) the median mark.

7. The mean height of 10 boys is 1·60 m and the mean height of 15
girls is 1·52 m. Find the mean height of the 25 boys and girls.

8.

mark	3	4	5
number of pupils	3	x	4

The table shows the number of pupils who scored marks 3, 4 or 5 in
a test. Given that the mean mark is 4·1, find x.

9. When two dice are thrown simultaneously, what is the probability
of obtaining the same number on both dice?

10. A bag contains 20 discs of equal size of which 12 are red, x are blue
and the rest are white.
  (a) If the probability of selecting a blue disc is $\frac{1}{4}$, find x.
  (b) A disc is drawn and then replaced. A second disc is drawn. Find
    the probability that neither disc is red.

11. Three dice are thrown. What is the probability that none of them
shows a 1 or a 6?

12. A coin is tossed four times. What is the probability of obtaining at
least three 'heads'?

13. A bag contains 8 balls of which 2 are red and 6 are white. A ball is
selected and not replaced. A second ball is selected. Find the
probability of obtaining:
  (a) two red balls,
  (b) two white balls,
  (c) one ball of each colour.

14. A bag contains x green discs and 5 blue discs. A disc is selected.
A second disc is drawn. Find, in terms of x, the probability of
selecting:
  (a) a green disc on the first draw,
  (b) a green disc on the first and second draws, if the first disc
    is replaced,
  (c) a green disc on the first and second draws, if the first disc
    is *not* replaced.

15. In a group of 20 people, 5 cannot swim. If two people are selected at random, what is the probability that neither of them can swim?

16. (a) What is the probability of winning the toss in five consecutive hockey matches?
    (b) What is the probability of winning the toss in all the matches in the FA cup from the first round to the final (i.e. 8 matches)?

17. Mr and Mrs Singh have three children. What is the probability that:
    (a) all the children are boys,
    (b) there are more girls than boys?
    (Assume that a boy is as likely as a girl.)

18. The probability that it will be wet today is $\frac{1}{6}$. If it is dry today, the probability that it will be wet tomorrow is $\frac{1}{8}$. What is the probability that both today and tomorrow will be dry?

19. Two dice are thrown. What is the probability that the *product* of the numbers on top is:
    (a) 12,   (b) 4,   (c) 11?

20. The probability of snow on January 1st is $\frac{1}{20}$. What is the probability that snow will fall on the next three January 1st?

Examination exercise 10B

1. The mass of each of 200 tea bags was checked by an inspector in a factory.
   The results are shown by the cumulative frequency curve.

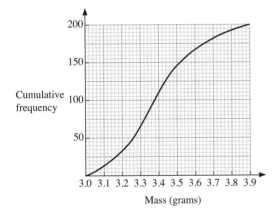

Mass (grams)

   Use the cumulative frequency curve to find
   (a) the median mass,   [1]
   (b) the interquartile range,   [2]
   (c) the number of tea bags with a mass greater than 3.5 grams.   [1]

Cambridge IGCSE Mathematics 0580
Paper 2 Q19 November 2007

2. (a) Each student in a class is given a bag of sweets.
   The students note the number of sweets in their bag.
   The results are shown in the table, where $0 \leqslant x < 10$.

Number of sweets	30	31	32
Frequency (number of bags)	10	7	x

   (i) State the mode.   [1]
   (ii) Find the possible values of the median.   [3]
   (iii) The mean number of sweets is 30.65.
      Find the value of x.   [3]

(b) The mass, m grams, of each of 200 chocolates is noted and the
   results are shown in the table.

Mass (m grams)	$10 < m \leqslant 20$	$20 < m \leqslant 22$	$22 < m \leqslant 24$	$24 < m \leqslant 30$
Frequency	35	115	26	24

   (i) Calculate an estimate of the mean mass of a chocolate.   [4]
   (ii) On a histogram, the height of the column for the
      $20 < m \leqslant 22$ interval is 11.5 cm.
      Calculate the heights of the other three columns.
      Do not draw the histogram.   [5]

Cambridge IGCSE Mathematics 0580
Paper 4 Q6 November 2008

3. The speeds (v kilometres/hour) of 150 cars passing a 50 km/h speed
limit sign are recorded.
A cumulative frequency curve to show the results is drawn below.

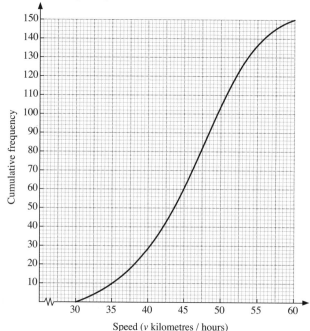

Speed (v kilometres / hours)

(a) Use the graph to find
   (i) the median speed, [1]
   (ii) the inter-quartile range of the speeds, [2]
   (iii) the number of cars travelling with speeds of more than
      50 km/h. [2]
(b) A frequency table showing the speeds of the cars is

Speed (v km/h)	$30 < v \leqslant 35$	$35 < v \leqslant 40$	$40 < v \leqslant 45$	$45 < v \leqslant 50$	$50 < v \leqslant 55$	$55 < v \leqslant 60$
Frequency	10	17	33	42	n	16

   (i) Find the value of n [1]
   (ii) Calculate an estimate of the mean speed. [4]
(c) **Answer this part of this question on a sheet of graph paper.**
Another frequency table for the same speeds is

Speed (v km/h)	$30 < v \leqslant 40$	$40 < v \leqslant 55$	$55 < v \leqslant 60$
Frequency	27	107	16

Draw an accurate histogram to show this information.
Use 2 cm to represent 5 units on the speed axis and 1 cm to
represent 1 unit on the frequency density axis (so that 1 cm^2
represents 2.5 cars). [5]

Cambridge IGCSE Mathematics 0580
Paper 4 Q7 June 2005

4. Rooms in a hotel are numbered from 1 to 19.
Rooms are allocated at random as guests arrive.
(a) What is the probability that the first guest to arrive is given a
room which is a prime number? (1 is not a prime number.) [2]
(b) The first guest to arrive is given a room which is a prime
number. What is the probability that the second guest to
arrive is given a room which is a prime number? [1]

Cambridge IGCSE Mathematics 0580
Paper 2 Q10 June 2005

5. A normal die, numbered 1 to 6, is rolled 50 times.

The results are shown in the frequency table.

Score	1	2	3	4	5	6
Frequency	15	10	7	5	6	7

(a) Write down the modal score.   [1]
(b) Find the median score.   [1]
(c) Calculate the mean score.   [2]
(d) The die is then rolled another 10 times.
    The mean score for the 60 rolls is 2.95.
    Calculate the mean score for the extra 10 rolls.   [3]

Cambridge IGCSE Mathematics 0580
Paper 4 Q2 June 2009

6. (a)

Grade	1	2	3	4	5	6	7
Number of students	1	2	4	7	4	8	2

The table shows the grades gained by 28 students in a history test.
    (i) Write down the mode.   [1]
    (ii) Find the median.   [1]
    (iii) Calculate the mean.   [3]
    (iv) Two students are chosen at random.
        Calculate the probability that they both gained grade 5.   [2]
    (v) From all the students who gained grades 4 or 5 or 6 or 7,
        two are chosen at random.
        Calculate the probability that they both gained grade 5.   [2]
    (vi) Students are chosen at random, one by one, from the
        original 28, until the student chosen has a grade 5.
        Calculate the probabilty that this is the third student
        chosen.   [2]
(b) Claude goes to school by bus.
    The probability that the bus is late is 0.1.
    If the bus is late, the probability that Claude is late to school is
    0.8.
    If the bus is not late, the probability that Claude is late to
    school is 0.05.
    (i) Calculate the probability that the bus is late and Claude is
        late to school.   [1]
    (ii) Calculate the probability that Claude is late to school.   [3]
    (iii) The school term lasts 56 days.
        How many days would Claude expect to be late?   [1]

Cambridge IGCSE Mathematics 0580
Paper 4 Q2 November 2007

7.

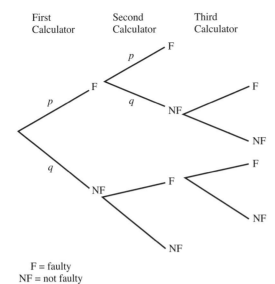

First
Calculator

Second
Calculator

Third
Calculator

F = faulty
NF = not faulty

The tree diagram shows a testing procedure on calculators, taken
from a large batch.
Each time a calculator is choosen at random, the probability that it
is faulty (F) is $\frac{1}{20}$.

(a) Write down the values of p and q. [1]

(b) Two calculators are chosen at random.
   Calculate the probability that
   (i) both are faulty, [2]
   (ii) **exactly one** is faulty. [2]

(c) If **exactly one** out of two calculators tested is faulty, then a third
   calculator is chosen at random.
   Calculate the probability that exactly one of the first two
   calculators is faulty **and** the third one is faulty. [2]

(d) The whole batch of calculators is rejected
   either if the first two chosen are both faulty
   or   if a third one needs to be chosen and it is faulty.
   Calculate the probability that the whole batch is rejected. [2]

(e) In one month, 1000 batches of calculators are tested in this way.
   How many batches are expected to be rejected? [1]

Cambridge IGCSE Mathematics 0580
Paper 4 Q8 June 2009

8. (a) All 24 students in a class are asked whether they like football and whether they like basketball.

Some of the results are shown in the Venn diagram below.

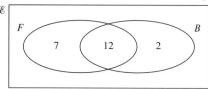

$\mathscr{E} = \{$students in the class$\}$.
$F = \{$students who like football$\}$.
$B = \{$students who like basketball$\}$.

  (i) How many students like both sports?   [1]
 (ii) How many students do not like either sport?   [1]
(iii) Write down the value of $n(F \cup B)$.   [1]
 (iv) Write down the value of $n(F' \cap B)$.   [1]
  (v) A student from the class is selected at random.
       What is the probability that this student likes basketball?   [1]
 (vi) A student who likes football is selected at random.
       What is the probability that this student likes basketball?   [1]
(b) Two students are selected at random from a group of 10 boys and 12 girls.

Find the probability that

  (i) they are both girls,   [2]
 (ii) one is a boy and one is a girl.   [3]

Cambridge IGCSE Mathematics 0580
Paper 4 Q4 November 2005

11 INVESTIGATIONS, PRACTICAL PROBLEMS, PUZZLES

William Shockley Every time you use a calculator you are making use of integrated circuits which were developed from the first transistor. The transistor was invented by William Shockley, working with two scientists, in 1947. The three men shared the 1956 Nobel Prize for physics. The story of the invention is a good example of how mathematics can be used to solve practical problems.

The first electronic computers did not make use of transistors or integrated circuits and they were so big that they occupied whole rooms themselves. A modern computer which can carry out just the same functions can be carried around in a briefcase.

11.1 Investigations

There are a large number of possible starting points for these investigations so it may be possible to allow students to choose investigations which appeal to them. On other occasions the same investigation may be set to a whole class.

Here are a few guidelines for you:

- If the set problem is too complicated try an easier case.
- Draw your own diagrams.
- Make tables of your results and be systematic.
- Look for patterns.
- Is there a rule or formula to describe the results?
- Can you *predict* further results?
- Can you *prove* any rules which you may find?

1. Opposite corners

Here the numbers are arranged in 10 columns.

1	2	3	4	5	6	7	8	9	10
11	12	13	14	15	16	17	18	19	20
21	22	23	24	25	26	27	28	29	30
31	32	33	34	35	36	37	38	39	40
41	42	43	44	45	46	47	48	49	50
51	52	53	54	55	56	57	58	59	60
61	62	63	64	65	66	67	68	69	70
71	72	73	74	75	76	77	78	79	80
81	82	83	84	85	86	87	88	89	90
91	92	93	94	95	96	97	98	99	100

In the 2×2 square

$$7 \times 18 = 126$$
$$8 \times 17 = 136$$

7	8
17	18

the difference between them is 10.

In the 3×3 square

$$12 \times 34 = 408$$
$$14 \times 32 = 448$$

12	13	14
22	23	24
32	33	34

the difference between them is 40.

Investigate to see if you can find any rules or patterns connecting the size of square chosen and the difference.

If you find a rule, use it to *predict* the difference for larger squares.

Test your rule by looking at squares like 8×8 or 9×9.

Can you *generalise* the rule?
[What is the difference for a square of size $n \times n$?]

Can you prove the rule?
Hint:
In a 3×3 square ...

x		?
?		?

What happens if the numbers are arranged in six columns or seven columns?

1	2	3	4	5	6
7	8	9	10	11	12
13	14	15	16	17	18
19					

1	2	3	4	5	6	7
8	9	10	11	12	13	14
15	16	17	18	19	20	21
22						

2. Scales

In the diagram we are measuring the mass of the package x using two masses.
If the scales are balanced, x must be 2 kg.

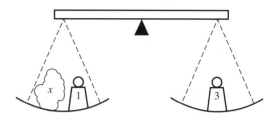

Show how you can measure all the masses from 1 kg to 10 kg using three masses: 1 kg, 3 kg, 6 kg.
It is possible to measure all the masses from 1 kg to 13 kg using a different set of three masses. What are the three masses?
It is possible to measure all the masses from 1 kg to 40 kg using four masses. What are the masses?

3. Buying stamps

You have one 1c, one 2c, one 5c and one 10c coin.
You can buy stamps of any value you like, but you must give the exact money.
How many different value stamps can you buy?
Suppose you now have one 1c, one 2c, one 5c, one 10c, one 20c, one 50c and one $1 note.
How many different value stamps can you buy now?

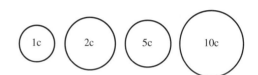

4. Frogs

This is a game invented by a French mathematician called Lucas.

Aim: To swap the positions of the discs so that they end up the other way round (with a space in the middle).
Rules 1. A disc can slide one square in either direction onto an empty square.
     2. A disc can hop over one adjacent disc of the other colour provided it can land on an empty square.

Example (a) Slide Ⓐ one square to the right.

    (b) Ⓑ hops over Ⓐ to the left.

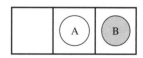

    (c) Slide Ⓐ one square to the right.

We took 3 moves.

1. Look at the diagram. What is the smallest number of moves needed for two discs of each colour?

2. Now try three discs of each colour. Can you complete the task in 15 moves?

3. Try four discs of each colour.
   Now look at your results and try to find a formula which gives the least number of moves needed for any number of discs x. It may help if you count the number of 'hops' and 'slides' separately.
4. Try the game with a different number of discs on each side. Say two reds and three blues. Play the game with different combinations and again try to find a formula giving the number of moves for x discs of one colour and y discs of another colour.

5. Triples

In this investigation a *triple* consists of three whole numbers in a definite order. For example, (4, 2, 1) is a triple and (1, 4, 2) is a different triple.

The three numbers in a triple do not have to be different. For example, (2, 2, 3) is a triple but (2, 0, 1) is not a triple because 0 is not allowed.

The *sum* of a triple is found by adding the three numbers together. So the sum of (4, 2, 1) is 7.

Investigate how many different triples there are with a given sum. See what happens to the number of different triples as the sum is changed.

If you find any pattern, try to explain why it occurs.

How many different triples are there whose sum is 22?

6. Mystic rose

Straight lines are drawn between each of the 12 points on the circle. Every point is joined to every other point. How many straight lines are there?

Suppose we draw a mystic rose with 24 points on the circle. How many straight lines are there?
How many straight lines would there be with n points on the circle?

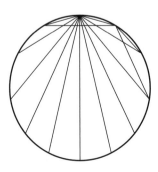

7. Knockout competition

Eight teams reach the 'knockout' stage of the World Cup.

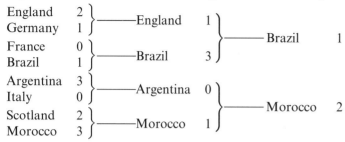

How would you organise a knockout competition if there were 12 teams? Or 15?

How many matches are played up to and including the final if there are:

(a) 8 teams,
(b) 12 teams,
(c) 15 teams,
(d) 23 teams,
(e) n teams?

In a major tournament like Wimbledon, the better players are seeded from 1 to 16. Can you organise a tournament for 32 players so that, if they win all their games:

(a) seeds 1 and 2 can meet in the final,
(b) seeds 1, 2, 3 and 4 can meet in the semi-finals,
(c) seeds 1, 2, 3, 4, 5, 6, 7, 8 can meet in the quarter-finals?

8. Discs

(a) You have five black discs and five white discs which are arranged in a line as shown.

We want to get all the black discs to the right-hand end and all the white discs to the left-hand end.

The only move allowed is to interchange two neighbouring discs.

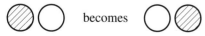 becomes

How many moves does it take?
How many moves would it take if we had fifty black discs and fifty white discs arranged alternately?

(b) Suppose the discs are arranged in pairs

 ... etc.

How many moves would it take if we had fifty black discs and fifty white discs arranged like this?
[Hint: In both cases work with a smaller number of discs until you can see a pattern.]

(c) Now suppose you have three colours, black, white and green, arranged alternately.

 ... etc.

You want to get all the black discs to the right, the green discs to the left and the white discs in the middle.
How many moves would it take if you have 30 discs of each colour?

9. Chess board

Start with a small board, just 4 × 4.

How many squares are there? [It is not just 16!]

How many squares are there on an 8 × 8 chess board?

How many squares are there on an $n \times n$ chess board?

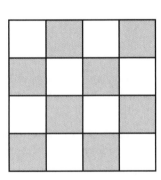

10. Area and perimeter

This is about finding different shapes in which the area is numerically equal to the perimeter.

This rectangle has an area of 10 square units and a perimeter of 14 units, so we will have to try another one.
There are some suggestions below but you can investigate shapes of your own choice if you prefer.

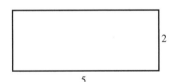

(a) Find rectangles with equal area and perimeter. After a while you can try adding on bits like this.

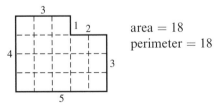

area = 18
perimeter = 18

(b) Suppose one dimension of the rectangle is fixed.
In this rectangle the length is 5 units.

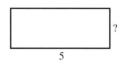

5

(c) Try right-angled triangles and equilateral triangles.
(d) Try circles, semi-circles and so on.
(e) How about three-dimensional shapes? Now we are looking for
cuboids, spheres, cylinders in which the volume is numerically
equal to the surface area.
(f) Can you find any connection between the square with equal
area and perimeter and the circle with equal area and perimeter?
How about the equilateral triangle with equal area and
perimeter?

11. Happy numbers (and more)

(a) Take the number 23.
Square the digits and add.

$$2 \quad 3$$
$$2^2 + 3^2 = 1 \quad 3$$
$$\qquad 1^2 + 3^2 = 1 \quad 0$$
$$\qquad\qquad 1^2 + 0^2 = 1$$

The sequence ends at 1 and we call 23 a 'happy' number.
Investigate for other numbers. Here are a few suggestions: 70,
85, 49, 44, 14, 15, 94.

(b) Now change the rule. Instead of squaring the digits we will cube
them.

$$2 \quad 1$$
$$2^3 + 1^3 = 0 \quad 9$$
$$\qquad 0^3 + 9^3 = 7 \quad 2 \quad 9$$
$$\qquad\qquad 7^3 + 2^3 + 9^3 = 1 \quad 0 \quad 8 \quad 0$$
$$\qquad\qquad\qquad 1^3 + 0^3 + 8^3 = 5 \quad 1 \quad 3$$
$$\qquad\qquad\qquad\qquad 5^3 + 1^3 + 3^3 = 153$$

And now we are stuck because 153 leads to 153 again.
Investigate for numbers of your own choice. Do any numbers
lead to 1?

12. Prime numbers

Write all the numbers from 1 to 104 in eight columns and draw a
ring around the prime numbers 2, 3, 5 and 7.

1	②	③	4	⑤	6	⑦	8
9	10	11	12	13	14	15	16
17	18	19	20	21	22	23	24
25							

If we cross out all the multiples of 2, 3, 5 and 7, we will be left with all the prime numbers below 104. Can you see why this works?

Draw *four* lines to eliminate the multiples of 2.
Draw *six* lines to eliminate the multiples of 3.
Draw *two* lines to eliminate the multiples of 7.
Cross out all the numbers ending in 5.

Put a ring around all the prime numbers less than 104.
[Check there are 27 numbers.]

Many prime numbers can be written as the sum of two squares. For example $5 = 2^2 + 1^2$, $13 = 3^2 + 2^2$. Find all the prime numbers in your table which can be written as the sum of two squares. Draw a red ring around them in the table.
What do you notice?
Check any 'gaps' you may have found.

Extend the table up to 200 and see if the pattern continues. In this case you will need to eliminate the multiples of 11 and 13 as well.

13. Squares

For this investigation you need either dotted paper or squared paper.

The shaded square has an area of 1 unit.
Can you draw a square, with its corners on the dots, with an area of 2 units?
Can you draw a square with an area of 3 units?
Can you draw a square with an area of 4 units?

Investigate for squares up to 100 units.

For which numbers x can you draw a square of area x units?

14. Painting cubes

The large cube below consists of 27 unit cubes.
All six faces of the large cube are painted green.

How many unit cubes have 3 green faces?
How many unit cubes have 2 green faces?
How many unit cubes have 1 green face?
How many unit cubes have 0 green faces?

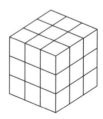

Suppose the large cube is $20 \times 20 \times 20$.
Answer the four questions above.

Answer the four questions for the cube which is $n \times n \times n$.

15. Final score

The final score in a football match was 3–2. How many different scores were possible at half-time?

Investigate for other final scores where the difference between the teams is always one goal [1–0, 5–4 etc]. Is there a pattern or rule which would tell you the number of possible half-time scores in a game which finished 58–57?

Suppose the game ends in a draw. Find a rule which would tell you the number of possible half-time scores if the final score was 63–63.

Investigate for other final scores [3–0, 5–1, 4–2 etc].

16. Cutting paper

The rectangle ABCD is cut in half to give two smaller rectangles.

Each of the smaller rectangles is mathematically similar to the large rectangle. Find a rectangle which has this property.

What happens when the small rectangles are cut in half? Do they have the same property?

Why is this a useful shape for paper used in business?

17. Matchstick shapes

(a) Here we have a sequence of matchstick shapes

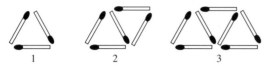

Can you work out the number of matches in the 10th member
of the sequence? Or the 20th member of the sequence?
How about the *n*th member of the sequence?

(b) Now try to answer the same questions for the patterns below.
Or you may prefer to design patterns of your own.

(i)

(ii)

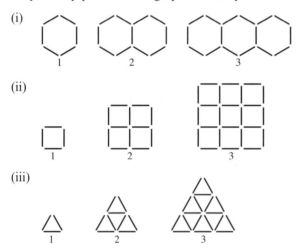

(iii)

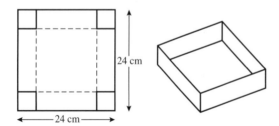

18. Maximum box

(a) You have a square sheet of card 24 cm by 24 cm.
You can make a box (without a lid) by cutting squares from the
corners and folding up the sides.

24 cm

24 cm

What size corners should you cut out so that the volume of the
box is as large as possible?
Try different sizes for the corners and record the results in a
table:

length of the side of the corner square (cm)	dimensions of the open box (cm)	volume of the box (cm^3)
1	22 × 22 × 1	484
2		
–		
–		

Now consider boxes made from different-sized cards:
15 cm × 15 cm and 20 cm by 20 cm.
What size corners should you cut out this time so that the
volume of the box is as large as possible?

Is there a connection between the size of the corners cut out
and the size of the square card?

(b) Investigate the situation when the card is not square.
Take rectangular cards where the length is twice the width
(20 × 10, 12 × 6, 18 × 9 etc).
Again, for the maximum volume is there a connection between
the size of the corners cut out and the size of the original
card?

19. Digit sum

Take the number 134.
Add the digits $1 + 3 + 4 = 8$.
The digit sum of 134 is 8.

Take the number 238.
$2 + 3 + 8 = 13$ [We continue if the sum is more than 9].
    $1 + 3 = 4$
The digit sum of 238 is 4.
Consider the multiples of 3:

Number	3	6	9	12	15	18	21	24	27	30	33	36
Digit sum	3	6	9	3	6	9	3	6	9	3	6	9

The digit sum is always 3, 6, or 9.
These numbers can be shown on a circle.
Investigate the pattern of the digit sums for multiples of:
(a) 2   (b) 5   (c) 6   (d) 7   (e) 8
(f) 9   (g) 11   (h) 12   (i) 13
Is there any connection between numbers where the pattern of the
digit sums is the same?
Can you (without doing all the usual working) predict what the
pattern would be for multiples of 43? Or 62?

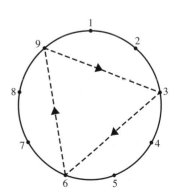

20. An expanding diagram

Look at the series of diagrams below.

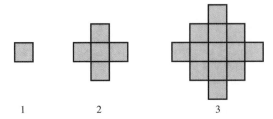

1   2   3

Each time new squares are added all around the outside of the previous diagram.

Draw the next few diagrams in the series and count the number of squares in each one.

How many squares are there in diagram number 15 or in diagram number 50?

What happens if we work in three dimensions? Instead of adding squares we add cubes all around the outside. How many cubes are there in the fifth member of the series or the fifteenth?

21. Fibonacci sequence

Fibonacci was the nickname of the Italian mathematician Leonardo de Pisa (A.D. 1170–1250). The sequence which bears his name has fascinated mathematicians for hundreds of years. You can if you like join the Fibonacci Association which was formed in 1963.

Here is the start of the sequence
1, 1, 2, 3, 5, 8, 13, 21, 34, 55, 89, 144, ...

There are no prizes for working out the next term!

The sequence has many interesting properties to investigate. Here are a few suggestions.

(a) Add three terms.
      $1 + 1 + 2, 1 + 2 + 3$, etc.
    Add four terms.
(b) Add squares of terms
      $1^2 + 1^2, 1^2 + 2^2, 2^2 + 3^2, ...$
(c) Ratios
      $\dfrac{1}{1} = 1, \dfrac{2}{1} = 2, \dfrac{3}{2} = 1.5, ...$
(d) In fours $\boxed{2\ \ 3\ \ 5\ \ 8}$
      $2 \times 8 = 16, 3 \times 5 = 15$
(e) In threes $\boxed{3\ \ 5\ \ 8}$
      $3 \times 8 = 24, 5^2 = 25$

(f) In sixes $\boxed{1 \ 1 \ 2 \ 3 \ 5 \ 8}$

square and add the first five numbers
$1^2 + 1^2 + 2^2 + 3^2 + 5^2 = 40$
$5 \times 8 = 40$.
Now try seven numbers from the sequence, or eight ...

(g) Take a group of 10 consecutive terms. Compare the sum of the 10 terms with the seventh member of the group.

22. Alphabetical order

A teacher has four names on a piece of paper which are in no particular order (say Smith, Jones, Biggs, Eaton). He wants the names in alphabetical order.

One way of doing this is to interchange each pair of names which are clearly out of order.
So he could start like this; S J B E
the order becomes   J S B E
He would then interchange S and B.

Using this method, what is the largest number of interchanges he could possibly have to make?

What if he had thirty names, or fifty?

23. Tiles

Gao counts the tiles by placing them in a pattern consisting of alternate black and white tiles. This one is five tiles across and altogether there are 13 tiles in the pattern.

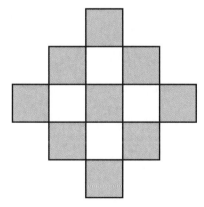

He makes the pattern so that there are always black tiles all around the outside. Draw the pattern which is nine tiles across. You should find that there are 41 tiles in the pattern.

How many tiles are there in the pattern which is 101 tiles across?

24. Diagonals

In a 4×7 rectangle the diagonal passes through 10 squares.

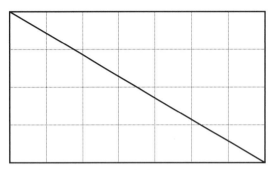

Draw rectangles of your own choice and count the number of squares through which the diagonal passes.

A rectangle is 640×250. How many squares will the diagonal pass through?

25. Biggest number

A calculator has the following buttons:

| + | − | × | ÷ | (|) | y^x | = |

Also the only digits buttons which work are the '1', '2' and '3'.

(a) You can press any button, but only once.
    What is the biggest number you can get?
(b) Now the '1', '2', '3' and '4' buttons are working.
    What is the biggest number you can get?
(c) Investigate what happens as you increase the number of digits which you can use.

26. What shape tin?

We need a cylindrical tin which will contain a volume of $600 \, \text{cm}^3$ of drink.
What shape should we make the tin so that we use the minimum amount of metal?
In other words, for a volume of $600 \, \text{cm}^3$, what is the smallest possible surface area?

What shape tin should we design to contain a volume of $1000 \, \text{cm}^3$?

Hint
Make a table.

r	h	A
2	?	?
3	?	?
⋮		

27. Find the connection

Work through the flow diagram several times, using a calculator.

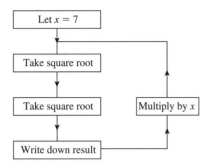

What do you notice?
Try different numbers for x (suggestions: 11, 5, 8, 27)
What do you notice?

What happens if you take the square root three times?

Suppose in the flow diagram you change
'Multiply by x' to 'Divide by x'. What happens now?

Suppose in the flow diagram you change
'Multiply by x' to 'Multiply by x^2'. What happens now?

28. Spotted shapes

For this investigation you need dotted paper. If you have not got any, you can make your own using a felt tip pen and squared paper.

The rectangle in Diagram 1 has 10 dots on the perimeter ($p = 10$) and 2 dots inside the shape ($i = 2$). The area of the shape is 6 square units ($A = 6$).

The triangle in Diagram 2 has 9 dots on the perimeter ($p = 9$) and 4 dots inside the shape ($i = 4$). The area of the triangle is $7\frac{1}{2}$ square units ($A = 7\frac{1}{2}$).

Draw more shapes of your own design and record the values for p, i and A in a table. Make some of your shapes more difficult like the one in Diagram 3.

Can you find a formula connecting p, i and A?
[Hint: $\frac{1}{2}i$, $\frac{1}{2}p$?]
Try out your formula with some more shapes to see if it always works.

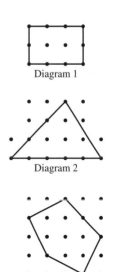

Diagram 1

Diagram 2

Diagram 3

29. Stopping distances

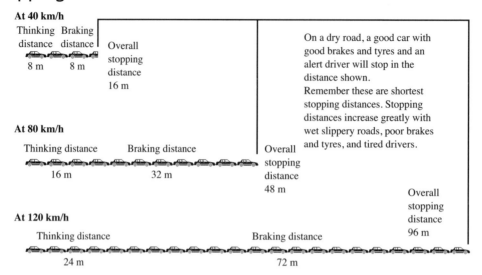

At 40 km/h

Thinking Braking
distance distance
8 m   8 m

Overall
stopping
distance
16 m

On a dry road, a good car with
good brakes and tyres and an
alert driver will stop in the
distance shown.
Remember these are shortest
stopping distances. Stopping
distances increase greatly with
wet slippery roads, poor brakes
and tyres, and tired drivers.

At 80 km/h

Thinking distance   Braking distance
16 m   32 m

Overall
stopping
distance
48 m

Overall
stopping
distance
96 m

At 120 km/h

Thinking distance   Braking distance
24 m   72 m

This diagram from the Highway Code gives the overall stopping
distances for cars travelling at various speeds.

What is meant by 'thinking distance'?
Work out the thinking distance for a car travelling at a speed of
90 km/h. What is the formula which connects the speed of the car
and the thinking distance?

(More difficult)
Try to find a formula which connects the speed of the car and the
overall stopping distance. It may help if you draw a graph of speed
(across the page) against *braking* distance (up the page).
What curve are you reminded of?

Check that your formula gives the correct answer for the overall
stopping distance at a speed of:
(a) 40 km/h   (b) 120 km/h.

30. Maximum cylinder

A rectangular piece of paper has a fixed perimeter of 40 cm.
It could for example be 7 cm × 13 cm.
This paper can make a hollow cylinder of height 7 cm or
of height 13 cm.
Work out the volume of each cylinder.

What dimensions should the paper have so that it can
make a cylinder of the maximum possible volume?

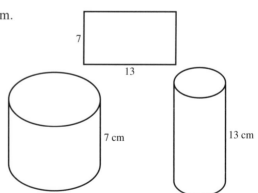

11.2 Practical problems

1. Timetabling

(a) Every year a new timetable has to be written for the school. We will look at the problem of writing the timetable for one department (mathematics). The department allocates the teaching periods as follows:

Upper 6	2 sets (at the same times); 8 periods in 4 doubles.
Lower 6	2 sets (at the same times); 8 periods in 4 doubles.
Year 5	6 sets (at the same times); 5 single periods.
Year 4	6 sets (at the same times); 5 single periods.
Year 3	6 sets (at the same times); 5 single periods.
Year 2	6 sets (at the same times); 5 single periods.
Year 1	5 mixed ability forms; 5 single periods not necessarily at the same times.

Here are the teachers and the maximum number of maths periods which they can teach.

A	33	
B	33	
C	33	
D	20	
E	20	
F	15	(must be years 5, 4, 3)
G	10	(must be years 2, 1)
H	10	(must be years 2, 1)
I	5	(must be year 3)

Furthermore, to ensure some continuity of teaching, teachers B and C must teach the U6 (Upper Sixth) and teachers A, B, C, D, E, F must teach year 5.

Here is a timetable form which has been started:

M	5				U6 B, C	U6 B, C		
Tu		5	U6 B, C	U6 B, C				
W				5				
Th					5	U6 B, C	U6 B, C	
F	U6 B, C	U6 B, C		5				

Your task is to write a complete timetable for the mathematics department subject to the restrictions already stated.

(b) If that was too easy, here are some changes.

U6 and L6 have 4 sets each (still 8 periods).
Two new teachers: J 20 periods maximum
                       K 15 periods maximum but cannot teach on
                          Mondays.

Because of games lessons: A cannot teach Wednesday afternoon
                                    B cannot teach Tuesday afternoon
                                    C cannot teach Friday afternoon.

Also: A, B, C and E must teach U6
       A, B, C, D, E, F must teach year 5.

For the pupils, games afternoons are as follows:
Monday year 2; Tuesday year 3; Wednesday year 5 L6, U6;
Thursday year 4; Friday year 1.

2. Hiring a car

You are going to hire a car for one week (seven days).

Which of the firms below should you choose?

Gibson car hire	Snowdon rent-a-car	Hav-a-car
$170 per week no charge up to 10 000 km	$10 per day 6·5c per km	$60 per week 500 km without charge 22c per km over 500 km

Work out as detailed an answer as possible.

3. Running a business

Mr Singh runs a small business making two sorts of steam cleaner:
the basic model B and the deluxe model D.
Here are the details of the manufacturing costs:

	model B	model D
Assembly time (in man-hours)	20 hours	30 hours
Component costs	$35	$25
Selling price	$195	$245

He employs 10 people and pays them each $160 for a 40-hour
week. He can spend up to $525 per week on components.

Speed Homework

Q1. Gary drove from London to Sheffield.
It took him 3 hours at an average speed of 80km/h.

Lyn drove from London to Sheffield.
She took 5 hours.

Assuming that Lyn   drove along the same roads as Gary   and did not take a break,

(a) work out Lyn's average speed from London to Sheffield.

(b) If Lyn did **not** drive along the same roads as Gary, explain how this could affect your answer to part (a).

Q2. A train travels from Madrid to Malaga at an average speed of 183 km/h.

The train leaves Madrid at 08 40   The train arrives at Malaga at 11 28

Work out the distance the train travels from Madrid to Malaga.

Q3. James and Peter cycled along the same 50 km route.

James took $2\frac{1}{2}$ hours to cycle the 50 km.

Peter started to cycle 5 minutes after James started to cycle.
Peter caught up with James when they had both cycled 15 km.

James and Peter both cycled at constant speeds.

Word out Peter's speed.

Q4. The distance from Fulbeck to Ganby is 10 miles.
The distance from Ganby to Horton is 18 miles.

Raksha is going to drive from Fulbeck to Ganby.
Then she will drive from Ganby to Horton.

Raksha leaves Fulbeck at 10 00
She drives from Fulbeck to Ganby at an average speed of 40mph.

Raksha wants to get to Horton at 10 35

Work out the average speed Raksha must drive at from Ganby to Horton.

Q5. Olly drove 56 km from Liverpool to Manchester.
He then drove 61 km from Manchester to Sheffield.

Olly's average speed from Liverpool to Manchester was 70 km/h.
Olly took 75 minutes to drive from Manchester to Sheffield.

(a) Work out Olly's average speed for his total drive from Liverpool to Sheffield.

Q6. On Monday, Tarek travelled by train from Manchester to London.

Tarek's train left Manchester at 08 35
It got to London at 11 05
The train travelled at an average speed of 110 miles per hour.

On Wednesday, Gill travelled by train from Manchester to London.

Gill's train also left at 08 35 but was diverted.
The train had to travel an extra 37 miles.
The train got to London at 11 35

Work out the difference between the average speed of Tarek's train and the average speed of Gill's train.

Q7. Sean drives from Manchester to Gretna Green.

He drives at an average speed of 50 mph for the first 3 hours of his journey.

He then has 150 miles to drive to get to Gretna Green.
Sean drives these 150 miles at an average speed of 30 mph.

Sean says,

"My average speed from Manchester to Gretna Green was 40 mph."

Is Sean right?
You must show how you get your answer.

Q8. A train travelled along a track in 110 minutes, correct to the nearest 5 minutes.

Jake finds out that the track is 270 km long.
He assumes that the track has been measured correct to the nearest 10 km.

(a) Could the average speed of the train have been greater than 160 km/h?

You must show how you get your answer.

The track was measured correct to the nearest 5 km.

(b) Explain how this could affect your decision in part (a).

(a) In one week the firm makes and sells six cleaners of each model. Does he make a profit?
[Remember he has to pay his employees for a full week.]
(b) What number of each model should he make so that he makes as much profit as possible? Assume he can sell all the machines which he makes.

4. How many of each?

A shop owner has room in her shop for up to 20 televisions. She can buy either type A for $150 each or type B for $300 each.
She has a total of $4500 she can spend and she must have at least 6 of each type in stock. She makes a profit of $80 on each television of type A and a profit of $100 on each of type B.

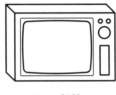

A cost $150

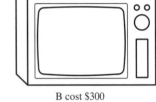

B cost $300

How many of each type should she buy so that she makes the maximum profit?

11.3 Puzzles and experiments

1. Cross numbers

(a) Copy out the cross number pattern.
(b) Fit all the given numbers into the correct spaces. Tick off the numbers from the lists as you write them in the square.

1.

2 digits	3 digits	4 digits	5 digits	6 digits
11	121	2104	14 700	216 841
17	147	2356	24 567	588 369
18	170	2456	25 921	846 789
19	174	3714	26 759	861 277
23	204	4711	30 388	876 452
31	247	5548	50 968	
37	287	5678	51 789	
58	324	6231	78 967	
61	431	6789	98 438	
62	450	7630		
62	612	9012		7 digits
70	678	9921		6 645 678
74	772			
81	774			
85	789			
94	870			
99				

2.

2 digits		3 digits	4 digits	5 digits	6 digits
12	47	129	2096	12 641	324 029
14	48	143	3966	23 449	559 641
16	54	298	5019	33 111	956 782
18	56	325	5665	33 210	
20	63	331	6462	34 509	
21	67	341	7809	40 551	
23	81	443	8019	41 503	
26	90	831	8652	44 333	*7 digits*
27	91	923		69 786	1 788 932
32	93			88 058	5 749 306
38	98			88 961	
39	99			90 963	
46				94 461	
				99 654	

2. Estimating game

This is a game for two players. On squared paper draw an answer grid with the numbers shown below.

Answer grid

891	7047	546	2262	8526	429
2548	231	1479	357	850	7938
663	1078	2058	1014	1666	3822
1300	1950	819	187	1050	3393
4350	286	3159	442	2106	550
1701	4050	1377	4900	1827	957

The players now take turns to choose two numbers from the question grid below and multiply them on a calculator.

Question grid

11	26	81
17	39	87
21	50	98

The game continues until all the numbers in the answer grid have been crossed out. The object is to get four answers in a line (horizontally, vertically or diagonally). The winner is the player with most lines of four.

A line of *five* counts as *two* lines of four.

A line of *six* counts as *three* lines of four.

3. The chess board problem

(a) On the 4×4 square below we have placed four objects subject to the restriction that nowhere are there two objects on the same row, column or diagonal.

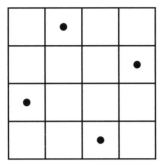

Subject to the same restrictions:
  (i) find a solution for a 5×5 square, using five objects,
 (ii) find a solution for a 6×6 square, using six objects,
(iii) find a solution for a 7×7 square, using seven objects,
(iv) find a solution for a 8×8 square, using eight objects.

It is called the chess board problem because the objects could be 'Queens' which can move any number of squares in any direction.

(b) Suppose we remove the restriction that no two Queens can be on the same row, column or diagonal. Is it possible to attack every square on an 8×8 chess board with less than eight Queens?

Try the same problem with other pieces like knights or bishops.

4. Creating numbers

Using only the numbers 1, 2, 3 and 4 once each and the operations $+, -, \times, \div, !$ create every number from 1 to 100.

You can use the numbers as powers and you must use all of the numbers 1, 2, 3 and 4.

[4! is pronounced 'four factorial' and means $4 \times 3 \times 2 \times 1$ (i.e. 24)
similarly $3! = 3 \times 2 \times 1 = 6$
            $5! = 5 \times 4 \times 3 \times 2 \times 1 = 120$]

Examples: $1 = (4 - 3) \div (2 - 1)$
             $20 = 4^2 + 3 + 1$
             $68 = 34 \times 2 \times 1$
            $100 = (4! + 1)(3! - 2!)$

5. Pentominoes

A pentomino is a set of five squares joined along their edges. Here are three of the twelve different pentomino designs.

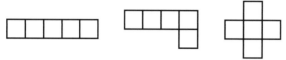

(a) Find the other nine pentomino designs to make up the complete set of twelve. Reflections or rotations of other pentominoes are not allowed.

(b) On squared paper draw an 8×8 square. It is possible to fill up the 8×8 square with the twelve different pentominoes together with a 2×2 square. Here we have made a possible start.

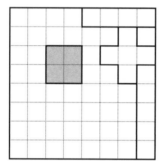

There are in fact many different ways in which this can be done.

(c) Now draw a 10×6 rectangle.

Try to fill up the rectangle with as many different pentominoes as you can. This problem is more difficult than the previous one but it is possible to fill up the rectangle with the twelve different pentominoes.

6. Calculator words

On a calculator work out $9508^2 + 192^2 + 10^2 + 6$. If you turn the calculator upside down and use a little imagination, you can see the word 'HEDGEHOG'.

Find the words given by the clues below.

1. $19 \times 20 \times 14 - 2 \cdot 66$ (not an upstanding man)

2. $(84 + 17) \times 5$ (dotty message)

3. $904^2 + 89\,621\,818$ (prickly customer)

4. $(559 \times 6) + (21 \times 55)$ (what a surprise!)

5. $566 \times 711 - 23\,617$ (bolt it down)

6. $\dfrac{9999 + 319}{8 \cdot 47 + 2 \cdot 53}$ (sit up and plead)

7. $\dfrac{2601 \times 6}{4^2 + 1^2}$; $(401 - 78) \times 5^2$ (two words) (not a great man)

8. $0.4^2 - 0.1^2$ (little Sidney)

9. $\dfrac{(27 \times 2000 - 2)}{(0.63 \div 0.09)}$ (not quite a mountain)

10. $(5^2 - 1^2)^4 - 14\,239$ (just a name)

11. $48^4 + 102^2 - 4^2$ (pursuits)

12. $615^2 + (7 \times 242)$ (almost a goggle)

13. $(130 \times 135) + (23 \times 3 \times 11 \times 23)$ (wobbly)

14. $164 \times 166^2 + 734$ (almost big)

15. $8794^2 + 25 \times 342.28 + 120 \times 25$ (thin skin)

16. $0.08 - (3^2 \div 10^4)$ (ice house)

17. $235^2 - (4 \times 36.5)$ (shiny surface)

18. $(80^2 + 60^2) \times 3 + 81^2 + 12^2 + 3013$ (ship gunge)

19. $3 \times 17 \times (329^2 + 2 \times 173)$ (unlimited)

20. $230 \times 230\frac{1}{2} + 30$ (fit feet)

21. $33 \times 34 \times 35 + 15 \times 3$ (beleaguer)

22. $0.32^2 + \frac{1}{1000}$ (Did he or didn't he?)

23. $(23 \times 24 \times 25 \times 26) + (3 \times 11 \times 10^3) - 20$ (help)

24. $(16^2 + 16)^2 - (13^2 - 2)$ (slander)

25. $(3 \times 661)^2 - (3^6 + 22)$ (pester)

26. $(22^2 + 29.4) \times 10$; $(3.03^2 - 0.02^2) \times 100^2$ (four words) (Goliath)

27. $1.25 \times 0.2^6 + 0.2^2$ (tissue time)

28. $(3^3)^2 + 2^2$ (wriggler)

29. $14 + (5 \times (83^2 + 110))$ (bigger than a duck)

30. $2 \times 3 \times 53 \times 10^4 + 9$ (opposite to hello, almost!)

31. $(177 \times 179 \times 182) + (85 \times 86) - 82$ (good salesman)

32. $6.2 \times 0.987 \times 1\,000\,000 - 860^2 + 118$ (flying ace)

33. $(426 \times 474) + (318 \times 487) + 22\,018$ (close to a bubble)

12 REVISION TESTS

Test 1

1. How many mm are there in 1 m 1 cm?

    A 1001
    B 1110
    C 1010
    D 1100

2. The circumference of a circle is 16π cm. The radius, in cm, of the circle is

    A 2
    B 4
    C $\dfrac{4}{\pi}$
    D 8

3. In the triangle below the value of cos x is

    A 0·8
    B 1·333
    C 0·75
    D 0·6

4. The line $y = 2x - 1$ cuts the x-axis at P. The coordinates of P are

    A $(0, -1)$
    B $(\frac{1}{2}, 0)$
    C $(-\frac{1}{2}, 0)$
    D $(-1, 0)$

5. The formula $b + \dfrac{x}{a} = c$ is rearranged to make x the subject. What is x?

    A $a(c - b)$
    B $ac - b$
    C $\dfrac{c - b}{a}$
    D $ac + ab$

6. The mean mass of a group of 11 men is 70 kg. What is the mean mass of the remaining group when a man of mass 90 kg leaves?

    A 80 kg
    B 72 kg
    C 68 kg
    D 62 kg

7. How many lines of symmetry has this shape?

    A 0
    B 1
    C 2
    D 4

8. In standard form the value of $200 \times 80\,000$ is

    A 16×10^{6}
    B $1\cdot6 \times 10^{9}$
    C $1\cdot6 \times 10^{7}$
    D $1\cdot6 \times 10^{8}$

9. The solutions of the equation $(x-3)(2x+1) = 0$ are

   A $-3, \frac{1}{2}$
   B $3, -2$
   C $3, -\frac{1}{2}$
   D $-3, -2$

10. In the triangle the size of angle x is

   A $35°$
   B $70°$
   C $110°$
   D $40°$

11. A man paid tax on $9000 at 30%. He paid the tax in 12 equal payments. Each payment was

   A $2·25
   B $22·50
   C $225
   D $250

12. The approximate value of $\dfrac{3·96 \times (0·5)^2}{97·1}$ is

   A $0·01$
   B $0·02$
   C $0·04$
   D $0·1$

13. Given that $\dfrac{3}{n} = 5$, then $n =$

   A 2
   B -2
   C $1\frac{2}{3}$
   D $0·6$

14. Cube A has side 2 cm. Cube B has side 4 cm.
$$\left(\frac{\text{Volume of B}}{\text{Volume of A}}\right) =$$

   A 2
   B 4
   C 8
   D 16

15. How many square tiles of side 50 cm will be needed to cover the floor shown?

   A 16
   B 32
   C 64
   D 84

6 m

2 m

4 m

2 m

16. The equation $ax^2 + x - 6 = 0$ has a solution $x = -2$. What is a?

   A 1
   B -2
   C $\sqrt{2}$
   D 2

17. Which of the following is/are correct?
   1. $\sqrt{0·16} = \pm 0·4$
   2. $0·2 \div 0·1 = 0·2$
   3. $\frac{4}{7} > \frac{3}{5}$

   A *1* only
   B *2* only
   C *3* only
   D *1* and *2*

18. How many prime numbers are there between 30 and 40?

A 0
B 1
C 2
D 3

19. A man is paid $600 per week after a pay rise of 20%. What was he paid before?

A $480
B $500
C $540
D $580

20. A car travels for 20 minutes at 45 km/h and then for 40 minutes at 60 km/h. The average speed for the whole journey is

A $52\frac{1}{2}$ km/h
B 50 km/h
C 54 km/h
D 55 km/h

21. The point $(3, -1)$ is reflected in the line $y = 2$. The new coordinates are

A $(3, 5)$
B $(1, -1)$
C $(3, 4)$
D $(0, -1)$

22. Two discs are randomly taken without replacement from a bag containing 3 red discs and 2 blue discs. What is the probability of taking 2 red discs?

A $\frac{9}{25}$
B $\frac{1}{10}$
C $\frac{3}{10}$
D $\frac{2}{5}$

23. The shaded area, in cm^2, is

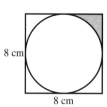

8 cm

8 cm

A $16 - 2\pi$
B $16 - 4\pi$
C $\dfrac{4}{\pi}$
D $64 - 8\pi$

24. Given the equation $5^x = 120$, the best approximate solution is $x =$

A 2
B 3
C 4
D 25

25. What is the sine of 45°?

1

45°

A 1
B $\frac{1}{2}$
C $\dfrac{1}{\sqrt{2}}$
D $\sqrt{2}$

Test 2

1. What is the value of the expression $(x-2)(x+4)$ when $x=-1$?

    A 9
    B -9
    C 5
    D -5

2. The perimeter of a square is $36\,\text{cm}$. What is its area?

    A $36\,\text{cm}^2$
    B $324\,\text{cm}^2$
    C $81\,\text{cm}^2$
    D $9\,\text{cm}^2$

3. AB is a diameter of the circle. Find the angle BCO.

    A $70°$
    B $20°$
    C $60°$
    D $50°$

4. The gradient of the line $2x+y=3$ is

    A 3
    B -2
    C $\frac{1}{2}$
    D $-\frac{1}{2}$

5. A firm employs 1200 people, of whom 240 are men. The percentage of employees who are men is

    A 40%
    B 10%
    C 15%
    D 20%

6. A car is travelling at a constant speed of $30\,\text{km/h}$. How far will the car travel in 10 minutes?

    A $\frac{1}{3}$ mile
    B $3\,\text{km}$
    C $5\,\text{km}$
    D $6\,\text{km}$

7. What are the coordinates of the point $(1,-1)$ after reflection in the line $y=x$?

    A $(-1,1)$
    B $(1,1)$
    C $(-1,-1)$
    D $(1,-1)$

8. $\frac{1}{3}+\frac{2}{5}=$

    A $\frac{2}{8}$
    B $\frac{3}{8}$
    C $\frac{3}{15}$
    D $\frac{11}{15}$

9. In the triangle the size of the largest angle is

    A $30°$
    B $90°$
    C $120°$
    D $80°$

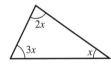

10. 800 decreased by 5% is

    A 795
    B 640
    C 760
    D 400

11. Which of the statements is (are) true?
    1. tan 60° = 2
    2. sin 60° = cos 30°
    3. sin 30° > cos 30°

    A *1* only
    B *2* only
    C *3* only
    D *2* and *3*

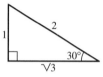

12. Given $a = \frac{3}{5}$, $b = \frac{1}{3}$, $c = \frac{1}{2}$ then

    A $a < b < c$
    B $a < c < b$
    C $a > b > c$
    D $a > c > b$

13. The *larger* angle between south-west and east is

    A 225°
    B 240°
    C 135°
    D 315°

14. Each exterior angle of a regular polygon with *n* sides is 10°; *n* =

    A 9
    B 18
    C 30
    D 36

15. What is the value of $1 - 0.05$ as a fraction?

    A $\frac{1}{20}$
    B $\frac{9}{10}$
    C $\frac{19}{20}$
    D $\frac{5}{100}$

16. Find the length *x*.

    A 5
    B 6
    C 8
    D $\sqrt{50}$

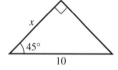

17. Given that $m = 2$ and $n = -3$, what is mn^2?

    A -18
    B 18
    C -36
    D 36

18. The graph of $y = (x - 3)(x - 2)$ cuts the *y*-axis at P.
    The coordinates of P are

    A (0, 6)
    B (6, 0)
    C (2, 0)
    D (3, 0)

19. $240 is shared in the ratio $2:3:7$. The largest share is

- **A** $130
- **B** $140
- **C** $150
- **D** $160

20. Adjacent angles in a parallelogram are $x°$ and $3x°$. The smallest angles in the parallelogram are each

- **A** $30°$
- **B** $45°$
- **C** $60°$
- **D** $120°$

21. When the sides of a square are increased by 10% the area is increased by

- **A** 10%
- **B** 20%
- **C** 21%
- **D** 15%

22. The volume, in cm³, of the cylinder is

- **A** 9π
- **B** 12π
- **C** 600π
- **D** 900π

23. A car travels for 10 minutes at 30 km/h and then for 20 minutes at 45 km/h. The average speed for the whole journey is

- **A** 40 km/h
- **B** $37\frac{1}{2}$ km/h
- **C** 20 km/h
- **D** 35 km/h

24. Four people each toss a coin. What is the probability that the fourth person will toss a 'tail'?

- **A** $\frac{1}{2}$
- **B** $\frac{1}{4}$
- **C** $\frac{1}{8}$
- **D** $\frac{1}{16}$

25. A rectangle 8 cm by 6 cm is inscribed inside a circle. What is the area, in cm², of the circle?

- **A** 10π
- **B** 25π
- **C** 49π
- **D** 100π

Test 3

1. The price of a T.V. changed from $240 to $300. What is the percentage increase?

- **A** 15%
- **B** 20%
- **C** 60%
- **D** 25%

2. Find the length x.

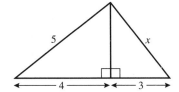

- **A** 6
- **B** 5
- **C** $\sqrt{44}$
- **D** $\sqrt{18}$

3. The bearing of A from B is 120°. What is the bearing of B from A?

    A 060°
    B 120°
    C 240°
    D 300°

4. Numbers m, x and y satisfy the equation $y = mx^2$.
When $m = \frac{1}{2}$ and $x = 4$ the value of y is

    A 4
    B 8
    C 1
    D 2

5. A school has 400 pupils, of whom 250 are boys. The ratio of boys to girls is

    A $5:3$
    B $3:2$
    C $3:5$
    D $8:5$

6. A train is travelling at a speed of 30 km per hour. How long will it take to travel 500 m?

    A 2 minutes
    B $\frac{3}{50}$ hour
    C 1 minute
    D $\frac{1}{2}$ hour

7. The approximate value of $\dfrac{9\cdot65 \times 0\cdot203}{0\cdot0198}$ is

    A 100
    B 10
    C 1
    D 180

8. Which point does *not* lie on the curve $y = \dfrac{12}{x}$?

    A $(6, 2)$
    B $(\frac{1}{2}, 24)$
    C $(-3, -4)$
    D $(3, -4)$

9. $t = \dfrac{c^3}{y}$, $y =$

    A $\dfrac{t}{c^3}$
    B $c^3 t$
    C $c^3 - t$
    D $\dfrac{c^3}{t}$

10. The largest number of 1 cm cubes which will fit inside a cubical box of side 1 m is

    A 10^3
    B 10^6
    C 10^8
    D 10^{12}

11. The shaded area in the Venn diagram represents

    A $A' \cup B$
    B $A \cap B'$
    C $A' \cap B$
    D $(A \cap B)'$

12. Which of the following has the largest value?

  A $\sqrt{100}$

  B $\sqrt{\dfrac{1}{0\cdot1}}$

  C $\sqrt{1000}$

  D $\dfrac{1}{0\cdot01}$

13. Two dice numbered 1 to 6 are thrown together and their scores are added. The probability that the sum will be 12 is

  A $\frac{1}{6}$

  B $\frac{1}{12}$

  C $\frac{1}{18}$

  D $\frac{1}{36}$

14. The length, in cm, of the minor arc is

  A 2π

  B 3π

  C 6π

  D $13\frac{1}{2}\pi$

15. Metal of mass 84 kg is made into 40 000 pins. What is the mass, in kg, of one pin?

  A $0\cdot0021$

  B $0\cdot0036$

  C $0\cdot021$

  D $0\cdot21$

16. What is the value of x which satisfies the simultaneous equations?
$$3x + y = 1$$
$$x - 2y = 5$$

  A -1

  B 1

  C -2

  D 2

17. What is the new fare when the old fare of $250 is increased by 8%?

  A $258

  B $260

  C $270

  D $281\cdot25

18. What is the area of this triangle?

  A $12x^2$

  B $15x^2$

  C $16x^2$

  D $30x^2$

19. What values of x satisfy the inequality $2 - 3x > 1$?

  A $x < -\frac{1}{3}$

  B $x > -\frac{1}{3}$

  C $x > \frac{1}{3}$

  D $x < \frac{1}{3}$

20. A right-angled triangle has sides in the ratio $5:12:13$. The tangent of the smallest angle is

   A $\frac{12}{5}$

   B $\frac{12}{13}$

   C $\frac{5}{13}$

   D $\frac{5}{12}$

21. The area of $\triangle ABE$ is $4\,cm^2$. The area of $\triangle ACD$ is

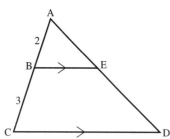

   A $10\,cm^2$

   B $6\,cm^2$

   C $25\,cm^2$

   D $16\,cm^2$

22. Given $2^x = 3$ and $2^y = 5$, the value of 2^{x+y} is

   A 15

   B 8

   C 4

   D 125

23. The probability of an event occurring is 0.35. The probability of the event *not* occurring is

   A $\dfrac{1}{0.35}$

   B 0.65

   C 0.35

   D 0

24. What fraction of the area of the rectangle is the area of the triangle?

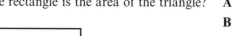

   A $\frac{1}{4}$

   B $\frac{1}{8}$

   C $\frac{1}{16}$

   D $\frac{1}{32}$

25. On a map a distance of $36\,km$ is represented by a line of $1.8\,cm$. What is the scale of the map?

   A $1:2000$

   B $1:20\,000$

   C $1:200\,000$

   D $1:2000\,000$

Test 4

1. What is the value of x satisfying the simultaneous equations
$$3x + 2y = 13$$
$$x - 2y = -1?$$

   A 7

   B 3

   C $3\frac{1}{2}$

   D 2

2. A straight line is 4·5 cm long. $\frac{2}{5}$ of the line is

    A 0·4 cm
    B 1·8 cm
    C 2 cm
    D 0·18 cm

3. The mean of four numbers is 12. The mean of three of the numbers is 13. What is the fourth number?

    A 9
    B 12·5
    C 7
    D 1

4. How many cubes of edge 3 cm are needed to fill a box with internal dimensions 12 cm by 6 cm by 6 cm?

    A 8
    B 18
    C 16
    D 24

For questions **5** to **7** use the diagram below.

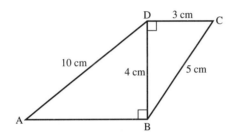

5. The length of AB, in cm, is

    A 6
    B $\sqrt{116}$
    C 8
    D $\sqrt{84}$

6. The sine of angle DCB is

    A 0·8
    B 1·25
    C 0·6
    D 0·75

7. The tangent of angle CBD is

    A 0·6
    B 0·75
    C 1·333
    D 1·6

8. The value of 4865·355 correct to 2 significant figures is

    A 4865·36
    B 4865·35
    C 4900
    D 49

9. What values of y satisfy the inequality $4y - 1 < 0$?

    A $y < 4$
    B $y < -\frac{1}{4}$
    C $y > \frac{1}{4}$
    D $y < \frac{1}{4}$

10. The area of a circle is $100\pi\,\mathrm{cm}^2$. The radius, in cm, of the circle is

   A 50
   B 10
   C $\sqrt{50}$
   D 5

11. If $f(x) = x^2 - 3$, then $f(3) - f(-1) =$

   A 5
   B 10
   C 8
   D 9

12. In the triangle BE is parallel to CD. What is x?

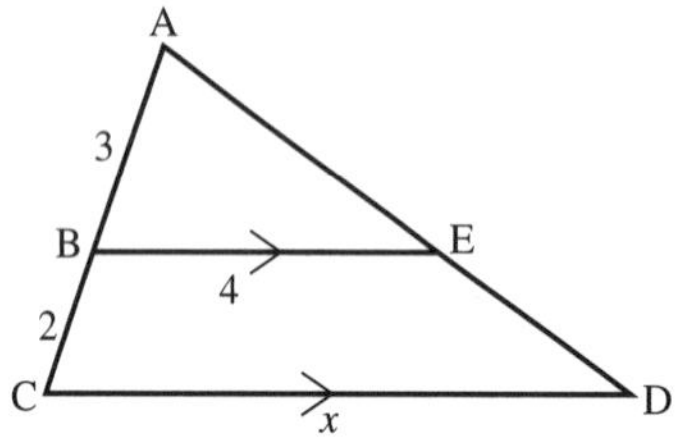

   A $6\frac{2}{3}$
   B 6
   C $7\frac{1}{2}$
   D $5\frac{3}{4}$

13. The cube root of 64 is

   A 2
   B 4
   C 8
   D 16

14. Given $\quad a + b = 10$
and $\quad a - b = 4$
then $\quad 2a - 5b =$

   A 0
   B -1
   C 1
   D 3

15. Given $16^x = 4^4$, what is x?

   A -2
   B $-\frac{1}{2}$
   C $\frac{1}{2}$
   D 2

16. What is the area, in m^2, of a square with each side $0.02\,\mathrm{m}$ long?

   A 0.0004
   B 0.004
   C 0.04
   D 0.4

17. I start with x, then square it, multiply by 3 and finally subtract 4. The final result is

   A $(3x)^2 - 4$
   B $(3x - 4)^2$
   C $3x^2 - 4$
   D $3(x - 4)^2$

18. How many prime numbers are there between 50 and 60?

   A 1
   B 2
   C 3
   D 4

19. What are the coordinates of the point (2, −2) after reflection in the line $y = -x$?

A (−2, 2)
B (2, −2)
C (−2, −2)
D (2, 2)

20. The area of a circle is $36\pi\,\text{cm}^2$. The circumference, in cm, is

A 6π
B 18π
C $12\sqrt{\pi}$
D 12π

21. The gradient of the line $2x - 3y = 4$ is

A $\frac{2}{3}$
B $1\frac{1}{2}$
C $-\frac{4}{3}$
D $-\frac{3}{4}$

22. When all three sides of a triangle are trebled in length, the area is increased by a factor of

A 3
B 6
C 9
D 27

23. $a = \sqrt{\left(\dfrac{m}{x}\right)}$

$x =$

A a^2m
B $a^2 - m$
C $\dfrac{m}{a^2}$
D $\dfrac{a^2}{m}$

24. A coin is tossed three times. The probability of getting three 'heads' is

A $\frac{1}{3}$
B $\frac{1}{6}$
C $\frac{1}{8}$
D $\frac{1}{16}$

25. A triangle has sides of length 5 cm, 5 cm and 6 cm. What is the area, in cm^2?

A 12
B 15
C 18
D 20

SPECIMEN PAPER 2

$$\begin{bmatrix} \text{Short-answer questions;} \\ \text{Extended level} \end{bmatrix}$$

TIME 1 h 30 m

Instructions to candidates

Answer **all** questions.
If working is needed for any question it must be shown.
The total of the marks for this paper is 70.
Electronic calculators should be used.
If the degree of accuracy is not specified in the question and if the answer is not exact, the answer should be given to three significant figures. Answers in degrees should be given to one decimal place.
For π, use either your calculator value or 3·142.
The number of marks is given in brackets [ ] at the end of each question or part question.

1. Find $\sqrt{\dfrac{16}{25}}$. [1]

2. How many minutes are there between 19:35 and midnight? [1]

3. For what range of values of x is $4x - 5 < 19$? [2]

4.

The road sign on the left stood at the top of a steep hill.
It was replaced by the road sign on the right.
(a) Explain why the two signs are equivalent. [1]
(b) If the old sign had been 1:7, what would the percentage on the new sign be? [1]

5. Find the value of $4ab$ when $a = 5 \times 10^4$ and $b = 7 \times 10^{-9}$.
Give your answer in standard form. [2]

6. Find the value of $9^{\frac{1}{2}} \times 125^{\frac{1}{3}} \times 4^0$. [2]

7. An armchair is advertised for sale at \$240. This is a 40% reduction on its original price. Work out the original price. [2]

8. (a) Change the following fractions to decimals, showing your full
calculator display in each case:

    (i) $\frac{9}{20}$ (ii) $\frac{4}{9}$ (iii) $\frac{33}{74}$ [1]

  (b) Which one of the above fractions is closest to $\dfrac{1}{\sqrt{5}}$?   [1]

9. (a) What is the gradient of the line $y = 8 - 3x$?   [1]

  (b) Find $\{(x, y) : y = 8 - 3x\} \cap \{(x, y) : x = 5\}$ [2]

10. To the nearest half metre, a room is 4 metres long and $2\frac{1}{2}$ metres
wide.

  (a) The actual length of the room is l metres.
Write down the upper and lower limits of l.   [1]

  (b) The actual area of the floor of the room is A square metres.
Calculate the upper and lower limits of A.   [2]

11. Solve the simultaneous equations:
$$3x + 2y = 10$$
$$2x - 3y = 11$$ [3]

12. A bicycle wheel has a radius of 40 cm.
How many times does it revolve during a journey of 10 km?
Give your answer to the nearest 100.
[For π, use either your calculator value or 3·142.]   [3]

13.

The map shows four cities on the north coast of Africa.

  (a) Use your protractor to find the bearing of

    (i) Tunis from Algiers,   [1]

    (ii) Casablanca from Tangier.   [1]

  (b) The scale of the map is 1 : 20 000 000.
Find the shortest distance in kilometres from Casablanca
to Tunis.   [1]

14. A hat of size N in Britain is equivalent to a hat of size C in
mainland Europe.

  (a) Find the value of $8N$ when N equals: (i) $6\frac{3}{4}$ (ii) $7\frac{1}{8}$ [2]

  (b) $C = 55$ when $N = 6\frac{3}{4}$ and $C = 58$ when $N = 7\frac{1}{8}$.
Write down a formula connecting C and N.   [1]

15. In 2011, twice as many books were borrowed from the school library as in 2010. The school librarian draws three possible diagrams to represent this.

(i) 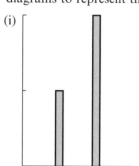   (ii) 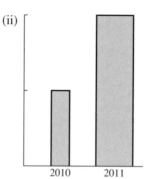   (iii)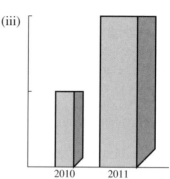

(a) Which diagram shows the information most fairly?   [1]

(b) What do the other two diagrams imply about the ratio $\dfrac{\text{number of books borrowed in 2011}}{\text{number of books borrowed in 2010}}$?   [2]

16. On graph paper, draw coordinate axes from -8 to $+8$ in both the x- and y-directions. Mark the point P(2, 0). Draw the triangle A with vertices (2, 1), (5, 2) and (5, 4).

(a) On the diagram enlarge triangle A with centre of enlargement P and scale factor -2.   [2]

(b) The area of triangle A is 3 square units. What is the area of the enlarged triangle?   [1]

17. y varies inversely with x.

(a) Write this statement as an equation in x, y and k, where k is a constant.   [1]

(b) If x decreases by 20%, find the percentage change in y.   [3]

18. $y = \dfrac{6}{x} + x$

(a) Find the values of y which correspond to values of x between 1 and 5 inclusive.   [2]

(b) On graph paper, draw an x-axis from 0 to 6 and a y-axis from 0 to 8. Plot the points you have found in part (a) on the graph paper, and join them up with a smooth curve.   [2]

19. Triangle A has vertices (0, 3), (2, 3) and (2, 5). Triangle B has vertices (5, 2), (7, 2) and (7, 4).

(a) Write down the vector of the translation which will map triangle A onto triangle B.   [1]

(b) On graph paper, draw triangle A and then rotate it anticlockwise through $90°$ about the point $(-2, 3)$. Label the rotated triangle C.   [1]

(c) Describe fully the rotation that will map triangle C onto triangle B.   [2]

20.

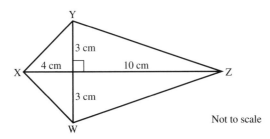

Not to scale

In the quadrilateral WXYZ, XZ is perpendicular to WY.
(a) Calculate the lengths of the four sides of quadrilateral
    WXYZ.   [2]
(b) Calculate the size of angle YZW.   [2]

21.

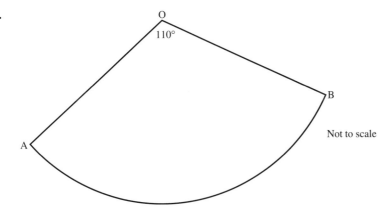

Not to scale

(a) Draw a sector OAB of a circle, centre O, radius 5 cm and
    angle AOB = 110°.   [1]
(b) Calculate the length of the arc AB.   [2]
(c) Calculate the area of the sector OAB.   [2]

22. The table shows the population of the world in thousands of
    millions, from 1750 to 1975.

year	1750	1815	1850	1900	1950	1975
population (thousands of millions)	0·65	1	1·4	2·3	3·7	4·4

(a) Using a scale of 2 cm to 50 years on the horizontal axis, and
    2 cm to 1 000 000 000 people on the vertical axis, plot the
    points from the table and join them with a smooth curve.   [1]
(b) Read off the world population in 1925.
    Write down your answer (i) in words or figures,   [1]
                            (ii) in standard form.   [1]
(c) By drawing a suitable tangent, estimate the rate of
    population growth in 1900.   [2]
(d) Extend the graph to estimate the population of the
    world in 2020.   [1]

23.

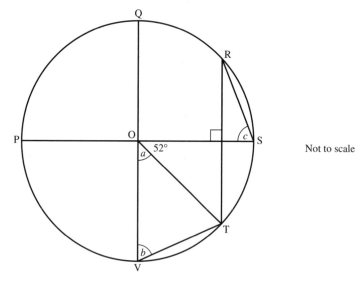

Not to scale

In the diagram, O is the centre of the circle.
QV is parallel to RT, PS is perpendicular to RT and angle
SOT = 52°.
Find the angles marked *a*, *b* and *c*.   [5]

SPECIMEN PAPER 4

[Structured questions; Extended level]

TIME 2 h 30 m

Instructions to candidates

Answer **all** questions.
All working must be clearly shown. Marks will be given for working which shows that you know how to solve the problem even if you get the answer wrong.

The number of marks is given in brackets [ ] at the end of each question or part question.
The total of the marks for this paper is 130.
Electronic calculators should be used.
If the degree of accuracy is not specified in the question, and if the answer is not exact, give the answer to three significant figures. Give answers in degrees to one decimal place.
For π, use either your calculator or a value of 3·142.

1. $r = \dfrac{2p^2}{q - 3}$

    (a) Find the value of r when (i) $p = 6$ and $q = 5$,   [1]
                                   (ii) $p = -4$ and $q = -1$.   [1]
    (b) Find the value of q when $p = 3$ and $r = 12$.   [2]
    (c) Find both possible values of p when $q = 8$ and $r = 10$.   [2]
    (d) The value of p is tripled and q remains unchanged.
        What effect does this have on the value of r?   [2]
    (e) Make p the subject of the formula.   [3]

2. Opposite sides of the hexagon in the diagram are parallel, and are in the ratio 2:1.
$\overrightarrow{AB} = \mathbf{x}$, $\overrightarrow{CD} = \mathbf{y}$, and $\overrightarrow{EF} = \mathbf{z}$.

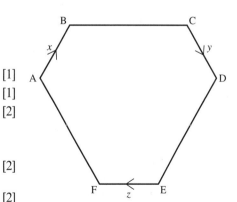

    (a) (i) Write down the vector $\overrightarrow{ED}$,   [1]
        (ii) Hence show that $\overrightarrow{EC} = 2\mathbf{x} - \mathbf{y}$.   [1]
        (iii) Find the vectors $\overrightarrow{AE}$ and $\overrightarrow{CA}$.   [2]
    (b) Write down in terms of $\mathbf{x}$, $\mathbf{y}$ and $\mathbf{z}$:
      $\overrightarrow{AE} + \overrightarrow{EC} + \overrightarrow{CA}$
      expressing your answer in its simplest form.   [2]
    (c) Write down a vector equation which follows from
      the result of part (b).   [2]
    (d) Use the above results to determine whether or
      not BE is parallel to CD.   [2]

3.

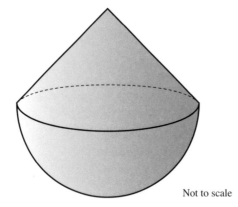

Not to scale

A glass paperweight consists of a cone mounted on a hemisphere.
The common radius (r) is 3 cm; the height of the cone (h) is 4 cm.
You are given:
the volume of a cone is $\frac{1}{3}\pi r^2 h$; the volume of a sphere is $\frac{4}{3}\pi r^3$;
the curved surface area of a cone is πrl (slant height l);
the surface area of a sphere is $4\pi r^2$.
(a) Calculate (i) the volume of the paperweight, [4]
          (ii) the surface area of the paperweight. [5]
(b) 1 cm^3 of the glass of which the paperweight is made weighs
    2·85 g. Calculate the mass of the paperweight. [2]

4. Answer the whole of this question on a sheet of graph paper.
Using a scale of 1 cm to represent 1 unit on each axis, draw a pair
of axes for $0 \leqslant x \leqslant 18$ and $0 \leqslant y \leqslant 14$.
(a) On your axes:
    (i) draw the line $y = 2x$, [2]
    (ii) mark the two points A(10, 0) and B(16, 5), [1]
    (iii) construct the locus of points which are equidistant
        from the points A and B, [3]
    (iv) construct the locus of points which are equidistant
        from the line $y = x$ and the x-axis, [3]
    (v) draw the circle which touches the x-axis at A, and
        which passes through B. [2]
(b) Which other line, already drawn, does the circle touch? [1]
(c) Draw the tangent to the circle at B, and write down the
    coordinates of the point at which it cuts the x-axis. [2]

5. (a) In triangle ABC, AB = 9 cm, BC = 7 cm and angle ABC = 128°.
    Calculate (i) the length of AC, [4]
            (ii) the area of triangle ABC. [3]
(b) The market place in Newark, Nottinghamshire is a rectangle
    PQRS. PQ = 105 m and QR = 65 m. In corner S stands the
    church. It is 40 m high.
    Work out the angle of elevation of the top of the church
    (i) from P, [2]
    (ii) from Q. [4]

6. A rectangle has length $(3x - 8)$ cm and width $(2x - 7)$ cm.
  (a) Write down and simplify an expression for the perimeter of
      the rectangle.   [2]
  (b) Write down an expression for the area of the rectangle.   [1]
  (c) If the area of the rectangle is 91 cm^2, show that
      $6x^2 - 37x - 35 = 0$ [3]
  (d) (i) Factorise $6x^2 - 37x - 35$.   [3]
      (ii) Solve the quadratic equation $6x^2 - 37x - 35 = 0$.   [3]
  (e) Write down the length and width of the rectangle when its
      area is 91 cm^2.   [2]

7. (a) On each of the first two holes on his golf course, a golfer can
       take 3, 4, 5, 6, 7 or 8 strokes. All outcomes are equally likely.
       Consider these two holes only.
       (i) Draw a possibility diagram, showing all his possible
           scores and totals.   [2]
       (ii) What is the probability that he takes a total of
            16 strokes?   [1]
       (iii) What is the probability that he takes a total of
            10 strokes?   [2]
       (iv) What is his most likely total?   [1]
   (b) If the weather is fine today, the probability that it will be fine
       tomorrow is 0·7.
       This and the other probabilities are shown in this matrix.

$$\begin{array}{c} \text{TOMORROW} \\ \begin{array}{cc} \text{fine} & \text{wet} \end{array} \end{array}$$

$$\text{TODAY} \quad \begin{array}{c} \text{fine} \\ \text{wet} \end{array} \begin{pmatrix} 0\cdot7 & 0\cdot3 \\ 0\cdot4 & 0\cdot6 \end{pmatrix}$$

The probability of the weather being fine on any one day is 0·6.
Copy and complete the tree diagram below, to represent all this
information.

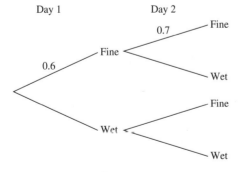

Calculate the probability of:
  (i) two fine days,   [2]
  (ii) a wet day followed by a fine day,   [2]
  (iii) one fine day and one wet day.   [2]

8. The table shows the number of hours of sunshine each day over a six-week period in a seaside town.

Number of hours of sunshine (N)	$0 \leqslant N \leqslant 2$	$2 < N \leqslant 4$	$4 < N \leqslant 6$	$6 < N \leqslant 8$	$8 < N \leqslant 10$	$10 < N \leqslant 12$
Frequency	9	8	6	2	4	13

(a) Work out an estimate of the mean number of hours of sunshine each day. [4]

(b) (i) Which is the modal class? [1]
   (ii) Find the median number of hours of sunshine. [2]

(c) Would you choose the mean, median or mode for publicity to attract visitors to the town? [1]

(d) Draw a histogram for the data, using the three class intervals $0 \leqslant N \leqslant 4$, $4 < N \leqslant 10$ and $10 < N \leqslant 12$. [5]

9. Graph paper must be used for the whole of this question.
A new model of bicycle is about to be marketed. It is estimated that if the selling price is fixed at $80, then 5000 bicycles will be sold; if it is fixed at $120, then only 3000 bicycles will be sold. These two points are plotted and connected with a straight line, as below:

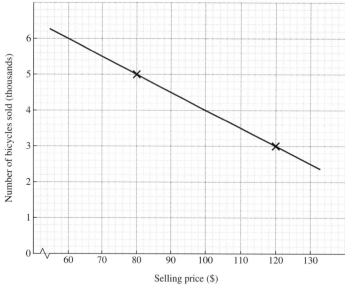

(a) Copy and complete the following table, by reading values from the graph and by calculation.

Selling price ($)	Number of bicycles sold	Sales revenue ($)
60		
70		
80	5000	400 000
90		
100		
110		
120	3000	360 000

[4]

(b) Using a horizontal scale of 2 cm to represent $10 (starting
at $60) and a vertical scale of 2 cm to represent $20 000,
draw the graph of selling price against sales revenue.   [5]
(c) (i) Use your graph to find what the selling price should be
in order to achieve the greatest sales revenue.   [2]
(ii) How many bicycles should be made at that price?   [2]

10. This is a Fibonacci sequence

$$1, 3, 4, 7, 11, 18, 29, 47, \ldots$$

(a) Write down the next three terms in the sequence.   [3]

(b) The matrix $\mathbf{M}$ is $\begin{pmatrix} 1 & 1 \\ 1 & 0 \end{pmatrix}$.

(i) 11 and 18 are two consecutive terms.
Pre-multiply them by $\mathbf{M}$, that is, work out

$$\begin{pmatrix} 1 & 1 \\ 1 & 0 \end{pmatrix}\begin{pmatrix} 18 \\ 11 \end{pmatrix}.$$ [2]

(ii) Choose another pair of consecutive terms and
pre-multiply them by $\mathbf{M}$.   [2]
(iii) Describe what happens when two consecutive terms in
the Fibonacci sequence are pre-multiplied by $\mathbf{M}$.   [2]
(c) (i) Work out $\mathbf{M}^2$.   [2]

(ii) Work out $\mathbf{M}^2\begin{pmatrix} 18 \\ 11 \end{pmatrix}$ [2]

(iii) Describe what has happened in (c) (ii).   [1]
(d) Show how this process will continue with $\mathbf{M}^3$ and $\mathbf{M}^4$.   [4]

STUDENT BOOK ANSWERS

1 Number

Exercise 1   *page 2*

1. 7·91	**2.** 22·22	**3.** 7·372	**4.** 0·066	**5.** 466·2
6. 1·22	**7.** 1·67	**8.** 1·61	**9.** 16·63	**10.** 24·1
11. 26·7	**12.** 3·86	**13.** 0·001	**14.** 1·56	**15.** 0·0288
16. 2·176	**17.** 0·02	**18.** 0·0001	**19.** 7·56	**20.** 0·7854
21. 360	**22.** 34 000	**23.** 18	**24.** 0·74	**25.** 2·34
26. 1620	**27.** 8·8	**28.** 1200	**29.** 0·00 175	**30.** 13·2
31. 200	**32.** 0·804	**33.** 0·8	**34.** 0·077	**35.** 0·0009
36. 0·01	**37.** 184	**38.** 20	**39.** 0·099	**40.** 3

Exercise 2   *page 2*

1. 20   **2.** 256; 65 536

3. $6^2 + 7^2 + 42^2 = 43^2$; $x^2 + (x+1)^2 + [x(x+1)]^2 = (x^2 + x + 1)^2$

4. (a) $54 \times 9 = 486$ (b) $57 \times 8 = 456$ (c) $52 \times 2 = 104$, $57 \times 2 = 114$ or $58 \times 3 = 174$

5. 5, 28; total 32   **7.** 12 units long, marks at 1, 4, 5 and 10   **8.** 21   **11.** 37

Exercise 3   *page 4*

1. $1\frac{11}{20}$	**2.** $\frac{11}{24}$	**3.** $1\frac{1}{2}$	**4.** $\frac{5}{12}$	**5.** $\frac{4}{15}$
6. $\frac{1}{10}$	**7.** $\frac{8}{15}$	**8.** $\frac{5}{42}$	**9.** $\frac{15}{26}$	**10.** $\frac{5}{12}$
11. $4\frac{1}{2}$	**12.** $1\frac{2}{3}$	**13.** $\frac{23}{40}$	**14.** $\frac{3}{40}$	**15.** $1\frac{7}{8}$
16. $1\frac{1}{12}$	**17.** $1\frac{1}{6}$	**18.** $2\frac{5}{8}$	**19.** $6\frac{1}{10}$	**20.** $9\frac{1}{10}$
21. $1\frac{9}{26}$	**22.** $\frac{1}{9}$	**23.** $\frac{2}{3}$	**24.** $5\frac{1}{4}$	**25.** $2\frac{2}{25}$

26. (a) $\frac{1}{2}$, $\frac{7}{12}$, $\frac{2}{3}$ (b) $\frac{2}{3}$, $\frac{3}{4}$, $\frac{5}{6}$ (c) $\frac{1}{3}$, $\frac{5}{8}$, $\frac{17}{24}$, $\frac{3}{4}$ (d) $\frac{5}{6}$, $\frac{8}{9}$, $\frac{11}{12}$

27. (a) $\frac{1}{2}$ (b) $\frac{3}{4}$ (c) $\frac{17}{24}$ (d) $\frac{7}{18}$ (e) $\frac{3}{10}$ (f) $\frac{5}{12}$

28. 5   **29.** 9   **30.** same

Exercise 4   *page 5*

1. 0·25	**2.** 0·4	**3.** 0·8	**4.** 0·75	**5.** 0·5
6. 0·375	**7.** 0·9	**8.** 0·625	**9.** 0·41̇6̇	**10.** 0·1̇6̇
11. 0·6̇	**12.** 0·83̇	**13.** 0·2̇85714̇	**14.** 0·4̇28571̇	**15.** 0·4̇
16. 0·45̇	**17.** 1·2	**18.** 2·625	**19.** 2·3̇	**20.** 1·7
21. 2·1875	**22.** 2·2̇85714̇	**23.** 2·8̇57142̇	**24.** 3·19	**25.** $\frac{1}{5}$
26. $\frac{7}{10}$	**27.** $\frac{1}{4}$	**28.** $\frac{9}{20}$	**29.** $\frac{9}{25}$	**30.** $\frac{13}{25}$
31. $\frac{1}{8}$	**32.** $\frac{5}{8}$	**33.** $\frac{21}{25}$	**34.** $2\frac{7}{20}$	**35.** $3\frac{19}{20}$
36. $1\frac{1}{20}$	**37.** $3\frac{1}{5}$	**38.** $\frac{27}{100}$	**39.** $\frac{7}{1000}$	**40.** $\frac{11}{100\,000}$
41. 0·58	**42.** 1·42	**43.** 0·65	**44.** 1·61	**45.** 0·07
46. 0·16	**47.** 3·64	**48.** 0·60	**49.** $\frac{4}{15}$, 0·33, $\frac{1}{3}$	**50.** $\frac{2}{7}$, 0·3, $\frac{4}{9}$

51. $\frac{7}{11}$, 0·705, 0·71   **52.** $\frac{5}{18}$, 0·3, $\frac{4}{13}$

Exercise 5   *page 6*

1. 3, 11, 19, 23, 29, 31, 37, 47, 59, 61, 67, 73
2. (a) 4, 8, 12, 16, 20   (b) 6, 12, 18, 24, 30   (c) 10, 20, 30, 40, 50
    (d) 11, 22, 33, 44, 55   (e) 20, 40, 60, 80, 100
3. 12 and 24   **4.** 15
5. (a) 1, 2, 3, 6   (b) 1, 3, 9   (c) 1, 2, 5, 10,
    (d) 1, 3, 5, 15   (e) 1, 2, 3, 4, 6, 8, 12, 24   (f) 1, 2, 4, 8, 16, 32
6. (a) Yes. Divide by 3, 5, 7, 11, 13, (i.e. odd prime numbers $< \sqrt{263}$)
    (b) No   (c) Prime numbers $< \sqrt{1147}$ **7.** 2, 3, 5, 41, 67, 89

Exercise 6   *page 6*

1. Rational: $\left(\sqrt{17}\right)^2$; 3·14; $\dfrac{\sqrt{12}}{\sqrt{3}}$; $3^{-1} + 3^{-2}$; $\frac{22}{7}$; $\sqrt{2\cdot25}$

2. (a) both irrational   (b) both rational
3. (a) 6π cm, irrational   (b) 6 cm, rational   (c) 36 cm^2, rational
    (d) 9π cm^2, irrational   (e) $36 - 9\pi$ cm^2, irrational
7. (a) No   (b) Yes e.g. $\sqrt{8} \times \sqrt{2} = 4$

Exercise 7   *page 7*

1. 18, 22   **2.** 30, 37   **3.** 63, 55   **4.** $-7, -12$ **5.** 21, 27   **6.** 2, -5
7. 16, 22   **8.** 16, 32   **9.** 25, 15   **10.** $-4, -10$ **11.** $-4, -3$ **12.** $7\frac{1}{2}, 3\frac{3}{4}$
13. $\frac{1}{3}, \frac{1}{9}$ **14.** 2, $\frac{2}{3}$ **15.** 32, 47   **16.** 840, 6720   **17.** 5, -1 **18.** 2, 1

Exercise 8   *page 8*

1. (a) $2n$ (b) $10n$ (c) $3n$ (d) $11n$ (e) $100n$ (f) n^2 (g) 10^n (h) n^3
2. $n^2 + 4$ **3.** 9, 11, 13, 15, 17
4. (a) 3, 4, 5, 6, 7   (b) 5, 10, 15, 20, 25   (c) 9, 19, 29, 39, 49   (d) 97, 94, 91, 88, 85
    (e) $1, \frac{1}{2}, \frac{1}{3}, \frac{1}{4}, \frac{1}{5}$ (f) 1, 4, 9, 16, 25

Exercise 9   *page 9*

1. $4n + 1$ **2.** $3n + 4$ **3.** $5n - 1$ **4.** $4n + 2$ **5.** $3n + 2$ **6.** $28 - 3n$
7. $5n$ **8.** 2^n **9.** $n(n + 2)$ **10.** $\dfrac{n}{n + 1}$ **11.** $7n$ **12.** n^2
13. $\dfrac{5}{n^2}$ **14.** $\dfrac{n + 2}{n}$ **15.** $4n - 1$ **16.** $2n + 3$ **17.** $9 - 2n$ **18.** $4n - 9$
19. (a) $2n + 6$ (b) $4n - 1$ (c) $5n + 3$
20. (a) $8n + 3$ (b) $2n + \frac{1}{2}$ (c) $3n - 10$
21. (a) $3n + 1$ (c) 3001

Exercise 10   *page 10*

1. (a) 8   (b) 8·17   (c) 8·17   **2.** (a) 20   (b) 19·6   (c) 19·62
3. (a) 20   (b) 20·0   (c) 20·04   **4.** (a) 1   (b) 0·815   (c) 0·81
5. (a) 311   (b) 311   (c) 311·14   **6.** (a) 0   (b) 0·275   (c) 0·28
7. (a) 0   (b) 0·00747   (c) 0·01   **8.** (a) 16   (b) 15·6   (c) 15·62
9. (a) 900   (b) 900   (c) 900·12   **10.** (a) 4   (b) 3·56   (c) 3·56
11. (a) 5   (b) 5·45   (c) 5·45   **12.** (a) 21   (b) 21·0   (c) 20·96
13. (a) 0   (b) 0·0851   (c) 0·09   **14.** (a) 1   (b) 0·515   (c) 0·52
15. (a) 3   (b) 3·07   (c) 3·07   **16.** 5·7   **17.** 0·8   **18.** 11·2
19. 0·1   **20.** 0·0   **21.** 11·1

Exercise 11   *page 11*

1. 195·5 cm   **2.** 36·5 kg   **3.** 3·25 kg   **4.** 95·55 m   **5.** 28·65 s
6. (a) 1·5, 2·5   (b) 2·25, 2·35   (c) 63·5, 64·5   (d) 13·55, 13·65
7. B   **8.** C   **9.**(a) Not necessarily   (b) 1 cm
10. (a) $16·5 \leqslant m < 17·5$ (b) $255·5 \leqslant d < 256·5$ (c) $2·35 \leqslant l < 2·45$ (d) $0·335 \leqslant m < 0·345$
     (e) $2·035 \leqslant v < 2·045$ (f) $11·95 \leqslant x < 12·05$ (g) $81·35 \leqslant T < 81·45$ (h) $0·25 \leqslant M < 0·35$
     (i) $0·65 \leqslant m < 0·75$ (j) $51\,500 \leqslant n < 52\,500$
11. No, max. card length 11·55 cm, min. envelope length 11·5 cm

Exercise 12   *page 13*

1. (a) 7·5, 8·5, 10·5 cm   (b) 26·5 cm   **2.** 46·75 cm^2
3. (a) 7   (b) 5   (c) 10   (d) 4   (e) 2   (f) 5   (g) 2   (h) 24
4. (a) 13   (b) 11   (c) 3   (d) 12·5
5. (a) 10·5   (ii) 4·3   **6.** (i) 13, 11   (ii) 3, 1   (iii) 0·8, 0·6
7. 56 cm^2 **8.** 55·7

Exercise 13   *page 14*

1. 70·56   **2.** 118·958   **3.** 451·62   **4.** 33 678·8   **5.** 0·6174
6. 1068   **7.** 19·53   **8.** 18 914·4   **9.** 38·72   **10.** 0·009 79
11. 2·4   **12.** 11   **13.** 41   **14.** 8·9   **15.** 4·7
16. 56   **17.** 0·0201   **18.** 30·1   **19.** 1·3   **20.** 0·31
21. 210·21   **22.** 294   **23.** 282·131   **24.** 35   **25.** 242

Exercise 14   *page 15*

1. 4×10^3 **2.** 5×10^2 **3.** 7×10^4 **4.** 6×10 **5.** $2·4 \times 10^3$
6. $3·8 \times 10^2$ **7.** $4·6 \times 10^4$ **8.** $4·6 \times 10$ **9.** 9×10^5 **10.** $2·56 \times 10^3$
11. 7×10^{-3} **12.** 4×10^{-4} **13.** $3·5 \times 10^{-3}$ **14.** $4·21 \times 10^{-1}$ **15.** $5·5 \times 10^{-5}$
16. 1×10^{-2} **17.** $5·64 \times 10^5$ **18.** $1·9 \times 10^7$ **19.** $1·1 \times 10^9$ **20.** $1·67 \times 10^{-24}$
21. $5·1 \times 10^8$ **22.** $2·5 \times 10^{-10}$ **23.** $6·023 \times 10^{23}$ **24.** 3×10^{10} **25.** $\$3·6 \times 10^6$

Exercise 15   *page 16*

1. $1·5 \times 10^7$ **2.** 3×10^8 **3.** $2·8 \times 10^{-2}$ **4.** 7×10^{-9} **5.** 2×10^6
6. 4×10^{-6} **7.** 9×10^{-2} **8.** $6·6 \times 10^{-8}$ **9.** $3·5 \times 10^{-7}$ **10.** 1×10^{-16}
11. 8×10^9 **12.** $7·4 \times 10^{-7}$ **13.** c, a, b **14.** 13   **15.** 16
16. (i) $9 \times 10^2, 4 \times 10^2$ (ii) $1 \times 10^8, 4 \times 10^7$
17. 50 min   **18.** 6×10^2 **19.** (a) 20·5 s   (b) $6·3 \times 10^{91}$ years

Exercise 16   *page 17*

1. 1 : 3   **2.** 1 : 6   **3.** 1 : 50   **4.** 1 : 1·6   **5.** 1 : 0·75
6. 1 : 0·375   **7.** 1 : 25   **8.** 1 : 8   **9.** 2·4 : 1   **10.** 2·5 : 1
11. 0·8 : 1   **12.** 0·02 : 1   **13.** $15, $25   **14.** $36, $84   **15.** 140 m 110 m
16. $18, $27, $72   **17.** 15 kg, 75 kg, 90 kg   **18.** 46 min, 69 min, 69 min
19. £39   **20.** 18 kg, 36 kg, 54 kg, 72 kg   **21.** $400, $1000, $1000, $1600
22. 5 : 3   **23.** $200   **24.** 3 : 7   **25.** $\frac{1}{7}x$ **26.** 6
27. 12   **28.** $120   **29.** 300 g   **30.** 625

Exercise 17   *page 18*

1. $1·68   **2.** $84   **3.** 6 days   **4.** $2\frac{1}{2}$ litres   **5.** 60 km
6. 119 g   **7.** $68·40   **8.** $2\frac{1}{4}$ weeks   **9.** 80 c   **10.** (a) 12; (b) 2100
11. 4   **12.** 5·6 days   **13.** $175   **14.** 540°   **15.** $1·20
16. 190 m   **17.** 1250   **18.** 11·2 h   **19.** 57·1 min   **20.** 243 kg
21. 12 days

Exercise 18   *page 20*

1. (a) €18·60   (b) £44·10   (c) 758 pesos   (d) 69·75 rupees   (e) 209·76 yen   (f) 0·27 dinars
2. (a) $537·63   (b) $3968·25   (c) $0·16   (d) $3000   (e) $33·04   (f) $663·72
3. £3·39   **4.** Cheaper in UK by $9·23   **5.** €494·67
6. Kuwait £325·86, France £495, Japan £706·14   **7.** €433·63

Exercise 19   *page 21*

1. (a) 70 m   (b) 16 m   (c) 3·55 m   (d) 108·5 m
2. (a) 5 cm   (b) 3·5 cm   (c) 0·72 cm   (d) 2·86 cm
3. (a) 450 000 cm   (b) 4500 m   (c) 4·5 km
4. 12·3 km   **5.** 4·71 km   **6.** 50 cm   **7.** 64 cm   **8.** 5·25 cm

Exercise 20   *page 22*

1. 40 m by 30 m; 12 cm^2; 1200 m^2 **2.** 1 m^2, 6 m^2 **3.** 0·32 km^2 **4.** 50 cm^2
5. 150 km^2 **6.** 0·75 hectares   **7.** 240 cm^2 **8.** 1 : 50 000

Exercise 21   *page 23*

1. (a) $\frac{3}{5}$ (b) $\frac{6}{25}$ (c) $\frac{7}{20}$ (d) $\frac{1}{50}$
2. (a) 25%   (b) 10%   (c) $87\frac{1}{2}$%   (d) $33\frac{1}{3}$%   (e) 72%   (f) 31%
3. (a) 0·36   (b) 0·28   (c) 0·07   (d) 0·134   (e) 0·6   (f) 0·875
4. (a) 45%; $\frac{1}{2}$; 0·6   (b) 4%; $\frac{6}{16}$; 0·38   (c) 11%; 0·111; $\frac{1}{9}$ (d) 0·3; 32%; $\frac{1}{3}$
5. (a) 85%   (b) 77·5%   (c) 23·75%   (d) 56%   (e) 10%   (f) 37·5%

Exercise 22   *page 24*

1. (a) $15   (b) 900 kg   (c) $2·80   (d) 125
2. $32   **3.** 13·2c   **4.** 52·8 kg
5. (a) $1·02   (b) $21·58   (c) $2·22   (d) $0·53
6. $243·28   **7.** $26 182   **8.** 96·8%   **9.** 77·5%   **10.** $71·48
11. 200   **12.** 29 000   **13.** 500 cm   **14.** $6·30   **15.** 400 kg
16. 325   **17.** $35·25   **18.** $8425·60

Exercise 23   *page 27*

1. (a) 25%, profit   (b) 25%, profit   (c) 10%, loss   (d) 20%, profit
    (e) 30%, profit   (f) 7·5%, profit   (g) 12%, loss   (h) 54%, loss
2. 28%   **3.** $44\frac{4}{9}$%   **4.** 46·9%   **5.** 12%   **6.** $5\frac{1}{3}$%
7. (a) $50   (b) $450   (c) $800   (d) $12·40
8. $50   **9.** $12   **10.** $5   **11.** 60c   **12.** $2200
13. 14·3%   **14.** 20%   **15.** 8 : 11   **16.** 21%   **17.** 20%

Exercise 24   *page 28*

1. (a) $216   (b) $115·50   (c) 2 years   (d) 5 years
2. $2295, $9045   **3.** 7·5%

Exercise 25   *page 29*

1. (a) $2180   (b) $2376·20   (c) $2590·06   **2.** (a) $5550   (b) $6838·16   (c) $8425·29
3. $13 107·96   **4.** (a) $36 465·19   (b) $40 202·87
5. (a) $9540   (b) $107 19·10   (c) $161 17·60
6. (a) $14 033·01   (b) $734·03   (c) $107 946·25
7. $9211·88   **8.** 8 years   **9.** 12 years   **10.** 13 years   **11.** $30 000 at 8%

Exercise 26   *page 31*

1. (a) $2\frac{1}{2}$ h   (b) $3\frac{1}{8}$ h   (c) 75 s   (d) 4 h
2. (a) 20 m/s   (b) 30 m/s   (c) $83\frac{1}{3}$ m/s   (d) 108 km/h   (e) 79·2 km/h
   (f) 1·2 cm/s   (g) 90 m/s   (h) 25 mph   (i) 0·03 miles per second
3. (a) 75 km/h   (b) 4·52 km/h   (c) 7·6 m/s   (d) 4×10^6 m/s   (e) $2·5 \times 10^8$ m/s
   (f) 200 km/h   (g) 3 km/h
4. (a) 110 000 m   (b) 10 000 m   (c) 56 400 m   (d) 4500 m   (e) 50 400 m
   (f) 80 m   (g) 960 000 m
5. (a) 3·125 h   (b) 76·8 km/h   **6.** (a) 4·45 h   (b) 23·6 km/h
7. 46 km/h   **8.** (a) 8 m/s   (b) 7·6 m/s   (c) 102·63 s   (d) 7·79 m/s
9. 1230 km/h   **10.** 3 h   **11.** 100 s   **12.** $1\frac{1}{2}$ minutes   **13.** 600 m
14. $53\frac{1}{3}$ s   **15.** 5 cm/s   **16.** 60 s   **17.** 120 km/h

Exercise 27   *page 32*

1. $\frac{7}{25}$, 0·28, 28%; $\frac{16}{25}$, 0·64, 64%; $\frac{5}{8}$, 0·625, $62\frac{1}{2}$%   **2.** 12·4 m
3. 3·08 kg   **4.** $56\frac{1}{4}$ km   **5.** $820   **6.** A
7. **(a)** 19:17   (b) 23:49   **8.** $36   **9.** 1·32

Exercise 28   *page 33*

1. $1·08 \times 10^9$ km   **2.** 167 days   **3.** 3 h 21 min   **4.** 17
5. (a) $4·95   (b) 25   (c) $11·75   **6.** 30 g zinc, 2850 g copper, 3000 g total

Exercise 29   *page 34*

1. 2 : 1   **2.** (a) 9·85   (b) 76·2   (c) 223 512   (d) 1678·1
3. 252 000   **4.** (a) 8   (b) 24   (c) 8   (d) 8
5. 0·18 s   **6.** $140   **7.** 5   **8.** THIS IS A VERY SILLY CODE   **9.** 29

Exercise 30   *page 36*

1. 3·041   **2.** 1460   **3.** 0·030 83   **4.** 47·98   **5.** 130·6
6. 0·4771   **7.** 0·3658   **8.** 37·54   **9.** 8·000   **10.** 0·6537
11. 0·037 16   **12.** 34·31   **13.** 0·7195   **14.** 3·598   **15.** 0·2445
16. 2·043   **17.** 0·3798   **18.** 0·7683   **19.** −0·5407   **20.** 0·070 40
21. 2·526   **22.** 0·094 78   **23.** 0·2110   **24.** 3·123   **25.** 2·230
26. 128·8   **27.** 4·268   **28.** 3·893   **29.** 0·6290   **30.** 0·4069
31. 9·298   **32.** 0·1010   **33.** 0·3692   **34.** 1·125   **35.** 1·677
36. 0·9767   **37.** 0·8035   **38.** 0·3528   **39.** 2·423   **40.** 1·639
41. 0·000 465 9   **42.** 0·3934   **43.** −0·7526   **44.** 2·454

Exercise 31   *page 37*

1. 40 000   **2.** 0·070 49   **3.** 405 400   **4.** 471·3   **5.** 20 810
6. $2·218 \times 10^6$ **7.** $1·237 \times 10^{-24}$ **8.** 3·003   **9.** 0·035 81   **10.** 47·40
11. −1748   **12.** 0·011 38   **13.** 1757   **14.** 0·026 35   **15.** 0·1651
16. 5447   **17.** 0·006 562   **18.** 0·1330   **19.** 0·4451   **20.** 0·036 16
21. 19·43   **22.** $1·296 \times 10^{-15}$ **23.** $5·595 \times 10^{14}$ **24.** $1·022 \times 10^{-8}$ **25.** 0·019 22
26. 0·9613

Exercise 32   *page 38*

1. (a) 1850, 1850, 12·5 (b) 4592, 4592, 14 (c) 50·4, 50·4, 63 (d) 31·6, 31·6, 221·2 (e) 42·3, 42·3, 384·93
  (f) 39·51, 39·51, 13{\cdot}71 (g) 21·2, 21·2, 95·4 (h) 42·4, 42·4 (i) 6·2449 . . . , 6·2449 . . . (j) 29·63,
  29·63
2. (a) A–T, B–P, C–S, D–R, E–Q
3. (a) 281   (b) 36   (c) 101:16
4. $1000
5. 6 times
6. (a) 5   (b) 100   (c) £3000   (d) 1   (e) 0·2   (f) 2   (g) 100   (h) £2000   (i) 400

Revision exercise 1A   *page 42*

1. (a) 185   (b) 150   (c) 40   (d) $\frac{11}{12}$ (e) $2\frac{4}{5}$ (f) $\frac{2}{5}$

2. 128 cm   **3.** $\frac{2}{5}$ **4.** $\frac{a}{b}$ **5.** (a) 0·0547   (b) 0·055   (c) $5·473 \times 10^{-2}$
6. 1·238   **7.** (a) 3×10^9 (b) $3·7 \times 10^4$ (c) $2·7 \times 10^{13}$
8. (a) $26   (b) 6 : 5   (c) 6   **9.** $75
10. (a) (i) 57·2%   (ii) $87\frac{1}{2}$%   (b) 40%   (c) 80 c   **11.** 5%
12. (a) $500   (b) $37\frac{1}{2}$%   **13.** $357·88   **14.** 3·05
15. (a) 2·4 km   (b) 1 km^2 **16.** (a) 300 m   (b) 60 cm   (c) 150 cm^2
17. (a) 1 : 50 000   (b) 1 : 4 000 000   **18.** (a) 22%   (b) 20·8%   (c) $240
19. (a) (i) 7 m/s   (ii) 200 m/s   (iii) 5 m/s   (b) (i) 144 km/h   (ii) 2·16 km/h
20. (a) 0·005 m/s   (b) 1·6 s   (c) 172·8 km   **21.** $33\frac{1}{3}$ km/h
22. (a) 3   (b) 10   (c) 1, 9   (d) 1, 8   (e) $m = 3, n = 9$
  (f) $p = 1, q = 3, r = 9, s = 8, t = 10$ **23.** About 3   **24.** $2·3 \times 10^9$
25. (a) 600   (b) 10 000   (c) 3   (d) 20
26. (a) 0·5601   (b) 3·215   (c) 0·6161   (d) 0·4743
27. (a) 0·340   (b) $4·08 \times 10^{-6}$ (c) 64·9   (d) 0·119
28. 33·1%

Examination exercise 1B   *page 45*

1. (a) $\frac{25}{32}$ (b) 0·781   **2.** $\frac{59}{37}$ or $1\frac{22}{37}$ or 1·59 (459. . . .)
3. (a) any irrational square root, π or e   (b) 61 or 67   **4.** (a) 0641   (b) $204
5. (a) −1·8   (b) 21   **6.** (a) 1240   (b) $(n + 4)^2 + 1$
7. (a) 21·5, 22·5   (b) 172   **8.** 75 000   76 200   **9.** 50·1225
10. 6 222 5000   **11.** $5·7 \times 10^{26}$
12. (a) 350, 250, 200   (b) 275   (c) 200   (d) 11 : 8 : 4   (e) 110·25
13. (a) $6000   (b) 12·5%   **14.** 20   **15.** (a) $2300   (b) $8·64
16. (a) 5   (b) 1   **17.** (a) 950 kg   (b) $405   (c) $0·43 or $0·426   (d) (i) $0·21   (ii) $0·28
19. (a) (ii) 80 200   (b) (ii) 40 200   (iii) 40 000   (c) (i) $n(2n + 1)$ (ii) n^2
20. (a) (i) $346·50   (ii) $350   (b) (i) 115   (ii) $430   (iii) 4·88%   (c) 55

2 Algebra 1

Exercise 1   *page 52*

1. $5°$
2. $-4°$
3. $-1°$
4. $4°$
5. $-4°$
6. $12°$
7. $-7°$
8. $-5°$
9. $-4°$
10. $0°$
11. (a) C
    (b) B
12. $-17\,\text{m}$

Exercise 2   *page 53*

1. 13
2. 211
3. -12
4. -31
5. -66
6. $6·1$
7. $9·1$
8. -35
9. $18·7$
10. -9
11. -3
12. 3
13. -2
14. -14
15. -7
16. 3
17. 181
18. $-2·2$
19. $8·2$
20. 17
21. 2
22. -6
23. -15
24. -14
25. -2
26. -12
27. -80
28. $-13·1$
29. $-4·2$
30. $12·4$
31. -7
32. 8
33. 4
34. -10
35. 11
36. 4
37. -20
38. 8
39. -5
40. -10
41. -26
42. -21

43. 8
44. 1
45. $-20·2$
46. -50
47. -508
48. -29
49. 0
50. -21
51. $-0·1$
52. -4
53. $6·7$
54. 1
55. -850
56. 4
57. 6
58. -4
59. -12
60. -31

Exercise 3   *page 53*

1. -8
2. 28
3. 12
4. 24
5. 18
6. -35
7. 49
8. -12
9. -2
10. 9
11. -4
12. 4
13. -4
14. 8
15. 70
16. -7
17. $\frac{1}{4}$
18. $-\frac{3}{5}$
19. $-0·01$
20. $0·0002$
21. 121
22. 6
23. -600
24. -1
25. -20
26. $-2·6$
27. -700
28. 18
29. -1000
30. 640
31. -6
32. -42
33. $-0·4$
34. $-0·4$
35. -200
36. -35
37. -2
38. $\frac{1}{2}$
39. $-\frac{1}{4}$
40. -90

Exercise 4   *page 54*

1. -10
2. 1
3. 12
4. -28
5. -2
6. 16
7. -3
8. 14
9. -28
10. 4
11. $-\frac{1}{6}$
12. 9
13. -30
14. 24
15. -1
16. -2
17. -30
18. 7
19. 3
20. 16
21. 93
22. 2400
23. 10
24. 1
25. -4
26. 48
27. -1
28. 0
29. -8
30. 170
31. -3
32. 1
33. 1
34. 0
35. 15
36. 5
37. $-2·4$
38. -180
39. 5
40. -994
41. 2
42. -48
43. 60
44. $-2·5$
45. -32
46. 0
47. $-0·1$
48. -16
49. $-4·3$
50. $-\frac{1}{16}$

Exercise 5   *page 54*

1. 21
2. $1·62$
3. 396
4. 650
5. $63·8$
6. 9×10^{12}
7. 800
8. $ac + ab - a^2$
9. $r - p + q$
10. $802; 4n + 2$
11. $2n + 6$

Exercise 6   page 56

1. 7   **2.** 13   **3.** 13   **4.** 22   **5.** 1   **6.** -1 **7.** 18   **8.** -4
9. -3 **10.** 37   **11.** 0   **12.** -4 **13.** -7 **14.** -2 **15.** -3 **16.** -8
17. -30 **18.** 16   **19.** -10 **20.** 0   **21.** 7   **22.** -6 **23.** -2 **24.** -7
25. -5 **26.** 3   **27.** 4   **28.** -8 **29.** -2 **30.** 2   **31.** 0   **32.** 4
33. -4 **34.** -3 **35.** -9 **36.** 4

Exercise 7   page 57

1. 9   **2.** 27   **3.** 4   **4.** 16   **5.** 36   **6.** 18
7. 1   **8.** 6   **9.** 2   **10.** 8   **11.** -7 **12.** 15
13. -23 **14.** 3   **15.** 32   **16.** 36   **17.** 144   **18.** -8
19. -7 **20.** 13   **21.** 5   **22.** -16 **23.** 84   **24.** 17
25. 6   **26.** 0   **27.** -25 **28.** -5 **29.** 17   **30.** $-1\frac{1}{2}$
31. 19   **32.** 8   **33.** 19   **34.** 16   **35.** -16 **36.** 12
37. 36   **38.** -12 **39.** 2   **40.** 11   **41.** -23 **42.** -26
43. 5   **44.** 31   **45.** $4\frac{1}{2}$

Exercise 8   page 57

1. -20 **2.** 16   **3.** -42 **4.** -4 **5.** -90 **6.** -160
7. -2 **8.** -81 **9.** 4   **10.** 22   **11.** 14   **12.** 5 or -5
13. 1 or -1 **14.** $\sqrt{5}$ **15.** 4   **16.** $-6\frac{1}{2}$ **17.** 54   **18.** 25
19. 4 or -4 **20.** 312   **21.** 45   **22.** 22   **23.** 14   **24.** -36
25. -7 **26.** 1 or -1 **27.** 901   **28.** -30 **29.** -5 **30.** $7\frac{1}{2}$
31. -7 **32.** $-\frac{3}{13}$ **33.** 7   **34.** -2 **35.** 0   **36.** $-4\frac{1}{2}$
37. 6 or -6 **38.** 2 or -2 **39.** 26   **40.** -9 **41.** $3\frac{1}{4}$ **42.** $-\frac{5}{6}$
43. 4   **44.** $2\frac{2}{3}$ **45.** $3\frac{1}{4}$ **46.** $-2\frac{1}{6}$ **47.** -13 **48.** 12
49. $1\frac{1}{3}$ **50.** $-\frac{5}{36}$

Exercise 9   page 59

1. $3x + 11y$ **2.** $2a + 8b$ **3.** $3x + 2y$ **4.** $5x + 5$ **5.** $9 + x$
6. $3 - 9y$ **7.** $5x - 2y - x^2$ **8.** $2x^2 + 3x + 5$ **9.** $-10y$ **10.** $3a^2 + 2a$
11. $7 + 7a - 7a^2$ **12.** $5x$ **13.** $\frac{10}{a} - b$ **14.** $\frac{5}{x} - \frac{5}{y}$ **15.** $\frac{3m}{x}$
6. $\frac{1}{2} - \frac{2}{x}$ **17.** $\frac{5}{a} + 3b$ **18.** $-\frac{n}{4}$ **19.** $7x^2 - x^3$ **20.** $2x^2$
21. $x^2 + 5y^2$ **22.** $-12x^2 - 4y^2$ **23.** $5x - 11x^2$ **24.** $\frac{8}{x^2}$ **25.** $5x + 2$
26. $12x - 7$ **27.** $3x + 4$ **28.** $11 - 6x$ **29.** $-5x - 20$ **30.** $7x - 2x^2$
31. $3x^2 - 5x$ **32.** $x - 4$ **33.** $5x^2 + 14x$ **34.** $-4x^2 - 3x$ **35.** $5a + 8$
36. $a + 9$ **37.** $ab + 4a$ **38.** $y^2 + y$ **39.** $2x - 2$ **40.** $6x + 3$
41. $x - 4$ **42.** $7x + 5y$ **43.** $4x^2 - 11x$ **44.** $2x^2 + 14x$ **45.** $3y^2 - 4y + 1$
46. $12x + 12$ **47.** $4ab - 3a + 14b$ **48.** $2x - 4$

Exercise 10   page 59

1. $x^2 + 4x + 3$ **2.** $x^2 + 5x + 6$ **3.** $y^2 + 9y + 20$ **4.** $x^2 + x - 12$
5. $x^2 + 3x - 10$ **6.** $x^2 - 5x + 6$ **7.** $a^2 - 2a - 35$ **8.** $z^2 + 7z - 18$

9. $x^2 - 9$ **10.** $k^2 - 121$ **11.** $2x^2 - 5x - 3$ **12.** $3x^2 - 2x - 8$

13. $2y^2 - y - 3$ **14.** $49y^2 - 1$ **15.** $9x^2 - 4$ **16.** $6a^2 + 5ab + b^2$

17. $3x^2 + 7xy + 2y^2$ **18.** $6b^2 + bc - c^2$ **19.** $-5x^2 + 16xy - 3y^2$ **20.** $15b^2 + ab - 2a^2$

21. $2x^2 + 2x - 4$ **22.** $6x^2 + 3x - 9$ **23.** $24y^2 + 4y - 8$ **24.** $6x^2 - 10x - 4$

25. $4a^2 - 16b^2$ **26.** $x^3 - 3x^2 + 2x$ **27.** $8x^3 - 2x$ **28.** $3y^3 + 3y^2 - 18y$

29. $x^3 + x^2y + x^2z + xyz$ **30.** $3za^2 + 3zam - 6zm^2$

Exercise 11   *page 60*

1. $x^2 + 8x + 16$ **2.** $x^2 + 4x + 4$ **3.** $x^2 - 4x + 4$ **4.** $4x^2 + 4x + 1$

5. $y^2 - 10y + 25$ **6.** $9y^2 + 6y + 1$ **7.** $x^2 + 2xy + y^2$ **8.** $4x^2 + 4xy + y^2$

9. $a^2 - 2ab + b^2$ **10.** $4a^2 - 12ab + 9b^2$ **11.** $3x^2 + 12x + 12$ **12.** $9 - 6x + x^2$

13. $9x^2 + 12x + 4$ **14.** $a^2 - 4ab + 4b^2$ **15.** $2x^2 + 6x + 5$ **16.** $2x^2 + 2x + 13$

17. $5x^2 + 8x + 5$ **18.** $2y^2 - 14y + 25$ **19.** $10x - 5$ **20.** $-8x + 8$

21. $-10y + 5$ **22.** $3x^2 - 2x - 8$ **23.** $2x^2 + 4x - 4$ **24.** $-x^2 - 18x + 15$

Exercise 12   *page 61*

1. 8   **2.** 9   **3.** 7   **4.** 10   **5.** $\frac{1}{3}$

6. 10   **7.** $1\frac{1}{2}$ **8.** -1 **9.** $-1\frac{1}{2}$ **10.** $\frac{1}{2}$

11. 35   **12.** 130   **13.** 14   **14.** $\frac{2}{3}$ **15.** $3\frac{1}{3}$

16. $-2\frac{1}{2}$ **17.** 3   **18.** $1\frac{1}{8}$ **19.** $\frac{3}{10}$ **20.** $-1\frac{1}{4}$

21. 10   **22.** 27   **23.** 20   **24.** 18   **25.** 28

26. -15 **27.** $\frac{99}{100}$ **28.** 0   **29.** 1000   **30.** $-\frac{1}{1000}$

31. 1   **32.** -7 **33.** -5 **34.** $1\frac{1}{6}$ **35.** 1

36. 2   **37.** -5 **38.** -3 **39.** $-1\frac{1}{2}$ **40.** 2

41. 1   **42.** $3\frac{1}{2}$ **43.** 2   **44.** -1 **45.** $10\frac{2}{3}$

46. $1\cdot1$ **47.** -1 **48.** 2   **49.** $2\frac{1}{2}$ **50.** $1\frac{1}{3}$

Exercise 13   *page 62*

1. $-1\frac{1}{2}$ **2.** 2   **3.** $-\frac{2}{5}$ **4.** $-\frac{1}{3}$ **5.** $1\frac{2}{3}$ **6.** 6

7. $-\frac{2}{5}$ **8.** $-3\frac{1}{5}$ **9.** $\frac{1}{2}$ **10.** -4 **11.** 18   **12.** 5

13. 4   **14.** 3   **15.** $2\frac{3}{4}$ **16.** $-\frac{7}{22}$ **17.** $\frac{1}{4}$ **18.** 1

19. 4   **20.** -11 **21.** $-7\frac{1}{3}$ **22.** $1\frac{1}{4}$ **23.** -5 **24.** 6

25. 3   **26.** 6   **27.** 2   **28.** 3   **29.** 4   **30.** 3

31. $10\frac{1}{2}$ **32.** 5   **33.** 2   **34.** -1 **35.** -17 **36.** $-2\frac{9}{10}$

37. $2\frac{10}{21}$ **38.** $\frac{1}{3}$ **39.** 14   **40.** 15

Exercise 14   *page 62*

1. $\frac{1}{4}$ **2.** -3 **3.** 4   **4.** $-7\frac{2}{3}$ **5.** -43

6. 11   **7.** $-\frac{1}{2}$ **8.** 0   **9.** 1   **10.** $-1\frac{2}{3}$

11. $\frac{1}{4}$ **12.** 0   **13.** $-\frac{6}{7}$ **14.** $1\frac{9}{17}$ **15.** $1\frac{22}{23}$

16. $\frac{2}{11}$ **17.** 4 cm   **18.** 5 m   **19.** 4

Exercise 15   *page 64*

1. $\frac{1}{3}$
2. $\frac{1}{5}$
3. $1\frac{2}{3}$
4. -3
5. $\frac{5}{11}$
6. -2
7. 6
8. $3\frac{3}{4}$
9. -7
10. $-7\frac{2}{3}$
11. 2
12. 3
13. 4
14. -2
15. -3
16. 3
17. $1\frac{5}{7}$
18. $4\frac{4}{5}$
19. 10
20. 24
21. 2
22. 3
23. 5
24. -4
25. $6\frac{3}{4}$
26. -3
27. 0
28. 3
29. 0
30. 1
31. 2
32. 3
33. 4
34. $\frac{3}{5}$
35. $1\frac{1}{8}$
36. -1
37. 1
38. 1
39. $\frac{1}{4}$
40. $-\frac{1}{3}$
41. $\frac{9}{10}$
42. 1
43. 2
44. $-\frac{1}{7}$
45. 2
46. 3

Exercise 16   *page 66*

1. $91, 92, 93$
2. $21, 22, 23, 24$
3. $57, 59, 61$
4. $506, 508, 510$
5. $12\frac{1}{2}$
6. $12\frac{1}{2}$
7. $11\frac{2}{3}$
8. $8\frac{1}{3}, 41\frac{2}{3}$
9. $1\frac{1}{4}, 13\frac{3}{4}$
10. $3\frac{1}{3}$ cm
11. 12 cm
12. 20
13. 5 cm
14. 7 cm
15. $18\frac{1}{2}, 27\frac{1}{2}$
16. $20°, 60°, 100°$
17. $45°, 60°, 75°$
18. 5
19. $6, 8$
20. $12, 24, 30$
21. $5, 15, 8$
22. $59\frac{2}{3}$ kg, $64\frac{2}{3}$ kg, $72\frac{2}{3}$ kg
23. $24, 22, 15$
24. $48, 12$
25. $40, 8$
26. 6
27. $168 \cdot 84 \text{ cm}^2$
28. 14
29. $\$45, \$31$
30. $\$21 \cdot 50$

Exercise 17   *page 69*

1. $\$3700$
2. 3
3. $1\frac{3}{7}$ m
4. $80°, 100°$
5. $30°, 60°, 90°, 120°, 150°, 270°$
6. $26, 58$
7. 2 km
8. 8 km
9. 400 m
10. 21
11. 23
12. $\$3600$
13. 15
14. 2 km
15. $7, 8, 9$
16. $2, 3, 4, 5$

Exercise 18   *page 71*

1. $x = 2, y = 1$
2. $x = 4, y = 2$
3. $x = 3, y = 1$
4. $x = -2, y = 1$
5. $x = 3, y = 2$
6. $x = 5, y = -2$
7. $x = 2, y = 1$
8. $x = 5, y = 3$
9. $x = 3, y = -1$
10. $a = 2, b = -3$
11. $a = 5, b = \frac{1}{4}$
12. $a = 1, b = 3$
13. $m = \frac{1}{2}, n = 4$
14. $w = 2, x = 3$
15. $x = 6, y = 3$
16. $x = \frac{1}{2}, z = -3$
17. $m = 1\frac{15}{17}, n = \frac{11}{17}$
18. $c = 1\frac{16}{23}, d = -2\frac{12}{23}$

Exercise 19   *page 71*

1. 1
2. -3
3. 2
4. 15
5. -12
6. -3
7. -2
8. -11
9. -21
10. 1
11. 0
12. 15
13. -10
14. 3
15. 6
16. -11
17. 2
18. 5
19. -19
20. -4
21. x
22. $-3x$
23. $4x$
24. $4y$
25. $9y$
26. $3x$
27. $-8x$
28. $4x$
29. $2x$
30. $3y$

Exercise 20   *page 78*

1. $x = 2, y = 4$
2. $x = 1, y = 4$
3. $x = 2, y = 5$
4. $x = 3, y = 7$
5. $x = 5, y = 2$
6. $a = 3, b = 1$
7. $x = 1, y = 3$
8. $x = 1, y = 3$
9. $x = -2, y = 3$
10. $x = 4, y = 1$
11. $x = 1, y = 5$
12. $x = 0, y = 2$
13. $x = \frac{5}{7}, y = 4\frac{3}{7}$
14. $x = 1, y = 2$
15. $x = 2, y = -3$
16. $x = 4, y = -1$
17. $x = 3, y = 1$
18. $x = 1, y = 2$
19. $x = 2, y = 1$
20. $x = -2, y = 1$
21. $x = 1, y = 2$
22. $a = 4, b = 3$
23. $x = -23, y = -78$
24. $x = 3, y = \frac{1}{2}$

25. $x = 4$, $y = 3$

26. $x = 5$, $y = -2$

27. $x = \frac{1}{3}$, $y = -2$

28. $x = 5\frac{5}{14}$, $y = \frac{2}{7}$

29. $x = 3$, $y = -1$

30. $x = 5$, $y = 0\cdot2$

Exercise 21   page 74

1. $5\frac{1}{2}$, $9\frac{1}{2}$

2. 6, 3 or $2\frac{2}{5}$, $5\frac{2}{5}$

3. 4, 10

4. $a = 2$, $c = 7$

5. $m = 4$, $c = -3$

6. $a = 1$, $b = -2$

7. $m = 1$c, $w = 3$c

8. TV \$200, video \$450

9. 7, 3

10. white 2g, brown $3\frac{1}{2}$g

11. 120 cm, 240 cm

12. 150 m 350 m

13. 2c × 15, 5c × 25

14. 10c × 14, 50c × 7

15. 20

16. man \$500, woman \$700

17. current 4 m/s, kipper 10 m/s

18. $\frac{5}{7}$

19. $\frac{3}{5}$

20. boy 10, mouse 3

21. 4, 7

22. $y = 3x - 2$

23. walks 4 m/s, runs 5 m/s

24. \$1×15, \$5×5

25. 36, 9

26. wind $4\frac{1}{2}$ knots, submarine $20\frac{1}{2}$ knots

27. $a = 1$, $b = 2$, $c = 5$

28. $y = 2x^2 - 3x + 5$

29. $y = x^2 + 3x + 4$

30. $y = x^2 + 2x - 3$

Exercise 22   page 76

1. $5(a + b)$

2. $7(x + y)$

3. $x(7 + x)$

4. $y(y + 8)$

5. $y(2y + 3)$

6. $2y(3y - 2)$

7. $3x(x - 7)$

8. $2a(8 - a)$

9. $3c(2c - 7)$

10. $3x(5 - 3x)$

11. $7y(8 - 3y)$

12. $x(a + b + 2c)$

13. $x(x + y + 3z)$

14. $y(x^2 + y^2 + z^2)$

15. $ab(3a + 2b)$

16. $xy(x + y)$

17. $2a(3a + 2b + c)$

18. $m(a + 2b + m)$

19. $2k(x + 3y + 2z)$

20. $a(x^2 + y + 2b)$

21. $xk(x + k)$

22. $ab(a^2 + 2b)$

23. $bc(a - 3b)$

24. $ae(2a - 5e)$

25. $ab(a^2 + b^2)$

26. $x^2y(x + y)$

27. $2xy(3y - 2x)$

28. $3ab(b^2 - a^2)$

29. $a^2b(2a + 5b)$

30. $ax^2(y - 2z)$

31. $2ab(x + b + a)$

32. $yx(a + x^2 - 2yx)$

Exercise 23   page 77

1. $(a + b)(x + y)$

2. $(a + b)(y + z)$

3. $(x + y)(b + c)$

4. $(x + y)(h + k)$

5. $(x + y)(m + n)$

6. $(a + b)(h - k)$

7. $(a + b)(x - y)$

8. $(m + n)(a - b)$

9. $(h + k)(s + t)$

10. $(x + y)(s - t)$

11. $(a - b)(x - y)$

12. $(x - y)(s - t)$

13. $(a - x)(s - y)$

14. $(h - b)(x - y)$

15. $(m - n)(a - b)$

16. $(x - z)(k - m)$

17. $(2a + b)(x + 3y)$

18. $(2a + b)(x + y)$

19. $(2m + n)(h - k)$

20. $(m - n)(2h + 3k)$

21. $(2x + y)(3a + b)$

22. $(2a - b)(x - y)$

23. $(x^2 + y)(a + b)$

24. $(m - n)(s + 2t^2)$

Exercise 24   page 77

1. $(x + 2)(x + 5)$

2. $(x + 3)(x + 4)$

3. $(x + 3)(x + 5)$

4. $(x + 3)(x + 7)$

5. $(x + 2)(x + 6)$

6. $(y + 5)(y + 7)$

7. $(y + 3)(y + 8)$

8. $(y + 5)(y + 5)$

9. $(y + 3)(y + 12)$

10. $(a + 2)(a - 5)$

11. $(a + 3)(a - 4)$

12. $(z + 3)(z - 2)$

13. $(x + 5)(x - 7)$

14. $(x + 3)(x - 8)$

15. $(x - 2)(x - 4)$

16. $(y - 2)(y - 3)$

17. $(x - 3)(x - 5)$

18. $(a + 2)(a - 3)$

19. $(a + 5)(a + 9)$

20. $(b + 3)(b - 7)$

21. $(x - 4)(x - 4)$

22. $(y + 1)(y + 1)$

23. $(y - 7)(y + 4)$

24. $(x - 5)(x + 4)$

25. $(x - 20)(x + 12)$

26. $(x - 15)(x - 11)$

27. $(y + 12)(y - 9)$

28. $(x - 7)(x + 7)$

29. $(x - 3)(x + 3)$

30. $(x - 4)(x + 4)$

Exercise 25   page 78

1. $(2x + 3)(x + 1)$

2. $(2x + 1)(x + 3)$

3. $(3x + 1)(x + 2)$

4. $(2x + 3)(x + 4)$

5. $(3x + 2)(x + 2)$

6. $(2x + 5)(x + 1)$

7. $(3x + 1)(x - 2)$

8. $(2x + 5)(x - 3)$

9. $(2x + 7)(x - 3)$

10. $(3x + 4)(x - 7)$

11. $(2x + 1)(3x + 2)$

12. $(3x + 2)(4x + 5)$

13. $(3x - 2)(x - 3)$ **14.** $(y - 2)(3y - 5)$ **15.** $(4y - 3)(y - 5)$ **16.** $(2y + 3)(3y - 1)$
17. $(2x - 5)(3x - 6)$ **18.** $(5x + 2)(2x + 1)$ **19.** $(6x - 1)(x - 3)$ **20.** $(4x + 1)(2x - 3)$
21. $(6x + 5)(2x - 1)$ **22.** $(16x + 3)(x + 1)$ **23.** $(2a - 1)(2a - 1)$ **24.** $(x + 2)(12x - 7)$
25. $(x + 3)(15x - 1)$ **26.** $(8x + 1)(6x + 5)$ **27.** $(16y - 3)(4y + 1)$ **28.** $(15x - 1)(8x + 5)$
29. $(3x - 1)(3x + 1)$ **30.** $(2a - 3)(2a + 3)$

Exercise 26   *page 78*

1. $(y - a)(y + a)$ **2.** $(m - n)(m + n)$ **3.** $(x - t)(x + t)$ **4.** $(y - 1)(y + 1)$
5. $(x - 3)(x + 3)$ **6.** $(a - 5)(a + 5)$ **7.** $(x - \frac{1}{2})(x + \frac{1}{2})$ **8.** $(x - \frac{1}{3})(x + \frac{1}{3})$
9. $(2x - y)(2x + y)$ **10.** $(a - 2b)(a + 2b)$ **11.** $(5x - 2y)(5x + 2y)$ **12.** $(3x - 4y)(3x + 4y)$

13. $\left(x - \dfrac{y}{2}\right)\left(x + \dfrac{y}{2}\right)$ **14.** $(3m - \frac{2}{3}n)(3m + \frac{2}{3}n)$ **15.** $(4t - \frac{2}{5}s)(4t + \frac{2}{5}s)$ **16.** $\left(2x - \dfrac{z}{10}\right)\left(2x + \dfrac{z}{10}\right)$

17. $x(x - 1)(x + 1)$ **18.** $a(a - b)(a + b)$ **19.** $x(2x - 1)(2x + 1)$
20. $2x(2x - y)(2x + y)$ **21.** $3x(2x - y)(2x + y)$ **22.** $2m(3m - 2n)(3m + 2n)$
23. $5(x - \frac{1}{2})(x + \frac{1}{2})$ **24.** $2a(5a - 3b)(5a + 3b)$ **25.** $3y(2x - z)(2x + z)$
26. $4ab(3a - b)(3a + b)$ **27.** $2a^3(5a - 2b)(5a + 2b)$ **28.** $9xy(2x - 5y)(2x + 5y)$
29. 161   **30.** 404   **31.** 4400   **32.** 2421   **33.** 4329
34. 0·75   **35.** 4·8   **36.** −2469   **37.** 0·0761
38. −10 900   **39.** 53·6   **40.** 0·000 005

Exercise 27   *page 79*

1. $-3, -4$ **2.** $-2, -5$ **3.** $3, -5$ **4.** $2, -3$ **5.** $2, 6$
6. $-3, -7$ **7.** $6, -1$ **8.** $5, -1$ **9.** $-7, 2$ **10.** $-\frac{1}{2}, 2$
11. $\frac{2}{3}, -4$ **12.** $1\frac{1}{2}, -5$ **13.** $\frac{2}{3}, 1\frac{1}{2}$ **14.** $\frac{1}{4}, 7$ **15.** $\frac{3}{5}, -\frac{1}{2}$
16. $7, 8$ **17.** $\frac{5}{6}, \frac{1}{2}$ **18.** $7, -9$ **19.** $-1, -1$ **20.** $3, 3$
21. $-5, -5$ **22.** $7, 7$ **23.** $-\frac{1}{3}, \frac{1}{2}$ **24.** $-1\frac{1}{4}, 2$ **25.** $13, -5$
26. $-3, \frac{1}{6}$ **27.** $\frac{1}{10}, -2$ **28.** $1, 1$ **29.** $\frac{2}{9}, -\frac{1}{4}$ **30.** $-\frac{1}{4}, \frac{3}{5}$

Exercise 28   *page 80*

1. $0, 3$ **2.** $0, -7$ **3.** $0, 1$ **4.** $0, \frac{1}{3}$ **5.** $4, -4$
6. $7, -7$ **7.** $\frac{1}{2}, -\frac{1}{2}$ **8.** $\frac{2}{3}, -\frac{2}{3}$ **9.** $0, -1\frac{1}{2}$ **10.** $0, 1\frac{1}{2}$
11. $0, 5\frac{1}{2}$ **12.** $\frac{1}{4}, -\frac{1}{4}$ **13.** $\frac{1}{2}, -\frac{1}{2}$ **14.** $0, \frac{5}{8}$ **15.** $0, \frac{1}{12}$
16. $0, 6$ **17.** $0, 11$ **18.** $0, 1\frac{1}{2}$ **19.** $0, 1$ **20.** $0, 4$
21. $0, 3$ **22.** $\frac{1}{2}, -\frac{1}{2}$ **23.** $1\frac{1}{3}, -1\frac{1}{3}$ **24.** $3, -3$ **25.** $0, 2\frac{2}{5}$
26. $\frac{1}{3}, -\frac{1}{3}$ **27.** $0, \frac{1}{4}$ **28.** $0, \frac{1}{6}$ **29.** $\frac{1}{4}, -\frac{1}{4}$ **30.** $0, \frac{1}{5}$

Exercise 29   *page 81*

1. $-\frac{1}{2}, -5$ **2.** $-\frac{2}{3}, -3$ **3.** $-\frac{1}{2}, -\frac{2}{3}$ **4.** $\frac{1}{3}, 3$
5. $\frac{2}{5}, 1$ **6.** $\frac{1}{3}, 1\frac{1}{2}$ **7.** $-0·63, -2·37$ **8.** $-0·27, -3·73$
9. $0·72, 0·28$ **10.** $6·70, 0·30$ **11.** $0·19, -2·69$ **12.** $0·85, -1·18$
13. $0·61, -3·28$ **14.** $-1\frac{2}{3}, 4$ **15.** $-1\frac{1}{2}, 5$ **16.** $3·56, -0·56$
17. $0·16, -3·16$ **18.** $-\frac{1}{2}, 2\frac{1}{3}$ **19.** $-\frac{1}{3}, -8$ **20.** $1\frac{2}{3}, -1$
21. $2·28, 0·22$ **22.** $-0·35, -5·65$ **23.** $-\frac{2}{3}, \frac{1}{2}$ **24.** $-0·58, 2·58$
25. $-2·69, 0·19$ **26.** $0·22, -1·55$ **27.** $-0·37, 5·37$ **28.** $-\frac{5}{6}, 1\frac{3}{4}$

29. $-\frac{7}{9}$, $1\frac{1}{4}$ **30.** $1\frac{2}{5}$, $-2\frac{1}{4}$ **31.** -4, $1\frac{1}{2}$ **32.** -3, $1\frac{2}{3}$
33. -2, $1\frac{2}{3}$ **34.** $-3\frac{1}{2}$, $\frac{1}{5}$ **35.** -3, $\frac{4}{5}$ **36.** $-8\frac{1}{2}$, 11

Exercise 30 *page 82*

1. -3, 2 **2.** -3, -7 **3.** $-\frac{1}{2}$, 2 **4.** 1, 4 **5.** $-1\frac{2}{3}$, $\frac{1}{2}$
6. $-0\cdot39$, $-4\cdot28$ **7.** $-0\cdot16$, $6\cdot16$ **8.** 3 **9.** 2, $-1\frac{1}{3}$ **10.** -3, -1
11. $0\cdot66$, $-22\cdot66$ **12.** -7, 2 **13.** $\frac{1}{4}$, 7 **14.** $-\frac{1}{2}$, $\frac{3}{5}$ **15.** 0, $3\frac{1}{2}$
16. $-\frac{1}{4}$, $\frac{1}{4}$ **17.** $-2\cdot77$, $1\cdot27$ **18.** $-\frac{2}{3}$, 1 **19.** $-\frac{1}{2}$, 2 **20.** 0, 3
21. (a) -1 (b) $0\cdot6258$ (c) $0\cdot5961$ (d) $0\cdot2210$

Exercise 31 *page 83*

1. $(x+4)^2 - 16$ **2.** $(x-6)^2 - 36$ **3.** $(x+\frac{1}{2}) - \frac{1}{4}$ **4.** $(x+2)^2 - 3$ **5.** $(x-3)^2$
6. $(x+1)^2 - 16$ **7.** $2(x+4)^2 - 27$ **8.** $2(x-2\cdot5)^2 - 6\cdot25$ **9.** $10 - (x-2)^2$ **10.** $4 - (x+1)^2$
11. (a) $x = \pm\sqrt{7} - 2$ (b) $x = 1\cdot5 \pm \sqrt{4\cdot25}$ (c) $x = \pm\sqrt{37} - 6$ **12.** $(x+3)^2 + 3$ requires finding $\sqrt{-1}$
13. $f(x) = (x+3)^2 + 3$ **14.** $g(x) = (x-3\cdot5)^2 - 12$ **15.** (a) 3 (b) $x = -2$ (c) $\frac{1}{3}$

Exercise 32 *page 85*

1. 8, 11 **2.** 11, 13 **3.** $12\,\text{cm}$ **4.** $6\,\text{cm}$
5. $x = 11$ **6.** $10\,\text{cm} \times 24\,\text{cm}$ **7.** $8\,\text{km}$ north, $15\,\text{km}$ east   **8.** $12\,\text{eggs}$
9. $13\,\text{eggs}$ **10.** 4 or -1 **11.** 2, 5 **12.** $\dfrac{40}{x}$ h, $\dfrac{40}{x-2}$ h, $10\,\text{km/h}$
13. $4\,\text{km/h}$ **14.** $60\,\text{km/h}$ **15.** $5\,\text{km/h}$ **16.** $157\,\text{km}$ **17.** $x = 2$
18. $x = 3$ or $9\cdot5$ **19.** $\frac{3}{4}$ **20.** $9\,\text{cm}$ or $13\,\text{cm}$

Revision exercise 2A *page 87*

1. (a) $-2\frac{1}{2}$ (b) $2\frac{2}{3}$ (c) 0, -5 (d) 2, -2 (e) -5, $2\frac{2}{3}$
2. (a) 14 (b) 18 (c) 28
3. (a) $(2x-y)(2x+y)$ (b) $2(x+3)(x+1)$ (c) $(2-3k)(3m+2n)$ (d) $(2x+1)(x-3)$
4. (a) $x = 3$, $y = -2$ (b) $m = 1\frac{1}{2}$, $n = -3$ (c) $x = 7$, $y = \frac{1}{2}$ (d) $x = -1$, $y = -2$
5. (a) 8 (b) 140 (c) 29 (d) 42 (e) 6 (f) -6
6. (a) $2x - 21$ (b) $(1-2x)(2a-3b)$ (c) 23
7. (a) 1 (b) $10\frac{1}{2}$ (c) 0, $3\frac{1}{2}$ (d) -3, -2 (e) 12
8. (a) $z(z-4)(z+4)$ (b) $(x^2+1)(y^2+1)$ (c) $(2x+3)(x+4)$ **9.** $\frac{7}{8}$
10. (a) $c = 5$, $d = -2$ (b) $x = 2$, $y = -1$ (c) $x = 9$, $y = -14$ (d) $s = 5$, $t = -3$
11. (a) $\frac{1}{2}$, $-\frac{1}{2}$ (b) $\frac{7}{11}$ (c) 3 (d) 0, 5
12. (a) $1\cdot78$, $-0\cdot28$ (b) $1\cdot62$, $-0\cdot62$ (c) $0\cdot87$, $-1\cdot54$ (d) $1\cdot54$, $-4\cdot54$
13. (a) $x = 9$ (b) $x = 10$
14. (a) 2 (b) -3 (c) 36 (d) 0 (e) 36 (f) 4
15. speed $= 5\,\text{km/h}$ **16.** $8\,\text{cm} \times 6\cdot5\,\text{cm}$ **17.** (a) -2, 4 (b) 16 (c) $6\cdot19$, $0\cdot81$
18. $-\frac{1}{5}$, 3 **19.** 8 **20.** $x = 13$ **21.** 21
22. 18 **23.** $6\,\text{cm}$ **24.** -4

Examination exercise 2B *page 89*

1. (a) 3 (b) 8 **2.** $(1, 3)$
3. (a) $13\cdot5$ (b) -1 and 4

4. $x = 10$, $y = 3$ **5.** $p = 2$, $q = -12$

6. (a) (i) $4x(x + 4)$ (b) $1 \cdot 1$

7. (a) (i) $(x + 4)(x - 5)$ (ii) $-4, 5$ (b) $-0 \cdot 55, 1 \cdot 22$

  (c) (i) $(m - 2n)(m + 2n)$ (ii) -12 (iii) $y = 20x + 5$ (iv) $n = \sqrt{\dfrac{m^2 - y}{4}}$ (d) (i) ± 4

    (ii) $n(m - 2n)(m + 2n)(m^2 + 4n^2)$

8. (a) (ii) $-9, 4$ (iii) $7 \cdot 81$ (b) (iii) $y = 4 \cdot 5$ (iv) $2 \cdot 55 \, \text{km/h}$

3 Mensuration

Exercise 1   *page 93*

1. $10 \cdot 2 \, \text{m}^2$ **2.** $22 \, \text{cm}^2$ **3.** $103 \, \text{m}^2$ **4.** $9 \, \text{cm}^2$
5. $31 \, \text{m}^2$ **6.** $6000 \, \text{cm}^2$ or $0 \cdot 6 \, \text{m}^2$ **7.** $26 \, \text{m}^2$ **8.** $18 \, \text{cm}^2$
9. $20 \, \text{cm}^2$ **10.** $13 \, \text{m}$ **11.** $15 \, \text{cm}$ **12.** $56 \, \text{m}$
13. $8 \, \text{m}, 10 \, \text{m}$ **14.** $12 \, \text{cm}$ **15.** 2500 **16.** 6 square units
17. 14 square units   **18.** 1849 **20.** $1100 \, \text{m}$

Exercise 2   *page 95*

1. $48 \cdot 3 \, \text{cm}^2$ **2.** $28 \cdot 4 \, \text{cm}^2$ **3.** $66 \cdot 4 \, \text{m}^2$ **4.** $3 \cdot 1 \, \text{cm}^2$
5. $18 \cdot 2 \, \text{cm}^2$ **6.** $12 \cdot 3 \, \text{cm}^2$ **7.** $2 \cdot 78 \, \text{cm}^2$ **8.** $36 \cdot 4 \, \text{m}^2$
9. $62 \cdot 4 \, \text{m}^2$ **10.** $30 \cdot 4 \, \text{m}^2$ **11.** $44 \cdot 9 \, \text{cm}^2$ **12.** $0 \cdot 28 \, \text{m}^2$
13. $63 \, \text{m}^2$ **14.** $70 \cdot 7 \, \text{m}^2$ **15.** $14 \, \text{m}^2$ **16.** $65 \cdot 8 \, \text{cm}^2$

17. $18 \cdot 1 \, \text{cm}^2$ **18.** $8 \cdot 0 \, \text{m}^2$ **19.** $14 \, \text{m}^2$ **20.** $52 \cdot 0 \, \text{cm}^2$
21. $124 \, \text{cm}^2$ **22.** $69 \cdot 8 \, \text{m}^2$ **23.** $57 \cdot 1 \, \text{cm}^2$ **24.** $10 \cdot 7 \, \text{cm}$
25. $50 \cdot 9°$ **26.** $4 \cdot 10 \, \text{m}$ **27.** $4 \cdot 85 \, \text{m}$ **28.** $7 \cdot 23 \, \text{cm}$
29. $60°; 23 \cdot 4 \, \text{cm}^2$ **30.** 292 **31.** $110 \, \text{cm}^2$
32. (a) $\dfrac{360°}{n}$ (b) $\dfrac{n}{2} \sin \dfrac{360°}{n}$ (c) $2 \cdot 6, 2 \cdot 94, 3 \cdot 1414, 3 \cdot 1416, 3 \cdot 1416$ as n increases, $A \rightarrow \pi$ **33.** $18 \cdot 7 \, \text{cm}$

Exercise 3   *page 98*

1. (a) $31 \cdot 4 \, \text{cm}$ (b) $78 \cdot 5 \, \text{cm}^2$ **2.** (a) $18 \cdot 8 \, \text{cm}$ (b) $28 \cdot 3 \, \text{cm}^2$
3. (a) $51 \cdot 4 \, \text{cm}$ (b) $157 \, \text{cm}^2$ **4.** (a) $26 \cdot 6 \, \text{cm}$ (b) $49 \cdot 1 \, \text{cm}^2$
5. (a) $26 \cdot 3 \, \text{cm}$ (b) $33 \cdot 3 \, \text{cm}^2$ **6.** (a) $25 \cdot 0 \, \text{cm}$ (b) $38 \cdot 5 \, \text{cm}^2$
7. (a) $35 \cdot 7 \, \text{cm}$ (b) $21 \cdot 5 \, \text{cm}^2$ **8.** (a) $50 \cdot 3 \, \text{cm}$ (b) $174 \, \text{cm}^2$
9. (a) $22 \cdot 0 \, \text{cm}$ (b) $10 \cdot 5 \, \text{cm}^2$ **10.** (a) $9 \cdot 42 \, \text{cm}$ (b) $6 \cdot 44 \, \text{cm}^2$
11. (a) $25 \cdot 1 \, \text{cm}$ (b) $25 \cdot 1 \, \text{cm}^2$ **12.** (a) $18 \cdot 8 \, \text{cm}$ (b) $12 \cdot 6 \, \text{cm}^2$

Exercise 4   *page 99*

1. $2 \cdot 19 \, \text{cm}$ **2.** $30 \cdot 2 \, \text{m}$ **3.** $2 \cdot 65 \, \text{km}$ **4.** $9 \cdot 33 \, \text{cm}$
5. $14 \cdot 2 \, \text{mm}$ **6.** $497\,000 \, \text{km}^2$ **7.** $21 \cdot 5 \, \text{cm}^2$
8. (a) $40 \cdot 8 \, \text{m}^2$ (b) 6 **9.** (a) $30;$ (b) $1508 \, \text{cm}^2$ (c) $508 \, \text{cm}^2$
10. 5305 **11.** 29 **12.** 970 **13.** (a) 80 (b) 7
14. $5 \cdot 39 \, \text{cm} \, (\sqrt{39})$ **15.** (a) $33 \cdot 0 \, \text{cm}$ (b) $70 \cdot 9 \, \text{cm}^2$ **16.** (a) $98 \, \text{cm}^2$ (b) $14 \cdot 0 \, \text{cm}^2$
17. $1 : 3 : 5$ **18.** $796 \, \text{m}^2$ **19.** $57 \cdot 5°$ **20.** Yes   **21.** $1 \cdot 716 \, \text{cm}$

Exercise 5   *page 103*

1. (a) $2 \cdot 09$ cm; $4 \cdot 19$ cm^2 (b) $7 \cdot 85$ cm; $39 \cdot 3$ cm^2 (c) $8 \cdot 20$ cm; $8 \cdot 20$ cm^2
2. $31 \cdot 9$ cm^2 **3.** $31 \cdot 2$ cm^2
4. (a) $7 \cdot 07$ cm^2 (b) $19 \cdot 5$ cm^2 **5.** (a) $85 \cdot 9°$ (b) $57 \cdot 3°$ (c) $6 \cdot 25$ cm
6. (a) 12 cm   (b) $30°$ **7.** (a) $3 \cdot 98$ cm   (b) $74 \cdot 9°$
8. (a) $30°$ (b) $10 \cdot 5$ cm   **9.** (a) 18 cm   (b) $38 \cdot 2°$
10. (a) 10 cm   (b) $43 \cdot 0°$ **11.** (a) $6 \cdot 14$ cm   (b) $27 \cdot 6$ m   (c) $28 \cdot 6$ cm^2
12. $14 \cdot 8$ km^2

Exercise 6   *page 106*

1. (a) $14 \cdot 5$ cm   (b) $72 \cdot 6$ cm^2 (c) $24 \cdot 5$ cm^2 (d) $48 \cdot 1$ cm^2
2. (a) $5 \cdot 08$ cm^2 (b) $82 \cdot 8$ m^2 (c) $5 \cdot 14$ cm^2
3. (a) $60°$, $9 \cdot 06$ cm^2 (b) $106 \cdot 3°$, $11 \cdot 2$ cm^2
4. 3 cm   **5.** $3 \cdot 97$ cm   **6.** $13 \cdot 5$ cm^2, 405 cm^3 **7.** $63 \cdot 6$ cm^2; $250 \cdot 5$ cm^2
8. 459 cm^2, 651 cm^2 **9.** $19 \cdot 6$ cm^2 **10.** $0 \cdot 313 r^2$
11. (a) $8 \cdot 37$ cm   (b) $54 \cdot 5$ cm   (c) $10 \cdot 4$ cm   **12.** $81 \cdot 2$ cm^2

Exercise 7   *page 108*

1. (a) 30 cm^3 (b) 168 cm^3 (c) 110 cm^3 (d) $94 \cdot 5$ cm^3 (e) 754 cm^3 (f) 283 cm^3
2. (a) 503 cm^3 (b) 760 m^3 (c) $12 \cdot 5$ cm^3
3. $3 \cdot 98$ cm   **4.** $6 \cdot 37$ cm   **5.** $1 \cdot 89$ cm   **6.** $5 \cdot 37$ cm
7. $9 \cdot 77$ cm   **8.** $7 \cdot 38$ cm   **9.** $12 \cdot 7$ m   **10.** $4 \cdot 24$ litres
11. 106 cm/s   **12.** 1570 cm^3, $12 \cdot 6$ kg   **13.** 3 : 4   **14.** cubes by 77 cm^3
15. No   **16.** $1 \cdot 19$ cm   **17.** 53 times   **18.** 191 cm

Exercise 8   *page 111*

1. $20 \cdot 9$ cm^3 **2.** 524 cm^3 **3.** 4189 cm^3 **4.** 101 cm^3 **5.** 268 cm^3 **6.** $4 \cdot 19 x^3$ cm^3
7. $0 \cdot 004\,19$ m^3 **8.** 3 cm^3 **9.** $93 \cdot 3$ cm^3 **10.** 48 cm^3 **11.** $92 \cdot 4$ cm^3 **12.** 262 cm^3
13. 235 cm^3 **14.** 415 cm^3 **15.** 5 m   **16.** $2 \cdot 43$ cm   **17.** $23 \cdot 9$ cm   **18.** 6 cm
19. $3 \cdot 72$ cm   **20.** $1 \cdot 93$ kg   **21.** 106 s   **22.** (a) 125   (b) 2744   (c) $2 \cdot 7 \times 10^7$
23. (a) $0 \cdot 36$ cm   (b) $0 \cdot 427$ cm   **24.** (a) $6 \cdot 69$ cm   (b) $39 \cdot 1$ cm
25. $10 \frac{2}{3}$ cm^3 **26.** $1 \cdot 05$ cm^3 **27.** 488 cm^3 **28.** 4 cm   **29.** $53 \cdot 6$ cm^3 **30.** $74 \cdot 5$ cm^3
31. $4 \cdot 24$ cm   **32.** 123 cm^3 **33.** $54 \cdot 5$ litres   **34.** (a) 16π (b) 8 cm   (c) 6 cm
35. 471 cm^3 **36.** 2720 cm^3 **37.** 943 cm^3 **38.** 5050 cm^3

Exercise 9   *page 115*

1. (a) 36π cm^2 (b) 72π cm^2 (c) 60π cm^2 (d) $2 \cdot 38\pi$ m^2
     (e) 400π m^2 (f) 65π cm^2 (g) 192π mm^2 (h) $10 \cdot 2\pi$ cm^2
     (i) $0 \cdot 000\,4\pi$ m^2 (j) 98π cm^2, 147π cm^2
2. $1 \cdot 64$ cm   **3.** $2 \cdot 12$ cm   **4.** $3 \cdot 46$ cm
5. (a) 3 cm   (b) 4 cm   (c) 5 cm   (d) $0 \cdot 2$ m   (e) 6 cm   (f) $0 \cdot 25$ cm   (g) 7 m
6. 303 cm^2 **7.** \$1178   **8.** \$3870   **9.** $94 \cdot 0$ cm^3
10. $44 \cdot 6$ cm^2 **11.** 675 cm^2 **12.** $1 \cdot 62 \times 10^8$ years   **13.** 377 cm^2
14. 20 cm, 10 cm   **15.** $71 \cdot 7$ cm^2 **16.** 147 cm^2

Revision exercise 3A   *page 117*

1. (a) $14 \, \text{cm}^2$ (b) $54 \, \text{cm}^2$ (c) $50 \, \text{cm}^2$ (d) $18 \, \text{m}^2$
2. (a) $56 \cdot 5 \, \text{m}$, $254 \, \text{m}^2$ (b) $10 \cdot 8 \, \text{cm}$ (c) $3 \cdot 99 \, \text{cm}$
3. (a) $9 \pi \, \text{cm}^2$ (b) $8 : 1$ 4. $3 \cdot 43 \, \text{cm}^2$, $4 \cdot 57 \, \text{cm}^2$
5. (a) $12 \cdot 2 \, \text{cm}$ (b) $61 \cdot 1 \, \text{cm}^2$
6. (a) $11 \cdot 2 \, \text{cm}$ (b) $10 \cdot 3 \, \text{cm}$ (c) $44 \cdot 7 \, \text{cm}^2$ (d) $31 \cdot 5 \, \text{cm}^2$ (e) $13 \cdot 2 \, \text{cm}^2$
7. $103 \cdot 1°$ 8. $9 \cdot 95 \, \text{cm}$ 9. (a) $905 \, \text{cm}^3$ (b) $5 \cdot 76 \, \text{cm}$
10. $8 \cdot 06 \, \text{cm}$ 11. $99 \cdot 5 \, \text{cm}^3$ 12. $333 \, \text{cm}^3$, $201 \, \text{cm}^3$ 13. $4 \, \text{cm}$
14. (a) $15 \cdot 6 \, \text{cm}^2$ (b) $93 \cdot 5 \, \text{cm}^2$ (c) $3741 \, \text{cm}^2$
15. $0 \cdot 370 \, \text{cm}$ 16. $104 \, \text{cm}^2$ 17. $5 \cdot 14 \, \text{cm}^2$ 18. $68c$ 19. 25 20. $20 \, \text{cm}^2$

Examination exercise 3B   *page 120*

1. (a) $45 \, 498 \, \text{km}$ (b) $7240 \, \text{km}$ 2. 23 3. $21 \cdot 3 \, \text{cm}^2$
4. $170 \, \text{cm}^2$ 5. (a) $14 \cdot 1 \, \text{cm}^2$ (b) $24 \cdot 8 \, \text{cm}$ 6. $314 \, \text{cm}^2$
7. (a) $6 \cdot 93 \, \text{cm}$ (b) $60 \cdot 6 \, \text{cm}^2$
8. (b) $42 \cdot 6 \, \text{kg}$ (c) $26 \cdot 4 \, \text{cm}$
    (d) (i) $0 \cdot 649 - 0 \cdot 651$ (ii) $5 \cdot 31 \, \text{cm}^2$ (iii) $501 \cdot 9 - 503\%$
9. (b) (i) $8 \cdot 49 \, \text{cm}$ (ii) $29 \cdot 0 \, \text{cm}$ (iii) $36 \, \text{cm}^2$ (iv) $695 \, \text{cm}^2$ (c) (i) $14 \cdot 5 \, \text{cm}$ (ii) $94 \cdot 8\%$

4 Geometry

Exercise 1   *page 125*

1. $95°$ 2. $49°$ 3. $100°$ 4. $77°$ 5. $129°$ 6. $95°$ 7. $a = 30°$
8. $e = 30°$, $f = 60°$ 9. $110°$ 10. $x = 54°$ 11. $a = 40°$
12. $a = 36°$, $b = 72°$, $c = 144°$, $d = 108°$ 13. $105°$
14. $a = 30°$, $b = 120°$, $c = 150°$ 15. $x = 20°$, $y = 140°$ 16. $a = 120°$, $b = 34°$, $c = 26°$
17. $a = 68°$, $b = 58 \cdot 5°$ 18. $25°$ 19. $44°$
20. $a = 30°$, $b = 60°$, $c = 150°$, $d = 120°$ 21. $a = 10°$, $b = 76°$ 22. $e = 71°$, $f = 21°$
23. $144°$ 24. $70°$ 25. $41°$, $66°$ 26. $46°$, $122°$ 27. $36°$

Exercise 2   *page 127*

1. $a = 72°$, $b = 108°$ 2. $x = 60°$, $y = 120°$ 3. $(n-2)180°$ 4. $110°$
5. $60°$ 6. $128 \frac{4}{7}°$ 7. 15 8. 12
9. 9 10. 18 11. 12 12. $36°$

Exercise 3   *page 129*

1. $a = 116°$, $b = 64°$, $c = 64°$ 2. $a = 64°$, $b = 40°$ 3. $x = 68°$
4. $a = 40°$, $b = 134$, $c = 134°$ 5. $m = 69°$, $y = 65°$ 6. $t = 48°$, $u = 48°$, $v = 42°$
7. $a = 118°$, $b = 100°$, $c = 62°$ 8. $a = 34°$, $b = 76°$, $c = 70°$, $d = 70°$ 9. $72°$, $108°$

Exercise 4   *page 130*

1. $10 \, \text{cm}$ 2. $4 \cdot 12 \, \text{cm}$ 3. $4 \cdot 24 \, \text{cm}$ 4. $12 \cdot 7 \, \text{cm}$ 5. $8 \cdot 72 \, \text{cm}$ 6. $5 \cdot 66 \, \text{cm}$
7. $6 \cdot 63 \, \text{cm}$ 8. $5 \, \text{cm}$ 9. $17 \, \text{cm}$ 10. $4 \, \text{cm}$ 11. $9 \cdot 85 \, \text{cm}$ 12. $7 \cdot 07 \, \text{cm}$
13. $3 \cdot 46 \, \text{m}$ 14. $40 \cdot 3 \, \text{km}$ 15. $13 \cdot 6 \, \text{cm}$ 16. $6 \cdot 34 \, \text{m}$ 17. $4 \cdot 58 \, \text{cm}$ 18. $84 \cdot 9 \, \text{km}$
19. $24 \, \text{cm}$ 20. $9 \cdot 80 \, \text{cm}$ 21. $5, 4, 3$; $13, 12, 5$; $25, 24, 7$; $41, 40, 9$; $61, 60, 11$
22. $x = 4 \, \text{m}$, $20 \cdot 6 \, \text{m}$ 23. $9 \cdot 49 \, \text{cm}$ 24. $18 \cdot 5 \, \text{km}$

Exercise 5   *page 133*

1. (a) 1, 1   (b) 1, 1   (c) 2, 2   (d) 2, 2   (e) 4, 4   (f) 0, 2   (g) 5, 5
(h) 0, 1   (i) 1, 1   (j) 0, 2   (k) 0, 2   (l) 0, 2   (m) ∞, ∞ (n) 0, 4
4. square 4, 4; rectangle 2, 2; parallelogram 0, 2; rhombus 2, 2; trapezium 0, 1; kite 1, 1;
equilateral triangle 3, 3; regular hexagon 6, 6
5. 34°, 56°   **6.** 35°, 35°   **7.** 72°, 108°, 80°   **8.** 40°, 30°, 110°   **9.** 116°, 32°, 58°
10. 55°, 55°   **11.** 26°, 26°, 77°   **12.** 52°, 64°, 116°   **13.** 70°, 40°, 110°   **14.** 54°, 72°, 36°
15. 60°, 15°, 75°, 135°

Exercise 6   *page 134*

1. 3   **2.** (a) 1   (b) 1   (c) 2   **3.** (a) Multiple answers   (b) 9
4. 2 planes of symmetry are shown.
There are another 2 formed by
joining the diagonals of the base to
the vertex of the pyramid.

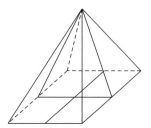

5. 4
6. (a) infinite number
(b) an infinite number running through the
tip of the cone to the centre of the base.

Exercise 7   *page 136*

1. $a = 2\frac{1}{2}$ cm, $e = 3$ cm   **2.** $x = 6$ cm, $y = 10$ cm   **3.** $x = 12$ cm, $y = 8$ cm
4. $m = 10$ cm, $a = 16\frac{2}{3}$ cm   **5.** $y = 6$ cm   **6.** $x = 4$ cm, $w = 1\frac{1}{2}$ cm
7. $e = 9$ cm, $f = 4\frac{1}{2}$ cm   **8.** $x = 13\frac{1}{3}$ cm, $y = 9$ cm   **9.** $m = 6$ cm, $n = 6$ cm
10. $m = 5\frac{1}{3}$ cm, $z = 4\frac{4}{5}$ cm   **11.** $v = 5\frac{1}{3}$ cm, $w = 6\frac{2}{3}$ cm   **12.** No
13. 2 cm, 6 cm   **14.** 16 m
15. (a) Yes   (b) No   (c) No   (d) Yes   (e) Yes   (f) No   (g) No   (h) Yes
18. 0·618; 1·618 : 1

Exercise 8   *page 138*

1. A and G; B and E.

Exercise 9   *page 140*

1. 16 cm^2 **2.** 27 cm^2 **3.** $11\frac{1}{4}$ cm^2 **4.** $14\frac{1}{2}$ cm^2 **5.** 128 cm^2 **6.** 12 cm^2
7. 8 cm   **8.** 18 cm   **9.** $4\frac{1}{2}$ cm   **10.** $7\frac{1}{2}$ cm   **11.** $2\frac{1}{2}$ cm   **12.** 6 cm
13. A = 32 cm^2 **14.** B = 27·9 cm^2 **15.** C = 40 cm^2 **16.** D = 225 cm^2
17. (a) $16\frac{2}{3}$ cm^2 (b) $10\frac{2}{3}$ cm^2 **18.** (a) 25 cm^2 (b) 21 cm^2
19. 8 cm^2 **20.** 6 cm   **21.** 24 cm^2
22. (a) $1\frac{4}{5}$ cm   (b) 3 cm   (c) 3 : 5   (d) 9 : 25
23. 150   **24.** 360   **25.** Less (for the same weight)

Exercise 10   *page 143*

1. 480 cm^3 **2.** 540 cm^3 **3.** 160 cm^3 **4.** 4500 cm^3 **5.** 81 cm^3
6. 11 cm^3 **7.** 16 cm^3 **8.** $85\frac{1}{3}$ cm^3 **9.** 4 cm   **10.** 21 cm
11. 4·6 cm   **12.** 9 cm   **13.** 6·6 cm   **14.** $4\frac{1}{2}$ cm   **15.** $168\frac{3}{4}$ cm^3
16. 106·3 cm^3 **17.** 12 cm   **18.** (a) 2 : 3   (b) 8 : 27   **19.** 8 : 125
20. $x_1^3 : x_2^3$ **21.** 54 kg   **22.** 240 cm^2 **23.** $9\frac{3}{8}$ litres   **24.** $2812\frac{1}{2}$ cm^2

Exercise 11   *page 147*

1. $a = 27°$, $b = 30°$
2. $c = 20°$, $d = 45°$
3. $c = 58°$, $d = 41°$, $e = 30°$
4. $f = 40°$, $g = 55°$, $h = 55°$
5. $a = 32°$, $b = 80°$, $c = 43°$
6. $x = 34°$, $y = 34°$, $z = 56°$
7. $43°$
8. $92°$
9. $42°$
10. $c = 46°$, $d = 44°$
11. $e = 49°$, $f = 41°$
12. $g = 76°$, $h = 52°$
13. $48°$
14. $32°$
15. $22°$
16. $a = 36°$, $x = 36°$

Exercise 12   *page 149*

1. $a = 94°$, $b = 75°$
2. $c = 101°$, $d = 84°$
3. $x = 92°$, $y = 116°$
4. $c = 60°$, $d = 45°$
5. $37°$
6. $118°$
7. $e = 36°$, $f = 72°$
8. $35°$
9. $18°$
10. $90°$
11. $30°$
12. $22\frac{1}{2}°$
13. $n = 58°$, $t = 64°$, $w = 45°$
14. $a = 32°$, $b = 40°$, $c = 40°$
15. $a = 18°$, $c = 72°$
16. $55°$
17. $e = 41°$, $f = 41°$, $g = 41°$
18. $8°$
19. $x = 30°$, $y = 115°$
20. $x = 80°$, $z = 10°$

Exercise 13   *page 150*

1. $a = 18°$
2. $x = 40°$, $y = 65°$, $z = 25°$
3. $c = 30°$, $e = 15°$
4. $f = 50°$, $g = 40°$
5. $h = 40°$, $i = 40°$
6. $n = 36°$
7. $k = 50°$, $m = 50°$, $n = 80°$, $p = 80°$
8. $n = 16°$, $p = 46°$
9. $x = 70°$, $y = 20°$, $z = 55°$

Exercise 14   *page 152*

1. $93°$
2. $36°$
7. $7·8$ cm
8. (a) $7·2$ cm   (b) $5·2$ cm   (c) $12·2$ cm   (d) $8·2$ cm
9. $10·4$ cm
10. $3·5$ cm
11. $6·6$ cm
12. $4·9$ cm, $7·8$ cm

Exercise 15   *page 153*

5. (a) A full circle   (b) An arc of a circle   (c) 60 circles
6.   R

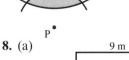

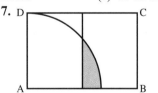

7. D

8. (a)

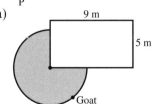

P

9 m

5 m

Goat

(b)

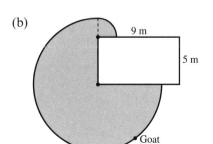

9 m

5 m

Goat

9. A line parallel to AB and 6 cm from AB

10. A spiral

11.

Exercise 16   *page 155*

1. (a), (b), (d)

2. (a) $a = 4$ cm, $x = 4$ cm, $y = 6$ cm   (b) 240 cm^3

3. (a) $a = 10$ cm, $b = 6$ cm, $c = 10$ cm, $d = 10$ cm   (b) 64 cm^3

4. (a) 168 mm^2 (b) 16 800 mm^3

5. $a = \sqrt{2}$, $b = \sqrt{2}$, $c = \sqrt{3}$, $x = \sqrt{2}$, $y = \sqrt{3}$

Revision exercise 4A   *page 156*

1. 80°

3. (a) 30°   (b) $22\frac{1}{2}$°   (c) 12

4. (a) 40°   (b) 100°

5. 4·12 cm

6. (i) 3 cm   (ii) 5·66 cm

7. (c) $2\frac{4}{5}$ cm

8. (b) 6 cm   **9.** $3\frac{2}{3}$ cm, $1\frac{1}{11}$ cm

10. 6 cm   **11.** 250 cm^3

12. (a) $3\frac{1}{3}$ cm   (b) 1620 cm^3

13. (a) 1 m^2 (b) 1000 cm^3

14. (a) 50°   (b) 128°

(c) $c = 50°$, $d = 40°$ (d) $x = 10°$, $y = 40°$

15. (a) 55°   (b) 45°

Examination exercise 4B   *page 158*

1. (a) 72° (b) 36°

2. (a) 5·66 cm   (b) 32·0 cm^2

3. (a) 320 cm^3 (b) 567 cm^2

4. (a) 55 cm by 40 cm   (b) $\dfrac{16}{25}$

5. $r = 24$ cm, $h = 36$ cm

6. (a) 58°   (b) 32° (c) 58° (d) 24°

7. (a) 54°   (b) 42° (c) 78°

8. arc of circle radius BD, centre B.

5 Algebra 2

Exercise 1   *page 163*

1. $\frac{5}{7}$ **2.** $\frac{7}{8}$ **3.** $5y$ **4.** $\frac{1}{2}$ **5.** 4   **6.** $\dfrac{x}{2y}$

7. 2   **8.** $\dfrac{a}{2}$ **9.** $\dfrac{2b}{3}$ **10.** $\dfrac{a}{5b}$ **11.** a **12.** $\frac{7}{8}$

13. $\dfrac{5 + 2x}{3}$ **14.** $\dfrac{3x + 1}{x}$ **15.** $\frac{32}{25}$ **16.** $\dfrac{4 + 5a}{5}$ **17.** $\dfrac{3}{4 - x}$ **18.** $\dfrac{b}{3 + 2a}$

19. $\dfrac{5x + 4}{8x}$ **20.** $\dfrac{2x + 1}{y}$ **21.** $\dfrac{x + 2y}{3xy}$ **22.** $\dfrac{6 - b}{2a}$ **23.** $\dfrac{2b + 4a}{b}$ **24.** $x - 2$

Exercise 2   *page 163*

1. $\dfrac{x + 2}{x - 3}$ **2.** $\dfrac{x}{x + 1}$ **3.** $\dfrac{x + 4}{2(x - 5)}$ **4.** $\dfrac{x + 5}{x - 2}$ **5.** $\dfrac{x + 3}{x + 2}$ **6.** $\dfrac{x + 5}{x - 2}$

7. $\dfrac{x + 2}{x}$ **8.** $\dfrac{3x}{x + 5}$ **9.** $\frac{1}{2}$ **10.** $\dfrac{3x}{x - 5}$ **11.** $\dfrac{3x - 5}{x}$ **12.** $\dfrac{x - 2}{x - 1}$

Exercise 3   *page 164*

1. $\frac{3}{5}$

2. $\frac{3x}{5}$

3. $\frac{3}{x}$

4. $\frac{4}{7}$

5. $\frac{4x}{7}$

6. $\frac{4}{7x}$

7. $\frac{7}{8}$

8. $\frac{7x}{8}$

9. $\frac{7}{8x}$

10. $\frac{5}{6}$

11. $\frac{5x}{6}$

12. $\frac{5}{6x}$

13. $\frac{23}{20}$

14. $\frac{23x}{20}$

15. $\frac{23}{20x}$

16. $\frac{1}{12}$

17. $\frac{x}{12}$

18. $\frac{1}{12x}$

19. $\frac{5x+2}{6}$

20. $\frac{7x+2}{12}$

21. $\frac{9x+13}{10}$

22. $\frac{1-2x}{12}$

23. $\frac{2x-9}{15}$

24. $\frac{-3x-12}{14}$

25. $\frac{3x+1}{x(x+1)}$

26. $\frac{7x-8}{x(x-2)}$

27. $\frac{8x+9}{(x-2)(x+3)}$

28. $\frac{4x+11}{(x+1)(x+2)}$

29. $\frac{-3x-17}{(x+3)(x-1)}$

30. $\frac{11-x}{(x+1)(x-2)}$

Exercise 4   *page 165*

1. $2\frac{1}{2}$

2. 3

3. $\frac{B}{A}$

4. $\frac{T}{N}$

5. $\frac{K}{M}$

6. $\frac{4}{y}$

7. $\frac{C}{B}$

8. $\frac{D}{4}$

9. $\frac{T+N}{9}$

10. $\frac{B-R}{A}$

11. $\frac{R+T}{C}$

12. $\frac{N-R^2}{L}$

13. $\frac{R-S^2}{N}$

14. 2

15. -7

16. $T-A$

17. $S-B$

18. $N-D$

19. $M-B$

20. $L-D^2$

21. $T-N^2$

22. $N+M-L$

23. $R-S-Z$

24. 7

25. $A+R$

26. $E+A$

27. $F+B$

28. F^2+B^2

29. $A+B+D$

30. A^2+E

31. $L+B$

32. $N+T$

33. 2

34. $4\frac{1}{2}$

35. $\frac{N-C}{A}$

36. $\frac{L-D}{B}$

37. $\frac{F-E}{D}$

38. $\frac{H+F}{N}$

39. $\frac{T+Z}{Y}$

40. $\frac{B+L}{R}$

41. $\frac{Q-m}{V}$

42. $\frac{n+a+m}{t}$

43. $\frac{s-t-n}{q}$

44. $\frac{t+s^2}{n}$

45. $\frac{c-b}{V^2}$

46. $\frac{r+6}{n}$

47. $\frac{s-d}{m}$

48. $\frac{t+b}{m}$

49. $\frac{j-c}{m}$

50. 2

51. $2\frac{2}{3}$

52. $\frac{C-AB}{A}$

53. $\frac{F-DE}{D}$

54. $\frac{a-hn}{h}$

55. $\frac{q+bd}{b}$

56. $\frac{n-rt}{r}$

57. $\frac{b+4t}{t}$

58. $\frac{z-St}{S}$

59. $\frac{s+vd}{v}$

60. $\frac{g-mn}{m}$

Exercise 5   *page 166*

1. 12

2. 10

3. BD

4. TB

5. RN

6. bm

7. 26

8. $BT+A$

9. $AN+D$

10. B^2N-Q

11. $ge + r$ **12.** $4\frac{1}{2}$ **13.** $\dfrac{DC - B}{A}$ **14.** $\dfrac{pq - m}{n}$ **15.** $\dfrac{vS + t}{r}$

16. $\dfrac{qt + m}{z}$ **17.** $\dfrac{bc - m}{A}$ **18.** $\dfrac{AE - D}{B}$ **19.** $\dfrac{nh + f}{e}$ **20.** $\dfrac{qr - b}{g}$

21. 4   **22.** -2 **23.** 2   **24.** $A - B$ **25.** $C - E$

26. $D - H$ **27.** $n - m$ **28.** $q - t$ **29.** $s - b$ **30.** $r - v$

31. $m - t$ **32.** 2   **33.** $\dfrac{T - B}{X}$ **34.** $\dfrac{M - Q}{N}$ **35.** $\dfrac{V - T}{M}$

36. $\dfrac{N - L}{R}$ **37.** $\dfrac{v^2 - r}{r}$ **38.** $\dfrac{w - t^2}{n}$ **39.** $\dfrac{n - 2}{q}$ **40.** $\frac{1}{4}$

41. $-\frac{1}{7}$ **42.** $\dfrac{B - DE}{A}$ **43.** $\dfrac{D - NB}{E}$ **44.** $\dfrac{h - bx}{f}$ **45.** $\dfrac{v^2 - Cd}{h}$

46. $\dfrac{NT - MB}{M}$ **47.** $\dfrac{mB + ef}{fN}$ **48.** $\dfrac{TM - EF}{T}$ **49.** $\dfrac{yx - zt}{y}$ **50.** $\dfrac{k^2m - x^2}{k^2}$

Exercise 6   page 167

1. $\frac{1}{2}$ **2.** $1\frac{2}{3}$ **3.** $\dfrac{B}{C}$ **4.** $\dfrac{T}{X}$ **5.** $\dfrac{M}{B}$

6. $\dfrac{n}{m}$ **7.** $\dfrac{v}{t}$ **8.** $\dfrac{n}{\sin 20°}$ **9.** $\dfrac{7}{\cos 30°}$ **10.** $\dfrac{B}{x}$

11. $6\frac{2}{3}$ **12.** $\dfrac{ND}{B}$ **13.** $\dfrac{HM}{N}$ **14.** $\dfrac{et}{b}$ **15.** $\dfrac{vs}{m}$

16. $\dfrac{mb}{t}$ **17.** $1\frac{1}{2}$ **18.** $3\frac{1}{3}$ **19.** $\dfrac{B - DC}{C}$ **20.** $\dfrac{Q + TC}{T}$

21. $\dfrac{V + TD}{D}$ **22.** $\dfrac{L}{MB}$ **23.** $\dfrac{N}{BC}$ **24.** $\dfrac{m}{cd}$ **25.** $\dfrac{tc - b}{t}$

26. $\dfrac{xy - z}{x}$ **27.** 1   **28.** $\frac{5}{6}$ **29.** $\dfrac{A}{C - B}$ **30.** $\dfrac{V}{H - G}$

31. $\dfrac{r}{n + t}$ **32.** $\dfrac{b}{q - d}$ **33.** $\dfrac{m}{t + n}$ **34.** $\dfrac{b}{d - h}$ **35.** $\dfrac{d}{C - e}$

36. $\dfrac{m}{r - e^2}$ **37.** $\dfrac{n}{b - t^2}$ **38.** $\dfrac{d}{mn - b}$ **39.** $\dfrac{M - Nq}{N}$ **40.** $\dfrac{Y + Tc}{T}$

41. $\dfrac{N - 2MP}{2M}$ **42.** $\dfrac{B - 6Ac}{6A}$ **43.** $\dfrac{K}{(C - B)M}$ **44.** $\dfrac{z}{y(y + z)}$ **45.** $\dfrac{m^2}{n - p}$

46. $\dfrac{q}{w - t}$

Exercise 7   page 168

1. 4   **2.** 24   **3.** 11   **4.** $B^2 - A$ **5.** $D^2 - C$

6. $H^2 + E$ **7.** $\dfrac{c^2 - b}{a}$ **8.** $a^2 + m$ **9.** $\dfrac{b^2 + t}{g}$ **10.** $b - r^2$

11. $d - t^2$ **12.** $b^2 + d$ **13.** $n - c^2$ **14.** $b - f^2$ **15.** $c - g^2$

16. $\dfrac{M - P^2}{N}$ **17.** $\dfrac{D - B}{A}$ **18.** $A^4 + D$ **19.** $\pm\sqrt{g}$ **20.** ± 4

21. $\pm\sqrt{B}$ **22.** $\pm\sqrt{(B - A)}$ **23.** $\pm\sqrt{(M + A)}$ **24.** $\pm\sqrt{(b - a)}$ **25.** $\pm\sqrt{(C - m)}$

26. $\pm\sqrt{(d - n)}$ **27.** $\pm\sqrt{\dfrac{n}{m}}$ **28.** $\pm\sqrt{\dfrac{b}{a}}$ **29.** $\dfrac{at}{z}$ **30.** $\pm\sqrt{\left(\dfrac{m + t}{a}\right)}$

31. $\pm\sqrt{(a - n)}$ **32.** $\pm\sqrt{40}$ **33.** $\pm\sqrt{(B^2 + A)}$ **34.** $\pm\sqrt{(x^2 - y)}$ **35.** $\pm\sqrt{(t^2 - m)}$

36. 8 **37.** $\dfrac{M^2 - A^2 B}{A^2}$ **38.** $\dfrac{M}{N^2}$ **39.** $\dfrac{N}{B^2}$ **40.** $a - b^2$

41. $\pm\sqrt{(a^2 - t^2)}$ **42.** $\pm\sqrt{(m - x^2)}$ **43.** $\dfrac{4}{\pi^2} - t$ **44.** $\dfrac{B^2}{A^2} - 1$ **45.** $\pm\sqrt{\left(\dfrac{C^2 + b}{a}\right)}$

46. $\pm\sqrt{\left(\dfrac{b^2 + a^2 x}{a^2}\right)}$ **47.** $\pm\sqrt{(x^2 - b)}$ **48.** $\pm\sqrt{(c - b)a}$ **49.** $\dfrac{c^2 - b^2}{a}$ **50.** $\pm\sqrt{\left(\dfrac{m}{a + b}\right)}$

Exercise 8   page 169

1. $3\frac{2}{3}$ **2.** 3 **3.** $\dfrac{D - B}{2N}$ **4.** $\dfrac{E + D}{3M}$ **5.** $\dfrac{2b}{a - b}$

6. $\dfrac{e + c}{m + n}$ **7.** $\dfrac{3}{x + k}$ **8.** $\dfrac{C - D}{R - T}$ **9.** $\dfrac{z + x}{a - b}$ **10.** $\dfrac{nb - ma}{m - n}$

11. $\dfrac{d + xb}{x - 1}$ **12.** $\dfrac{a - ab}{b + 1}$ **13.** $\dfrac{d - c}{d + c}$ **14.** $\dfrac{M(b - a)}{b + a}$ **15.** $\dfrac{n^2 - mn}{m + n}$

16. $\dfrac{m^2 + 5}{2 - m}$ **17.** $\dfrac{2 + n^2}{n - 1}$ **18.** $\dfrac{e - b^2}{b - a}$ **19.** $\dfrac{3x}{a + x}$ **20.** $\dfrac{e - c}{a - d}$ or $\dfrac{c - e}{d - a}$

21. $\dfrac{d}{a - b - c}$ **22.** $\dfrac{ab}{m + n - a}$ **23.** $\dfrac{s - t}{b - a}$ or $\dfrac{t - s}{a - b}$ **24.** $2x$ **25.** $\dfrac{v}{3}$

26. $\dfrac{a(b + c)}{b - 2a}$ **27.** $\dfrac{5x}{3}$ **28.** $-\dfrac{4z}{5}$ **29.** $\dfrac{mn}{p^2 - m}$ **30.** $\dfrac{mn + n}{4 + m}$

Exercise 9   page 170

1. $-\left(\dfrac{by + c}{a}\right)$ **2.** $\pm\sqrt{\left(\dfrac{e^2 + ab}{a}\right)}$ **3.** $\dfrac{n^2}{m^2} + m$ **4.** $\dfrac{a - b}{1 + b}$ **5.** $3y$

6. $\dfrac{a}{e^2 + c}$ **7.** $-\left(\dfrac{a + lm}{m}\right)$ **8.** $\dfrac{t^2 g}{4\pi^2}$ **9.** $\dfrac{4\pi^2 d}{t^2}$ **10.** $\pm\sqrt{\dfrac{a}{3}}$

11. $\pm\sqrt{\left(\dfrac{t^2 e - ba}{b}\right)}$ **12.** $\dfrac{1}{a^2 - 1}$ **13.** $\dfrac{a + b}{x}$ **14.** $\pm\sqrt{(x^4 - b^2)}$ **15.** $\dfrac{c - a}{b}$

16. $\dfrac{a^2 - b}{a + 1}$ **17.** $\pm\sqrt{\left(\dfrac{G^2}{16\pi^2} - T^2\right)}$ **18.** $-\left(\dfrac{ax + c}{b}\right)$ **19.** $\dfrac{1 + x^2}{1 - x^2}$ **20.** $\pm\sqrt{\left(\dfrac{a^2 m}{b^2 + n}\right)}$

21. $\dfrac{P - M}{E}$ **22.** $\dfrac{RP - Q}{R}$ **23.** $\dfrac{z - t^2}{x}$ **24.** $(g - e)^2 - f$ **25.** $\dfrac{4np + me^2}{mn}$

Exercise 10   page 172

1. (a) $S = ke$ (b) $v = kt$ (c) $x = kz^2$ (d) $y = k\sqrt{x}$
    (e) $T = k\sqrt{L}$ (f) $C = kr$ (g) $A = kr^2$ (h) $V = kr^3$

2. (a) 9   (b) $2\frac{2}{3}$ **3.** (a) 35   (b) 11

5.

x	1	3	4	$5\frac{1}{2}$
z	4	12	16	22

6.

r	1	2	4	$1\frac{1}{2}$
V	4	32	256	$13\frac{1}{2}$

7.

h	4	9	25	$2\frac{1}{4}$
w	6	9	15	$4\frac{1}{2}$

8. (a) 18   (b) 2   **9.** (a) 42   (b) 4   **10.** $333\frac{1}{3}$ N/cm^3

11. 180 m; 2 s   **12.** 675 J; $\sqrt{\frac{4}{3}}$ cm   **13.** 4 cm; 49 h

14. $15\frac{5}{8}$ h   **15.** 9000 N; 25 m/s   **16.** $15^4 : 1\,(50\,625 : 1)$

Exercise 11   *page 174*

1. (a) $x = \dfrac{k}{y}$ (b) $s = \dfrac{k}{t^2}$ (c) $t = \dfrac{k}{\sqrt{q}}$ (d) $m = \dfrac{k}{w}$ (e) $z = \dfrac{k}{t^2}$

2. (a) 1   (b) 4   **3.** (a) $2\frac{1}{2}$ (b) $\frac{1}{2}$

4. (a) 36   (b) ±4 **5.** (a) 1·2   (b) ±2

6. (a) 16   (b) ±10 **7.** (a) 6   (b) 16

8. (a) $\frac{1}{2}$ (b) $\frac{1}{20}$

9.

y	2	4	1	$\frac{1}{4}$
z	8	4	16	64

10.

t	2	5	20	10
y	25	4	$\frac{1}{4}$	1

11.

x	1	4	256	36
r	12	6	$\frac{3}{4}$	2

12. (a) 6   (b) 50   **13.** (a) 0·36   (b) 6

14. $k = 100$, $n = 3$ **15.** $k = 12$, $n = 2$

x	1	2	4	10
z	100	$12\frac{1}{2}$	1·5625	$\frac{1}{10}$

v	1	4	36	10 000
y	12	6	2	$\frac{3}{25}$

16. 2·5 m^3; 200 N/m^2 **17.** 3 h; 48 men   **18.** 2 days; 200 days   **19.** 6 cm

Exercise 12   *page 177*

1. 3^4 **2.** $4^2 \times 5^3$ **3.** 3×7^3 **4.** $2^3 \times 7$ **5.** 10^{-3}

6. $2^{-2} \times 3^{-3}$ **7.** $15^{\frac{1}{2}}$ **8.** $3^{\frac{1}{3}}$ **9.** $10^{\frac{1}{5}}$ **10.** $5^{\frac{3}{2}}$

11. x^7 **12.** y^{13} **13.** z^4 **14.** z^{100} **15.** m

16. e^{-5} **17.** y^2 **18.** w^6 **19.** y **20.** x^{10}

21. 1   **22.** w^{-5} **23.** w^{-5} **24.** x^7 **25.** a^8

26. k^3 **27.** 1   **28.** x^{29} **29.** y^2 **30.** x^6

31. z^4 **32.** t^{-4} **33.** $4x^6$ **34.** $16y^{10}$ **35.** $6x^4$

36. $10y^5$ **37.** $15a^4$ **38.** $8a^3$ **39.** 3   **40.** $4y^2$

41. $\frac{5}{2}y$ **42.** $32a^4$ **43.** $108x^5$ **44.** $4z^{-3}$ **45.** $2x^{-4}$

46. $\frac{5}{2}y^5$ **47.** 1   **48.** $21w^{-3}$ **49.** $2n^4$ **50.** $2x$

Exercise 13   *page 177*

1. 27   **2.** 1   **3.** $\frac{1}{9}$ **4.** 25   **5.** 2

6. 4   **7.** 9   **8.** 2   **9.** 27   **10.** 3

11. $\frac{1}{3}$ **12.** $\frac{1}{2}$ **13.** 1   **14.** $\frac{1}{5}$ **15.** 10

16. 8   **17.** 32   **18.** 4   **19.** $\frac{1}{9}$ **20.** $\frac{1}{8}$

21. 18   **22.** 10   **23.** 1000   **24.** $\frac{1}{1000}$ **25.** $\frac{1}{9}$

26. 1   **27.** $1\frac{1}{2}$ **28.** $\frac{1}{25}$ **29.** $\frac{1}{10}$ **30.** $\frac{1}{4}$

31. $\frac{1}{4}$ **32.** $100\,000$ **33.** 1 **34.** $\frac{1}{32}$ **35.** $0{\cdot}1$

36. $0{\cdot}2$ **37.** $1{\cdot}5$ **38.** 1 **39.** 9 **40.** $1\frac{1}{2}$

41. $\frac{3}{10}$ **42.** 64 **43.** $\frac{1}{100}$ **44.** $1\frac{2}{3}$ **45.** $\frac{1}{100}$

46. 1 **47.** 100 **48.** 6 **49.** 750 **50.** -7

Exercise 14   *page 178*

1. $25x^4$ **2.** $49y^6$ **3.** $100a^2b^2$ **4.** $4x^2y^4$ **5.** $2x$ **6.** $\dfrac{1}{9y}$

7. x^2 **8.** $\dfrac{x^2}{2}$ **9.** 1 **10.** $\dfrac{2}{x}$ **11.** $36x^4$ **12.** $25y$

13. $16x^2$ **14.** $27y$ **15.** 25 **16.** 1 **17.** 49 **18.** 1

19. $8x^6y^3$ **20.** $100x^2y^6$ **21.** $\dfrac{3x}{2}$ **22.** $\dfrac{2}{x}$ **23.** x^3y^5 **24.** $12x^3y^2$

25. $10y^4$ **26.** $3x^3$ **27.** $x^3y^2z^4$ **28.** x **29.** $3y$ **30.** $27x^{\frac{3}{2}}$

31. $10x^3y^5$ **32.** $32x^2$ **33.** $\frac{5}{2}x^2$ **34.** $\dfrac{9}{x^2}$ **35.** $2a^2$ **36.** $a^3b^3c^2$

37. (a) 2^5 (b) 2^7 (c) 2^6 (d) 2^0

38. (a) 3^{-3} (b) 3^{-4} (c) 3^{-1} (d) 3^{-2}

39. 16 **40.** $\frac{1}{4}$ **41.** $\frac{1}{6}$ **42.** 1 **43.** $16\frac{1}{8}$ **44.** $\frac{3}{8}$

45. $\frac{1}{4}$ **46.** $\frac{5}{256}$ **47.** $1\frac{1}{16}$ **48.** 0 **49.** $\frac{1}{4}$ **50.** $\frac{1}{4}$

51. 3 **52.** 4 **53.** -1 **54.** -2 **55.** 3 **56.** 3

57. 1 **58.** $\frac{1}{5}$ **59.** 0 **60.** -4 **61.** 2 **62.** -5

63. 1 **64.** $\frac{1}{18}$ **65.** (a) $3{\cdot}60$ (b) $5{\cdot}44$

Exercise 15   *page 179*

1. $<$ **2.** $>$ **3.** $>$ **4.** $=$

5. $<$ **6.** $<$ **7.** $=$ **8.** $>$

9. $<$ **10.** $>$ **11.** $<$ **12.** $>$

13. $>$ **14.** $>$ **15.** $=$ **16.** F

17. F   **18.** T   **19.** F   **20.** F

21. T   **22.** T   **23.** F   **24.** F

25. $x > 13$ **26.** $x < -1$ **27.** $x < 12$ **28.** $x \leqslant 2\frac{1}{2}$

29. $x > 3$ **30.** $x \geqslant 8$ **31.** $x < \frac{1}{4}$ **32.** $x \geqslant -3$

33. $x < -8$ **34.** $x < 4$ **35.** $x > -9$ **36.** $x < 8$

37. $x > 3$ **38.** $x \geqslant 1$ **39.** $x < 1$ **40.** $x > 2\frac{1}{3}$

Exercise 16   *page 180*

1. $x > 5$ **2.** $x \leqslant 3$ **3.** $x > 6$ **4.** $x \geqslant 1$

5. $x < 1$ **6.** $x < -3$ **7.** $x > 0$ **8.** $x > 4$

9. $x > 2$ **10.** $x < -3$ **11.** $1 < x < 4$ **12.** $-2 \leqslant x \leqslant 5$

13. $1 \leqslant x < 6$ **14.** $0 \leqslant x < 5$ **15.** $-1 \leqslant x \leqslant 7$ **16.** $0{\cdot}2 < z < 2$

17. $x > 80$ **18.** $x > 10$ **19.** $x < -2{\cdot}5$ **20.** $0 < x < 4$

21. $5 \leqslant x \leqslant 9$ **22.** $-1 < x < 4$ **23.** $5{\cdot}5 \leqslant x \leqslant 6$ **24.** $\frac{1}{2} < x < 8$

25. $-8 < x < 2$ **26.** $\{1, 2, 3, 4, 5, 6\}$ **27.** $\{7, 11, 13, 17, 19\}$ **28.** $\{2, 4, 6, 8, 10\}$

29. $\{4, 9, 16, 25, 36, 49\}$ **30.** $\{5, 10\}$ **31.** $\{-4, -3, -2, -1\}$ **32.** $\{2, 3, 4, \ldots 12\}$
33. $\{1, 4, 9\}$ **34.** $\{2, 3, 5, 7, 11\}$ **35.** $\{2, 4, 6, \ldots 18\}$ **36.** $n = 5$
37. $x = 7$ **38.** $y = 5$ **39.** $4 < z < 5$ **40.** $4 < p < 5$
41. $\frac{1}{2}$ (or other values) **42.** $1, 2, 3, \ldots 14$ **43.** 19 **44.** $\frac{1}{2}$ (or other values)
45. 19
46. (a) $-3 \leqslant x < 6$ (b) $-2 < x < 2$ (c) $-3 \leqslant x \leqslant 2$ (d) $-3 \leqslant x < 7$ **47.** 17

Exercise 17   *page 182*

1. $x \leqslant 3$ **2.** $y \geqslant 2\frac{1}{2}$ **3.** $1 \leqslant x \leqslant 6$
4. $x < 7, y < 5$ **5.** $y \geqslant x$ **6.** $x + y < 10$
7. $x < 8, y > -2, y < x$ **8.** $x \geqslant 0, y \geqslant x - 1, x + y \leqslant 7$ **9.** $y \geqslant 0, y \leqslant x + 2, x + y \leqslant 6$

10. 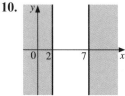 **11.**   **12.** 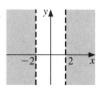 **13.**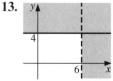

14. 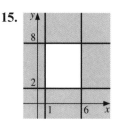 **15.** 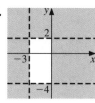 **16.**   **17.**

18. 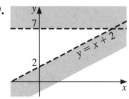 **19.** 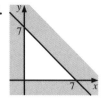 **20.** 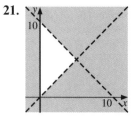 **21.**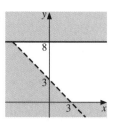

22. 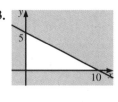 **23.** 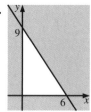 **24.** 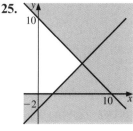 **25.**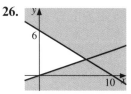

26. 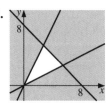 **27.**

Exercise 18   *page 184*

1. (a) maximum value $= 26$ at $(6, 5)$ (b) minimum value $= 12$ at $(3, 3)$
2. (a) maximum value $= 25$ at $(8, 3)$ (b) minimum value $= 9$ at $(7, 2)$
3. (a) maximum value $= 40$ at $(20, 0)$ (b) minimum value $= 112$ at $(14, 8)$
4. $(3, 3), (4, 2), (4, 3), (4, 4), (5, 1), (5, 2), (6, 0)$
5. $(0, 6), (0, 7), (0, 8), (1, 5), (1, 6), (2, 4)$ 6. $(3, 2), (2, 3), (2, 4), (2, 5), (1, 4), (1, 5), (0, 5), (0, 6)$
7. $(2, 4), (2, 5), (2, 6), (2, 7), (2, 8), (2, 9), (2, 10), (3, 6), (3, 7), (3, 8)$
8. (a) $(6, 7), (7, 7), (8, 6), (7, 6), (6, 8)$ (b) 7 defenders, 6 forwards have lowest wage bill ($190 000)
9. (a) 18   (b) €2·60   10. (a) 14   (b) $(7, 7)$, $154000   (c) $(13, 3)$, $190 000
11. (a) 10   (b) 250, $(4, 7)$ 12. (a) $x \geqslant 6, y \geqslant 6, x + y \leqslant 20, 1{\cdot}5x + 3y \leqslant 45$
  (b) $(6, 12)$, $14   (c) $(10, 10)$, $18
13. $(14, 11)$, $1360   14. (a) $(9, 7)$, $44   (b) $(15, 5)$, $45

Revision exercise 5A   *page 187*

1. (a) $\dfrac{9x}{20}$ (b) $\dfrac{7}{6x}$ (c) $\dfrac{5x - 2}{6}$ (d) $\dfrac{5x + 23}{(x - 1)(x + 3)}$

2. (a) $(x - 2)(x + 2)$ (b) $\dfrac{3}{x + 2}$
3. (a) $s = t(r + 3)$ (b) $r = \dfrac{s - 3t}{t}$ (c) $t = \dfrac{s}{r + 3}$

4. (a) $z = x - 5y$ (b) $m = \dfrac{11}{k + 3}$ (c) $z = \dfrac{T^2}{C^2}$

5. (a) 50   (b) 50   6. (a) 16   (b) ± 4
7. (a) (i) 3   (ii) 4   (iii) $\frac{1}{4}$ (b) (i) 4   (ii) 0

8. (a) 9, 10   (b) 2, 3, 4, 5   9. $\dfrac{t^2}{k^2} - 5$ 10. $\dfrac{z + 2}{z - 3}$

11. (a) $\frac{3}{5}$ (b) $\dfrac{k(1 - y)}{y}$ 12.

13. (a) $1\frac{5}{6}$ (b) 0·09   14. 21
15. (a) $\dfrac{5 + a^2}{2 - a}$ (b) $-\left(\dfrac{cz + b}{a}\right)$ (c) $\dfrac{a^2 + 1}{a^2 - 1}$

16. (a) $\dfrac{7}{2x}$ (b) $\dfrac{3a + 7}{a^2 - 4}$ (c) $\dfrac{x - 8}{x(x + 1)(x - 2)}$

17. $p = \dfrac{10t^2}{s}$ 18. 6 or 7   19. $y \geqslant 2, x + y \leqslant 6, y \leqslant 3x$

20. $x \geqslant 0, y \geqslant x - 2, x + y \leqslant 7, y \geqslant 0$ 22. (a) 512   (b) 6 h   (c) 2^{21}

Examination exercise 5B   *page 189*

1. $\dfrac{x^2 - 6x + 25}{4(x - 3)}$ 2. $\dfrac{-18}{(2x + 3)(x - 3)}$ 3. (a) $0{\cdot}8^2$ (b) $0{\cdot}8^{-1}$ 4. $9x^2$

5. (a) 0   (b) 0·2   (c) 0·6   6. (a) $3x^2$ (b) -6
7. $\dfrac{2}{c}$ 8. $x < 23{\cdot}5$ 9. $y = [8(1 - x)]^2$
10. (a) $x^2(a + b)$ (b) $x = \pm \sqrt{\dfrac{(p^2 + d^2)}{(a + b)}}$ 11.   1·25   12. 0·128

13. (a) $y \propto \dfrac{1}{x^2}$ or $y = \dfrac{k}{x^2}$ (b) 30 (c) 3·46 (d) 4·93 (e) divided by 4

(f) increases by 25% (g) $x = \sqrt{\dfrac{120}{y}}$

14. (a) $x + y \leqslant 12,\ x \geqslant 4$ (d) (i) \$18 [from (4,4)] (ii) \$27 [from (6,6)]

15. (a) (i) 439·8 to 440 cm^2 (ii) $h = \dfrac{A - 2\pi r^2}{2\pi r}$ (iii) 3·99 to 4·01 (iv) 9·77 to 9·78

(b) (i) 134 (ii) $\dfrac{x}{45}$ (iii) $\dfrac{x - 75}{48}$ (iv) $x = 3915$

6 Trigonometry

Exercise 2 *page 194*

1. 4·54	**2.** 3·50	**3.** 3·71	**4.** 6·62	**5.** 8·01
6. 31·9	**7.** 45·4	**8.** 4·34	**9.** 17·1	**10.** 13·2
11. 38·1	**12.** 3·15	**13.** 516	**14.** 79·1	**15.** 5·84
16. 2·56	**17.** 18·3	**18.** 8·65	**19.** 11·9	**20.** 10·6
21. 119	**22.** 10·1	**23.** 3·36 cm	**24.** 4·05 cm	**25.** 4·10 cm
26. 11·7 cm	**27.** 9·48 cm	**28.** 5·74 cm	**29.** 9·53 cm	**30.** 100 m
31. 56·7 m	**32.** 16·3 cm	**33.** 0·952 cm	**34.** 8·27 m	

Exercise 3 *page 196*

1. 5, 5·55	**2.** 13·1, 27·8	**3.** 34·6, 41·3	**4.** 20·4, 11·7	**5.** 94·1, 94·1
6. 15·2, 10, 6·43	**7.** 4·26	**8.** 3·50	**9.** 26·2	**10.** 8·82

11. (a) 17·4 cm (b) 11·5 cm (c) 26·5 cm **12.** (a) 6·82 cm (b) 6·01 cm (c) 7·31 cm

Exercise 4 *page 198*

1. 36·9°	**2.** 44·4°	**3.** 48·2°	**4.** 60°	**5.** 36·9°	**6.** 50·2°
7. 29·0°	**8.** 56·4°	**9.** 38·9°	**10.** 43·9°	**11.** 41·8°	**12.** 39·3°
13. 60·3°	**14.** 50·5°	**15.** 13·6°	**16.** 34·8°	**17.** 60·0°	**18.** 42·0°
19. 36·9°	**20.** 51·3°	**21.** 19·6°	**22.** 17·9°	**23.** 32·5°	**24.** 59·6°
25. 54·8°	**26.** 46·3°				

Exercise 5 *page 201*

1. 19·5°	**2.** 4·1 m	**3.** (a) 26·0 km	(b) 23·4 km
4. (a) 88·6 km	(b) 179·3 km	**5.** 4·1 m	**6.** 8·6 m
7. (a) 484 km	(b) 858 km	(c) 985 km, 060·6°	
8. 954 km, 133°	**9.** 56·3°	**10.** 35·5°	**11.** 71·6°
12. 91·8°	**13.** 180 m	**14.** 36·4°	**15.** 10·3 cm
16. 9·51 cm	**17.** 71·1°	**18.** 67·1 m	**19.** 138 m
20. 83·2 km	**21.** 60°	**22.** 13·9 cm	**23.** Yes
24. 11·1 m; 11·1 s; 222 m		**25.** 4·4 m	

Exercise 6 *page 203*

1. 100 m	**2.** 89 n miles	**3.** 103 km
4. 99 km; 024°	**5.** 9190 km/h; 255°	**6.** 11 km

Exercise 7   *page 204*

1. (a) 13 cm   (b) 13·6 cm   (c) 17·1°
2. (a) 4·04 m   (b) 38·9°   (c) 11·2 m   (d) 19·9°
3. (a) 8·49 cm   (b) 8·49 cm   (c) 10·4 cm   (d) 35·3°   (e) 35·3°
4. (a) 10 m   (b) 7·81 m   (c) 9·43 m   (d) 70·2°
5. (a) 14·1 cm   (b) 18·7 cm   (c) 69·3°   (d) 29·0°   (e) 41·4°
6. (a) 4·47 m   (b) 7·48 m   (c) 63·4°   (d) 74·5°   (e) 53·3°
7. 10·8 cm; 21·8°
8. (a) $h\tan 65°$ or $\dfrac{h}{\tan 25°}$ (b) $h\tan 57°$ or $\dfrac{h}{\tan 33°}$ (c) 22·7 m
9. 22·6 m   10. 55·0 m   11. 7·26 m   12. 43·3°

Exercise 8   *page 208*

3. 162°   4. 153°   5. (a) 140°   (b) 110°   (c) 50°
6. 110°   7. 135°
8. 160°   9. 82°   10. 58°, 122°   11. 20·5°, 159·5°
12. 60°, 120°   13. (a) 41°, 139°   (b) 30°, 150°
14. (a) (i) 16°, 111°   (ii) 153°   (b) 2·2   (c) 63°

Exercise 9   *page 210*

1. 6·38 m   2. 12·5 m   3. 5·17 cm   4. 40·4 cm   5. 7·81 m, 7·10 m
6. 3·55 m, 6·68 m   7. 8·61 cm   8. 9·97 cm   9. 8·52 cm   10. 15·2 cm
11. 35·8°   12. 42·9°   13. 32·3°   14. 37·8°   15. 35·5°, 48·5°
16. 68·8°, 80·0°   17. 64·6°   18. 34·2°   19. 50·6°   20. 39·1°
21. 39·5°   22. 21·6°

Exercise 10   *page 212*

1. 6·24   2. 6·05   3. 5·47   4. 9·27   5. 10·1
6. 8·99   7. 5·87   8. 4·24   9. 11·9   10. 154
11. 25·2°   12. 78·5°   13. 115·0°   14. 111·1°   15. 24·0°
16. 92·5°   17. 99·9°   18. 38·2°   19. 137·8°   20. 34·0°
21. 60·2°   22. $\widehat{B} = 112·2°$ 23. 94·55°   24. 51·2°

Exercise 11   *page 214*

1. 6·7 cm   2. 10·8 m   3. 35·6 km   4. 25·2 m
5. 38·6°, 48·5°, 92·9°   6. 40·4 m   7. 9·8 km; 085·7°
8. (a) 29·6 km   (b) 050·5°   9. (a) 10·8 m   (b) 72·6°   (c) 32·6°
10. 378 km, 048·4°   11. (a) 62·2°   (b) 2·33 km   12. 9·64 m
13. 8·6°

Revision exercise 6A   *page 215*

1. (a) 45·6°   (b) 58·0°   (c) 3·89 cm   (d) 33·8 m
2. (a) 1·75   (b) 60·3°   3. (a) 12·7 cm   (b) 5·92 cm   (c) 36·1°
4. 5·39 cm   5. (a) 220°   (b) 295°
6. 0·335 m
7. (a) 6·61 cm   (b) 12·8 cm   (c) 5·67 cm   8. (a) 86·9 cm   (b) 53·6 cm   (c) 133 cm

9. 52·4 m   **10.** (a)   14·1 cm (b)   35·3°   (c) 35·3°
11. (a) 6·63 cm   (b) 41·8°   **12.** (a)   11·3 cm (b)   8·25 cm   (c) 55·6°
13. 45·2 km, 33·6 km   **14.** 73·4°   **15.** 8·76 m, 9·99 m
16. 0·539   **17.** 4·12 cm, 9·93 cm   **18.** 26·4°

Examination exercise 6B   *page 218*

1. 0·276 m   **2.** (a) 121 m (b) (i) 280°   (ii) 069°   **3.** (a) 232°   (b) 175·4°
4. (b) 9·60 to 9·603 m   **5.** (a) (i) 60° (ii) 13 km   (b) (i) 145° (ii) 61·4°   (iii) 15·3 km
  (c) 139 to 140 km² **6.** 45, 135   **7.**   7·94
8. (a) 24·7 m   (b) 11·5 m   **9.** (a) 11·3°   (b) 233°
10. (a) 2   (b) 30 cm³   (c) 45°   (d) 37·5°   (e) 4·92 to 4·93 cm

7 Graphs

Exercise 1   *page 224*

For questions **1** to **10** end points of lines are given.

1. $(-3, -5)$ and $(3, 7)$ **2.** $(-3, -13)$ and $(3, 5)$ **3.** $(-3, -7)$ and $(3, 5)$
4. $(-2, 10)$ and $(4, 4)$ **5.** $(-2, 14)$ and $(4, 2)$ **6.** $(-3, 1)$ and $(3, 4)$
7. $(-3, -15)$ and $(3, 3)$ **8.** $(-3, 2\frac{1}{2})$ and $(3, 5\frac{1}{2})$ **9.** $(-2, -7)$ and $(4, 5)$
10. $(-2, 18)$ and $(4, 0)$ **11.** $(0, 0), (1, 4), (1·6, 1·6)$ **12.** $(0, 1), (2\frac{1}{4}, 1), (4\frac{1}{2}, 10)$
13. $(-2, -6), (1·25, 3·75), (4·5, 0·5)$ **14.** $(-1·5, 1·5)(0·67, 8), (3·5, 8), (3·5, -3·5)$
15. $(4, -2), (0·33, 5·33), (-2·28, -5·14)$ **16.** $(-2, 3), (0·6, 8·2), (2·5, 2·5), (1·33, 1·33)$
17. (a) $560   (b) 2400 km   **18.** (a) 3·4   (b) 3 h 20 m
19. (a) $440   (b) 42 km/h   **20.** (a) $4315   (b) 26 000 km

Exercise 2   *page 226*

1. $1\frac{1}{2}$ **2.** 2   **3.** 3   **4.** $1\frac{1}{2}$ **5.** $\frac{1}{2}$ **6.** $-\frac{1}{6}$
7. -7 **8.** -1 **9.** 4   **10.** -4 **11.** 5   **12.** $-1\frac{3}{7}$
13. 6   **14.** 0   **15.** 0   **16.** infinite   **17.** infinite   **18.** -8
19. $5\frac{1}{3}$ **20.** 0   **21.** $\dfrac{b-d}{a-c}$ or $\dfrac{d-b}{c-a}$ **22.** $\dfrac{n+b}{m-a}$ **23.** $\dfrac{2f}{a}$
24. -4 **25.** 0   **26.** $-\dfrac{6d}{c}$ **27.** (a) $-1\frac{1}{5}$ (b) $\frac{1}{10}$ (c) $\frac{4}{5}$
28. (a) infinite   (b) $-\frac{3}{10}$ (c) $\frac{3}{10}$ **29.** $3\frac{1}{2}$
30. (a) $\dfrac{n+4}{2m-3}$ (b) $n = -4$ (c) $m = 1\frac{1}{2}$ **31.** (b) 7·2   (c) $(4, 4)$
32. (b) yes, $PQ = PR$ (c) $(3, 2·5)$

Exercise 3   *page 227*

1. 1, 3   **2.** 1, -2 **3.** 2, 1   **4.** 2, -5 **5.** 3, 4
6. $\frac{1}{2}$, 6   **7.** 3, -2 **8.** 2, 0   **9.** $\frac{1}{4}$, -4 **10.** -1, 3
11. -2, 6   **12.** -1, 2   **13.** -2, 3   **14.** -3, -4 **15.** $\frac{1}{2}$, 3
16. $-\frac{1}{3}$, 3   **17.** 4, -5 **18.** $1\frac{1}{2}$, -4 **19.** 10, 0   **20.** 0, 4

Exercise 4   *page 228*

1. $y = 3x + 7$ **2.** $y = 2x - 9$ **3.** $y = -x + 5$ **4.** $y = 2x - 1$
5. $y = 3x + 5$ **6.** $y = -x + 7$ **7.** $y = \frac{1}{2}x - 3$ **8.** $y = 2x - 3$
9. $y = 3x - 11$ **10.** $y = -x + 5$ **11.** $y = \frac{1}{3}x - 4$

Exercise 5   *page 228*

1. A: $y = 3x - 4$ B: $y = x + 2$ **2.** C: $y = \frac{2}{3}x - 2$ D: $y = -2x + 4$
3. (a) $y = 2x + 5$ (b): $y = -x + 3$ **4.** (a) $y = 3x + 1$ B: $y = x - 2$

Exercise 6   *page 230*

1.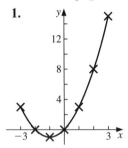

2.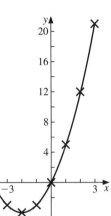

3.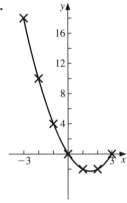

4.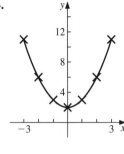

5.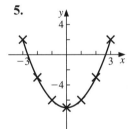

6.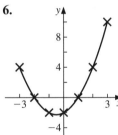

7.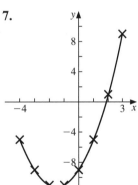

8.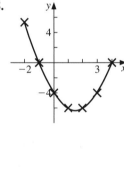

9.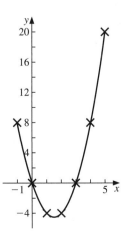

10.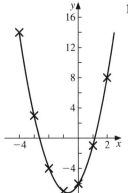

11.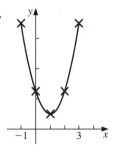

12.

13.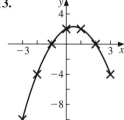

14.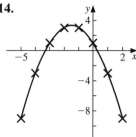

15.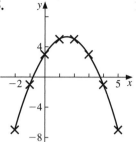

16.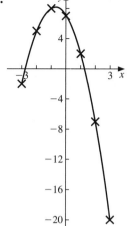

17.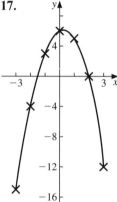

18.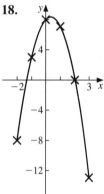

19.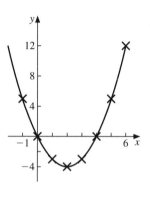

20.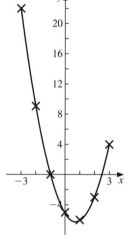

Exercise 7 *page 231*

1. (a) 4 (b) 8 (c) 10·6 2. (a) 3 (b) −5 (c) 1·5
3. (a) 7·25 (b) −2 (c) −0·8, 3·8 15. (a) 0·75 (b) 1·23
16. (a) 3·13 (b) 3·35 17. (a) −2·45 (b) 1·4
18. (a) 5 (b) 10·1 9c) −1·25 20. (a) 245 (b) 41 (c) $25 < x < 67$

Exercise 8 *page 233*

1. (a) $10\cdot7\,\text{cm}^2$ (b) $1\cdot7\,\text{cm} \times 53\,\text{cm}$ (c) $12\cdot25\,\text{cm}^2$
  (d) $3\cdot5\,\text{cm} \times 3\cdot5\,\text{cm}$ (e) square
2. $15\,\text{m} \times 30\,\text{m}$ 3. (a) 2·5 s (b) 31·3 m (c) $2 < t < 3$
4. (a) 108 m/s (b) 1·4 s (c) $2\cdot3 < t < 3\cdot6$ 6. 3·3
10. (d) $1\cdot41 \pm 0\cdot02$ (e) $1\cdot15 \leqslant x \leqslant 1\cdot3\ (\pm0\cdot02)$

Exercise 9 *page 236*

1. (a) (i) 40 km (ii) 24 km (iii) 72 km (iv) 8 km
  (b) (i) 40 miles (ii) 35 miles (iii) 10 miles (iv) 20 miles
2. (a) (i) €28 (ii) €112 (iii) €70 (b) (i) £40 (ii) £60 (iii) £100 (c) £110

3. (a) 180   (b) $C = 0.2x + 35$

4. (a) 30 litres   (b) 6 km/litre; 30 km/litre   (c) $6\frac{2}{3}$ km/litre; 6 litres

5. (a) 2000   (b) 200   (c) $1.6 \leqslant x \leqslant 2.4$

6. (a) Yes   (b) No   (c) About \$250–\$270

Exercise 10 *page 239*

1. (a) -0.4, 2.4 (b) -0.8, 3.8 (c) -1, 3 (d) -0.4, 2.4 **2.** -0.3, 3.3

3. 0.6, 3.4 **4.** 0.3, 3.7

5. (a) $y = 3$ (b) $y = -2$ (c) $y = x + 4$ (d) $y = x$ (e) $y = 6$

6. (a) $y = 6$ (b) $y = 0$ (c) $y = 4$ (d) $y = 2x$ (e) $y = 2x + 4$

7. (a) $y = -4$ (b) $y = 2x$ (c) $y = x - 2$ (d) $y = -3$ (e) $y = 2$

8. (a) $y = 5$ (b) $y = 2x$ (c) $y = 0.2$ (d) $y = 3 - x$ (e) $y = 3$

9. (a) $y = 0$ (b) $y = -2\frac{1}{2}$ (c) $y = -8x$ (d) $y = -3$ (e) $y = -5\frac{1}{2}x$

10. (a) -1.65, 3.65 (b) -1.3, 2.3 (c) -1.45, 3.45

11. (a) 1.7, 5.3 (b) 0.2, 4.8 **12.** (a) -3.3, 0.3 (b) -4.6, -0.4

13. (a) -2.35, 0.85 (b) -2.8, 1.8

14. (a) (i) -0.4, 2.4 (ii) -0.5, 2 (b) $-1.3 < x < 2.8$

15. (a) 3.4, -5.4 (b) 2.4, 7.6 (c) ± 4.2

16. (a) ± 3.7 (c) ± 2.8 **17.** (a) 1.75 (b) 0, ± 1.4

18. (a) $1.6 < x < 7.4$ (b) 6.9

19. (a) 2.6 (b) 0.45 (c) 0.64 (d) 5.7

20. (a) -1.6, 0.6 (b) $-\frac{1}{2}$, 1

Exercise 11 *page 243*

1. (a) 45 min   (b) 09 : 15   (c) 60 kmh   (d) 100 km/h   (e) 57·1 km/h

2. (a) 09 : 15   (b) 64 km/h   (c) 37·6 km/h   (d) 47 km   (e) 80 km/h

3. 11 : 05   **4.** 12 : 42   **5.** 12 : 35   **6.** $1\frac{1}{8}$ h   **7.** 1 h

8. (a) (i) B   (ii) A   (b) 8 s to 18 s   (c) About 15 s

     (d) About 9 s   (e) B   (f) A

Exercise 12 *page 245*

1. (a) $1\frac{1}{2}$ m/s^2 (b) 675 m   (c) $11\frac{1}{4}$ m/s

2. (a) 600 m   (b) 20 m/s   (c) 225 m   (d) -2 m/s^2

3. (a) 600 m   (b) $387\frac{1}{2}$ m   (c) 0 m/s^2

4. (a) 20 m/s   (b) 750 m

5. (a) 8 s   (b) 496 m   (c) 12·4 m/s

6. (a) 30 m/s   (b) $-2\frac{1}{7}$ m/s^2 (c) 20 s

7. (a) 15 m/s   (b) $2\frac{1}{4}$ m/s^2

8. (a) 40 m/s   (b) 10 s

9. (a) 50 m/s   (b) 20 s

10. (a) 20 m/s   (b) 20 s

Exercise 13 *page 247*

1. 225 m   **2.** 60 m   **3.** $\frac{2}{3}$ km

4. 10 s   **5.** 3 min   **6.** 50 m

7. 18·75 m   **8.** 1·39 km   **9.** 250 m

10. Yes. Stopping distance $= 46.5$ m

11. 94 375 m

12. (a) 0.8 m/s^2 (b) 670 m   **13.** (a) 0.35 m/s^2 (b) 260 m

Revision exercise 7A   page 248

1. (a) $y = x - 7$ (b) $y = 2x + 5$ (c) $y = -2x + 10$ (d) $y = \dfrac{x+1}{2}$

2. (a) 2 (b) 1 (c) $-3\frac{1}{2}$ (d) 0 (e) 10

3. (a) 2, -7 (b) -4, 5 (c) $\frac{1}{2}$, 4 (d) $-\frac{1}{2}$, 5 (e) -2, 12 (f) $-\frac{2}{3}$, 8

4. A : $y = 6$; B : $y = \frac{1}{2}x - 3$; C : $y = 10 - x$; D : $y = 3x$

5. A : $4y = 3x - 16$; B : $2y = x - 8$; C : $2y + x = 8$; D : $4y + 3x = 16$

6. (a) $y = 2x - 3$ (b) $y = 3x + 4$ (c) $y = 10 - x$ (d) $y = 7$

7. (a) A(0, -8), (4, 0) (b) 2 (c) $y = 2x - 8$

8. 25 sq. units   9. -3 10. 219

12. (a) $y = 3x$ (b) $y = 0$ (c) $y = 11 - x$ (d) $y = 5x$

13. (a) 1·56, -2·56 (b) ± 2·24 (c) ± 2·65

14. (a) 0·84, 4·15 (b) $0.65 < x < 3.85$ (c) 3·3

15. (a) 9·2 (b) 0·6 (c) 1·4 (d) 1·65

16. (a) 0·3 m/s^2 (b) 1050 m (c) 40 s

17. (a) 30 m/s (b) 600 m

Examination exercise 7B   page 251

1. (a) $y = 2x - 6$ (b) (3, 0)   2. (a) $y = \dfrac{8 - 3x}{2}$.   (b) $-\dfrac{3}{2}$ (c) (0, 4)

3. $y = \dfrac{1}{2}x + 5$ 4. (a) 1000, 1400, 1960, 2744, 3842 (c) 3·2 or 3·3

5. (a) (i) 3   (ii) -4·25 to -4 (b) (i) -1·6, 2·0, 8·6 to 8·63   (ii) 9·2

   (c) -9, 3   (d) $0 < x < 6$ (e) (i) $y = 1 - x$ (ii) $y = 3$

6. (a) $p = 5(.04)$, $q = 0$, $r = 8$·66 (c) (i) -2·95 to -2·6, -0·75 to -0·6, 0·5 to 0·6

   (ii) $a = 3$ $b = -1$ (d) -4·5 to -3

7. (a) 1·05 m/s^2 (b) 3360 m   (c) 18·7 m/s

8. (a) (i) 10·625 hours (iii) 10 hours 37 mins 30 secs   (b) (i) 0608   (ii) 78·7 km/h

   (c) (i) increasing (more slowly)   (ii) decreasing   (iii) 12·5 m/s^2 (iv) 170 m

   (v) areas above and below broken line are approximately equal   (vi) 61·2 km/h

9. (a) 0·9, -10·1   (c) any integer $\geqslant 1$ (e) (i) -0·45 to -0·3, 0·4 to 0·49, 2·9 to 2·99

   (ii) $2x^3 - 6x^2 + 1 = 0$ (f) (i) tangent drawn with gradient ≈ 2

8 Sets, Vectors and Functions

Exercise 1   page 257

1. (a) 8 (b) 3 (c) 4 (d) 18 (e) 7

2. (a) 9 (b) 5 (c) 4 (d) 20 (e) 31

3. (a) 8 (b) 3 (c) 3 (d) 2 (e) 18 (f) 0

4. (a) 59 (b) 11 (c) 5 (d) 40 (e) 11 (f) 124

5. (a) 120 (b) 120 (c) 490 (d) 80 (e) 40 (f) 10 (g) 500

Exercise 2   page 259

1. (a) {5, 7} (b) {1, 2, 3, 4, 5, 6, 7, 8, 9, 11, 13} (c) 5 (d) 11

   (e) true (f) true (g) false (h) true

2. (a) {2, 3, 5, 7}   (b) {1, 2, 3, … 9}   (c) 4   (d) ∅   (e) false
   (f) true   (g) false   (h) true
3. (a) {2, 4, 6, 8, 10}   (b) {16, 18, 20}   (c) ∅   (d) 15   (e) 11
   (f) 21   (g) false   (h) false   (i) true   (j) true
4. (a) {1, 3, 4, 5}   (b) {1, 5}   (c) 1   (d) {1, 5}   (e) {1, 3, 5, 10}
   (f) 4   (g) true   (h) false   (i) true
5. (a) 4   (b) 3   (c) {b, d}   (d) {a, b, c, d, e}   (e) 5
   (f) 2
6. (a) 2   (b) 4   (c) {1, 2, 4, 6, 7, 8, 9}   (d) {7, 9}
   (e) {1, 2, 4}   (f) {1, 2, 4, 7, 9}   (g) {1, 2, 4, 6, 8}   (h) {6, 7, 8, 9}   (i) {1, 2, 4, 7, 9}

Exercise 3   *page 260*

1. (a) 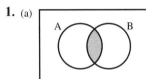   (b) 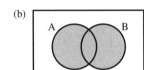   (c)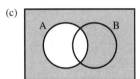

(d) 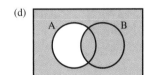   (e) 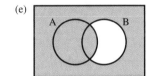   (f)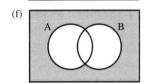

2. (a) 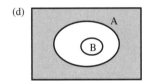   (b) 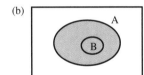   (c)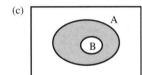

(d)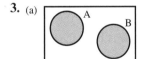

3. (a)    (b) 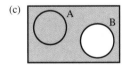   (c)    (d)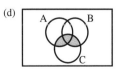

4. (a) 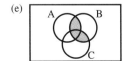   (b) 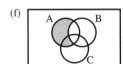   (c) 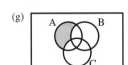   (d)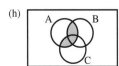

(e) 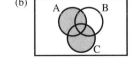   (f) 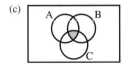   (g)   (h)

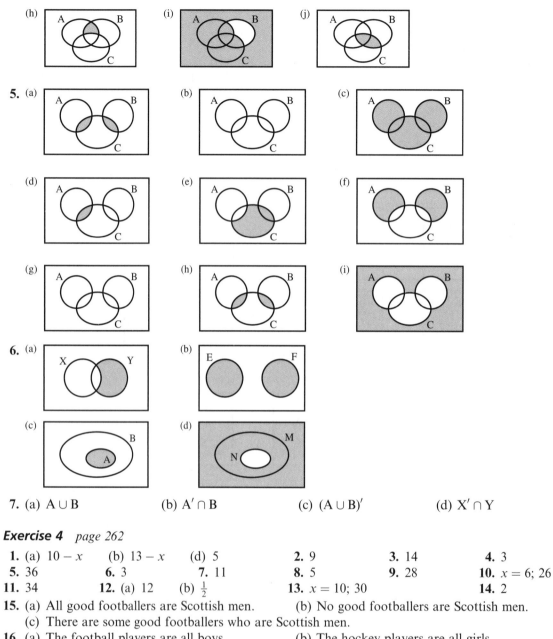

7. (a) $A \cup B$ (b) $A' \cap B$ (c) $(A \cup B)'$ (d) $X' \cap Y$

Exercise 4   *page 262*

1. (a) $10 - x$ (b) $13 - x$ (d) 5   **2.** 9   **3.** 14   **4.** 3

5. 36   **6.** 3   **7.** 11   **8.** 5   **9.** 28   **10.** $x = 6$; 26

11. 34   **12.** (a) 12   (b) $\frac{1}{2}$ **13.** $x = 10$; 30   **14.** 2

15. (a) All good footballers are Scottish men.   (b) No good footballers are Scottish men.

     (c) There are some good footballers who are Scottish men.

16. (a) The football players are all boys.   (b) The hockey players are all girls.

     (c) There are some people who play both football and hockey.

     (d) There are no boys who play hockey.   (e) $B \cap F = \varnothing$ (f) $H \cup F = \mathscr{E}$

17. (a) $S \cap T = \varnothing$ (b) $F \subset T$ (c) $S \cap F \neq \varnothing$

     (d) All living creatures are either spiders, animals that fly or animals which taste nice.

     (e) Animals which taste nice are all spiders.

18. (a) All tigers who believe in fairies also believe in Eskimos.

     (b) All tigers who believe in fairies or Eskimos are in hospital.

     (c) There are no tigers in hospital who believe in Eskimos.

     (d) $H \subset T$ (e) $T \cap X \neq \varnothing$

19. (a) There are no good bridge players called Peter.
   (b) All school teachers are either called Peter, are good bridge players or are women.
   (c) There are some women teachers called Peter.
   (d) $W \cap B = \varnothing$ (e) $B \subset (W \cap P)$

Exercise 5   *page 266*

1. d	**2.** 2c	**3.** 3c	**4.** 3d	**5.** 5d
6. 3c	**7.** −2d	**8.** −2c	**9.** −3c	**10.** −c
11. c + d	**12.** c + 2d	**13.** 2c + d	**14.** 3c + d	**15.** 2c + 2d
16. 2c + 3d	**17.** 2c − d	**18.** 3c − d	**19.** −c + 2d	**20.** −c + 3d
21. −c + d	**22.** −c − 2d	**23.** −2c − 2d	**24.** −3c − 6d	**25.** −2c + 3d
26. c + 6d	**27.** $\overrightarrow{QI}$	**28.** $\overrightarrow{QU}$	**29.** $\overrightarrow{QH}$	**30.** $\overrightarrow{QB}$
31. $\overrightarrow{QF}$	**32.** $\overrightarrow{QJ}$	**33.** $\overrightarrow{QZ}$	**34.** $\overrightarrow{QL}$	**35.** $\overrightarrow{QE}$
36. $\overrightarrow{QX}$	**37.** $\overrightarrow{QW}$	**38.** $\overrightarrow{QK}$		

39. (a) −a   (b) b − a   (c) −b   (d) a + b
40. (a) a + b   (b) a − 2b   (c) −a + b   (d) −a − b
41. (a) −a − b   (b) 3a − b   (c) 2a − b   (d) −2a + b
42. (a) a − 2b   (b) a − b   (c) 2a   (d) −2a + 3b
43. (a) −3a + 2b   (b) 3a − b   (c) 4a − b   (d) −4a + b
44. (a) 2a − c   (b) 2a − c   (c) 3a   (d) a + b + c   (e) −3a − b
45. (a) b − c   (b) 2b + 2c   (c) a + 2b + 2c   (d) −a − b   (e) c − a − b
46. (a) a + c   (b) −a + c   (c) a + b + c   (d) b − c   (e) −a + 2c

Exercise 6   *page 268*

1. (a) a   (b) −a + b   (c) 2b   (d) −2a   (e) −2a + 2b
   (f) −a + b   (g) a + b   (h) b   (i) −b + 2a   (j) −2b + a
2. (a) a   (b) −a + b   (c) 3b   (d) −2a   (e) −2a + 3b
   (f) $-a + \frac{3}{2}b$ (g) $a + \frac{3}{2}b$ (h) $\frac{3}{2}b$ (i) −b + 2a   (j) −3b + a
3. (a) 2a   (b) −a + b   (c) 2b   (d) −3a   (e) −3a + 2b
   (f) $-\frac{3}{2}a + b$ (g) $\frac{3}{2}a + b$ (h) $\frac{1}{2}a + b$ (i) −b + 3a   (j) −2b + a
4. (a) $\frac{1}{2}a$ (b) −a + b   (c) 4b   (d) $-\frac{3}{2}a$ (e) $-\frac{3}{2}a + 4b$
   (f) $-a + \frac{8}{3}b$ (g) $\frac{1}{2}a + \frac{8}{3}b$ (h) $-\frac{1}{2}a + \frac{8}{3}b$ (i) $\frac{3}{2}a - b$ (j) a − 4b
5. (a) 5a   (b) b − a   (c) $\frac{3}{2}b$ (d) −6a   (e) $\frac{3}{2}b - 6a$
   (f) b − 4a   (g) 2a + b   (h) a + b   (i) 6a − b   (j) $a - \frac{3}{2}b$
6. (a) 4a   (b) b − a   (c) 3b   (d) −5a   (e) 3b − 5a
   (f) $\frac{3}{4}b - \frac{5}{4}a$ (g) $\frac{15}{4}a + \frac{3}{4}b$ (h) $\frac{11}{4}a + \frac{3}{4}b$ (i) 5a − b   (j) a − 3b
7. $\frac{1}{2}s - \frac{1}{2}t$ **8.** $\frac{1}{3}a + \frac{2}{3}b$ **9.** a + c − b   **10.** 2m + 2n
11. (a) b − a   (b) b − a   (c) 2b − 2a   (d) b − 2a   (e) b − 2a
   (f) 2b − 3a
12. (a) y − z   (b) $\frac{1}{2}y - \frac{1}{2}z$ (c) $\frac{1}{2}y + \frac{1}{2}z$ (d) $-x + \frac{1}{2}y + \frac{1}{2}z$ (e) $-\frac{2}{3} + \frac{1}{3}y + \frac{1}{3}z$
   (f) $\frac{1}{3}x + \frac{1}{3}y + \frac{1}{3}z$

Exercise 7   *page 272*

1.   **2.**   **3.**   **4.** 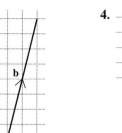   **5.**

6. 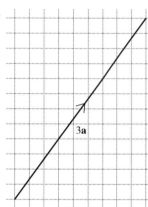   **7.**    **8.** 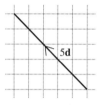   **9.**

10. 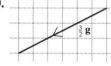   **11.**   **12.**

13. $\begin{pmatrix} -11 \\ 9 \end{pmatrix}$ **14.** $\begin{pmatrix} -1 \\ -4 \end{pmatrix}$ **15.** $\begin{pmatrix} 4 \\ 8 \end{pmatrix}$ **16.** $\begin{pmatrix} 4 \\ 3 \end{pmatrix}$ **17.** $\begin{pmatrix} 8 \\ -7 \end{pmatrix}$ **18.** $\begin{pmatrix} 18 \\ -1 \end{pmatrix}$ **19.** $\begin{pmatrix} 7 \\ -4 \end{pmatrix}$ **20.** $\begin{pmatrix} -18 \\ -16 \end{pmatrix}$

21. $\begin{pmatrix} 13 \\ 19 \end{pmatrix}$ **22.** $\begin{pmatrix} 8 \\ 5 \end{pmatrix}$ **23.** $\begin{pmatrix} 10 \\ -13 \end{pmatrix}$ **24.** $\begin{pmatrix} 17 \\ 35 \end{pmatrix}$ **25.** $\begin{pmatrix} 4 \\ 8 \end{pmatrix}$ **26.** $\begin{pmatrix} 4 \\ 3 \end{pmatrix}$ **27.** $\begin{pmatrix} -1 \\ 1 \end{pmatrix}$ **28.** $\begin{pmatrix} -16 \\ 3 \end{pmatrix}$

29. $\begin{pmatrix} -2\frac{1}{2} \\ -3 \end{pmatrix}$ **30.** $\begin{pmatrix} -3\frac{1}{2} \\ \frac{1}{2} \end{pmatrix}$ **31.** $\begin{pmatrix} -3 \\ -7 \end{pmatrix}$ **32.** $\begin{pmatrix} -2 \\ 12 \end{pmatrix}$ **33.** $\begin{pmatrix} 4 \\ 8 \end{pmatrix}$ **34.** $\begin{pmatrix} -5\frac{1}{2} \\ 4\frac{1}{2} \end{pmatrix}$ **35.** $\begin{pmatrix} 0 \\ 0 \end{pmatrix}$ **36.** $\begin{pmatrix} -2\frac{1}{2} \\ -6 \end{pmatrix}$

37. (b) **l** and **s**; **n** and **r**; **p** and **t**; **m** and **q**

38. (a) true   (b) true   (c) true   (d) true   (e) false   (f) false

39. (a) 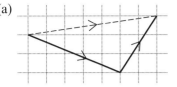   (b)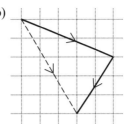

40. (a)    (b)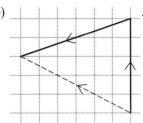

(c) 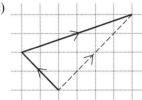   (d)

Exercise 8   *page 273*

1. $\begin{pmatrix} 2 \\ -2 \end{pmatrix}$ **2.** $\begin{pmatrix} 6 \\ -2 \end{pmatrix}$ **3.** (b) $\begin{pmatrix} 0 \\ 3 \end{pmatrix}$; $\begin{pmatrix} -5 \\ -5 \end{pmatrix}$ **4.** (b) $\begin{pmatrix} 4 \\ 2 \end{pmatrix}$; $\begin{pmatrix} -7 \\ 0 \end{pmatrix}$

5. (a) $\begin{pmatrix} 3 \\ -3 \end{pmatrix}$ (b) $\begin{pmatrix} 1\frac{1}{2} \\ -1\frac{1}{2} \end{pmatrix}$ (c) $\begin{pmatrix} 3\frac{1}{2} \\ 3\frac{1}{2} \end{pmatrix}$; M$(3\frac{1}{2}, 3\frac{1}{2})$

6. (a) $\begin{pmatrix} -1 \\ 6 \end{pmatrix}$ (b) $\begin{pmatrix} -\frac{1}{2} \\ 3 \end{pmatrix}$ (c) $\begin{pmatrix} 5\frac{1}{2} \\ -4 \end{pmatrix}$; M$(5\frac{1}{2}, -4)$

7. (a) (i) $\begin{pmatrix} -6 \\ 3 \end{pmatrix}$ (ii) $\begin{pmatrix} -2 \\ 1 \end{pmatrix}$ (iii) $\begin{pmatrix} 2 \\ 3 \end{pmatrix}$ (b) (i) $\begin{pmatrix} 0 \\ -9 \end{pmatrix}$ (ii) $\begin{pmatrix} 0 \\ -3 \end{pmatrix}$ (iii) $\begin{pmatrix} -2 \\ 2 \end{pmatrix}$

8. $\begin{pmatrix} 1 \\ -2 \end{pmatrix}$ or $\begin{pmatrix} -1 \\ 2 \end{pmatrix}$ **9.** $\begin{pmatrix} 0 \\ 2 \end{pmatrix}$ or $\begin{pmatrix} 0 \\ -2 \end{pmatrix}$

10. (a) $\mathbf{q} - \mathbf{p}$ (b) $\mathbf{q} + 2\mathbf{p}$ (c) $\mathbf{p} + \mathbf{q}$

11. (a) $\begin{pmatrix} 1 \\ -3 \end{pmatrix}$ (b) $\begin{pmatrix} -1 \\ 3 \end{pmatrix}$ (c) $\begin{pmatrix} 3 \\ 1 \end{pmatrix}$ (d) $\begin{pmatrix} -3 \\ -1 \end{pmatrix}$

Exercise 9   *page 275*

1. 5   **2.** $\sqrt{17}$ **3.** 13   **4.** 3   **5.** 5

6. $\sqrt{45}$ **7.** $\sqrt{74}$ **8.** $\sqrt{208}$ **9.** 10   **10.** $\sqrt{89}$

11. (a) $\sqrt{320}$ (b) no   **12.** (a) $\sqrt{148}$ (b) no

13. $\sqrt{29}$ **14.** $\sqrt{26}$ **15.** $\sqrt{10}$

16. (a) 5   (b) $n = \pm 4$ **17.** (a) 13   (b) $m = \pm 13$

18. (a) 5   (b) $p = 0$ **19.** (a) 9   (b) 6   (c) 5

20. (a) 30   (b) 5   (c) $\sqrt{50}$ (d) 4

Exercise 10   *page 277*

1. (a) $2\mathbf{a}$; $3\mathbf{b}$ (b) $-\mathbf{b} + \mathbf{a}$ (c) $-3\mathbf{b} + 2\mathbf{a}$ (d) $4\mathbf{a} - 3\mathbf{b}$
    (e) $4\mathbf{a} - 6\mathbf{b}$ (f) $\overrightarrow{EC} = 2\overrightarrow{ED}$

2. (a) $2\mathbf{b}$; $\frac{5}{2}\mathbf{a}$ (b) $-\mathbf{a} + \mathbf{b}$ (c) $-\frac{5}{2}\mathbf{a} + 2\mathbf{b}$ (d) $-5\mathbf{a} + 6\mathbf{b}$
    (e) $-\frac{15}{2}\mathbf{a} + 6\mathbf{b}$ (f) $\overrightarrow{XC} = 3\overrightarrow{XY}$

3. (a) $-\mathbf{b} + \mathbf{a}$; $-3\mathbf{b} + 3\mathbf{a}$ (b) $-2\mathbf{b} + \frac{3}{2}\mathbf{a}$ (c) $-\frac{1}{2}\mathbf{a}$; $-2\mathbf{b} + \frac{3}{2}\mathbf{a}$

4. (a) $-\mathbf{a} + \mathbf{b}$; $-\frac{2}{3}\mathbf{a} + \frac{2}{3}\mathbf{b}$ (b) $\frac{1}{2}\mathbf{a}$; $-\frac{1}{6}\mathbf{a} + \frac{2}{3}\mathbf{b}$ (c) $-\frac{1}{2}\mathbf{a} + 2\mathbf{b}$ (d) $\overrightarrow{MX} = 3\overrightarrow{MP}$

5. (a) $-\mathbf{b} + \mathbf{a}$; $-3\mathbf{a} + \mathbf{b}$ (b) $-\frac{3}{2}\mathbf{a} + \frac{1}{2}\mathbf{b}$ (c) $(k - \frac{3}{2})\mathbf{a} + (\frac{1}{2} - k)\mathbf{b}$ (d) $k = \frac{3}{2}$

6. (a) $-\mathbf{a} + \mathbf{b}$ (b) $-\frac{1}{4}\mathbf{a} + \frac{1}{4}\mathbf{b}$ (c) $\mathbf{a} + (m - 1)\mathbf{b}$ (d) $m = \frac{4}{3}$

7. (a) $-\mathbf{c} + \mathbf{d}$ (b) $-\frac{1}{5}\mathbf{c} + \frac{1}{5}\mathbf{d}$ (c) $\frac{4}{5}\mathbf{c} + \frac{1}{5}\mathbf{d}$ (d) $\mathbf{c} + (n - 1)\mathbf{d}$
    (e) $n = \frac{5}{4}$

8. (a) $-\mathbf{a} + \mathbf{b}$; $-\frac{1}{2}\mathbf{a} + \frac{1}{2}\mathbf{b}$; $\frac{1}{2}\mathbf{a} + \frac{1}{2}\mathbf{b}$ (b) $\frac{1}{3}\mathbf{a} + \frac{1}{3}\mathbf{b}$ (c) $-\frac{2}{3}\mathbf{a} + \frac{1}{3}\mathbf{b}$
    (d) $-\mathbf{a} + \frac{1}{2}\mathbf{b}$ (e) $m = \frac{2}{3}$

9. (a) $-\mathbf{a} + \mathbf{b}$ (b) $\frac{1}{2}\mathbf{b}$ (c) $-\mathbf{a} + \mathbf{c}$ (d) $-\frac{1}{2}\mathbf{a} + \frac{1}{2}\mathbf{c}$
    (e) $\frac{1}{2}\mathbf{a} + \frac{1}{2}\mathbf{c}$ (f) $-\frac{1}{2}\mathbf{b} + \frac{1}{2}\mathbf{a} + \frac{1}{2}\mathbf{c}$ (g) $\mathbf{a} + \mathbf{c} = \mathbf{b}$

10. (a) $-\mathbf{b} + \mathbf{a}$ (b) $m\mathbf{a} + (1 - m)\mathbf{b}$ (c) $4\mathbf{a} + 2\mathbf{b}$ (d) $n = \frac{1}{6}$, $m = \frac{2}{3}$

11. (a) $-\mathbf{c} + \mathbf{d}$; $-\frac{1}{4}\mathbf{c} + \frac{1}{4}\mathbf{d}$; $\frac{3}{4}\mathbf{c} + \frac{1}{4}\mathbf{d}$ (b) $-\mathbf{c} + \frac{1}{2}\mathbf{d}$

    (c) $(1 - h)\mathbf{c} + \dfrac{h}{2}\mathbf{d}$ (d) $(1 - h)\mathbf{c} + \dfrac{h}{2}\mathbf{d} = k;\ \dfrac{3}{4}\mathbf{c} + \dfrac{k}{4}\mathbf{d};\ h = \frac{2}{5}$, $k = \frac{4}{5}$

Exercise 11   *page 280*

1. (a) 5, 10, 1   (b) 21, 101, -29

2. $x \to \boxed{\times 5} \to \boxed{+4} \to 5x + 4$

3. $x \to \boxed{-4} \to \boxed{\times 3} \to 3(x - 4)$

4. $x \to \boxed{\times 2} \to \boxed{+7} \to \boxed{\text{square}} \to (2x + 7)^2$

5. $x \to \boxed{\times 5} \to \boxed{+9} \to \boxed{\div 4} \to \dfrac{5x + 9}{4}$

6. $x \to \boxed{\times -3} \to \boxed{\text{subtract from 4}} \to \boxed{\div 5} \to \dfrac{4 - 3x}{5}$

7. $x \to \boxed{\text{square}} \to \boxed{\times 2} \to \boxed{+1} \to 2x^2 + 1$

8. $x \to \boxed{\text{square}} \to \boxed{\times 3} \to \boxed{\div 2} \to \boxed{+5} \to \dfrac{3x^2}{2} + 5$

9. $x \to \boxed{\times 4} \to \boxed{-5} \to \boxed{\text{square root}} \to \sqrt{(4x - 5)}$

10. $x \rightarrow \boxed{\text{square}} \rightarrow \boxed{+10} \rightarrow \boxed{\text{square root}} \rightarrow \boxed{\times 4} \rightarrow 4\sqrt{(x^2 + 10)}$

11. $x \rightarrow \boxed{\times 3} \rightarrow \boxed{\text{subtract from 7}} \rightarrow \boxed{\text{square}} \rightarrow (7 - 3x)^2$

12. $x \rightarrow \boxed{\times 3} \rightarrow \boxed{+1} \rightarrow \boxed{\text{square}} \rightarrow \boxed{\times 4} \rightarrow \boxed{+5} \rightarrow 4(3x + 1)^2 + 5$

13. $x \rightarrow \boxed{\text{square}} \rightarrow \boxed{\text{subtract from 5}} \rightarrow 5 - x^2$

14. $x \rightarrow \boxed{\text{square}} \rightarrow \boxed{+1} \rightarrow \boxed{\text{square root}} \rightarrow \boxed{\times 10} \rightarrow \boxed{+6} \rightarrow \boxed{\div 4} \rightarrow \dfrac{10\sqrt{(x^2 + 1)} + 6}{4}$

15. $x \rightarrow \boxed{\text{cube}} \rightarrow \boxed{\div 4} \rightarrow \boxed{+1} \rightarrow \boxed{\text{square}} \rightarrow \boxed{\text{subtract 6}} \rightarrow \left(\dfrac{x^3}{4} + 1\right)^2 - 6$

16. (a) $-9, 11, \frac{1}{2}$ (b) $0\cdot8, -2\cdot7, \frac{1}{80}$ (c) $4, 1\cdot2, 36$

17. (a) 0 (b) 6 (c) 12

18. (a) 10 (b) $\frac{1}{2}$ (c) 2

19. (a) $6, 24, 6$ (b) $0, \sqrt{2}, \sqrt{6}$ (c) $-6, 6, 9\frac{3}{4}$

20. (a) ±3 (b) ±3 (c) ±2 (d) ±6

21. (a) $10, 21$ (b) $111, 411, 990, 112$

22. (a) 7 (b) 10 (c) 5 (d) 14
     (e) 7 (f) 7

23. (a) 3 (b) 6 (c) 8 (d) 10

24. (a) 11 (b) 17 (c) 7

25. (a) 5 (b) 17 (c) $1\frac{1}{2}$ (d) 3

26. $a = 3, b = 5$ **27.** $a = 2, b = -5$ **28.** $a = 7, b = 1$

Exercise 12 *page 283*

1. (a) $x \mapsto 4(x + 5)$ (b) $x \mapsto 4x + 5$ (c) $x \mapsto (4x)^2$ (d) $x \mapsto 4x^2$
     (e) $x \mapsto x^2 + 5$ (f) $x \mapsto 4(x^2 + 5)$ (g) $x \mapsto [4(x + 5)]^2$

2. (a) $-2\cdot5$ (b) $\pm\sqrt{\frac{5}{3}}$

3. (a) $x \mapsto 2(x - 3)$ (b) $x \mapsto 2x - 3$ (c) $x \mapsto x^2 - 3$ (d) $x \mapsto (2x)^2$
     (e) $x \mapsto (2x)^2 - 3$ (f) $x \mapsto (2x - 3)^2$

4. (a) 2 (b) 11 (c) 6 (d) 2
     (e) 1 (f) 64

5. (a) -3 (b) 2 (c) $1\frac{1}{2}$ (d) 5

6. (a) $x \rightarrow 2(3x - 1) + 1$ (b) $x \rightarrow 3(2x + 1) - 1$ (c) $x \rightarrow 2x^2 + 1$ (d) $x \rightarrow (3x - 1)^2$
     (e) $x \rightarrow 2(3x - 1)^2 + 1$ (f) $x \rightarrow 3(2x^2 + 1) - 1$

7. (a) 11 (b) 9 (c) 11 (d) 14
     (e) 81 (f) -1

8. (a) 2 (b) $0, 2$ (c) $\pm\sqrt{2}$

9. $x \mapsto \dfrac{x + 2}{5}$ **10.** $x \mapsto \dfrac{x}{5} + 2$ **11.** $x \mapsto \dfrac{x}{6} - 2$ **12.** $x \mapsto \dfrac{3x - 1}{2}$

13. $x \mapsto \dfrac{4x}{3} + 1$ **14.** $x \mapsto \dfrac{x - 2}{6}$ **15.** $x \mapsto \dfrac{2x - 24}{5}$ **16.** $x \mapsto \dfrac{x - 3}{-7}$

17. $x \mapsto \dfrac{3x - 12}{-5}$ **18.** $x \mapsto 10 - 3x$ **19.** $x \mapsto \dfrac{4(5x + 3) + 1}{2}$ **20.** $x \mapsto \dfrac{7x - 30}{6}$

21. $x \mapsto 20x - 164$

22. (a) $\sqrt{x}$ (b) log (c) $x!$ (d) x^2 (e) $\dfrac{1}{x}$ (f) tan (g) $\dfrac{1}{x}$ (h) $\sqrt{}$ (i) cos

(j) log or ln (k) tan (l) $x!$ (m) cos (n) sin (o) cos (p) $\dfrac{1}{x}$ (q) $x!$ (r) log

Revision exercise 8A page 285

1. (a) $\{5\}$ (b) $\{1, 3, 5, 6, 7\}$ (c) $\{2, 4, 6, 7, 8\}$ (d) $\{2, 4, 8\}$ (e) $\{1, 2, 3, 4, 5, 8\}$ **2.** 32

3. (a) 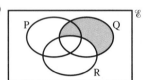 (b) 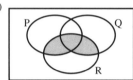 (c)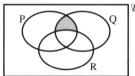

4. (a) $(A \cup B)' \cap C$ (b) $(A \cup B) \cap C'$

5. (a) (i) $S \subset T$ (ii) $S \cap M' \neq \varnothing$ (b) There are no women on the train over 25 years old.

6. (a) 4 (b) 11 (c) 17 **7.** (a) $\mathbf{r} - \mathbf{p}$ (b) $\frac{1}{2}\mathbf{p} - \frac{1}{2}\mathbf{p}$ (c) $\frac{1}{2}\mathbf{r} + \frac{1}{2}\mathbf{p}$

8. (a) 5 (b) $\sqrt{68}$ (c) $\sqrt{41}$ **9.** $n = 2, m = -15$

10. (a) $\begin{pmatrix} -1 \\ 4 \end{pmatrix}$ (b) $\begin{pmatrix} 4 \\ 4 \end{pmatrix}$ (c) $\begin{pmatrix} 6 \\ -4 \end{pmatrix}$ (d) $\begin{pmatrix} 2 \\ 2 \end{pmatrix}$

11. (a) $\mathbf{a} - \mathbf{c}$ (b) $\frac{1}{2}\mathbf{a} + \mathbf{c}$ (c) $\frac{1}{2}\mathbf{a} - \frac{1}{2}\mathbf{c}$ CA is parallel to NM and CA $= 2$ NM.

12. $m = 3, n = 2$ **13.** (a) $\begin{pmatrix} -3 \\ 2 \end{pmatrix}$ (b) $\begin{pmatrix} -1\frac{1}{2} \\ 1 \end{pmatrix}$ (c) $\begin{pmatrix} 1\frac{1}{2} \\ 3 \end{pmatrix}$

14. (a) -5 (b) 0 (c) -3 (d) $8; ff : x \mapsto 4x - 9$

15. $f^{-1} : x \mapsto \dfrac{(x - 4)}{3}; h^{-1} : x \mapsto 5x + 2$ (a) 3 (b) $5\frac{1}{3}$ **16.** (a) 3 (b) 0, 5

Examination exercise 8B page 287

1. (a) 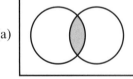 (b) 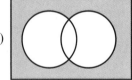 (c)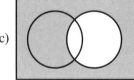

2. (a) 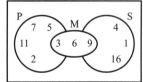 (b) 4 **3.** (a) ϕ (b) $\mathscr{E}$ $\notin \mathscr{C} \not\ni \mathfrak{A}$

4. (a) $2\mathbf{a} - \mathbf{g}$ (b) $\dfrac{5}{2}\mathbf{a} + \dfrac{1}{2}\mathbf{g}$ **5.** (b) $\dfrac{1}{2}\mathbf{a} + \dfrac{1}{3}\mathbf{b}$ (c) $-\dfrac{1}{2}\mathbf{a} + \dfrac{2}{3}\mathbf{b}$

6. (a) (i) $\mathbf{c} - \mathbf{d}$ (ii) $\mathbf{d} - 0.5\,\mathbf{c}$ (iii) $1.5\,\mathbf{c} - \mathbf{d}$

(b) (i) $120°$ (ii) 27.7 cm^2 (iii) 13.9 cm (iv) 166 to 167 cm^2

7. (a) -14 (b) $2x^3 - 6x^2 + 12x - 9$ (c) $\dfrac{x + 1}{2}$

8. (a) 8 (b) $\dfrac{5-x}{3}$ (c) 8

9. (a) 1 (b) 0 (c) $\dfrac{\tan x - 6}{2}$

10. (a) 2 (b) $\dfrac{2+2x}{2x-1}$ (c) $\dfrac{3}{x-1}$ (d) 256 (e) -3

9 Matrices and Transformations

Exercise 1 *page 292*

1. $\begin{pmatrix} 2 & 4 \\ 4 & 2 \end{pmatrix}$ **2.** $\begin{pmatrix} 1 & 6 & -1 \\ 7 & -10 & 6 \end{pmatrix}$ **3.** — **4.** $\begin{pmatrix} -4 & 2 \\ 0 & 0 \end{pmatrix}$

5. $(8 \quad 10)$ **6.** $\begin{pmatrix} 0 & 15 \\ 3 & -6 \end{pmatrix}$ **7.** $\begin{pmatrix} 0 & -3 \\ 1 & 0 \\ -9 & -5 \end{pmatrix}$ **8.** $\begin{pmatrix} 4 & 3 \\ 7 & 6 \end{pmatrix}$

9. — **10.** $\begin{pmatrix} 6 & 7 \\ 5 & 0 \end{pmatrix}$ **11.** $\begin{pmatrix} -1 & 9 \\ -2 & -1 \\ 25 & 10 \end{pmatrix}$ **12.** $\begin{pmatrix} 1 & -5\frac{1}{2} \\ \frac{1}{2} & 4 \end{pmatrix}$

13. $\begin{pmatrix} -1 & 12 \\ 4 & 7 \end{pmatrix}$ **14.** $\begin{pmatrix} 15 & 20 \\ -4 & -9 \end{pmatrix}$ **15.** $\begin{pmatrix} 5 & -10 \\ 2 & 7 \end{pmatrix}$ **16.** $\begin{pmatrix} 3 & 14 \\ -2 & 9 \end{pmatrix}$

17. $\begin{pmatrix} 12 \\ 13 \end{pmatrix}$ **18.** $\begin{pmatrix} 5 \\ 13 \end{pmatrix}$ **19.** $\begin{pmatrix} 14 & 1 \\ -32 & -13 \end{pmatrix}$ **20.** $\begin{pmatrix} 8 & -27 \\ 23 & -2 \end{pmatrix}$

21. $\begin{pmatrix} 8 & -27 \\ 23 & -2 \end{pmatrix}$ **22.** — **23.** — **24.** $\begin{pmatrix} 16 & 20 \\ 4 & 5 \\ 12 & 15 \end{pmatrix}$

25. $\begin{pmatrix} 30 & 40 \\ -8 & -18 \end{pmatrix}$ **26.** $\begin{pmatrix} 7 \\ 36 \end{pmatrix}$ **27.** $\begin{pmatrix} 12 & 15 \\ 4 & 5 \end{pmatrix}$ **28.** (17)

29. $\begin{pmatrix} -18 & 14 & -6 \\ 26 & -26 & 8 \end{pmatrix}$ **30.** $\begin{pmatrix} 1 & -6 \\ 18 & 13 \end{pmatrix}$ **31.** $\begin{pmatrix} -107 & -84 \\ 252 & 61 \end{pmatrix}$ **32.** —

33. $\begin{pmatrix} -9 & 13 & -17 \\ 3 & -4 & 5 \\ 0 & -7 & 14 \end{pmatrix}$ **34.** $\begin{pmatrix} 59 \\ -21 \end{pmatrix}$ **35.** $\begin{pmatrix} 1 & 5 & 1 \\ 3 & -11 & 0 \\ 22 & -20 & 7 \end{pmatrix}$ **36.** $\begin{pmatrix} 45 & -140 \\ -28 & 101 \end{pmatrix}$

37. $x = 6, y = 3, z = 0$ **38.** $x = 4, y = 5, z = 7, w = -5, v = 0$
39. $a = 4, b = 13, c = 23, d = 2$ **40.** $x = 1, y = 4$
41. $m = 5, n = -\frac{1}{3}$ **42.** $p = 3, q = -1$
43. $x = 1, y = 2, z = -1, w = -2$ **44.** $y = 2, z = -1, x = 1, w = -2$
45. $a = -3, e = 4, k = 2$ **46.** $m = 3, n = 5, p = 3, q = 3$
47. $x = 2\frac{2}{3}$ **48.** $k = \pm 1$
49. (a) $k = 2$ (b) $m = 8$ **50.** (a) $n = 3$ (b) $q = 9$

Exercise 2 *page 294*

1. $\begin{pmatrix} 1 & -1 \\ -3 & 4 \end{pmatrix}$ **2.** $\begin{pmatrix} 5 & -2 \\ -2 & 1 \end{pmatrix}$ **3.** $\frac{1}{2}\begin{pmatrix} 2 & -4 \\ -1 & 3 \end{pmatrix}$ **4.** $\frac{1}{3}\begin{pmatrix} 1 & -2 \\ -1 & 5 \end{pmatrix}$ **5.** $\frac{1}{2}\begin{pmatrix} 2 & 2 \\ 1 & 2 \end{pmatrix}$

6. $\frac{1}{5}\begin{pmatrix} 2 & 3 \\ 1 & 4 \end{pmatrix}$ **7.** $\frac{1}{8}\begin{pmatrix} 3 & -1 \\ 2 & 2 \end{pmatrix}$ **8.** $\frac{1}{6}\begin{pmatrix} 4 & 3 \\ -2 & 0 \end{pmatrix}$ **9.** $\frac{1}{5}\begin{pmatrix} -3 & 2 \\ -1 & -1 \end{pmatrix}$ **10.** No inverse

11. $\frac{1}{14}\begin{pmatrix} 4 & 2 \\ -1 & 3 \end{pmatrix}$ **12.** $-\frac{1}{5}\begin{pmatrix} 1 & -1 \\ -2 & -3 \end{pmatrix}$ **13.** $-\frac{1}{5}\begin{pmatrix} -4 & 3 \\ -1 & 2 \end{pmatrix}$ **14.** $\frac{1}{7}\begin{pmatrix} 1 & 0 \\ 5 & 7 \end{pmatrix}$ **15.** $-\frac{1}{6}\begin{pmatrix} -4 & -1 \\ 2 & 2 \end{pmatrix}$

16. $\frac{1}{2}\begin{pmatrix} 3 & -4 \\ -1 & 2 \end{pmatrix}$ **17.** $-\frac{1}{2}\begin{pmatrix} 1 & 0 \\ -3 & -2 \end{pmatrix}$ **18.** $\begin{pmatrix} -1 & 3 \\ 0 & -3 \end{pmatrix}$ **19.** $\begin{pmatrix} 3 & 2 \\ 1 & 3 \end{pmatrix}$ **20.** $\begin{pmatrix} 2 & -3 \\ 2 & 4 \end{pmatrix}$

21. $\frac{2}{3}\begin{pmatrix} 1 & -1 \\ 2 & 1 \end{pmatrix}$ **22.** (a) $\begin{pmatrix} 5 & -11 \\ -1 & 3 \end{pmatrix}$ (b) $\frac{1}{2}\begin{pmatrix} 1 & 3 \\ 0 & 2 \end{pmatrix}$ (c) $\frac{1}{2}\begin{pmatrix} 3 & 1 \\ 1 & 1 \end{pmatrix}$ **23.** $\begin{pmatrix} 1 & -3 \\ 4 & 0 \end{pmatrix}$

24. $\begin{pmatrix} 4 \\ 3 \end{pmatrix}$ **25.** 4; 1; $1\frac{1}{2}$ **26.** $x = -2$ **27.** (a) 42 (b) 14

28. (a) (i) 3 (ii) 2 (b) 9550 (c) The total cost of 50 floor tiles and 300 wall tiles.

Exercise 3 *page 297*

1. 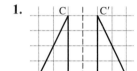 **2.** 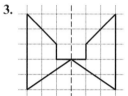 **3.** **4.**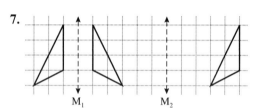

5. **6.** 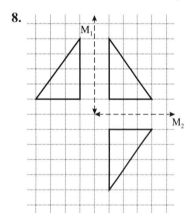 **7.**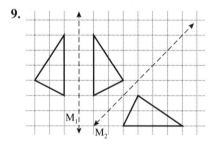

8.

9.

Exercise 4 *page 298*

1. (c) (i) $(-6, 8)$ (ii) $(6, -4)$ (iii) $(8, 6)$
2. (c) (i) $(8, 8)$ (ii) $(8, -6)$ (iii) $(-8, 6)$
3. (c) (i) $(3, -1)$ (ii) $(4, 2)$ (iii) $(-1, 1)$
4. (b) (i) $y = 1$ (ii) $y = x$ (iii) $y = -x$ (iv) $y = 2$
5. (f) $(1, -1), (-3, -1), (-3, -3)$
6. (f) $(8, -2), (6, -6), (6, -6)$

Exercise 5　*page 299*

1.

2.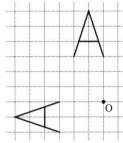

3.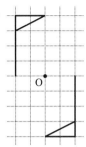

4.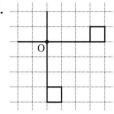

5.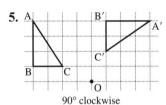

90° clockwise

6.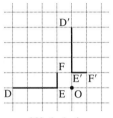

90° clockwise

7.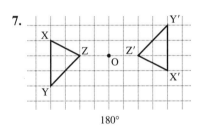

180°

8.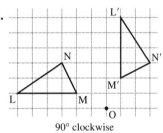

90° clockwise

Exercise 6　*page 300*

1. (a) A′(3, −1) B′(6, −1) C′(6, −3)　　　　(b) D′(3, −3) E′(3, −6) F′(1, −6)
   (c) P′(−7, −4) Q′(−5, −4) R′(−5, −1)

2. (b) (i) (4, −1), (7, −1), (7, −3)　　(ii) (−1, −4), (−1, −7), (−3, −7)
   (iii) (−4, 1), (−7, 1), (−7, 3)

3. (c) (−2, 1), (−2, −1), (1, −2)

4. (e) (−5, 2), (−5, 6), (−3, 5)

5. (b) (i) 90° anticlockwise, centre (0, 0)　　(ii) 180°, centre (2, 1)
   (iii) 90° clockwise, centre (2, 0)　　(iv) 180°, centre $(3\frac{1}{2}, 2\frac{1}{2})$
   (v) 90° anticlockwise, centre (6, 1)　　(vi) 90° clockwise, centre (1, 3)

6. (e) (i) 180°, centre $(\frac{1}{2}, \frac{1}{2})$　　(ii) 90° anticlockwise, centre (−2, 4)

Exercise 7　*page 302*

1. (a) $\begin{pmatrix} 7 \\ 3 \end{pmatrix}$　　(b) $\begin{pmatrix} 0 \\ -9 \end{pmatrix}$　　(c) $\begin{pmatrix} 9 \\ 10 \end{pmatrix}$　　(d) $\begin{pmatrix} -10 \\ 3 \end{pmatrix}$

   (e) $\begin{pmatrix} -1 \\ 13 \end{pmatrix}$　　(f) $\begin{pmatrix} 10 \\ 0 \end{pmatrix}$　　(g) $\begin{pmatrix} -9 \\ -4 \end{pmatrix}$　　(h) $\begin{pmatrix} -10 \\ 0 \end{pmatrix}$

2. (5, 2)　　3. (5, 6)　　4. (8, −5)　　5. (0, 6)　　6. (4, −7)　　7. (−3, 4)
8. (−3, −5)　　9. (−1, −8)　　10. (5, 2)　　11. (−2, 1)

Exercise 8   *page 304*

1. 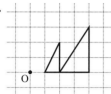   **2.** 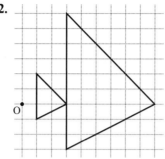   **3.**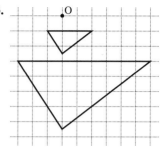

4.    **5.** 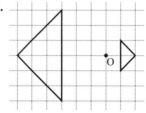   **6.**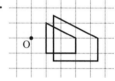

7. (4, 8), (8, 4), (10, 10)
10. (1, 4), (7, 8), (11, 2)
13. (5, 4), −2
16. (3, 4), (3, 3), (5, 3)
19. (6, 5), $(3\frac{1}{2}, 5)$, $(3\frac{1}{2}, 3)$

8. (3, 6), (7, 2), (9, 8)
11. (2, 1), +3
14. (6, 6), −1
17. (3, 7), (1, 7), (3, 3)

9. (1, 1), (10, 4), (4, 7)
12. (11, 9), $\frac{1}{2}$
15. $(\frac{1}{2}, 1)$, $(6\frac{1}{2}, 1)$, $(\frac{1}{2}, 5)$
18. (10, 7), (6, 7), (6, 5)

Exercise 9   *page 305*

1. (a) (i)   Rotation 90° clockwise, centre (0, −2)   (ii)   Reflection in $y = x$

(iii) Translation $\begin{pmatrix} 3 \\ 7 \end{pmatrix}$ (iv) Enlargement, scale factor 2, centre (−5, 5)

(v)   Translation $\begin{pmatrix} -7 \\ -3 \end{pmatrix}$ (vi) Reflection in $y = x$

2. (a) Rotation 90°clockwise, centre (4, −2)   (b) Translation $\begin{pmatrix} 8 \\ 2 \end{pmatrix}$
(c) Reflection in $y = x$
(d) Enlargement, scale factor $\frac{1}{2}$, centre (7, −7)   (e) Rotation 90° anticlockwise, centre (−8, 0)
(f) Enlargement, scale factor 2, centre (−1, −9)   (g) Rotation 90° anticlockwise, centre (7, 3)

3. (a) Enlargement, scale factor $1\frac{1}{2}$, centre (1, −4)   (b) Rotation 90° clockwise, centre (0, −4)

(c) Reflection in $y = -x$ (d) Translation $\begin{pmatrix} 11 \\ 10 \end{pmatrix}$

(e) Enlargement, scale factor $\frac{1}{2}$, centre (−3, 8)   (f) Rotation 90° anticlockwise, centre $(\frac{1}{2}, 6\frac{1}{2})$
(g) Enlargement, scale factor 3, centre (−2, 5)

Exercise 10   *page 308*

1. (2, −3)   **2.** (5, −1)   **3.** (6, 4)   **4.** (4, −6)   **5.** (0, 0)
6. (−6, 4)   **7.** (3, −2)   **8.** (3, 2)   **9.** (−2, 3)   **10.** (0, 0)

11. $(-3, 2)$ **12.** $(-3, -2)$ **13.** $(-3, -2)$ **14.** $(-6, -4)$ **15.** $(6, 0)$
16. $(-2, 3)$ **17.** $(0, 4)$ **18.** $(6, 8)$ **19.** $(3, 2)$ **20.** $(0, 0)$

Exercise 11   *page 308*

1. (a) $(-4, 4)$ (b) $(2, -2)$ (c) $(0, 0)$ (d) $(0, 4)$ (e) $(0, 0)$
2. (a) $(-2, 5)$ (b) $(-4, 0)$ (c) $(2, -2)$ (d) $(1, -1)$
3. (a) reflection in y-axis   (b) rotation $180°$, centre $(-2, 2)$
   (c) rotation $90°$ clockwise, centre $(2, 2)$
4. (a) rotation $90°$ anticlockwise, centre $(0, 0)$ (b) translation $\begin{pmatrix} -2 \\ 5 \end{pmatrix}$

   (c) rotation $90°$ anticlockwise, centre $(2, -4)$
   (d) rotation $90°$ anticlockwise, centre $(-\frac{1}{2}, 3\frac{1}{2})$
5. (a) rotation $90°$ anticlockwise, centre $(2, 2)$ (b) enlargement, scale factor $\frac{1}{2}$, centre $(8, 6)$
   (c) rotation $90°$ clockwise, centre $(-\frac{1}{2}, -3\frac{1}{2})$
6. $\mathbf{A}^{-1}$: reflection in $x = 2$
   $\mathbf{B}^{-1}$: B
   $\mathbf{C}^{-1}$: translation $\begin{pmatrix} 6 \\ -2 \end{pmatrix}$
   $\mathbf{D}^{-1}$: D
   $\mathbf{E}^{-1}$: E
   $\mathbf{F}^{-1}$: translation $\begin{pmatrix} -4 \\ -3 \end{pmatrix}$
   $\mathbf{G}^{-1}$: $90°$ rotation anticlockwise, centre $(0, 0)$
   $\mathbf{H}^{-1}$: enlargement, scale factor 2, centre $(0, 0)$
7. (a) $(4, 0)$ (b) $(-6, -1)$ (c) $(-2, -2)$ (d) $(2, -2)$ (e) $(6, 2)$
8. (a) $(1, -6)$ (b) $(4, -2)$ (c) $(2, 7)$ (d) $(4, -6)$ (e) $(2, -4)$
9. (a) $(-1, -2)$ (b) $(8, 2)$ (c) $(4, -6)$ (d) $(0, -3)$

10. (b) rotation, $180°$, centre $(4, 0)$ (c) translation $\begin{pmatrix} 12 \\ -4 \end{pmatrix}$

Exercise 12   *page 310*

1. (a) $(-2, 2), (-2, 6). (-4, 6)$ (b) $(-2, 2), (-6, 2), (-6, 4)$ (c) $(2, -2), (6, -2), (6, -4)$
   (d) $(2, 2), (2, 6), (4, 6)$ (e) $(1, 1), (3, 1), (3, 2)$
2. (a) enlargement: scale factor 2, centre $(0, 0)$ (b) enlargement: scale factor $-\frac{1}{2}$, centre $(0, 0)$
   (c) reflection in $y = -x$ (d) enlargement: scale factor -2, centre $(0, 0)$
3. A : reflection in x-axis   B : reflection in y-axis   C : reflection in $y = x$
   D : rotation, $-90°$, centre $(0, 0)$ E : reflection in $y = -x$ F : rotation, $180°$, centre $(0, 0)$
   G : rotation, $+90°$, centre $(0, 0)$ H : identity (no change)
4. (b) ratio $= 4 : 1$
5. rotation $45°$ anticlockwise, centre $(0, 0)$; enlargement scale factor $\sqrt{2}$ $(1·41)$
6. rotation $26·6°$ clockwise; enlargement scale factor $\sqrt{5}$ $(2·24)$
7. (d) rotation $90°$ clockwise, centre $(0, 0)$
8. $y = 2x$ **9.** $y - 3x$
10. (c) $OB = \sqrt{20}$, $OB' = 3\sqrt{20}$ (d) $36·9°$ (e) rotation $36·9°$; enlargement scale factor 3

11. (a) $\begin{pmatrix} 0 & -1 \\ 1 & 0 \end{pmatrix}$ (b) $\begin{pmatrix} -1 & 0 \\ 0 & -1 \end{pmatrix}$ (c) $\begin{pmatrix} 0·866 & -0·5 \\ 0·5 & 0·866 \end{pmatrix}$ (d) $\begin{pmatrix} 0 & 1 \\ -1 & 0 \end{pmatrix}$

   (e) $\begin{pmatrix} 0·5 & -0·866 \\ 0·866 & 0·5 \end{pmatrix}$ (f) $\begin{pmatrix} -0·866 & -0·5 \\ 0·5 & -0·866 \end{pmatrix}$ (g) $\begin{pmatrix} 0·707 & -0·707 \\ 0·707 & 0·707 \end{pmatrix}$ (h) $\begin{pmatrix} 0·6 & -0·8 \\ 0·8 & 0·6 \end{pmatrix}$

12. (a) $+90°$ (b) $+36·9°$ (c) $-60°$ (d) $-53·1°$

Exercise 13   *page 313*

1. (a) reflection in $y = x - 1$
   (b) reflection in $y = 1$
   (c) rotation $-90°$, centre $(2, -2)$
   (d) enlargement, scale factor 3, centre $(2, -1)$

2. (c) $\begin{pmatrix} 0 & -1 \\ -1 & 0 \end{pmatrix}$ (d) $\mathbf{BA} \equiv \mathbf{A}$ then $\mathbf{B}$

3. (c) $\begin{pmatrix} 2 & 2 \\ 0 & 2 \end{pmatrix}$ (d) $\mathbf{BA} \equiv \mathbf{A}$ then $\mathbf{B}$

4. (a) $\begin{pmatrix} \frac{1}{2} & 0 \\ 0 & \frac{1}{2} \end{pmatrix}$
   (b) $\begin{pmatrix} 1 & -2 \\ 0 & 1 \end{pmatrix}$
   (c) $\frac{1}{3}\begin{pmatrix} 1 & 0 \\ 0 & 3 \end{pmatrix}$

5. (a) $(14, 3)$
   (b) $m = 3, n = \frac{1}{2}$
   (c) $h = 1, k = -2$

6. (a) $(-2, 2)$
   (c) $\begin{pmatrix} 2 \\ -2 \end{pmatrix}$
   (d) $\begin{pmatrix} 2 & 0 \\ 0 & 2 \end{pmatrix}\begin{pmatrix} x \\ y \end{pmatrix} + \begin{pmatrix} 2 \\ -2 \end{pmatrix}$

7. (a) $x = 2$
   (c) $\begin{pmatrix} 4 \\ 0 \end{pmatrix}$
   (d) $\begin{pmatrix} -1 & 0 \\ 0 & 1 \end{pmatrix}\begin{pmatrix} x \\ y \end{pmatrix} + \begin{pmatrix} 4 \\ 0 \end{pmatrix}$

8. (a) $\begin{pmatrix} 2 & 0 \\ 0 & 2 \end{pmatrix}\begin{pmatrix} x \\ y \end{pmatrix} + \begin{pmatrix} -1 \\ -3 \end{pmatrix}$
   (b) $\begin{pmatrix} 2 & 0 \\ 0 & 2 \end{pmatrix}\begin{pmatrix} x \\ y \end{pmatrix} + \begin{pmatrix} -\frac{1}{2} \\ -1 \end{pmatrix}$
   (c) $\begin{pmatrix} 0 & 1 \\ 1 & 0 \end{pmatrix}\begin{pmatrix} x \\ y \end{pmatrix} + \begin{pmatrix} -3 \\ 3 \end{pmatrix}$
   (d) $\begin{pmatrix} -1 & 0 \\ 0 & -1 \end{pmatrix}\begin{pmatrix} x \\ y \end{pmatrix} + \begin{pmatrix} 3 \\ 5 \end{pmatrix}$
   (e) $\begin{pmatrix} 1 & 0 \\ 0 & -1 \end{pmatrix}\begin{pmatrix} x \\ y \end{pmatrix} + \begin{pmatrix} 0 \\ 2 \end{pmatrix}$
   (f) $\begin{pmatrix} 0 & 1 \\ -1 & 0 \end{pmatrix}\begin{pmatrix} x \\ y \end{pmatrix} + \begin{pmatrix} 4 \\ 0 \end{pmatrix}$

Exercise 14   *page 315*

1. rotation: $+90°$, centre $(0, 0)$
2. reflection in y-axis
3. reflection in $y = -x$
4. reflection in $y = x$
5. enlargement: scale factor 2, centre $(0, 0)$
6. enlargement: scale factor $\frac{1}{2}$, centre $(0, 0)$
7. enlargement: scale factor -2, centre $(0, 0)$
8. enlargement: scale factor $-\frac{1}{2}$, centre $(0, 0)$

9. $\begin{pmatrix} 0 & -1 \\ 1 & 0 \end{pmatrix}$
10. $\begin{pmatrix} 0 & 1 \\ 1 & 0 \end{pmatrix}$
11. $\begin{pmatrix} -1 & 0 \\ 0 & -1 \end{pmatrix}$
12. $\begin{pmatrix} -1 & 0 \\ 0 & -1 \end{pmatrix}$

13. $\begin{pmatrix} 3 & 0 \\ 0 & 3 \end{pmatrix}$
14. $\begin{pmatrix} 0 & -1 \\ -1 & 0 \end{pmatrix}$
15. $\begin{pmatrix} -2 & 0 \\ 0 & -2 \end{pmatrix}$
16. $\begin{pmatrix} 1 & 0 \\ 0 & -1 \end{pmatrix}$

17. $\begin{pmatrix} 0 & 1 \\ -1 & 0 \end{pmatrix}$
18. $\begin{pmatrix} \frac{1}{2} & 0 \\ 0 & \frac{1}{2} \end{pmatrix}$

Exercise 15   *page 317*

1. $x = 0$ (y-axis)
2. $x = 0$ (the y-axis)
3. (d) shear with shear factor 2, invariant line $x = 0$ (the y-axis)
4. $y = 0$
5. (a) stretch: parallel to x-axis, scale factor 3
   (b) shear: x-axis invariant
   (c) shear: y-axis invariant
   (d) stretch: parallel to y-axis, scale factor 2
6. stretch: parallel to x-axis, scale factor 2
7. stretch: parallel to x-axis, scale factor 3
8. stretch: parallel to y-axis, scale factor 2
9. stretch: parallel to x-axis, scale factor $1\frac{1}{2}$
10. shear: x-axis invariant
11. stretch: parallel to x-axis, scale factor -2
12. stretch: parallel to y-axis, scale factor 3
13. stretch: parallel to x-axis, scale factor $\frac{1}{2}$
14. $y = -x$
15. $y = -x$

Revision exercise 9A   *page 318*

1. (a) $\begin{pmatrix} 6 & 4 \\ 2 & 8 \end{pmatrix}$ (b) $\begin{pmatrix} 4 & -1 \\ 1 & 2 \end{pmatrix}$ (c) $\begin{pmatrix} 1\frac{1}{2} & 1 \\ \frac{1}{2} & 2 \end{pmatrix}$ (d) $\begin{pmatrix} -3 & 13 \\ -1 & 11 \end{pmatrix}$ (e) $\begin{pmatrix} 1 & 3 \\ 0 & 4 \end{pmatrix}$

2. (a) $\begin{pmatrix} -9 & -1 \\ 5 & 1\frac{1}{3} \end{pmatrix}$ (b) $\begin{pmatrix} 12 & 6 \\ 4 & 2 \end{pmatrix}$ (c) $\begin{pmatrix} 9 & -2 \\ 2 & -7 \end{pmatrix}$

3. (a) (14); $\begin{pmatrix} -1 & -15 \\ 3 & 15 \end{pmatrix}$ (b) $X = \begin{pmatrix} 1 & 3 \\ 2 & 4 \end{pmatrix}$ 4. $\frac{1}{13}\begin{pmatrix} 5 & 1 \\ -3 & 2 \end{pmatrix}$

5. $x = 3$; $-\frac{1}{9}\begin{pmatrix} -1 & -2 \\ -3 & 3 \end{pmatrix}$ 6. $h = 4$, $k = -4$

7. (a) $a = \pm 4$ (b) $a = \pm 3$ 8. (a) $(4, -1)$ (b) $(4, 1)$ (c) $(-3, 2)$
9. (a) A′(−3, −1) B′(1, −1) C′(−3, −7)   (b) A′(2, −2) B′(6, −2) C′(2, −8)
   (c) A′(1, 1) B′(2, 1) C′(1, −$\frac{1}{2}$)   (d) A′(4, 2) B′(3, 2) C′(4, 3$\frac{1}{2}$)
   (e) A′(−2, 2) B′(−6, 2) C′(−2, 8)
10. (a) reflection in *y*-axis   (b) reflection in $y = x$ (c) rotation, −90°, centre (0, 0)
    (d) reflection in $y = -x$ (e) rotation, 180°, centre (0, 0)   (f) rotation, −90°, centre (0, 0)
11. (a) reflection in $x = \frac{1}{2}$ (b) reflection in $y = -x$ (f) rotation, 180°, centre (1,1)
12. (a) (−1, −3)   (b) (−1, 3)   (c) (6, 2)   (d) (−3, 1)   (e) (−2, 6)   (f) (0, 2)
13. (a) (−1, 2)   (b) (1, −2)   (c) (10, −2)   (d) (6, −2)   (e) (−10, 2)   (f) (12, 2)
14. (a) (−2, 5)   (b) (−4, −3)   (c) rotation, +90°, centre (0, 0)
15. (a) rotation, +90°, centre (0, 0)   (b) reflection in *x*-axis   (c) rotation, 180°, centre (0, 0)
    (d) rotation, −90°, centre (0, 0)   (e) reflection in $y = -x$
16. (a) reflection in $y = x$ (b) reflection in *y*-axis
    (c) enlargement, scale factor 3, centre (0, 0)

17. (a) $\begin{pmatrix} -1 & 0 \\ 0 & -1 \end{pmatrix}$ (b) $\begin{pmatrix} 1 & 0 \\ 0 & -1 \end{pmatrix}$ (c) $\begin{pmatrix} 4 & 0 \\ 0 & 4 \end{pmatrix}$ (d) $\begin{pmatrix} 0 & -1 \\ -1 & 0 \end{pmatrix}$ (e) $\begin{pmatrix} 0 & 1 \\ -1 & 0 \end{pmatrix}$

18. (a) enlargement, scale factor 2, centre (0, 0); translation $\begin{pmatrix} 5 \\ -2 \end{pmatrix}$
    (b) (11, −4)   (c) (3, −1)   (d) (1, 3)

19. (a) $\begin{pmatrix} 0 & 1 \\ 1 & 0 \end{pmatrix}$ (b) $\begin{pmatrix} -1 & 0 \\ 0 & 1 \end{pmatrix}$ (c) $\begin{pmatrix} 0 & 1 \\ -1 & 0 \end{pmatrix}$
    (d) $\begin{pmatrix} 0 & -1 \\ 1 & 0 \end{pmatrix}$ **AB** ≡ rotation − 90°, centre (0, 0)
        AB ≡ rotation + 90°, centre (0, 0)

Examination exercise 9B   *page 322*

1. (a) $\begin{pmatrix} 2x + 12 & 3x + 6 \\ 14 & 15 \end{pmatrix}$ (b) 5   2. $a = 3$, $b = 4$

3. $\frac{1}{2}\begin{pmatrix} 5 & -3 \\ 4 & -2 \end{pmatrix}$ 4. $\begin{pmatrix} 13 & 21 \\ 21 & 34 \end{pmatrix}$ 5. $\begin{pmatrix} -11 \\ -11 \\ -14 \end{pmatrix}$

6. (a) (i) triangle (−1, −2) (−1, −3) (−3, −2)   (ii) Reflection in $y = -x$ (b) $\begin{pmatrix} 0 & -1 \\ 1 & 0 \end{pmatrix}$
7. (a) (i) (5, 3) (ii) (3, 5)   (b) $\begin{pmatrix} 0 & 1 \\ 1 & 0 \end{pmatrix}$ (c) TM(Q) = (k, k)   (d) $\begin{pmatrix} 0 & 1 \\ 1 & 0 \end{pmatrix}$
   (e) (i) $\begin{pmatrix} 0 & 1 \\ -1 & 0 \end{pmatrix}$ (ii) Rotation, centre (0, 0), 90° clockwise

8. (c) (i) (5, −7) (8, −7) (8, −5) (ii) (−4, 2) (−7, 2) (−7, 4) (iii) (−2, −2) (−5, −2) (−5, −4)

(d) (i) (3, 2) (7·5, 2) (7·5, 4) (ii) $\dfrac{1}{15}\begin{pmatrix} 1 & 0 \\ 0 & 1\cdot5 \end{pmatrix}$ (iii) Stretch, y-axis invariant, factor $\dfrac{2}{3}$

9. (c) Triangle (1, 2) (1, 5) (3, 3) (d) (−2, 1) (−5, 1) (−3, 3) (e) Reflection in y-axis

(f) (i) (2, −1) (5, −4) (3, 0) (ii) Shear, y-axis invariant (iii) $\begin{pmatrix} 1 & 0 \\ 1 & 1 \end{pmatrix}$

10 Statistics and Probability

Exercise 1 *page 327*

1. (a) Squash (b) 160 (c) 10
3. (a) $3000 (b) $4000 (c) $6000 (d) $11 000
4. red 50°; green 70°; blue 110°; yellow 40°; pink 90°
5. eggs 270°; milk 12°; butter 23·4°; cheese 54°; salt/pepper 0·6°
6. (a) A 60°; B 100°; C 60°; D 140°; E 0°
  (b) A 50°; B 75°; C 170°; D 40°; E 25°
  (c) A 48·5°; B 76·2°; C 62·3°; D 96·9°; E 76·2°
7. 18°, 54°, 54°, 234° **8.** 80°, 120°, 160° **9.** $x = 8$ **10.** 100
11. (a) 22·5% (b) $x = 45°$, $y = 114°$ **12.** (a) 144° (b) posters 5%, cinema 1%
13. Area of second apple looks much more than twice area of first.
14. Vertical axis starts at 130.

Exercise 3 *page 332*

Frequency densities for histograms are given here.

1. 1, 1·4, 0·3
3. 0·5, 1·4, 2, 1, 0·2
5. 0·55, 1·8, 0·7, 0·25
7. 8, 11, 6, 4, 1, 0·5

2. 1, 0·6, 1·2, 1·7, 1·3, 0·25
4. 1·8, 9·2, 7, 1·3, 0·8
6. 1·6, 3·4, 2·8, 0·73

Exercise 4 *page 336*

1. (a) mean = 6; median = 5; mode = 4. (b) mean = 9; median = 7; mode = 7.
  (c) mean = 6·5; median = 8; mode = 9. (d) mean = 3·5; median = 3·5; mode = 4.
2. (a) mean = 7·82; median = 8; mode = 8. (b) mean = 5; median = 4; mode = 4.
  (c) mean = 2·1; median = 2·5; mode = 4. (d) mean = $\frac{13}{18}$; median = $\frac{1}{2}$; mode = $\frac{1}{2}$.
3. 78 kg **4.** 35·2 cm **5.** (a) 2 (b) 9
6. (a) 20·4 m (b) 12·8 m (c) 1·66 m **7.** 55 kg **8.** 12
9. mean = 17, median = 3. The median is more representative.
10. the median.
11. many answers (e.g. 4, 4, 6, 10, 11)
12. 3·38
13. 3·475
14. (a) mean = 3·025; median = 3; mode = 3. (b) mean = 17·75; median = 17; mode = 17.
  (c) mean = 3·38; median = 4; mode = 4.
15. (a) 5·17 (b) 5

16. (i) 9   (ii) 9   (iii) 15   **17.** (i) 5   (ii) 10   (iii) 10
18. 12   **19.** $3\frac{2}{3}$ **20.** 4·68
21. $\dfrac{ax + by + cz}{a + b + c}$

Exercise 5   *page 339*

1. (a)

number of words	frequency f	midpoint x	fx
1–5	6	3	18
6–10	5	8	40
11–15	4	13	52
16–20	2	18	36
21–25	3	23	69
Totals	20	—	215

(b) 10·75   **2.** (a) 68·25   **3.** 68·25   **4.** 3·77

5. (a) 181 cm
(b) The raw data is unavailable and an assumption has been made with the midpoint of each interval.
(c) 180–90 cm

Exercise 6   *page 344*

2. (a) strong positive correlation   (b) no correlation
(c) weak negative correlation
3. 11   **4.** 9   **5.** no correlation   **6.** (c) about 26   (d) about 46
7. (b) 33 m.p.g.   (c) 63 m.p.h.

Exercise 7   *page 347*

1. (a) 47   (b) 30, 63   (c) 33   (d) 37   (e) 23
2. (a) 32   (b) 26, 43   (c) 17   (d) 30   (e) 84
3. (a) 20 kg   (b) 10·5 kg   **4.** (a) 80·5 cm   (b) 22 cm
5. (a) 71 s   (b) 20 s   **6.** (a) 45   (b) 14
7. (a) 36·5 g   (b) 20 g   (c) 25
8. (a) 26   (b) 25·2   (c) 26·1

Exercise 8   *page 351*

1. (a) $\frac{1}{6}$ (b) $\frac{1}{2}$ (c) $\frac{1}{2}$ (d) $\frac{1}{3}$
2. (a) $\frac{1}{4}$ (b) $\frac{1}{2}$
3. (a) (i) $\frac{3}{5}$ (ii) $\frac{2}{5}$ (b) (i) $\frac{5}{9}$ (ii) $\frac{4}{9}$
4. (a) $\frac{1}{11}$ (b) $\frac{2}{11}$ (c) 0   (d) $\frac{1}{11}$
5. (a) $\frac{1}{8}$ (b) $\frac{3}{8}$ (c) $\frac{1}{8}$ (d) $\frac{7}{8}$
6. (a) $\frac{5}{11}$ (b) $\frac{7}{22}$ (c) $\frac{15}{22}$ (d) $\frac{17}{22}$
7. (a) $\frac{1}{2}$ (b) $\frac{3}{25}$ (c) $\frac{9}{100}$ (d) $\frac{2}{25}$
8. (a) $\frac{1}{12}$ (b) $\frac{1}{36}$ (c) $\frac{5}{18}$ (d) $\frac{1}{6}$ (e) $\frac{1}{6}$; most likely total = 7.
9. (a) $\frac{1}{12}$ (b) $\frac{5}{36}$ (c) $\frac{2}{3}$ (d) $\frac{1}{12}$ (e) $\frac{1}{36}$
10. (a) 1   (b) 0   (c) 1   (d) 0
11. (a) $\frac{3}{8}$ (b) $\frac{1}{16}$ (c) $\frac{15}{16}$ (d) $\frac{1}{4}$

12. $\frac{271}{1000}$

13. $\frac{x}{12}$, 3

14. (a) $\frac{1}{144}$ (b) $\frac{1}{18}$ (c) $\frac{1}{72}$; head, tail and total of 7.

15. (a) $\frac{1}{216}$ (b) $\frac{1}{72}$ (c) $\frac{1}{8}$ (d) $\frac{5}{108}$ (e) $\frac{5}{72}$ (f) $\frac{1}{36}$

Exercise 9   *page 354*

1. (a) $\frac{1}{2}$ (b) $\frac{1}{2}$ (c) $\frac{1}{4}$ **2.** (a) $\frac{1}{16}$ (b) $\frac{25}{144}$

3. (a) $\frac{1}{121}$ (b) $\frac{9}{121}$ **4.** (a) $\frac{1}{288}$ (b) $\frac{1}{72}$

5. (a) $\frac{1}{125}$ (b) $\frac{1}{125}$ (c) $\frac{1}{10\,000}$ (d) $\frac{3}{500}$ (e) $\frac{3}{500}$

6. (a) $\frac{1}{9}$ (b) $\frac{4}{27}$ (c) $\frac{4}{9}$

Exercise 10   *page 355*

1. (a) $\frac{49}{100}$ (b) $\frac{9}{100}$ **2.** (a) $\frac{9}{64}$ (b) $\frac{15}{64}$ **3.** (a) $\frac{7}{15}$ (b) $\frac{1}{15}$

4. (a) $\frac{2}{9}$ (b) $\frac{2}{15}$ (c) $\frac{1}{45}$ (d) $\frac{14}{45}$

5. (a) $\frac{1}{12}$ (b) $\frac{1}{6}$ (c) $\frac{1}{3}$ (d) $\frac{2}{9}$

6. (a) $\frac{1}{216}$ (b) $\frac{125}{216}$ (c) $\frac{25}{72}$ (d) $\frac{91}{216}$

7. (a) $\frac{1}{6}$ (b) $\frac{1}{30}$ (c) $\frac{1}{30}$ (d) $\frac{29}{30}$

8. (a) $\frac{27}{64}$ (b) $\frac{1}{64}$

9. (a) 6   (b) $\frac{1}{3}$

10. (a) $\frac{1}{64}$ (b) $\frac{27}{64}$ (c) $\frac{9}{64}$ (d) $\frac{27}{64}$; Sum $= 1$

11. (a) $\frac{3}{20} \times \frac{2}{19} \times \frac{1}{18} \left(= \frac{1}{1140} \right)$ (b) $\frac{1}{4} \times \frac{4}{19} \times \frac{1}{16} \left(= \frac{1}{114} \right)$

      (c) $\left(\frac{3}{20} \times \frac{5}{19} \times \frac{12}{18} \right) \times 6$ (d) $\frac{5}{20} \times \frac{4}{19} \times \frac{3}{18} \times \frac{2}{17}$

12. (a) $\dfrac{1}{10\,000}$ (b) $\dfrac{523}{10\,000}$ (c) $\dfrac{9^4}{10^4}$

13. (a) $\dfrac{x}{x+y}$ (b) $\dfrac{x(x-1)}{(x+y)(x+y-1)}$ (c) $\dfrac{2xy}{(x+y)(x+y-1)}$ (d) $\dfrac{y(y-1)}{(x+y)(x+y-1)}$

14. (a) $\frac{1}{5}$ (b) $\frac{18}{25}$ (c) $\frac{1}{20}$ (d) $\frac{2}{25}$ (e) $\frac{77}{100}$

15. $\frac{3}{10}$ **16.** $\frac{9}{140}$ **17.** (a) 5   (b) $\frac{1}{64}$

18. (a) $\frac{3}{5}$ (b) $\frac{1}{3}$ (c) $\frac{2}{15}$ (d) $\frac{2}{21}$ (e) $\frac{1}{7}$ (f) $\frac{1}{35}$

19. (a) $\frac{1}{220}$ (b) $\frac{1}{22}$ (c) $\frac{3}{11}$ (d) 5   (e) 300

20. (a) $\dfrac{10 \times 9}{1000 \times 999}$ (b) $\dfrac{990 \times 989}{1000 \times 999}$ (c) $\dfrac{2 \times 10 \times 990}{1000 \times 999}$

21. (a) $\frac{3}{20}$ (b) $\frac{7}{20}$ (c) $\frac{1}{2}$

22. (a) 0·00781   (b) 0·511   **23.** (a) $\frac{21}{506}$ (b) $\frac{455}{2024}$ (c) $\frac{945}{2024}$

Revision exercise 10A   *page 359*

1. $162°$ **2.** 41·7%   **3.** $54°$

4. (a) 84   (b) 19·2   **5.** (a) 3·4   (b) 3   (c) 3

6. (a) 5·45   (b) 5   (c) 5   **7.** 1·552 m   **8.** 3

9. $\frac{1}{6}$ **10.** (a) 5   (b) $\frac{4}{25}$ **11.** $\frac{8}{27}$ **12.** $\frac{5}{16}$

13. (a) $\frac{1}{28}$ (b) $\frac{15}{28}$ (c) $\frac{3}{7}$

14. (a) $\dfrac{x}{x+5}$ (b) $\left(\dfrac{x}{x+5}\right)^2$; $\dfrac{x}{x+5}$, $\dfrac{x(x-1)}{(x+5)(x+4)}$

15. $\frac{1}{19}$ **16.** (a) $\frac{1}{32}$ (b) $\frac{1}{256}$ **17.** (a) $\frac{1}{8}$ (b) $\frac{1}{2}$

18. $\frac{35}{48}$ **19.** (a) $\frac{1}{9}$ (b) $\frac{1}{12}$ (c) 0   **20.** $\dfrac{1}{20^3}$

Examination exercise 10B   *page 361*

1. (a) 3·365 to 3·375 (b) 0·26 to 0·27 (c) 55, 56 or 57
2. (a) (i) 30   (ii) 30, 30·5, 31   (iii) 3   (b) (i) 20·93 or 20·9   (ii) 2·6, 0·7, 0·8
3. (a) (i) 46·5 km/h   (ii) 9·5   (iii) 48   (b) (i) $n = 32$ (ii) 46·4 km/h
4. (a) $\dfrac{8}{19}$ (b) $\dfrac{7}{18}$
5. (a) 1   (b) 2·5   (c) 2·96   (d) 2·9
6. (a) (i) 6   (ii) 4·5   (iii) 4·54   (iv) $\dfrac{1}{63}$ (v) $\dfrac{1}{35}$ (vi) $\dfrac{92}{819}$

   (b) (i) 0·08   (ii) 0·125   (iii) 7

7. (a) $p = \dfrac{1}{20}$ (b) $q = \dfrac{19}{20}$ (b) (i) $\dfrac{1}{400}$ (ii) $\dfrac{38}{400}$ (c) $\dfrac{38}{8000}$ (d) $\dfrac{58}{8000}$ (e) 7·25

8. (a) (i) 12   (ii) 3   (iii) 21   (iv) 2   (v) $\dfrac{14}{24}$ (vi) $\dfrac{12}{19}$

   (b) (i) $\dfrac{132}{462}$ (ii) $\dfrac{240}{462}$ $\left(\text{i.e. } \dfrac{10}{22} \times \dfrac{12}{21}\right)$

11 Investigations, Practical Problems, Puzzles

11.1 Investigations   *page 367*

Note: It must be emphasised that the *process* of obtaining reliable results is far more important than these few results. It is not suggested that 'obtaining a formula' is the only aim of these coursework tasks. The results are given here for some of the investigations merely as a check for teachers or students working on their own.

It is not possible to summarise the enormous number of variations which students might think of for themselves. Obviously some original thoughts will be productive while many others will soon 'dry up'.

1. With the numbers written in c columns, the difference for a $(n \times n)$ square is $(n-1)^2 \times c$.

8. (a) For n blacks and n whites, number of moves $= \dfrac{n(n+1)}{2}$.

   (b) For n blacks and n whites, number of moves $= \dfrac{n(n+2)}{2}$.

   (c) For n of each colour, number of moves $= \frac{3}{2}n(n+1)$.

9. 4×4: There are 30 squares (i.e. $16 + 9 + 4 + 1$)
   8×8: There are 204 squares ($64 + 49 + 36 + 25 + 16 + 9 + 4 + 1$)

$n \times n$: Number of squares $= 1^2 + 2^2 + 3^2 + \ldots + n^2$

$$= \frac{n}{6}(n+1)(2n+1)$$

This could be found using method of differences or from the standard result for $\sum_1^n r^2$.

14. For $n \times n \times n$: 3 green faces $= 8$
    2 green faces $= 12(n-2)$
    1 green face $= 6(n-2)^2$
    0 green face $= (n-2)^3$

18. For a square card, corner cut out $= \frac{1}{6}$ (size of card).

   For a rectangle $a \times 2a$, corner cut out $\cong \dfrac{a}{4 \cdot 732}$.

20. For diagram number n, number of squares $= 2n^2 - 2n + 1$.

21. (a) Another Fibonacci sequence.
    (b) Terms are alternate terms of original sequence.
    (c) Ratio tends towards $1 \cdot 618$ (to 4 s.f.), the 'Golden Ratio'
    (d) (first $\times$ fourth) $=$ (second $\times$ third) $+ 1$
    (e) (first $\times$ third) $\pm 1 =$ (second)^2, alternating $+$ and $-$
    (f) sum of 10 terms $= 11 \times$ seventh term.
    (g) For six terms $a\,b\,c\,d\,e\,f$ let
    $x = e \times f - (a^2 + b^2 + c^2 + d^2 + e^2)$

The numbers in the first difference column are the squares of the terms in the original Fibonacci sequence.

	x	first difference
first six	0	1
second six	1	1
third six	2	4
fourth six	6	9
fifth six	15	25
sixth six	40	64
seventh six	104	169
eighth six	279	

22. For n names, maximum possible number of interchanges $= \dfrac{n(n-1)}{2}$.

24. Consider three cases of rectangles $m \times n$
    (a) m and n have no common factor. Number of squares $= m + n - 1$.
       e.g. 3×7, number $= 3 + 7 - 1 = 9$
    (b) n is a multiple of m. Number of squares $= n$ e.g. 3×12, number $= 12$
    (c) m and n share a common factor a
       so $m \times n = a(m' + n')$
       number of squares $= a(m' + n' - 1)$
       e.g. $640 \times 250 = 10(64 \times 25)$, number of squares $= 10(64 + 25 - 1) = 880$.

26. For smallest surface area, height of cylinder $= 2 \times$ radius.

27. (a) square root twice, 'multiply by x' gives cube root of x.
    (b) square root three times, 'multiply by x' gives seventh root of x.
    (c) square root twice, 'divide by x' gives $1/(\sqrt[3]{x})$
    (d) square root twice, 'multiply by x^2' gives $(\sqrt[3]{x})^2$

28. Pick's theorem: $A = i + \frac{1}{2}p - 1$

12 Revision tests

Test 1   page 390

1. C	**2.** D	**3.** D	**4.** B	**5.** A
6. C	**7.** A	**8.** C	**9.** C	**10.** B
11. C	**12.** A	**13.** D	**14.** C	**15.** C
16. D	**17.** A	**18.** C	**19.** B	**20.** D
21. A	**22.** C	**23.** B	**24.** B	**25.** C

Test 2   page 393

1. B	**2.** C	**3.** A	**4.** B	**5.** D
6. C	**7.** A	**8.** D	**9.** B	**10.** C
11. B	**12.** D	**13.** A	**14.** D	**15.** C
16. D	**17.** B	**18.** A	**19.** B	**20.** B
21. C	**22.** D	**23.** A	**24.** A	**25.** B

Test 3   page 395

1. D	**2.** D	**3.** D	**4.** B	**5.** A
6. C	**7.** A	**8.** D	**9.** D	**10.** B
11. C	**12.** D	**13.** D	**14.** B	**15.** A
16. B	**17.** C	**18.** A	**19.** D	**20.** D
21. C	**22.** A	**23.** B	**24.** B	**25.** D

Test 4   page 398

1. B	**2.** B	**3.** A	**4.** C	**5.** D
6. A	**7.** B	**8.** C	**9.** D	**10.** B
11. C	**12.** A	**13.** B	**14.** B	**15.** D
16. A	**17.** C	**18.** B	**19.** B	**20.** D
21. A	**22.** C	**23.** C	**24.** C	**25.** A

Specimen Paper 2

page 402

1. $\frac{4}{5}$ **2.** 265 min   **3.** $x < 6$

4. (a) $\frac{1}{8} \times 100 = 12\frac{1}{2}\%$ (b) $14(\cdot 3)\%$ **5.** $1 \cdot 4 \times 10^{-3}$ **6.** 15   **7.** \$400

8. (a) $0 \cdot 45$, $0 \cdot 444\,444\,444$, $0 \cdot 445\,945\,945$ (b) $\frac{33}{74}$

9. (a) -3 (b) -7 **10.** (a) $4 \cdot 25$ m, $3 \cdot 75$ m   (b) $11 \cdot 7 \, \text{m}^2$, $8 \cdot 44 \, \text{m}^2$

11. $x = 4$, $y = -1$ **12.** 4000

13. (a) (i) $090°$ (ii) $224°$ (b) 1760 km   **14.** (a) (i) 54   (ii) 57   (b) $C = 8N + 1$

15. (a) Diagram (i)   (b) Diagram (ii) 4:1; Diagram (iii) $8 \cdot 1$

16. (a) A triangle with vertices $(2, -2)$, $(-4, -4)$ and $(-4, -8)$ (b) 12 square units

17. (a) $xy = k$ or $y = \dfrac{k}{x}$ (b) y increases by 25%

18. (a)

x	1	2	3	4	5
y	7	5	5	5·5	6·2

(b)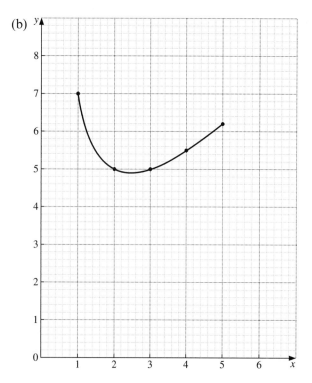

19. (a) $\begin{pmatrix} 5 \\ -1 \end{pmatrix}$ (b) C has vertices $(-2, 5)$, $(-2, 7)$ and $(-4, 7)$

(c) Rotation $90°$ clockwise about the point $(0, 0)$

20. (a) $XY = XW = 5\,cm$, $YZ = WZ = 10\cdot4\,cm$ (b) $\angle YZW = 33\cdot4°$

21. (b) $9\cdot60\,cm$ (c) $24\cdot0\,cm^2$

22. (b) (i) $2\,800\,000\,000$ or $2\cdot8$ thousand million   (ii) $2\cdot8 \times 10^9$

(c) approximately 20 million per year   (d) approximately 7 thousand million

23. $a = 38°$, $b = 71°$, $c = 64°$

Specimen Paper 4

page 407

1. (a) (i) 36   (ii) -8 (b) $4\frac{1}{2}$ (c) $+5, -5$

(d) $\times 9$ (e) $p = \sqrt{\dfrac{r(q-3)}{2}}$

2. (a) (i) $2x$ (ii) $\overrightarrow{EC} = \overrightarrow{ED} + \overrightarrow{DC}$ (iii) $\overrightarrow{AE} = 2\mathbf{y} - \mathbf{z}$ $\overrightarrow{CA} = 2\mathbf{z} - \mathbf{x}$

(b) $\mathbf{x} + \mathbf{y} + \mathbf{z}$

(c) $\mathbf{x} + \mathbf{y} + \mathbf{z} = 0$

(d) $\overrightarrow{BE} = 2\mathbf{y} - \mathbf{x} - \mathbf{z} = 2\mathbf{y} + \mathbf{y}$ (using (c)) $= 3\mathbf{y}$

BE and CD are parallel.

3. (a) (i) $94\cdot2\,cm^3$ (ii) $104\,cm^2$

(b) $268\,g$

4. (a)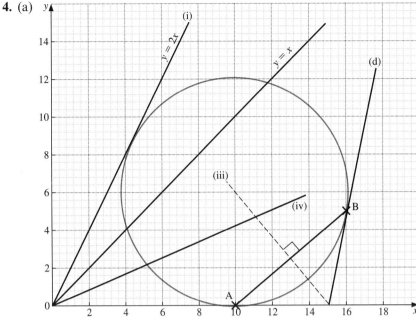

(b) $y = 2x$

(c) $(15, 0)$

5. (a) (i) $14 \cdot 4$ cm   (ii) $24 \cdot 8$ cm^2 (b) (i) $31 \cdot 6°$ (ii) $17 \cdot 9°$

6. (a) $10x - 30$ (b) $(3x - 8)(2x - 7)$

(d) (i) $(6x + 5)(x - 7)$ (ii) $x = -\frac{5}{6}$ or 7

(e) length 13 cm, width 7 cm

7. (a) (ii) $\frac{1}{36}$ (iii) $\frac{5}{36}$ (iv) 11   (b) (i) $0 \cdot 42$ (ii) $0 \cdot 16$ (iii) $0 \cdot 34$

8. (a) $6 \cdot 10$ hours   (b) (i) $10 < N \leqslant 12$ (ii) 5 h 20 m

(d)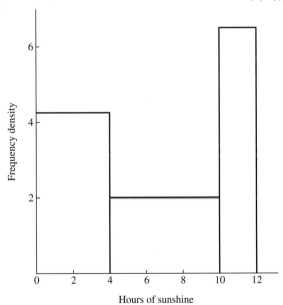

9. (a)

Selling price	Number sold	Sales revenue ($)
60	6000	360 000
70	5500	385 000
80	5000	400 000
90	4500	405 000
100	4000	400 000
110	3500	385 000
120	3000	360 000

(c) (i) $90   (ii) 4500

10. (a) 76, 123, 199

(b) (i) $\begin{pmatrix} 29 \\ 18 \end{pmatrix}$ (c) (i) $\begin{pmatrix} 2 & 1 \\ 1 & 1 \end{pmatrix}$ (ii) $\begin{pmatrix} 47 \\ 29 \end{pmatrix}$

Subject Index